Effects of Resource Distribution on Animal–Plant Interactions

A Gambian Epauleted Fruit Bat, *Epomophorus gambianus,* feeds on fruits of the wild fig, *Ficus sycomorus,* in East Africa. Adapted from a photograph by Merlin D. Tuttle, Bat Conservation International. © Merlin D. Tuttle. The illustration is by Ali Partridge. The editors acknowledge the help of Dr. Gary Bateman (curator) and Dr. Tad Theimer for loan of museum specimens of the bat from the Vertebrate Collection, Department of Biological Sciences, Northern Arizona University.

Effects of Resource Distribution on Animal–Plant Interactions

Edited by

Mark D. Hunter
Department of Entomology
Pennsylvania State University
University Park, Pennsylvania

Takayuki Ohgushi
Shiga Prefectural
Junior College
Shiga, Japan

Peter W. Price
Department of Biological Science
Northern Arizona University
Flagstaff, Arizona

Academic Press, Inc.
Harcourt Brace Jovanovich, Publishers
San Diego   New York   Boston   London   Sydney   Tokyo   Toronto

Copyright © 1992 by ACADEMIC PRESS, INC.

All Rights Reserved.

No part of this publication may be reproduced or transmitted in any form or by any means, electronic or mechanical, including photocopy, recording, or any information storage and retrieval system, without permission in writing from the publisher.

Academic Press, Inc.
San Diego, California 92101

United Kingdom Edition published by
Academic Press Limited
24–28 Oval Road, London NW1 7DX

Library of Congress Cataloging-in-Publication Data

Effects of resource distribution on animal–plant interactions /
    [edited by] Mark D. Hunter, Takayuki Ohgushi, Peter W. Price.
        p.   cm.
    Includes bibliographical references and index.
    ISBN 0-12-361955-6
    1. Animal–plant relationships.   2. Variation (Biology)
I. Hunter, Mark D.   II. Ohgushi, Takayuki.   III. Price, Peter W.
QH549.5.E34   1992
574.5'24--dc20   91-24301
                                                                      CIP

PRINTED IN THE UNITED STATES OF AMERICA
91 92 93 94   9 8 7 6 5 4 3 2 1

Contents

Section I
Phenotypic and Genotypic Variation in Plants and Animals

Section II
Resource Distribution, Reproduction, and Population Dynamics

5. Nectar Distributions, Pollinator Behavior, and Plant Reproductive Success
Beverly J. Rathcke

6. Plant Resources as the Mechanistic Basis for Insect Herbivore Population Dynamics
Peter W. Price

7. Factoring Natural Enemies into Plant Tissue Availability to Herbivores
Jack C. Schultz

8. Resource Limitation on Insect Herbivore Populations
Takayuki Ohgushi

Section III
Resource Distribution and Patterns in Animal–Plant Communities

9. Bottom-Up versus Top-Down Regulation of Vertebrate Populations: Lessons from Birds and Fish
James R. Karr, Michele Dionne, and Isaac Schlosser

10. Interactions within Herbivore Communities Mediated by the Host Plant: The Keystone Herbivore Concept
Mark D. Hunter

11. Loose Niches in Tropical Communities: Why Are There So Few Bees and So Many Trees?
David W. Roubik

Section IV
Evolutionary Responses to the Distribution of Resources

Preface

One major component of the complexity of natural biological systems is heterogeneity. Even the most superficially homogeneous communities of animals and plants exhibit considerable spatial and temporal variability when they are examined closely. Although heterogeneity has been recognized in biological systems for hundreds of years, incorporating much of this variability into our understanding of the natural world is in its infancy. Until relatively recently, for example, fields such as population and community ecology have essentially ignored variation in time and space in their conceptual and theoretical models.

The purpose of this volume is to explore the importance of natural variability in one field of biology—animal–plant interactions. Specifically, we argue that our understanding of population, community, and evolutionary level interactions between plants and the animals that depend on them are better understood when heterogeneity is taken into account. The causes and consequences of variability among plants as resources for animals (Chapters 4, 5, 6, 7, 9, 10, 14) and variability among animals as both resources for (Chapters 4, 11, 12, 13) and exploiters of plants (Chapters 2, 3, 4, 8) link all the contributions in this volume together. We suggest that many of the complex interactions we perceive in natural animal–plant systems arise from the superimposing of herbivore variation upon plant variation.

Indeed, variability among the individual herbivores within a species is a recurrent theme in this volume, and we explore the consequences of that variability for plant pollination and reproductive success (Chapters 5 and 13), animal population dynamics (Chapters 2 and 8), and animal social systems (Chapter 3). The authors of these chapters express the view that individual variation among herbivores within a species has important ramifications at population, community, and evolutionary levels, and we predict that major advances in animal–plant biology will result from the synthesis of studies at these different levels of organization.

We think this volume should also stimulate more focused debate on the relative strengths of bottom-up versus top-down effects in terrestrial food webs. To what extent is pattern in plant-feeding species and communities generated by plants and vegetation compared to carnivore effects? This theme runs through the volume. Some chapters take a balanced approach by illustrating the tight interactive linkages between plant resources, herbivores, and carnivores (Chapters 1, 7, 9, 10). Other chapters either debate the validity of strong bottom-up effects that generate pattern (Chapters 4, 6) or discuss the important role of resource display and variation in the evolution of life histories of animals exploiting plants (Chapter 12). Coupling such variation provided by plants to the necessarily opportunistic exploitation by pollinators provides a constantly shifting set of relationships in bee communities (Chapter 11). In spite of the different approaches taken in these chapters, the logic of a building-block approach, with plants as the autotrophic foundation, generates a frame of reference that provides considerable comparative power across diverse systems.

Another theme in this volume is inevitably the necessity of integrating the abiotic environment into the understanding of plant and animal interactions. This is most explicitly addressed in Chapter 14, but many of the chapters invoke the importance of physical factors, not only in defining plant distribution and abundance, but as major mediating components in linkages among trophic levels.

Although we have organized the chapters in this volume to reflect a natural hierarchy of organization, from the individual to the community and evolutionary level, there is considerable overlap between sections. We interpret this as an encouraging sign of an emerging synthesis of ideas. Section I considers genetic and phenotypic variability among animals and plants; Section II explores their reproduction and population dynamics; Section III investigates patterns in animal–plant communities; and Section IV describes interactions between animals and plants developed over evolutionary time.

We use the term "herbivore" in its broadest sense, to include frugivores, seed predators, pollen and nectar feeders, as well as animals that consume leaves and other vegetative parts. We have tried, where possible, to consider the consequences of heterogeneity for animal–plant interactions in a variety of animal taxa, including birds, bats, rodents, fish, and insects and to draw generalizations where they emerge.

After a short introduction, the remaining chapters in this volume contain brief literature reviews, descriptions of the author(s) current research interests, and novel hypotheses or research directions. This "where we've been, where we are, and where we ought to go" approach is designed to appeal to the widest possible audience. We anticipate a readership of informed undergraduates in the biological sciences who wish to survey a particular field, and graduate students who wish to incorporate natural heterogeneity into the design of their research projects, as well as professional animal–plant ecologists who will test some of the particular hypotheses presented here. To this end, we have made the language in this book as "jargon-free" as possible and accessible to biologists with a wide range of interests and experience. Since explicit consideration of heterogeneity in natural systems is a young and developing field, we hope that readers will contact the authors directly with questions, comments, and ideas.

The original idea for this book grew out of the Fifth International Congress of Ecology 1990, Yokohama, Japan, and although we invited several other thors to contribute, we would like to thank the organizing committee for bringing the core group together. Special thanks is due to Professor Shoichi Kawano, who suggested the topic and who extended his warmest hospitality.

All chapters in this volume have been peer reviewed by colleagues outside this project. We would like to extend our gratitude to the following for their help, guidance, and criticism: Heidi Appel, Alan Berryman, Cathy Bach, James Cresswell, Hugh Dingle, Niles Eldredge, Stan Faeth, Peter Feinsinger, Ted Floyd, Doug Futuyma, Fred Gould, Brad Hawkins, David Inouye, Masao Ito, Lorrie Klosterman, Jill Landsberg, William Lidicker, Tom Martin, Judy Myers, Mary Power, Mark Rausher, Shoichi Sakagami, Ellen Simms, John Thompson, Peter Turchin, Nick Waser, and Allan Watt.

1

Natural Variability in Plants and Animals

Mark D. Hunter
Department of Entomology
Pennsylvania State University
University Park, Pennsylvania

Peter W. Price
Department of Biological Science
Northern Arizona University
Flagstaff, Arizona

I. Introduction

If the world were one continuous carpet of vegetation, equal in all ways as a resource for animals, there would be no need for this book. In recent years, however, more texts have begun to stress the variability of plants as a food source and as shelter for the animals that they support (Rosenthal and Janzen, 1979; Denno and McClure, 1983; Strong, Lawton and Southwood, 1984). From these and other sources has come the realization that we must consider the distribution of quality as well as quantity of plants and plant parts if we are to develop our understanding of plant-herbivore interactions. Herbivores, too, exhibit significant genetic and phenotypic variability, both within and between species. Variable herbivores, therefore, are likely to exert ecological and evolutionary pressures on plants.

Variability is hierarchical. Most ecologists would agree that variation among different species of plants has a considerable influence on the distribution and abundance of herbivores. We know much less about the effects of heterogeneity among plant populations and individuals on the animals that depend upon them. Our level of understanding is poorer still when it comes to the converse; we know surprisingly little about the effects of variation among herbivore populations and individuals on plant ecology and evolution. However, we might expect that many of the complex patterns we see in animal-plant systems arise from interactions between heterogeneous plants and heterogeneous herbivores.

Figure 1 is a simplified representation of the major factors that generate variability in plants and herbivores. We feel that this model is applicable at the individual, population, or species level. In the model, plant heterogeneity arises from variability among herbivores, climate, soils, plant pathogens, decomposers, and symbionts. We have ignored genetic mutation because we assume, for the sake of simplicity, that the prevalence of particular plant genotypes reflects natural selection acting through the other factors mentioned above.

Likewise, herbivore heterogeneity arises from variability among plants, natural enemies (including pathogens), climate, and symbionts. A more complex (and realistic) model might represent the relative strengths of these forces by the thickness of the arrows on Figure 1. However, we simply do not have sufficient information to do this for any natural system of which we are aware. That mycorrhizae (Siqueira *et al.*, 1989) and plant pathogens (Power, 1991) can influence plant heterogeneity is almost certain, but the relative strength of their effects compared to, for example, climate is unkown.

One purpose of this book is to investigate the "thickness of the lines" between herbivores and plants on Figure 1. We want to explore the consequences of variability among plants for herbivores and the consequences of variability among herbivores for plants. Currently we know more about the former than the latter, and the contents of this book necessarily reflect that unfortunate bias. Even excluding variability among herbivores, there is remarkably little known about the effects of defoliation, especially below ground (Reichman and Smith, 1991), on plant population dynamics and community structure (Crawley, 1989). We would like to make a plea for considerable expansion in this area of research.

We see a strong parallel between our current inability to determine the relative thickness of the lines on Figure 1, and the debate between the strengths of "bottom-up" and "top-down" forces in natural populations and communities. There remains considerable disagreement in terrestrial systems, for example, about the relative roles of resource distribution and natural enemies in determining herbivore population dynamics and community structure (Hairston *et al.*, 1960; Lawton and Strong, 1981; Faeth, 1987; Price, 1990). Part of this disagreement may arise from a lack of appreciation for the degree of variability at lower trophic levels, and how it can interact with species at higher trophic levels (Price *et al.*, 1980; Kareiva and Sahakian, 1990). A simple genetic mutation in pea plants, for example, can determine the efficacy of a natural enemy on its herbivorous prey (Kareiva and Sahakian, 1990).

We would emphasize, therefore, that variability at lower trophic levels can have cascading effects up the trophic system. This is not meant to deny the power of "top-down" processes in many communities and several chapters in this volume consider such effects. Rather, we wish to reestablish the fundamental role of energy flow up through the system as the template

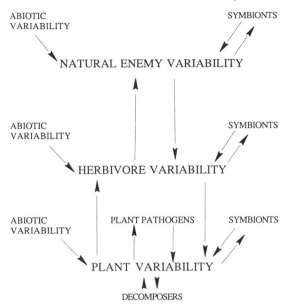

Figure 1 Major ecological forces that generate variability among plants and animals in natural systems. Both abiotic and biotic forces can "cascade" up and down the trophic web so that the action of climate on a natural enemy complex, for example, can influence the distribution and abundance of plants and even decomposers.

upon which all species interactions, "top-down" or "bottom-up" necessarily take place. A balanced view on the relative roles of "top-down" and "bottom-up" forces has emerged in aquatic systems (Carpenter *et al.*, 1985, 1987; McQueen *et al.*, 1986, 1989; Schindler, 1978), and we hope to stimulate a similar synthesis in terrestrial systems by emphasizing how variability in plants can influence interactions at higher trophic levels.

Having listed major factors that generate variability in plants and animals (Fig. 1), we will end this Introduction by reviewing briefly some of the known causes and consequences of variability among primary producers. We will leave it to other authors in this volume to describe the causes and consequences of variability among herbivores, and to consider the benefits of integrating a knowledge of variability at both trophic levels for our understanding of animal-plant interactions.

A. Sources and Patterns of Plant Variability

1. Presence or Absence of the Host Plant

Of critical importance to specialist herbivores, and of varying importance to oligophagous and polyphagous herbivores, is the presence or absence of a particular plant species. Even at this fundamental level, there is a continuum of spatial scales to be considered—the factors that make the pedunculate oak *Quercus robur* L. widely distributed in Europe and absent from North America are not the same factors that, for example, determine its presence or absence in a particular woodland in Scotland. There are probably a hierarchy of factors that reduce a plant's distribution from its "potential" to its "realized" distribution ranging from climate, topography, and elevation, to dispersal, soil variability, microclimate, competition, and herbivory. Bartholomew (1958) considered that the biotic or abiotic factor to which an organism has least adaptability or over which it has least control will ultimately determine its distribution. This slight modification of Liebig's Law of the Minimum (Liebig, 1840; Odum, 1959) is a rather static picture and, in reality, the forces that determine plant distribution are dynamic in ecological and evolutionary time. The generation of light gaps in tropical forests by the death of canopy trees, for example, plays a critical role in the regeneration of some tree species (Denslow *et al.*, 1990 but see Welden *et al.*, 1991). Plant succession, of course, describes a turnover of species with time and adaptation, too, can lead to the occurrence of plants in novel or hostile environments (Antonovics *et al.*, 1971). The more dynamic view of vegetation, and therefore resources for herbivores, was certainly fostered by Loucks' (1970) view of periodic perturbation as an essential generator and preserver of heterogeneity, and acted as a precursor to the now active study of patch dynamics (e.g. Pickett and White, 1985).

Given the considerable variability in patterns of plant distribution, the consequences of these patterns for herbivores are understandably complex. The relationships among climate, plant distribution, and plant-herbivore interactions are poorly documented, although there is growing evidence that the evolution of host plant choice may be influenced by biogeographic patterns and plant hybrid zones (Whitham, 1989; Boecklen and Spellenberg, 1990; Scriber and Lederhouse (Chapter 14)). Although the use of island biogeography theory to explain the numbers of herbivores associated with individual plants, plant populations, and plant species has received some criticism (Kuris *et al.*, 1980), there is good evidence that plant distribution and abundance can influence the biomass and species richness of herbivores associated with primary producers (Southwood, 1960, 1961; Root, 1973; Strong, 1979). Other studies, however, have suggested that a diversity of host plant species may actually be important to particular herbivores. For example, some monkeys are known to include plant species of apparent low quality in their diet, perhaps in order to accumulate specific nutrients (Oates *et al.*, 1980).

The influence of plant succession on resource use by herbivores has received more attention, and effects on plant palatability, herbivore specialization, and arthropod community structure have been described (Reader and Southwood, 1981; Futuyma, 1976; Haering and Fox, 1987). Given that natural perturbations can generate patches of vegetation in varying stages of succession, we should not be surprised to find that disturbances such as flooding (Power *et al.*, 1985) and drought (Young and Smith, 1987) can have significant indirect impacts on animal populations and communities. When perturbations influence patterns of plant succession, the consequences for animals need not be restricted to herbivores. Pianka (1989), for example, has suggested that the maintenance of species diversity in lizard communities in Australia is dependent upon regeneration of spinifex grassland, which is dominated by fire.

Within a habitat, the spatial distribution of plants is often critical to plant-herbivore interactions. For example, in temperate forests the probability of a plant's nearest neighbor being a conspecific is many times higher than it would be in the majority of tropical forests. As a consequence, herbivore dispersal strategies, their population dynamics, host plant choice, and plant defensive strategies can vary dramatically with the diversity of the plant community (Gilbert, 1975, 1979; Crawley, 1983). The density of particular plant species within a habitat, either because they are unusually beneficial or unusually deleterious, can be of enormous importance in determining patterns of herbivore population change. The presence of oak species (particularly chestnut oak) is strongly associated with outbreaks of the gypsy moth in the north eastern United States (Houston and Valentine, 1977; Doane and McManus, 1981).

2. Temporal Variability

Plants as resources for animals can vary greatly during a year. Annual plants may be absent for extended periods, and specialist herbivores have evolved life cycles accordingly. The availability of plant parts is often highly seasonal: pollen, fruits, seeds, buds, and the leaves of deciduous trees are examples. The migration of ungulates (Sinclair, 1985), the narrow larval feeding period of some phytophagous insects (Feeny, 1970; West, 1985), and the spawning of marine invertebrates (Starr *et al.*, 1990) can all be viewed as adaptations that exploit a temporally variable plant resource. Seasonal variation in the quantity or quality of plants and plant parts can have direct effects on the growth, reproduction, and movement of herbivores (Feeny, 1970; Rockwood, 1974; Haukioja and Niemela, 1979; Ramachandran, 1987) or indirect effects by their interaction with climate or natural enemies (Price *et al.*, 1980; Schultz, 1983; Hunter, 1987). The quality of plants as a resource for animals can vary over a much shorter time period too. Diurnal changes in flower availability are common, and continual changes in the nectar supply of some species are known to influence the behavior and foraging strategy of pollinator species (Gill and Wolf, 1975).

The consequences of seasonal variability in the quantity and quality of plants and plant parts has been investigated most thoroughly for phytophagous insects (e.g. Strong *et al.*, 1984). Many studies suggest that, in general, foliage quality declines with age due to decreasing water content and increasing toughness (Feeny, 1970; Wint, 1983). The effects of leaf age on leaf chemistry, particularly the concentrations of secondary plant compounds, appear more variable with different studies describing seasonal increases, decreases, and complex changes over time (Feeny, 1976; Schultz *et al.*, 1982; Lindroth, 1989).

Separating the often confounded effects of toughness, water content, and chemistry remains difficult, but profound effects of seasonally variable foliage quality on herbivore population dynamics and resource use patterns are common (Rhoades, 1985). Narrow phenological windows of suitability of host plants can influence herbivore populations within (Satchell, 1962; Hunter, 1990) and between (Varley *et al.*, 1973; Phillipson and Thompson, 1983) years, presumably through their effects on herbivore survivorship, growth rate, and fecundity (Wint, 1983; Schroeder, 1986; Raupp *et al.*, 1988). Seasonal variation in host plant quality may also influence herbivores over evolutionary time by providing opportunities for seasonal specialists (Mattson, 1980) and by influencing dietary breadth (Hunter, 1990).

3. Variation among Individual Plants

Variation among individuals within a plant species and population adds further resource variability to the environment encountered by herbivores. There are three major factors that generate this variability: plant age, plant genotype, and the influence of the environment. These, of course, are interacting variables—the influence of an impoverished soil on plant reproduction, for example, may vary with age and genotype.

Plants of different ages are often differentially susceptible to herbivores (e.g. Martin, 1966; Niemela *et al.*, 1980; Kearsley and Whitham, 1989; Price *et al.*, 1990), and individuals of an unsuitable age class may be "transparent" to foraging animals. Since some plants do not begin reproduction until they reach a particular age or size, they cannot act as a food resource for nectarivores, frugivores, or seed predators until that time. Plant age may also interact with other trophic levels to influence animals that depend upon primary producers. For example, the abundance of parasitoids and predators of some phytophagous insects varies with the age of the host plant (Gagne and Martin, 1968; Munster-Swendsen, 1980).

The relative contributions of genotype and environment to plant variability are often difficult to untangle. The genetics of plant resistance, although exploited within artificial crop ecosystems, is not well understood in natural plant communities (Edmunds and Alstad, 1978). This is a rapidly developing field and a number of authors have demonstrated clear differences in herbivore performance on plants of different genotype (Bergman and Tingey, 1979; Price *et al.*, 1980) and interactions between plant genotype and other trophic levels (Weis *et al.*, 1985; Weis and Gorman, 1990).

Environmental impacts on plant resource quality have been demonstrated more frequently than genetic effects, probably because they are more easily manipulated. Light level and soil type can influence the carbon/

nitrogen balance (Bryant *et al.*, 1983) and secondary chemistry of the foliage of many plant species (e.g. Larsson *et al.*, 1986; Bryant, 1987) and environmental effects on plant growth and reproduction are well documented. Herbivores themselves exert an environmental influence on plants that can change the quality and quantity of resource for themselves and other animals. Wound-induced changes in leaf chemistry and structure are now well documented (Green and Ryan, 1972; Haukioja and Niemela, 1977; Hunter, 1987; Rossiter *et al.*, 1988) and can influence the population dynamics and community structure of animals that utilize plants.

Whatever the relative contributions of age, genotype, and environment, individual plants within a species are not replicates of each other. Between-plant variation can influence the foraging strategy of herbivores since risk from natural enemies or the environment may vary among host plant individuals (Schultz *et al.*, 1990). Conversely, the degree to which herbivores are clumped on their hosts may influence their population dynamics (Cook and Hubbard, 1977; Hassell *et al.*, 1987; Elkinton *et al.*, 1990). Variable host plant quality can also influence the growth and fecundity of herbivores and the phenotype of offspring that they produce (Rossiter *et al.*, 1988; Rossiter, 1991).

4. Variation within Individual Plants

At any given point in time, individual plants exhibit variability in their tissues, presenting a patchwork of resource quality to would-be consumers. Within-plant variation in resource quality has received increasing attention by animal-plant biologists (Denno and McClure, 1983; Whitham, 1986; Kimmerer and Potter, 1987; Craig *et al.* 1989), and the sources of within-plant variability are probably similar to those operating between plants— tissue age, tissue genotype (the prevalence of somatic mutation awaits further investigation), and the influence of the environment. Sun and shade leaves, for example, can differ in their chemistry and suitability to herbivores, and wound-induced changes in leaf quality can be as powerful among leaves within a plant as they are between individual plants (Hunter, 1987).

Consequently, variability within plants can affect herbivores in ways similar to variability among plants, influencing their foraging strategies (Barbosa and Greenblatt, 1979; Claridge, 1986), population dynamics (Bultman and Faeth, 1988) and offspring quality (Rossiter, 1991). The same tissue type can vary chemically (Claridge, 1986; Bultman and Faeth, 1988) and physically (Myyasi *et al.*, 1976; Hunter, 1987) within one plant, exerting influence both directly and indirectly on herbivore biology.

There is a dynamic mosaic of plant resource quality and quantity in ecological and evolutionary time and, at the interface of this mosaic and the rest of the biotic and the abiotic environment, animals that use plants must find food, avoid natural enemies, and reproduce. The following chapters describe both a) the influence of plant resource heterogeneity on herbivore population quality, population dynamics, and community structure, and b) the variability within and among herbivore species that impacts plant ecology and evolution.

References

Antonovics, J., Bradshaw, A. D., and Turner, R. G. (1971). Heavy metal tolerance in plants. *Adv. Ecol. Res.* **7**, 1–85.

Barbosa, P., and Greenblatt, J. (1979). Effects of leaf age and position on larval preferences of the fall webworm, *Hyphantria curea* (Lepidoptera: Arctiidae). *Can. Ent.* **111**, 381–383.

Bartholomew, G. A. (1958). The role of physiology in the distribution of terrestrial vertebrates. *In:* "Zoogeography" (C. L. Hubbs, ed.) *Amer. Assoc. Adv. Sci. Pub.* 151, Washington D.C.

Bergman, J. M., and Tingey, W. M. (1979). Aspects of interaction between plant genotypes and biological control. *Bull. Entomol. Soc. Amer.* **25**, 275–279.

Boecklen, W. J., and Spellenberg, R. (1990). Structure of herbivore communities in two oak (*Quercus* spp.) hybrid zones. *Oecologia* **85**, 92–100.

√ Bryant, J. P. (1987). Feltleaf willow-snow shoe hair interactions: Plant carbon/nutrient balance and flood plain succession. *Ecology* **68**, 1319–1327.

Bryant, J. P., Chapin, F. S., and Klein, D. R. (1983). Carbon/nutrient balance of boreal plants in relation to vertebrate herbivory. *Oikos* **40**, 357–368

Bultman, T. L., and Faeth, S. H. (1988). Abundance and mortality of leaf miners on artificially shaded Emory oak. *Ecol. Entomol.* **13**, 31–42.

Carpenter, S. R., Kitchell, J. F., and Hodgson, J. R. (1985). Cascading trophic interactions and lake productivity. *Bioscience* **35**, 634–639.

Carpenter, S. R., Kitchell, J. F., Hodgson, J. R., Cochran, P. A., Elser, J. J., Elser, M. M., Lodge, D. M., Kretchmer, D., He, X., and von Ende, C. N. (1987). Regulation of lake primary productivity by food web structure. *Ecology* **68**, 1863–1876.

Claridge, D. W. (1986). The distribution of a typhlocybine leafhopper, *Ribautiana ulmi* (Homoptera: Cicadellidae) on a specimen of wych elm tree. *Ecol. Entomol.* **11**, 31–39.

Cook, R. M., and Hubbard, S. F. (1977). Adaptive searching strategies in insect parasites. *J. Anim. Ecol.* **46**, 115–175.

Craig, T. P., Itami, J. K., and Price, P. W. (1989). A strong relationship between oviposition preference and larval performance in a shoot-galling sawfly. *Ecology* **70**, 1691–1699.

Crawley, M. J. (1983). "Herbivory, the dynamics of animal–plant interactions." Blackwell, Oxford.

Crawley, M. J. (1989). Insect herbivores and plant population dynamics. *Ann. Rev. Entomol.* **34**, 531–564.

Denno, R. F., and McClure, M. S. (1983). "Variable plants and herbivores in natural and managed systems." Academic Press, New York.

Denslow, J. S., Schultz, J. C., Vitousek, P. M., and Strain, B. R. (1990). Growth responses of tropical shrubs to treefall gap environments. *Ecology* **71**, 165–179.

Doane, C. C., and McManus, M. L. (1981). The gypsy moth: Research toward integrated pest management. USDA Forest Service Technical Bulletin 1584.

Edmunds, G. F., and Alstad, D. N. (1978). Coevolution in insect herbivores and conifers. *Science* **199**, 941–945.

Elkinton, J. S., Gould, J. R., Ferguson, C. S., Liebhold, A. M., and Wallner, W. E. (1990). Experimental manipulation of gypsy moth density to assess impact of natural enemies. *In:* "Population Dynamics of Forest Insects" (A. D. Watt, S. R. Leather, M. D. Hunter, and N. A. C. Kidd, eds.). Intercept, Andover.

Faeth, S. H. (1987). Indirect interactions between seasonal herbivores via leaf chemistry and structure. *In:* "Chemical mediation of coevolution" (K. Spencer, ed.). AIBS Symposium.

Feeny, P. (1970). Seasonal changes in oak leaf tannins and nutrients as a cause of spring feeding by winter moth caterpillars. *Ecology* **51**, 565–581.

Feeny, P. (1976). Plant apparency and chemical defense. *Rec. Adv. Phytochem.* **10**, 1–40.

Futuyma, D. J. (1976). Food plant specialisation and environmental predictability in Lepidoptera. *Am. Nat.* **110**, 285–292.

Gagne, W. C., and Martin, J. L. (1968). The insect ecology of red pine plantations in central Ontario V. The Coccinellidae (Coleoptera). *Can. Ent.* **100**, 835–846.

Gilbert, L. E. (1975). Ecological consequences of a coevolved mutualism between butterflies and plants. *In:* "Coevolution of animals and plants" (L. E. Gilbert and P. H. Raven, eds.). University of Texas Press, Austin.

Gilbert, L. E. (1979). Development of theory in the analysis of insect–plant interactions. *In:* "Analysis of Ecological Systems" (D. J. Horn, G. R. Stairs and R. D. Mitchell, eds.). Ohio State University Press, Columbus.

Gill, F. B., and Wolf, L. L. (1975). Economics of feeding territoriality in the golden-winged sunbird. *Ecology* **56**, 333–345.

Green, T. R., and Ryan, C. A. (1972). Wound-induced proteinase inhibitor in plant leaves: a possible defence mechanism against insects. *Science* **175**, 776–777.

Haering, R., and Fox, B. J. (1987). Short-term coexistence and long-term competitive displacement of two dominant species of *Iridomyrmex:* the successional response of ants to regenerating habitats. *J. Anim. Ecol.* **56**, 495–507.

Hairston, N. G., Smith, F. E., and Slobodkin, L. B. (1960). Community structure, population control and competition. *Am. Nat.* **44**, 421–425.

Hassell, M. P., Southwood, T. R. E., and Reader, P. M. (1987). The dynamics of the viburnum whitefly (*Aleurotrachelus jelinekii* Fraunf.): a case study on population regulation. *J. Anim. Ecol.* **56**, 283–300.

Haukioja, E., and Niemela, P. (1977). Retarded growth of a geometrid larva after mechanical damage to leaves of its host tree. *Annales zoologici Fennici* **14**, 48–52.

Haukioja, E., and Niemela, P. (1979). Birch leaves as a resource for herbivores: seasonal occurence of increased resistence in foliage after mechanical damage of adjacent leaves. *Oecologia* **39**, 151–159.

Houston, D. R., and Valentine, H. T. (1977). Comparing and predicting forest stand susceptibility to gypsy moth. *Can. J. For. Res.* **7**, 447–461.

Hunter, M. D. (1987). Opposing effects of spring defoliation on late season oak caterpillars. *Ecol. Entomol.* **12**, 373–382.

Hunter, M. D. (1990). Differential susceptibility to variable plant phenology and its role in competition between two insect herbivores on oak. *Ecol. Entomol.* **15**: 401–408.

Kareiva, P., and Sahakian, R. (1990). Tritrophic effects of a single architectural mutation in pea plants. *Nature* **345**, 433–434.

Kearsley, M. J. C., and Whitham, T. G. (1989). Developmental changes in resistance to herbivory: Implications for individuals and populations. *Ecology* **70**, 422–434.

Kimmerer, T. W., and Potter, D. A. (1987). Nutritional quality of specific leaf tissues and selective feeding by a specialist leafminer. *Oecologia* **71**, 548–551.

Kuris, A. M., Blaustein, A. R., and Alio, J. J. (1980). Hosts as islands. *Am. Nat.* **116**, 570–586.

Larsson, S., Wiren, A., Lundgren, L., and Ericsson, T. (1986). Effects of light and nutrient stress on leaf phenolic chemistry in *Salix dasyclados* and susceptibility to *Galerucella lineola* (Coleoptera). *Oikos* **47**, 205–210.

Lawton, J. H., and Strong, D. R. (1981). Community patterns and competition in folivorous insects. *Am. Nat.* **118**, 317–338.

Liebig, J. (1840). "Chemistry and its application to agriculture and physiology." Taylor and Walton, London.

Lindroth, R. L. (1989). Biochemical detoxification: mechanism of differential tiger swallowtail tolerance to phenolic glycosides. *Oecologia* **81,** 219–224.

Loucks, O. L. (1970). Evolution and diversity, efficiency, and community stability. *Amer. Zool.* **10,** 17–25.

Martin, J. L. (1966). The insect ecology of red pine plantations in central Ontario IV. The crown fauna. *Can. Ent.* **98,** 10–27.

Mattson, W. J. (1980). Herbivory in relation to plant nitrogen content. *Ann. Rev. Ecol. Syst.* **11,** 119–161.

Mayyasi, A. M., Pulley, P. E., Coulson, R. N., DeMichele, D. W., and Foltz, J. L. (1976). A mathematical description of the within tree distribution of the various developmental stages of *Dendroctonus frontalis Zimm.* (Coleoptera: Scolytidae). *Res. Pop. Ecol.* **18,** 135–145.

McQueen, D. J., Johannes, M. R. S., Post, J. R., Stewart, T. J., and Lean, D. R. S. (1989). Bottom-up and top-down impacts on freshwater pelagic community structure. *Ecol. Monogr.* **59,** 289–309.

McQueen, D. J., Post, J. R., and Mills, E. L. (1986). Trophic relationships in freshwater pelagic systems. *Can. J. Fish. Aquat. Sci.* **43,** 1571–1581.

Munster-Swendsen, M. (1980). The distribution in time and space of parasitism in *Epinotia tedella* (Cl) (Lepidoptera: Tortricidae). *Ecol. Entomol.* **5,** 373–383.

Niemela, P., Tuomi, J., and Haukioja, E. (1980). Age-specific resistance in trees: Defoliation of tamaracks (*Larix laricina*) by larch budmoth (*Zeiraphera improbana*) (Lepidoptera: Tortricidae). *Rep. Kevo. Subarctic Res. Stat.* **16,** 49–57.

Oates, J. F., Waterman, P. A., and Choo, G. M. (1980). Food selection by the south Indian leaf-monkey, *Presbytis johnii*. *Oecologia* **45,** 45–56.

Odum, E. P. (1959). "Fundamentals of ecology." 2nd ed. W. B. Saunders, Philadelphia.

Phillipson, J., and Thompson, D. J. (1983). Phenology and intensity of phyllophage attack on *Fagus sylvatica* in Wytham Woods, Oxford. *Ecol. Entomol.* **8,** 315–330.

Pianka, E. R. (1989). Desert lizard diversity: additional comments and some data. *Am. Nat.* **134,** 344–364.

Pickett, S. T. A., and White, P. S. (1985). "The ecology of natural disturbance and patch dynamics." Academic Press, New York.

Power, A. G. (1991). Virus spread and vector dynamics in genetically diverse plant populations. *Ecology* **72,** 232–241.

Power, M. E., Mathews, W. J., and Stewart, A. J. (1985). Grazing minnows, piscivorous bass, and stream algae: dynamics of a strong interaction. *Ecology* **66,** 1448–1456.

Price, P. W. (1990). Evaluating the role of natural enemies in latent and eruptive species: New approaches in life table construction. *In:* "Population Dynamics of Forest Insects" (A. D. Watt, S. R. Leather, M. D. Hunter, and N. A. C. Kidd, eds.). Intercept, Andover.

Price, P. W., Bouton, C. E., Gross, P., McPheron, B. A., Thompson, J. N., and Weis, A. E. (1980). Interactions among three trophic levels: influence of plants on interactions between insect herbivores and natural enemies. *Ann. Rev. Ecol. Syst.* **11,** 41–65.

Price, P. W., Cobb, N., Craig, T. P., Fernandes, G. W., Itami, J. K., Mopper, S., and Preszler, R. W. (1990). Insect herbivore population dynamics on trees and shrubs: New approaches relevant to latent and eruptive species and life table development. *In:* "Insect-Plant Interactions" (E. A. Bernays, ed.) . Vol. 2. CRC Press, Boca Raton, Florida.

Ramachandran, R. (1987). Influence of host plants on the wind dispersal and the survival of an Australian geometrid caterpillar. *Ent. Exp. Appl.* **44,** 289–294.

Raupp, M. J., Werren, J. H., and Sadof, C. S. (1988). Effects of short-term phenological changes in leaf suitability on the survivorship, growth, and development of gypsy moth (Lepidoptera: Lymantriidae) larvae. *Environ. Entomol.* **17,** 316–319.

Reader, P. M., and Southwood, T. R. E. (1981). The relationship between palatability to invertebrates and the successional status of a plant. *Oecologia* **51,** 271–275.

Reichman, O. J., and Smith, S. C. (1991). Responses to simulated leaf and root herbivory by a biennial. *Ecology* **72,** 116–124.

Rhoades, D. F. (1985). Offensive–defensive interactions between herbivores and plants: Their relevance to herbivore population dynamics and community theory. *Amer. Nat.* **125,** 205–238.

Rockwood, L. L. (1974). Seasonal changes in the susceptibility of *Crescentia alata* leaves to the flea beetle, *Oedionychus* sp. *Ecology* **55,** 142–148.

Root, R. B. (1973). Organization of plant–arthropod association in simple and diverse habitats: The fauna of collards (*Brassica oleracea*). *Ecol. Monogr.* **43,** 95–124.

Rosenthal, G. A., and Janzen, D. H. (1979). "Herbivores: Their interaction with secondary plant metabolites." Academic Press, New York.

Rossiter, M. C. (1991). Environmentally-based maternal effects: A hidden force in insect population dynamics? *Oecologia* **87,** 288–294.

Rossiter, M. C., Schultz, J. C., and Baldwin, I. T. (1988). Relationships among defoliation, red oak phenolics, and gypsy moth growth and reproduction. *Ecology* **69,** 267–277.

Satchell, J. E. (1962). Resistance in oak (*Quercus* spp.) to defoliation by *Tortrix viridana* L. in Roudsea Wood National Nature Reserve. *Ann. Appl. Biol.* **50,** 431–442.

Schindler, D. W. (1978). Factors regulating phytoplankton production and standing crop in the world's freshwaters. *Limnol. Oceanogr.* **23,** 478–486.

Schroeder, L. A. (1986). Changes in tree leaf quality and growth performance of Lepidopteran larvae. *Ecology* **67,** 1628–1636.

Schultz, J. C. (1983). Habitat selection and foraging tactics of caterpillars in heterogeneous trees. *In:* "Variable Plants and Herbivores in Natural and Managed Systems" (R. F. Denno and M. S. McClure, eds.). Academic Press, New York.

Schultz, J. C., Foster, M. A., and Montgomery, M. E. (1990). Host plant-mediated impacts of a baculovirus on gypsy moth populations. *In:* "Population Dynamics of Forest Insects" (A. D. Watt, S. R. Leather, M. D. Hunter, and N. A. C. Kidd, eds.). Intercept, Andover.

Schultz, J. C., Nothnagle, P. J., and Baldwin, I. T. (1982). Individual and seasonal variation in leaf quality of two northern hardwood tree species. *Am. J. Bot.* **69,** 753–759.

Sinclair, A. R. E. (1985). Does interspecific competition or predation shape the African ungulate community? *J. Anim. Ecol.* **54,** 899–918.

Siqueira, J. O., Colozzifilho, A., and Deoliveira, E. (1989). Occurence of vesicular-arbuscular *Mycorrhizae* in agroecosystems and natural ecosystems of Minas-Gerais State. *Pesquisa Agropecuaria Brasileira* **24,** 1499–1506.

Southwood, T. R. E. (1960). The abundance of the Hawaiian trees and the number of their associated insect species. *Proc. Hawaiian Ent. Soc.* **17,** 299–303.

Southwood, T. R. E. (1961). The number of species of insect associated with various trees. *J. Animal Ecol.* **30,** 1–8.

Starr, M., Himmelman, J. H., and Therriault, J. C. (1990). Direct coupling of marine invertebrate spawning with phytoplankton blooms. *Science* **247,** 1071–1074.

Strong, D. R., Jr. (1979). Biogeographic dynamics of insect–host plant communities. *Ann. Rev. Entomol.* **24,** 89–119.

Strong, D. R., Lawton, J. H., and Southwood, T. R. E. (1984). "Insects on Plants. Community patterns and mechanisms." Blackwell, Oxford.

Varley, G. C., Gradwell, G. R., and Hassell, M. P. (1973). Insect Population Ecology. An analytical approach. Blackwell, Oxford.

Weis, A. E., Abrahamson, W. G., and McCrea, K. D. (1985). Host gall size and oviposition success by the parasitoid *Eurytoma gigantea*. *Ecol. Entomol.* **10,** 341–348.

Weis, A. E., and Gorman, W. L. (1990). Measuring selection on reaction norms: An exploration of the *Eurosta-Solidago* system. *Evolution* **44,** 820–831.

Welden, C. W., Hewett, S. W., Hubbell, S. P., and Foster, R. B. (1991) Sapling survival, growth, and recruitment: relationship to canopy height in a neotropical forest. *Ecology* **72**, 35–50.

West, C. (1985). Factors underlying the late seasonal appearance of the lepidopterous leaf mining guild on oak. *Ecol. Entomol.* **10**, 111–120.

Whitham, T. G. (1986). Costs and benefits of territoriality: Behavioral and reproductive release by competing aphids. *Ecology* **67**, 139–147.

Whitham, T. G. (1989). Plant hybrid zones as sinks for pests. *Science* **244**, 1490–1493.

Wint, G. R. W. (1983). The role of alternative host plant species in the life of a polyphagous moth, *Operophtera brumata* (Lepidoptera: Geometridae). *J. Anim. Ecol.* **52**, 439–450.

Young, T. P., and Smith, A. P. (1987). Alpine herbivory on Mount Kenya. *In* "Tropical Alpine Environments: Plant Form and Function" (P. Rundel, ed.), Springer-Verlag, Berlin.

2

The Impact of Resource Variation on Population Quality in Herbivorous Insects: A Critical Aspect of Population Dynamics

MaryCarol Rossiter

Department of Entomology
Pennsylvania State University
University Park, Pennsylvania

I. Introduction

A. The Relationship between Individual Quality, Population Quality, and Population Dynamics

An important, but often overlooked feature of population biology is that environmental variation does not affect an herbivore population *per se*. It affects the individuals that make up that population. The contribution of an individual to the response of the population (e.g., mortality or fecundity) is determined by the interaction of its genotype and the environment it encounters. In theoretical models, the cumulative phenotypic effect of gene–environment interactions is represented by the average response of the population to each environmental feature included in the model. The use of average response to characterize an interaction between herbivore and environment is parsimonious, concise, and practical. However, it may not be a realistic representation if the variation that contributes to the average response is, itself, the premise for alternative developmental histories that have an additional impact on population growth (or decline) not otherwise expressed in the model.

Gene–environment interactions are responsible for qualitative features of an individual (i.e., life history expression) and for qualitative features of a population (i.e., average life-history expression of the group). The form of the gene–environment interactions, captured by population quality variables, can provide explicit proximal causes for the population dynamics of a species as well as ultimate causes for the population dynamics by virtue of their ability to alter the nature of feedback loops between the herbivore population and its environment.

I shall use general systems theory (Milsum, 1968; Berryman, 1981) as the framework to demonstrate that population quality should be included in the development of population dynamics models, whether the goal is to uncover general ecological processes or predict the population growth of a particular species. In the absence of information on the *nature* (i.e., the underlying biological mechanism) of a population's variation in physiological and behavioral adjustments to resource variation, there may be a great reduction in the utility of parameters typically used in the prediction of herbivore population dynamics–population size (N) and a population's average response to resources or natural enemies (e.g., based on predator–prey, or host quality–fecundity relationships). In this chapter, I shall argue that population quality is a critical variable in the ecological and evolutionary fate of herbivore populations. By extension, population quality is critical to community composition and stability whenever an herbivore population has a significant effect on the fate of other species in the community.

To understand fully the impact of resource variation on the population dynamics of an herbivore and its position in the community, we need, first, to consider the impact of resource variation on the individual. This bottom-up approach argues strongly for the need to consider the basic biology of the organism (genetics, development, behavior) when developing hypotheses about population and community-level phenomena. *Individual quality* describes the effect of the environment on the expression of a genotype with respect to the success (i.e., fitness) of the individual or lineage. By extension, *population quality* describes the cumulative impact of individual quality on success (i.e., growth or decline) of a population. Quality does not imply superiority; it merely recognizes that individuals and populations can differ by virtue of innate constitution and experience.

Throughout the history of population studies, mathematical ecologists have stated their awareness of both the importance and the omission of population quality in describing and predicting the fate of populations (e.g., Berryman, 1981; Lomnicki, 1988; Getz and Haight, 1989). With respect to humans, the economist and Nobel laureate T. W. Schultz (1980) thinks the omission arises from a reliance on the quantitative theory of populations, owing considerably to Malthus who "could not have anticipated the substitution by parents of quality for quantity of children" (p. 18). In the study of insects, W. G. Wellington was insightful, and also humorously indignant, that individuals of a population were viewed merely as participants of a count and cast into uniformity, a condition emblazoned with the title "monolithic lump of protoplasm" (Wellington, 1977, p. 2).

Population quality has been slighted for good reason. First, it is not immediately clear what should be measured, that is, which gene–environment interactions alter the probability of survival or reproduction. Second, the effort required to characterize the interaction between environmental heterogeneity and population quality can be staggering (Montgomery, 1990). Fortunately, recent advances in biotechnology and statistical and computing capabilities make such characterization feasible, and the process will become more efficient as collaboration between the sciences of organismal biology and population biology increases (e.g., see Calow and Sibley, 1990).

B. Thesis and Organization of Chapter

It is the aim of this chapter to demonstrate that the inclusion of population quality variables in models of herbivore population dynamics can improve their heuristic value and predictive power. This improvement will support the development of successful herbivore control programs that minimize environmental hazard and the development of conservation programs aimed at the preservation of a particular herbivore taxon or entire communities in endangered habitats.

To establish the importance of population quality in herbivore population biology, I have focused primarily on temperate forest insects that experience outbreak. In Section II, I describe the relevance of population quality to population dynamics in terms of general systems theory. This is followed by empirical information that highlights the pitfalls of omitting population quality factors in theoretical and empirical studies of population dynamics. In Section III, the criteria for measuring population quality are provided, followed by a discussion of the conceptual and logistic difficulties of such measurements. Section IV focuses on the contribution of resource variation to the expression of population quality. The resource emphasized is food quality, critical in its own right and often the mediator in other ecological and autecological forces on the herbivore. Environmentally-based maternal effects provide a most remarkable example of the influence of resource variation on population quality. Environmentally-based maternal effects occur when the environmental experience of the parent(s) produces a phenotypic alteration in the offspring. This phenomenon is documented for a number of herbivore species. In Section V, the logic used in the development of the Maternal-Effects Hypothesis of Outbreak is presented with theoretical and empirical support. The chapter ends with a general approach to testing the hypothesis.

II. Relevance of Population Quality to Population Dynamics

A. A General Systems Theory Perspective

General systems theory provides an excellent framework to investigate the dynamic features of herbivore population behavior (see Berryman, 1981, 1989). To explain the participation of population quality in population dynamics, I shall apply the basic concepts of general systems theory (as outlined in Berryman, 1981) with a simple herbivore example. The components of the herbivore system represented in Figure 1 include herbivore number (population size), herbivore population quality, natural enemy number, food quality, and weather. Population quality variables refer to any features of the herbivore's biology that have the potential to influence population growth; they include genetic, physiological, and behavioral characteristics.

In Figure 1, two types of input variables are represented. Exogenous effects (squares) are input variables that experience no feedback from the system (e.g., weather, cross-generational phenomena). State variables (circles) are input variables that experience feedback in the system (e.g., parasite density influences and is influenced by herbivore density). Input variables

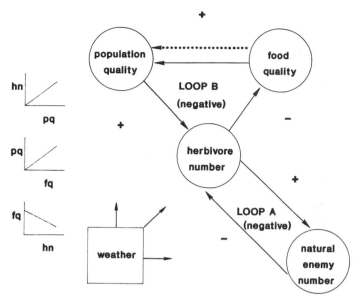

Figure 1 General systems diagram of a simple herbivore system; the sign of each interaction reflects the form of the relationship (see graphs) between state variables (circles); solid arrows represent interaction processes; dashed arrows indicate a time lag in the interaction; exogenous variables (squares) can influence each state variable, but with no feedback effect. See text for amplification.

should be the most sensitive indicators of change in population growth (e.g., important mortality agents identified by life table analysis).

In response to input, the value of a state variable can be modified. The process that links stimulus (e.g., food quality) and response (e.g., population quality) is represented by an arrow (e.g., gene–environment interaction). Interactive processes can occur between a state variable and exogenous effects (herbivore number and weather), between state variables (herbivore and natural enemy number), or within a state variable (intraspecific competition). Feedback occurs when a stimulus is fed back to its origin through one or more interactions. When state variables mutually influence one another, they are involved in a positive or negative feedback loop. The net effect of the feedback loop can be determined by multiplying the signs of the component processes. When the product of the loop is negative, state variables tend to return to their original condition; this encourages stability in the dynamics of the system (e.g., feedback loops A and B). When the product of the feedback loop is positive, state variables move in the same direction as the initial stimulus. This results in destabilization.

Although no positive feedback loop is shown in Figure 1, a modification to the relationship between population size and food quality can produce one. Let us hypothesize that there is a threshold effect involved in the influence of population size on food quality. Once a density threshold is crossed, the sign of the interaction changes from positive to negative. This changes the sign of the feedback loop to a positive one, a condition that promotes destabilization. An example of this is found in the bark beetle, *Dendroctonus ponderosae*. Once population size crosses a threshold (set by local ecological factors), cooperative behavior (aggregation) leads to the loss of defensive response in the host plant (Raffa, 1988). The next point to be made about the general systems model is that a time delay in a feedback loop (even a negative one) can cause destabilization in the system (see dashed arrow in loop B). The magnitude of the time-lag effect and speed with which it is transmitted will determine the dynamic behavior of a system.

B. Support for the Inclusion of Population Quality in Models of Population Dynamics

Proximal causes for the initiation of an outbreak in natural herbivore populations are largely unknown. Empirical and theoretical work indicates that outbreak can occur when a deterministic or stochastic event increases the availability of a limiting resource as occurs in agricultural and forestry monocultures (Risch, 1987) or increases population number through immigration (Rankin and Singer, 1984). Outbreak can also occur when action of the regulatory agent(s) is absent or impeded (e.g., after introduction of foreign species [McClure, 1988) or through application of pesticides (Huffaker and Messenger, 1976)]. Escape from a regulating agent is commonly invoked as the *cause* of outbreak in undisturbed systems, although there are very few data sets to support this hypothesis. Royama (1977) showed that the statistical methods used to correlate number of herbivores and regulating agents cannot distinguish whether escape from natural enemies is the cause or the consequence of an outbreak.

Extensive empirical efforts to describe correlations between population size and environmental features of the herbivore's ecology, such as weather (Martinat, 1987), habitat and host plant availability (Redfearn and Pimm, 1987), and abundance of natural enemies (Price, 1987), have not yielded an answer as to the cause of outbreak. In few cases can these correlated factors, acting alone or together, be used reliably to predict population dynamics. Two interpretations are possible. First, the correlation between the value of some environmental input variable and population size is not the cause of but the consequence of outbreak. This "chicken and egg" conundrum persists because most studies of outbreak populations occur after population release from low density, and the environmental conditions preceding outbreak are lost (Hunter *et al.*, 1991). A second interpretation is that the environmental input variable is critical to population destabilization *only* when the value for average herbivore quality predisposes the population to accelerated growth. If the latter holds, outbreaks will be initiated only when certain environmental and population quality conditions are met simultaneously. In this chapter, I shall argue that *population quality is a critical component of population dynamics because it constitutes the baseline from which the impact of environment is determined.*

Mathematical models show that outbreak can result from a time delay in the response of a population to some density-dependent factor (Caswell, 1972; May *et al.*, 1974; Berryman, 1978, 1981). Theoreticians have suggested that features of population quality (genetic, physiological, or behavioral traits) which produce time lags can provoke population fluctuations (May, 1975; Berryman, 1987). However, little is known empirically about the impact of population quality on the population growth of herbivorous insect species, despite a respectable history of verbal support for its importance (e.g., Wellington, 1957, 1977; Uvarov, 1961; Leonard, 1970; Capinera, 1979; Berryman, 1981, 1988; Rhoades, 1983; Barbosa and Baltensweiler, 1987; Haukioja and Neuvonen, 1987; Mitter and Schneider, 1987; Lomnicki, 1988).

To extend my argument that the inclusion of population quality is critical to understanding herbivore population dynamics, let us consider evidence from forest insect species that experience eruptive or cyclic outbreaks. The mechanisms that induce noticeable fluctuations in population size may be similar regardless of the magnitude or regularity of the fluctuations, but detection of variation in population quality may be easier in species which are extreme in their fluctuations.

A number of eruptive forest pest species have been extensively studied in order to develop a method of predicting outbreak (e.g., see Berryman, 1988). Overall, these efforts have provided a wealth of correlative data on population size and the state of the environment, but they infrequently lead to the reliable prediction of population dynamics (but see Berryman *et al.*, 1990). For example, the spruce budworm, *Choristoneura fumiferana*, experiences outbreaks correlated with a dramatic increase in the availability of high-quality food (Morris, 1963; Kimmins, 1971). Under the condition of even-age monocultures of over-mature balsam fir trees and dry weather, the population will usually (but not always, see Morris, 1963) undergo an outbreak. In even-aged stands of young trees, outbreak will not occur. Less clear are the circumstances of outbreak under intermediate conditions such as a mixed species stand or a monoculture in a stable age distribution where outbreaks occur with less frequency (Ghent, 1958; Rhoades, 1983). Escape from bird predation has been invoked as the agent of population release (Morris *et al.*, 1958; Buckner, 1966; Ludwig *et al.*, 1978), but it is unclear whether the decrease in bird predation is the cause or consequence of outbreak (Crawley, 1983). Watt (1963) hypothesized that outbreak was the result of either escape from natural enemies or the result of greater vigor at intermediate densities, in other words, the result of a shift in population quality.

The factor assumed to cause outbreak in some bark beetles, e.g.,

Dendroctonus rufipennis, is increased availability of high-quality food due to reduction in the spruce defensive response (resinosis) in diseased or senescent trees (Coulson, 1979). However, availability of high quality (noninduced) food alone is not always associated with population growth. Beetles from an outbreak population were allowed to colonize test logs in which the normal defensive response of the tree was absent. This subpopulation showed the same reduced fecundity as conspecifics in living trees as the outbreak declined (McCambridge and Knight, 1972). From this, I conclude that initial condition of population quality can be of greater importance to population growth than is current environmental quality.

For the gypsy moth, *Lymantria dispar,* much effort has been expended on the development of methods to predict population growth for purposes of control (Doane and McManus, 1981). From the simplest models that depend on egg mass density to complex models that include many parameters to describe the action of environmental factors on population dynamics (e.g., natural enemies, food quality), the ability to predict outbreak is still very poor (Chapters 3, 4 in Doane and McManus, 1981; Valentine, 1983; Elkinton *et al.,* 1990). In a review of gypsy moth population dynamics, Elkinton and Liebhold (1990) conclude that the reason for failure of regulatory agents at population release is unknown, but that changes affecting fecundity and early larval survival may be involved. This strongly suggests that the status of population quality is critical to release from its low-density equilibrium. In this species, variation in population quality is expressed in fecundity, development time, hatch phenology, and susceptibility to toxins (Rossiter, 1987, 1991a,b; Rossiter *et al.,* 1990).

For cyclic rather than eruptive herbivores, outbreak occurs at predictable intervals. For these species, efforts have focused on discovering the reason for cycles. The general theory of outbreak recognizes that the response of any state variable has the potential to generate outbreaks through effects on the feedback structure of the system (Berryman, 1981). From empirical efforts there is no consensus on the proximal causes of herbivore cycles, although hypotheses involving weather, migration, food availability, and natural enemies have been well tested (many examples in Berryman, 1988). With a few exceptions, population quality variables have not been considered. The most notable exception concerns noninsect herbivores—the cycling microtine species. The value for average population quality, measured as degree of agressive behavior, was found to shift as population density changed (Krebs and Myers, 1974). Chitty (1960, 1967) hypothesized that microtine cycles were driven by changes in gene frequencies for this population quality trait. Stenseth (1981) tested this hypothesis with a theoretical model and found it implausible. However, Chitty's hypothesis and, consequently, Stenseth's model, did not include any interaction between environmental variables (proximal extrinsic factors) and population quality (proximal intrinsic factors). Stenseth (1981) concludes that the hypothesis seems more plausible if the interaction between intrinsic and extrinsic factors is included. This amplified hypothesis has not been tested.

A search for the cause of the 9-year population cycles of the larch budmoth, *Zeiraphera diniana,* and the autumnal moth, *Epirrata autumnata,* has been extensive. Most aspects of the species' natural history and many aspects of the biology are well documented. For both species, deteriorating food quality during outbreak and population decline are correlated with reduction of survival and fecundity in the current and subsequent generations (Fischlin and Baltensweiler, 1979; Baltensweiler and Fischlin, 1988; Haukioja *et al.,* 1988). Baltensweiler and Fischlin (1988) report the results of a simulation model of larch budmoth dynamics, which included a time delay in the negative feedback between food quality and population size; this loop represented the long-term induced defensive responses of the host plant. The model accurately describes the 9-year cycles. However, the authors state that the mechanism by which food quality influences dynamics is unknown; they consider the interaction between food quality and genotype as one likely candidate. For the autumnal moth, a qualitative model of population growth also showed that a time delay in the negative feedback loop between food quality and population size is very important to population dynamics (Haukioja *et al.,* 1988). These authors also suggested that a shift in population quality may be involved (see Section IV,D).

Douglas-fir tussock moth, *Orgyia pseudotsugata,* also experiences 9-year cycles. Factors causing numerical change have been identified only during the outbreak phase, and none of these factors is recognized to affect numerical change in non-outbreak populations (Mason and Overton, 1983). The proximal cause of these cycles is unknown, but numerical interactions with the host plant or with predators or parasitoids are considered likely candidates (Berryman *et al.,* 1990).

There exists one model developed exclusively to evaluate the role of population quality in the dynamics of herbivorous insects. A stochastic model was developed for the tent caterpillar, *Malacosoma pluviale,* to assess the effects of changing population quality on population growth (Wellington *et al.,* 1975). Values for population quality parameters were based on earlier results showing that behavioral and developmental traits were associated with survival and reproductive success, and these changes in population quality were correlated with changes in population size (Wellington, 1957, 1960, 1964). In simulations, colonies (families) were removed under many harvesting schemes ranging from intense, random removal to very limited removal of the highest quality colonies. Wellington *et al.* (1975) found that colony quality rather than number of colonies removed had, by far, the most pronounced effect on population growth. They also found that environmental factors that generated stress enhanced the effect of the harvesting scheme.

III. Criteria for Measurement of Population Quality

To include herbivore quality in the study of population dynamics, criteria must be developed for what to measure and how to measure it. What to measure is probably more elusive because natality and mortality can be significantly altered by an array of life-history traits whose expression is dependent on the environment. Where extensive data on the natural history, biology, and population dynamics of a species exist, synthetic thinking should identify the most likely aspect of population quality to consider.

A. Sources of Phenotypic Variation in Population Quality

The literature holds many examples of variation in population quality, but few are identified as such. Within a generation, quality variation is evident in reports of phenotypic variation in life-history traits. In this type of study, there is often an emphasis on cases of obvious morphological, developmental, and behavioral polymorphisms [e.g., phase change in grasshoppers (Nolte, 1974), variation in diapause (Mousseau and Roff, 1989), adult dispersal (Zera *et al.*, 1983), and propagule size (Capinera, 1979)]. The source of variation for complex life-history traits has just begun to be characterized with the application of quantitative genetics techniques to natural populations. By partitioning phenotypic variation into its causal components, these techniques are useful in deciphering the impact of resource variation on expression of the herbivore phenotype as well as providing an insight into the underlying mechanisms of gene–environment interaction.

Most empirical studies of population dynamics provide data on the numerical response of a population, thereby joining variation due to genes, environment, and gene–environment interaction under one umbrella. To understand how environmental factors such as food quality or weather can alter population quality, we need to know the extent that phenotype (e.g., reproductive output) is set by genetics, by environment, and by the interaction of the two. At the population level, sources of phenotypic variation can be identified and quantified with quantitative genetics methods. To date, this approach has been used successfully for a few herbivore species (e.g., Rausher, 1984; Via, 1984; Hare and Kennedy, 1986; Pashley, 1988; Boonstra and Boag, 1987; Rossiter, 1987). I shall describe a procedure called full-sib analysis (Falconer, 1981) with a hypothetical example and discuss the results to illustrate the utility of measuring components of phenotypic variance.

In this hypothetical example, larval development time was selected as the population quality trait to be measured for several reasons:

1. some life table analyses indicate that mortality by larval parasitoids is a key factor in population regulation (Crawley, 1983);
2. field work demonstrates that the longer the larval period, the greater the probability of parasitism (Cheng, 1970; Price *et al.*, 1980); and

3. development time can be greatly influenced by food quality; by virtue of this gene–environment interaction, development time is a population quality variable that is capable of modifying the magnitude of the parasitoid response to herbivore number.

The experimental design requires 30 mothers selected at random from the source population. Offspring from each mother are randomly divided across three host treatments (H1, H2, H3) and measured for the length of the larval period. [For the sake of simplicity, we shall assume that all familial variation is additive genetic variation, an assumption that can be tested with half-sib analysis (Falconer, 1981).] The sources of phenotypic variation in development time are evaluated with analysis of variance, which partitions the variance into these components: (1) *genetic:* variation arising from genetic differences among families, measured within each host environment and then summed across all hosts; (2) *plasticity:* variation arising from differences among siblings in their response to different host environments (summarized as the family's *norm of reaction*) and then summed across all families; and (3) *gene–environment interactions:* variation arising from differences among families in their norm of reaction.

Figure 2 a–e presents a series of results illustrating the need to understand the interaction between environmental and population quality factors. For clarity in this discussion, the population response is distilled to that of three representative families. In Figure 2a, there is no genetic variation for development time. Within each family, development time is the same regardless of host environment, and so there is no plasticity for development time. In Figure 2b, there is no genetic variation for development time, but significant plasticity. This means that the host environment alone will set the mean development time for the population. In Figure 2c, genetic variation for development time is expressed among families within each of three host environments, but no plasticity is expressed. Under the conditions of Figures 2a and 2c, genetic composition alone sets the mean development time for the population. In Figure 2d, genetic variation for development time is expressed among families within each of three host environments, and within each family, plasticity is expressed. In Figure 2e, genetic variation for development time is expressed in H2 and H3 but not in H1, and 2 of 3 families express plasticity across host environments. This model is very simple and ignores several complications provided by the natural world, such as variable host quality over space and time. In any case, the point is to show that *the expression of population quality is a function of the initial condition of the population and the environment that it encounters.*

B. Estimating Population Quality Parameters

For inclusion of population quality variables in population dynamics models, the best starting point is the least complex distillation of information about gene-environment interactions. A single value representing average

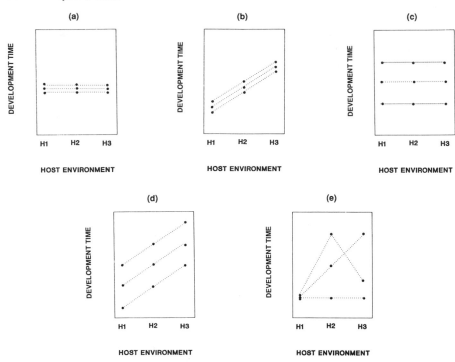

Figure 2 The genetic and plasticity status for development time in five hypothetical populations (a–e), each of which experiences three host environments. For clarity, the response of each population is distilled to that of three representative families. Norm of reaction plots show a population with (a) no genetic variation and no plasticity; (b) no genetic variation but plasticity expressed by all families; (c) genetic variation expressed in all host environments and plasticity expressed by all families; and (e) genetic variation expressed only in host environments 2 and 3 and plasticity expressed in 2 of 3 families.

population quality (PQ_{ave}) can be calculated as

$$PQ_{ave} = \frac{1}{n} \sum_{i=1}^{n} P_i(\text{LHT}_i) \tag{1}$$

which is the sum of the mean values for the herbivore life history trait (LHT) in each environment ($i = 1\ldots n$), divided by n environments; the influence of environmental heterogeneity is accounted for by weighting the mean LHT value by (P_i), the proportional contribution of each environment to the total environment.

The capacity for behavioral and physiological adjustment to environmental variation is set by the genotype that mediates the interaction between structural genes, regulatory genes, hormones, and the environment. Estimates of the realized phenotype, whether at the individual or population level, may require at least some knowledge of the mechanism of gene–environment interaction in order to improve the accuracy of a population quality component in a dynamics model. For example, if the response of the population quality trait to a range of environments is non-linear or, changes direction because of a developmental threshold effect, then gene–environment interactions are capable of changing the magnitude of a feedback effect or removing the feedback loop from the system through loss of density dependence. This point is elaborated in Section V with an example involving negative maternal effects, a phenomenon by which the environment produces opposing effects in parents and offspring.

C. The Time Scale for Measuring Population Quality

Although most genetic and evolutionary aspects of population biology focus on population variation, there has been limited application of these results to the study of population dynamics. Part of the problem lies in the time scale implicitly accepted by ecological and evolutionary biologists. With an evolutionary viewpoint, researchers evaluate the action of selection or the consequences of selection on a trait or suite of traits that confer fitness and generate an adaptive life history strategy. Although the quality of a population can change as a plastic response, the genetic structure of the population can also change over time and influence population dynamics. Moreover, the nature of the plastic response can change through long-term genetic change.

The search for cases in which genetic change underlies population dynamics, first discussed by Pimentel (1961), has been limited. For herbivores, the focus has been on highly specialized and relatively sessile herbivores suspected to be under very strong directional selection for host utilization. Scale insects, for example, appear to exhibit genetic change associated with population outbreak (Alstad and Edmunds, 1983; Wainhouse and Gate, 1988). In most plant–herbivore systems, greater mobility and the lack of extreme specialization would reduce the intensity of selection, making the measurement of genetic change even less tractable. Moreover, the use of traditional allozyme analysis would be inappropriate whenever selection occurs on regulatory rather than structural genes. If genetic change occurs in association with population cycles, that is, on an ecological time scale, we will need to develop methods more sensitive to these changes.

With an ecological viewpoint, the focus is usually on the immediate effects of the environment on phenotype or population size rather than on fitness, let alone genetic change. This is not surprising since one of the greatest challenges in biology is to define mechanisms that connect genotype to phenotype and phenotype to fitness (Dykhuizen and Dean, 1990), and then to discern the role of a genotype in greater interactive phenomena–population structure and dynamics and community construction and stability.

Work on cycling microtine rodents provides the earliest and one of the few attempts to link herbivore genotype with population level phenomena (see Section II,B). Recently, Boonstra and Boag (1987) applied quantitative genetics methods on offspring of field-captured meadow voles, reared under homogeneous conditions. They found that variation in population quality, measured in terms of developmental traits, had no additive genetic component indicating that short-term genetic selection was not possible. Most of the variation was assigned to nonadditive genetic components, which include the effects of dominance, epistasis, and the common environment, that is, maternal effects. While this may further convince some that there is no premise for including genetic considerations in the study of population dynamics, their work clearly indicates that long-term selection can produce a life-history strategy that equips a taxon with plastic responses to deal with resource variation. At the population level, plasticity could be the source of changing population quality seen in microtine populations. Plasticity is, perhaps, the basis of the opportunistic strategy seen in so many forest insect pest species.

While there may be no need to invoke short-term genetic change in population fluctuations, long-term genetic change may produce a life-history strategy that is based on a selected suite of plastic responses to changes in density or environmental quality, a *quality-alteration trait*. The evolution of a quality-alteration trait has theoretical support (Hastings and Caswell, 1979; Cooper and Kaplan, 1982; Wallace, 1982; Caswell, 1983; Lacey *et al.*, 1983; Smith-Gill, 1983) as well as empirical support (Henrich and Travis, 1988). I suggest that population cycles may reflect shifts in population quality that are formatted by the expression of quality-alteration traits. Group selection or Larmarckianism need not be invoked for the evolution of such a trait if success within a lineage is based on the ability for recurrent plastic alteration of life history. The evolution of a quality-alteration trait would occur, most likely, at the level of the regulatory genes.

IV. Resource Variation Effects Population Quality

A. Relationship between Food Quality and Population Quality

Food-quality variation is not an isolated component of resource variation because food is often a mediator (i.e., involved in feedback loops) of other environmental components of an herbivore system. An exogenous environmental effect such as weather can modify the system response by changing the magnitude of interactions between food quality and other state variables through, for example, the alteration of budburst phenology or host plant chemistry (Andrewartha and Birch, 1954; Holliday, 1977; Hunter, 1990). Food quality can also be influenced by interaction with other state variables such as conspecific population size (intraspecific competition) and the population size of other feeding members of the community (interspecific competition) (Lance *et al.*, 1986; Hunter, 1987; Hunter and Willmer, 1989). The alteration of food quality by such interactions has been considered by theo-reticians and empiricists in attempts to determine the relationship between variable food quality and herbivore population dynamics. There is a consensus that food quality is important to herbivore population dynamics. The central tenet of several hypotheses is that herbivore population growth is critically influenced by food quality, and that food-quality variation is influenced by site, weather, and by damage history (Haukioja and Hakala, 1975; White, 1978; Rhoades, 1979, 1983; Haukioja, 1980; Mattson and Haack, 1987). The empirical and theoretical tests of these hypotheses have been equivocal (e.g., Valentine, 1983; Fowler and Lawton, 1985; Haukioja and Neuvonen, 1987; Edelstein-Keshet and Rausher, 1989; Montgomery and Wallner, 1988). Some of the empirical contradictions may dissolve if the relationship between population quality and food quality is recognized as a critical factor in population growth. If the herbivore's response to nutrients or secondary compounds varies with its metabolic state, whether genetically or environmentally induced, then it will be difficult to get a clear picture of the role of food quality on population dynamics without accounting for the quality of the herbivore.

B. Population Quality Shifts Associated with Density Changes

The best-characterized shifts in population quality between generations come from correlative studies on the relationship between population density and expression of life-history traits. In review papers, Peters and Barbosa (1977) and Barbosa and Baltensweiler (1987) collected many examples in which changes in density were associated with changes in population quality as measured by shifts in body coloration, diet breadth, development time, susceptibility to disease, nutrient metabolism, survival, activity level, dispersal, and migratory behavior.

Some interesting cases include that of the mountain pine beetle, *Dendroctonus ponderosae*, which exhibits a density-correlated shift in the behavior related to oviposition site selection (Raffa and Berryman, 1983). Raffa (1988) hypothesized that the shift in population quality may be involved in its release to high densities. The tent caterpillar, *M. pluviale*, exhibits a shift in several population quality traits—larval and adult activity levels, fecundity, and dispersal behavior—as the density changes from one generation to the next. Wellington (1960, 1964) hypothesized that the deterioration in population quality was related to the rate of habitat deterioration. Myers (1990) hypothesized that the regional synchrony in tent caterpillar decline was related to a shift in population quality, possibly based on selection for pathogen resistance. The gypsy moth, *L. dispar*, exhibits density-correlated shifts in fecundity (Campbell, 1967; Richerson *et al.*, 1978) and oviposition site selection (Skaller, 1985). Finally, several herbivore species (*M. pluviale* and *Z. diniana*) exhibit a decline in vigor that is unrelated to disease or parasitism and that is initiated during high density, and continues even after a return to low density (Wellington, 1960; Day and Baltensweiler, 1972).

C. The Influence of Maternal Effects on Population Quality

For insects, maternal effects are the result of resource provisioning by one generation for the next, and the transfer occurs via the egg. These resource-based maternal effects are the product of gene–environment interactions experienced in the parental generation and expressed on a time delay. To date, most work on maternal effects in plant–herbivore systems focuses on nonnutritional factors such as the impact of photoperiod, age, and crowding in the maternal generation on diapause state, polymorphism, or sexuality in the offspring generation (Mousseau and Dingle, 1991). Little is known of the biochemical mechanisms involved, but transfer of hormones from mother to offspring has been demonstrated in locust and aphid species (Mousseau and Dingle, 1991). There are considerably fewer species for which the role of nutritional factors on the expression and translation of maternal effects has been addressed. I shall describe some of these studies in the context of their importance to the issue of population quality. Paternal effects have been demonstrated for several insects (e.g., Hoffmann and Harshman, 1985; Giesel, 1988), but the extent of their influence in most herbivore systems is unstudied. Rather than use the term *parental* effects, I shall use the more familiar term maternal effects, although the intergenerational effects described below may involve paternal effects.

Maternal effects have the potential to alter population size through a time-delayed impact on population quality traits that influence mortality and natality. Maternal effects can influence the value of offspring population quality traits such as phenology, mobility, migration, growth rate, resistance to physiological stress, and fecundity of offspring (Mousseau and Dingle, 1991). Offspring population quality traits of particular interest include

1. the length of the early prefeeding–dispersal period when inability to locate acceptable food results in death (Andrewartha and Birch, 1954; Cain *et al.*, 1985; Ramachandran, 1987; Hunter, 1990);
2. development time that can translate to probability of escape from natural enemies through alteration of exposure time (Cheng, 1970; Moran and Hamilton, 1980; Price *et al.*, 1980); and
3. development time that can influence diet quality (and, consequently, fecundity) because the nutritional value of foliage in temperate deciduous forests drops as the season progresses (Feeny, 1970; Schultz *et al.*, 1982; Schroeder, 1986; Rossiter *et al.*, 1988).

The following examples demonstrate the influence of maternal effects on population quality traits known to influence mortality and natality. For each of these species, maternal effects that influence population quality traits provide a time delay in the population's response to the environment. They may be an important proximal cause of outbreak dynamics (see Section V).

1. Resource-Based Maternal Effects in the Gypsy Moth

In the course of my own research, I discovered that population quality has the potential for significant impact on herbivore population dynamics. Using the gypsy moth as a model system, I found that variation in food quality experienced by the parental generation produced significant life-history variation in the next generation. The parental host species and the interaction between parent and offspring host accounted for 24% and 25%, respectively, of the explained variation in daughters' development time (Rossiter, 1991a). In another experiment, nutritional quality of red oak in the parental diet was correlated with offspring life history; offspring attained greater pupal weights and fecundities when their mothers fed on trees with higher damage levels. When mothers experienced greater condensed tannin levels, sons had lower pupal weights and daughters had a shorter prefeeding stage, a trait associated with the length of time available for host location through windborne dispersal (Rossiter, 1991a). In a third study, I found that egg quality was related to pesticide resistance. Individuals from eggs provisioned first along the ovariole were twice as resistant as siblings from eggs provisioned last (Rossiter *et al.*, 1990). In a fourth study, I found that individuals from larger eggs hatch earlier (both sexes) and daughters develop faster and become heavier pupae (Rossiter, 1991b). Egg weight has both a genetic and maternal effects component (M.C. Rossiter and D. Cox-Foster, unpublished data, 1990). Ongoing research (unpublished) shows that food quality in the parental generation accounts for over 40% of the variation in longevity of neonates held under starvation, a condition that mimics the pre-feeding, windborne movement phase of neonates; during this phase, larvae perish if acceptable food is not encounted. Longevity under starvation is critical because inclement weather, asynchrony with budbreak, and the passive nature of windborne travel can impede food location. The mortality level attributable to neonate death from nutritional stress in the wild is unknown. However, the greatest level of mortality—40 to 70% of the starting population—occurs sometime during the early instar period (Gould et. al, 1990).

2. Resource-Based Maternal Effects in the Western Tent Caterpillar

In this species, maternal effects are evident in the correlation between the activity phenotype and position of the egg in the mass, with first yolked/laid eggs giving rise to individuals with markedly different orientation and behavioral response to light, higher activity levels, and faster developmental rates compared with siblings from last yolked/laid eggs (Wellington, 1957, 1965). Within a population, families differ in their average activity phenotye (Wellington, 1957, 1964). Between generations, average activity phenotype (a population quality trait) varies in relation to population density with average activity level decreasing over each generation of an outbreak, then increasing in the year *following* a return to low density (Wellington, 1960).

3. Resource-Based Maternal Effects in other Herbivore Species

Autumnal moth larvae (*E. autumnata*) from parents reared on poor-quality food (defoliated in the previous year) were four times as likely to survive on poor-quality food (defoliated in current year) than were larvae from parents reared on undefoliated control trees (Haukioja and Neuvonen, 1987). These authors also reported the work of Jeker (1981), which demonstrated a similar response in *Agelastica alni*. In this species, offspring from mothers reared on foliage with little or no damage showed a 35% drop in fecundity when reared on damaged trees compared to siblings reared on control trees. By contrast, offspring of mothers reared on damaged foliage showed no difference in fecundity on control or damaged trees.

For the spruce budworm, *Choristoneura* spp., maternal effects have been invoked as the source of a quality shift in offspring of a single mating. As a consequence of differential provisioning along the ovariole, egg weight varied within a family and was associated with survival and growth characteristics into adulthood (Campbell, 1962). Harvey (1977, 1983, 1985) found that average egg weight varied among families within populations and that egg weight was significantly correlated with larval survival under warm overwintering temperature stress.

For the fall webworm, *Hyphantria cunea*, poor nutrition in the parental diet affected egg viability and early larval establishment (Morris, 1967). For the tobacco budworm, *Heliothis virescens*, nutritional quality of the maternal diet influenced offspring growth and ability to handle nutritional stress (Gould, 1988). In locust species, the degree of maternal crowding influenced egg lipid reserves, color, activity level, and development time in offspring (Hardie and Lees, 1985).

V. The Presence of Time-Delayed Effects on Population Growth

From a general systems perspective, maternal effects provide a particularly interesting example of a time-delayed effect of the environment of the parental generation (t) on the population size of the offspring generation (N at t + 1) through alteration of offspring survival, and on the population size of the subsequent generation (N at t + 2...) through alteration of fecundity in the offspring generation. Maternal effects provide a proximal cause for the presence of delayed density-dependent regulation in theoretical models, which include time lags (e.g., Berryman, 1978, 1987; Turchin, 1990). Turchin (1990) used temporal records from a single location for each of 14 forest outbreak insect species to test a theoretical model of delayed density-dependent regulation with time lags of t + 1 and t + 2. He found that eight species, some cyclic and some eruptive, gave strong evidence for delayed density-dependent regulation. Prior to this study, regulation was considered to be density-independent for seven of these eight species. The results clearly demonstrate that time lags can be an important phenomenon in the population dynamics of herbivores.

What are the proximal causes for time lags in herbivore systems? Berryman (1978, 1987) included environmental variables as well as population quality variables as potential sources of the delayed response. To date, most applications of this hypothesis have focused on environmental factors, including food quality. This is reasonable because food quality can be a state variable (involved in a feedback loop) whenever increased defoliation results in the deterioration of food quality (short-term induction) (e.g., Haukioja and Neuvonen, 1987; Karban, 1987; Rossiter *et al.*, 1988). Where induction effects are cross-generational, researchers have suggested that long-term induction may provide the proximal cause for a time delay and, consequently, destabilization of the system (e.g., Haukioja *et al.*, 1988; Mason and Wickman, 1988; Montgomery and Wallner, 1988; Baltensweiler and Fischlin, 1988). However, the time delay introduced by long-term induction should act to stabilize, rather than destabilize the population, under the assumption that herbivore quality remains unchanged. This interpretation is supported by the theoretical work of Edelstein-Keshet and Rausher (1989) on the ability of host induction effects to cause herbivore population destabilization. According to their model of mobile, nonselective herbivores, induction effects can, of their own accord, stabilize population growth over a wide range of conditions. Only in the presence of an Allee effect can host induction effects generate significant fluctuations. An Allee effect occurs when a shift in the population's growth potential occurs after some critical density threshold is crossed (Allee, 1931).

A. The Maternal-Effects Hypothesis of Population Outbreak

The maternal-effects component of population quality provides a delayed density-dependent effect through the interaction of population quality and food quality of previous generations. I hypothesize that the development of an outbreak begins when (1) the maternal-effects contribution to population quality enhances survival (and, hence, population size) to a point that (2) the numerical response of natural enemies is impaired, causing escape from the negative density-dependent interaction between herbivore and natural enemy. As a consequence, positive density-dependent growth, characteristic of outbreak development, begins. *This hypothesis pivots on the contribution of a time-delayed shift in population quality to interference in the negative feedback loop between herbivore and natural enemy.* There is interference with the numerical response of natural enemies because herbivore number in the previous generation(s) does not predict the current availability. Escape from natural enemies has been invoked frequently as the cause of population outbreak (e.g., discussed in Crawley, 1983). However, empiricists have been unable to determine whether escape is the cause or consequence of the positive density-dependent growth which is characteristic of an outbreak. The Maternal Effects Hypothesis provides an alternative explanation for the association between escape from natural enemies and the development of an outbreak.

B. Action of Maternal Effects from a General Systems Perspective

As its foundation, the Maternal-Effects Hypothesis uses the theoretical prediction (Berryman, 1978, 1987) that the presence of a delayed density-dependent response provides an ultimate (systems level) cause of insect outbreak. The Maternal-Effects Hypothesis suggests a proximal cause (i.e., the specific cause within a given system) for the presence of the delayed density-dependent response. Consider the operation of the Maternal-Effects Hypothesis from a general systems perspective. The system presented in Figure 3 includes two feedback loops: A involves natural enemies and herbivore number; B involves food quality, population quality, and herbivore number. *Although both loops are negative and capable of promoting stability, the magnitude of the food-quality loop (B) in this example is insufficient to stabilize population growth in the absence of the natural enemy loop (A).* Three generations are represented, the grandparental generation (t-2) in Figure 3a, the parental generation (t-1) in Figure 3b, and the current (t) generation in Figure 3c. The involvement of three generations is not critical to the hypothesis. Rather, the essential time period will be the one needed to generate maternal effects of sufficient strength to permit escape from natural enemies.

In Figure 3a, generation t-2 has a food-quality experience (e.g., induction of host defense), which generates maternal effects that enhance the population quality in generation t-1 (e.g., through changes in offspring dispersal, growth, behavior, and physiological resistance parameters). As a consequence, herbivore number in generation t-1 is altered through a time-lag effect. This causes a discrepancy in the herbivore number *expected* by the natural enemies in t-1 (which is based on herbivore number in t-2) and the realized herbivore number in t-1. As a consequence, the magnitude of the negative feedback effect on herbivore number (loop A) is reduced.

Generation t-1 (Fig. 3b) has a food-quality experience and response similar to that in t-2. Maternal effects alter the population quality and, consequently, the herbivore number in generation t. The discrepancy between the herbivore number expected by natural enemies and the realized herbivore number in generation t is now greater than that of the previous generation. Loop A (natural enemies) remains negative until generation t (Fig. 3c) when the enhancement of herbivore number (through alteration of population quality via maternal effects) is great enough to eliminate the density-dependent relationship between herbivore number and natural enemy number. In general systems terms, feedback loop A is temporarily disfunctional and the population is destabilized. The immediate effect is the

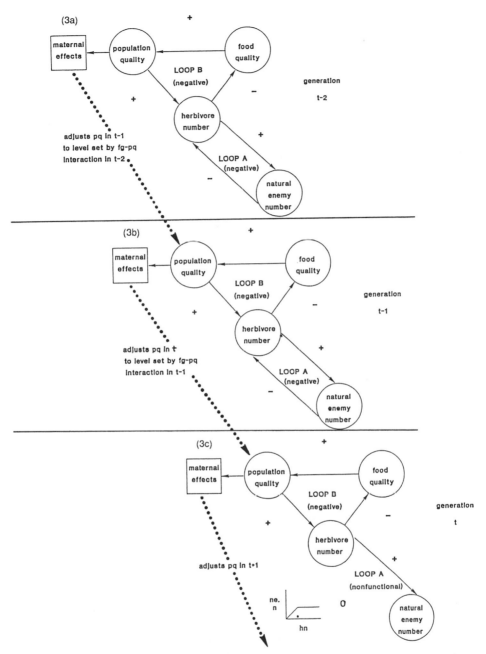

Figure 3 Events responsible for the initiation of an herbivore outbreak as set forth by the Maternal-Effects Hypothesis. Maternal effects act as exogenous variables (squares) and are the vehicle for delivery of time-delayed effects of the interaction between food quality (FQ) and population quality (PQ) in the previous generation(s). Graph in bottom panel, (3c) shows the loss of density-dependence between herbivore number (HN) and natural enemy number (NE.N) due to maternal effects enhancement of PQ.

promotion of positive density-dependent growth that will be enhanced or tempered by current environmental conditions (e.g., food quality, weather, and pathogens).

Population quality may be reduced via maternal effects as the population attains outbreak status or begins its decline if deteriorating food quantity or quality or disease alter the nongenetic parental contribution. Reduced population quality is commonplace during population decline (e.g., Myers, 1990), and quality can remain low for at least one generation after the return to low density (e.g., Wellington, 1960).

This proposed mechanism need not be limited to the time scale used above. The number of generations required to cross the critical population quality threshold where maternal effects are great enough to interfere with the natural enemy response may vary by species, geographic location, and the stochasticity of weather. However, for cyclic species that exhibit regional outbreaks, there may be a predictable program of plastic changes in quality (i.e., the quality-alteration trait produces a predictable sequence of metabolic shifts over generations once its expression is initiated), which yield a predictable sequence of changes in population growth rate, thereby giving rise to periodicity.

C. The Biology of Maternal Effects

There are two plausible sources of cross-generational shifts in population quality. The first is rapid short-term directional selection for host-utilization ability in a genetically variable population. This hypothesis has been tested (Stenseth, 1981; Boonstra and Boag, 1987) but is, as yet, unsupported. The other mechanism for a cross-generational shift in population quality is based on the directed expression of phenotypic plasticity, that is, the presence of a quality alteration trait (Section III,C). Such a trait produces metabolic adjustments in response to resource variation. Where maternal effects are involved, the environment dictates the metabolic state of the parent, and the metabolic state of the parent determines the quality and quantity of the contribution to offspring (e.g., egg nutrients, hormone precursors).

The state of the environment can dictate both the magnitude and direction of the maternal effect. The impact of the environment on one generation can be directly or inversely correlated with its impact on the next generation, giving rise to positive and negative maternal effects. With a negative maternal effect, the environmental quality experienced by the parental generation is directly correlated with parental response but inversely correlated with offspring response. Although rarely investigated, negative maternal effects are known for several species. In the gypsy moth, increased defoliation in the parental generation is correlated with reduced pupal weight and fecundity in mothers but increased pupal weight in offspring (Rossiter, 1991a). For autumnal moth, offspring whose parents were debilitated by poor food quality were four times as likely to survive poor

Table 1 Intrinsic and Extrinsic Features of Herbivore Species Suited for Testing the Population Quality Hypothesis of Outbreak

Outbreak herbivore species	Evidence of population quality shift between generations	Evidence of maternal effects	Evidence of defoliation-induced shift in food quality[a]	Evidence of delayed density dependence
Lymantria dispar	Yes[5,22]	Yes[18-20]	Yes[21]	Yes[12,23]
Epirrata autumnata	Yes[7]	Yes[8]	Yes[9]	Yes[9]
Orgyia pseudotsugata	Yes[10]	?	Yes[10]	Yes[10,23]
Choristoneura fumiferana	?	Yes[4,6]	Yes[11]	No[23]
Malacosoma californicum pluviale	Yes[25,26]	Yes[24,27]	Yes[14,28]	?
Zeirapha diniana	Yes[1]	Possibly[2]	Yes[3]	Yes[23]
Dendroctonus ponderose	Yes[16]	?	Yes[15]	[b]
Hyphantria cunea	Yes[13]	Yes[13]	Yes[17]	No[23]

[a] Effect can be positive or negative.
[b] Exhibits Allee effect.

References:

1. Baltensweiler (1971);
2. Baltensweiler et al. (1977);
3. Baltensweiler & Fischlin (1988);
4. Campbell (1962);
5. Campbell (1967);
6. Harvey (1985);
7. Haukioja & Neuvonen (1985);
8. Haukioja & Neuvonen (1987);
9. Haukioja et al. (1988);
10. Mason & Wickman (1988);
11. Mattson et al. (1988);
12. Montgomery & Wallner (1988);
13. Morris (1967);
14. Myers & Williams (1984);
15. Raffa (1988);
16. Raffa & Berryman (1983);
17. Rhoades (1983);
18. Rossiter (1991a);
19. Rossiter (1991b);
20. Rossiter et al. (1990);
21. Rossiter et al. (1988);
22. Skaller (1985);
23. Turchin (1990);
24. Wellington (1957);
25. Wellington (1960);
26. Wellington (1964);
27. Wellington (1965);
28. Williams & Myers (1984).

food quality (Haukioja and Neuvonen, 1987). Negative maternal effects have also been found in *Orchesella cincta*, a soil arthropod (Janssen *et al.*, 1988).

Negative maternal effects in herbivores may represent an adaptive strategy that allows the regulation of offspring quality in response to resource variation. Presence of a quality-alteration trait involving negative maternal effects could produce the superficial appearance of self-regulation at the population level whenever population lineages share the same plasticity response for altering offspring quality.

D. Testing the Maternal-Effects Hypothesis

There are several approaches for testing the Maternal-Effects Hypothesis. The first is to develop a formal theoretical model of population growth that includes terms which describe the average effect of the environment in t-1 on survival and fecundity in t, and the density-dependent relationship between herbivore number and natural enemy number. Second, parameters must be set after identification of outbreak species that exhibit delayed density-dependence or are known to express environmentally-based maternal effects or cross-generational shifts in population quality (see Table 1). To use the model for prediction in a particular herbivore system, the relationship between the critical feature of the parental environment that influences maternal effects expression (e.g., food quality) and the population-quality trait altered by the presence of the maternal effect (e.g., ability to withstand starvation in early development) must be identified. In addition, information on the population size of herbivores and natural enemies, as well as the average values of the pertinent population quality trait(s) and environmental quality trait(s) must be gathered in each generation over a succession of years ranging from low-density conditions through the development of an outbreak.

The Maternal-Effects Hypothesis can also be falsified with manipulative field experiments in which the relationship between environmental quality and population quality is disrupted and the change in population growth is compared to that of a control. Possible treatment effects include transplanting herbivore populations to locations that differ in environmental quality (e.g., Myers, 1990), or augmenting food quality or natural enemy number in low-density populations, that is, before the initiation of an outbreak. Whether developing or testing a theoretical model, or using a manipulative experimental approach, the biology of the individual should provide the starting point.

Acknowledgments

I thank A. Berryman, F. Gould, M. Hunter, J. Myers, P. Turchin, and two anonymous reviewers for insightful comments on the manuscript. I am especially grateful to A. Berryman and P. Turchin for discussions critical to the formalization of my ideas. This work was supported by USDA Competitive Grant No. 89-37250-4590.

References

Allee, W. C. (1931). "Animal Aggregations: A Study in General Sociology." Univ. of Chicago Press, Chicago, Illinois.

Alstad, D. N., and Edmunds, G. F., Jr. (1983). Adaptation, host specificity, and gene flow in the black pineleaf scale. *In* "Variable Plants and Herbivores in Natural and Managed Systems" (R.F. Denno and M. S. McClure, eds.), pp. 413–426. Academic Press, New York.

Andrewartha, H. G., and Birch, L. C. (1954). "The Distribution and Abundance of Animals." University of Chicago Press, Chicago, Illinois.

Baltensweiler, W. (1971). The relevance of changes in the composition of larch budmoth populations for the dynamics of its numbers. *Proc. Adv. Study Inst. Dyn. Numbers Popul.* 1970:208–219.

Baltensweiler, W., and Fischlin, A. (1988). The larch budmoth in the Alps. *In* "Dynamics of Forest Insect Populations: Patterns, Causes, Implications" (A. A. Berryman, ed.), pp. 331–352. Plenum, New York.

Baltensweiler, W., Benz, G., Bovey, P., and DeLuicchi, V. (1977). Dynamics of larch budmoth populations. *Annu. Rev. Entomol.* **22,** 79–100.

Barbosa, P., and Baltensweiler, W. (1987). Phenotypic plasticity and herbivore outbreaks. *In* "Insect Outbreaks" (P. Barbosa and J. C. Schultz, eds.), pp. 469–504. Academic Press, New York.

Berryman, A. A. (1978). Population cycles of the Douglas-fir tussock moth (Lepidoptera: Lymantriidae): The time-delay hypothesis. *Can. Entomol.* **110,** 513–518.

Berryman, A. A. (1981). "Population Systems. A General Introduction." Plenum, New York.

Berryman, A. A. (1987). The theory and classification of outbreaks. *In* "Insect Outbreaks" (P. Barbosa and J. C. Schultz, eds.), pp. 3–30. Academic Press, New York.

Berryman, A. A. (ed.) (1988). "Dynamics of Forest Insect Populations: Causes, Implications." Plenum, New York.

Berryman, A. A. (1989). The conceptual foundations of ecological dynamics. *Bull. Ecol. Soc. Amer.* **70,** 230–236.

Berryman, A. A., Millstein, J. A., and Mason, R. R. (1990). Modelling Douglas-fir tussock moth population dynamics: The case for simple theoretical models. *In* "Population Dynamics of Forest Insects" (A. D. Watt, S. R. Leather, M. D. Hunter, and N. A. Kidd, eds.), pp. 369–380. Intercept, Andover, Hampshire, UK.

Boonstra, R., and Boag, P. T. (1987). A test of the Chitty hypothesis: Inheritance of life-history traits in meadow voles *Microtus pennsylvanicus. Evolution* **41,** 929–947.

Buckner, C. H. (1966). The role of invertebrate predators in the biological control of forest insects. *Annu. Rev. Entomol.* **11,** 449–470.

Cain, M. L., Eccleston, J., and Kareiva, P. M. (1985). The influence of food plant dispersion on caterpillar searching success. *Ecol. Entomol.* **10,** 1–7.

Calow, P., and Sibley, R. M. (1990). A physiological basis of population processes: Ecotoxicological implications. *Funct. Ecol.* **4,** 283–288.

Campbell, I. M. (1962). Reproductive capacity in the genus *Choristoneura* Led. (Lepidoptera: Tortricidae). I. Quantitative inheritance and genes as controllers of rates. *Can. J. Genet. Cytol.* **4,** 272–288.

Campbell, R. W. (1967). The analysis of numerical change in gypsy moth populations. *For. Sci. Monogr. 15,* Soc. of American Foresters, Bethesda, Maryland.

Capinera, J. L. (1979). Qualitative variation in plants and insects: Effect of propagule size on ecological plasticity. *Am. Nat.* **114,** 350–361.

Caswell, H. (1972). A simulation study of a time lag population model. *J. Theor. Biol.* **34,** 419–439.

Caswell, H. (1983). Phenotypic plasticity in life-history traits: Demographic effects and evolutionary consequences. *Am. Zool.* **23,** 35–46.

Cheng, L. (1970). Timing the attack by *Lyphia dubia* Fall. (Diptera: Tachinidae) on the winter moth *Operophtera brumata* (L.) (Lepidoptera: Geometridae) as a factor affecting parasite success. *J. Anim. Ecol.* **39,** 313–320.

Chitty, D. (1960). Population processes in the vole and their relevance to general theory. *Can. J. Zool.* **38,** 99–113.

Chitty, D. (1967). The natural selection of self-regulatory behaviour in animal populations. *Proc. Ecol. Soc. Aust.* **2,** 51–78.

Cooper, W. S., and Kaplan, R. H. (1982). Adaptive "coin-flipping": A decision-theoretic examination of natural selection for random individual variation. *J. Theor. Biol.* **94,** 135–151.

Coulson, R. N. (1979). Population dynamics of bark beetles. *Annu. Rev. Entomol.* **24,** 417–447.

Crawley, M. J. (1983). "Herbivory: The Dynamics of Animal-Plant Interactions." Univ. of California Press, Berkeley, California.

Day, K. R., and Baltensweiler, W. (1972). Change in proportion of larval colour-types of the larchform *Zeiraphera diniana* when reared on two media. *Entomol. Exp. Appl.* **15,** 287–298.

Doane, C. C., and McManus, M. L. (eds.) (1981). "The Gypsy Moth: Research Toward Integrated Pest Management." *For. Serv. Tech. Bull. 1584,* U.S.D.A., Washington, D.C.

Dykhuizen, D. E., and Dean, A. M. (1990). Enzyme activity and fitness: Evolution in solution. *Trends Ecol. Evol.* **5,** 257–263.

Edelstein-Keshet, L., and Rausher, M. D. (1989). The effects of inducible plant defenses on herbivore populations. I. Mobile herbivores in continuous time. *Am. Nat.* **133,** 787–810.

Elkinton, J. S., and Liebhold, A. M. (1990). Population dynamics of gypsy moth in North America. *Annu. Rev. Entomol.* **35,** 571–596.

Elkinton, J. S., Gould, J. R., Ferguson, C. S., Liebhold, A. M., and Wallner, W. E. (1990). Experimental manipulation of gypsy moth density to assess impact of natural enemies. *In* "Population Dynamics of Forest Insects" (A. D. Watt, S. R. Leather, M. D. Hunter, and N. A. Kidd, eds.), pp. 275–287. Intercept, Andover, Hampshire, UK.

Falconer, D. S. (1981). "Introduction to Quantitative Genetics," 2nd Ed. Longman, New York.

Feeny, P. (1970). Seasonal changes in oak leaf tannins and nutrients as a cause of spring feeding by winter moth caterpillars. *Ecology* **51,** 565–581.

Fischlin, A., and Baltensweiler, W. (1979). Systems analysis of the larch bud moth system. Part I. The larch–larch bud moth relationship. *Mitt. Schweiz. Entomol. Ges.* **52,** 273–289.

Fowler, S. V., and Lawton, J. H. (1985). Rapidly induced defenses and talking trees: The devil's advocate position. *Am. Nat.* **126,** 181–195.

Freeland, W. J. (1974). Vole cycles: Another hypothesis. *Am. Nat.* **108,** 238–245.

Getz, W. M., and Haight, R. G. (1989). "Population Harvesting." Princeton University Press, Princeton, New Jersey.

Ghent, A. W. (1958). Studies of regeneration in forest stands devastated by the spruce budworm. II. Age, height, growth and related studies of balsam fir seedlings. *For. Sci.* **4,** 135–146.

Giesel, J. T. (1988). Effects of parental photoperiod on development time and density sensitivity of progeny of *Drosophila melanogaster. Evolution* **42,** 1348–1350.

Gould, F. (1988). Stress specificity of maternal effects in *Heliothis virescens* (Boddie) (Lepidoptera: Noctuidae) larvae. *Mem. Entomol. Soc. Can.* **146,** 191–197.

Gould, J. R., Elkinton, J. S., and Wallner, W. E. (1990). Density-dependent suppression of experimentally created gypsy moth, *Lymantria dispar* (Lepidoptera: Lymantriidae) populations by natural enemies. *J. Anim. Ecol.* **59,** 213–233.

Hardie, J., and Lees, A. D. (1985). Endocrine control of polymorphism and polyphenism. *In* "Comprehensive Insect Physiology, Biochemistry, and Pharmacology" (G. A. Kerkut, and L. I. Gilbert, eds.), Vol. 8, pp. 441–490. Pergamon Press, Oxford, England.

Hare, J. D., and Kennedy, G. G. (1986). Genetic variation in plant–insect associations: Survival of *Leptinotarsa decemlineata* populations on *Solanum carolinense. Evolution* **40,** 1031–1043.

Harvey, G. T. (1977). Mean weight and rearing performance of successive egg clusters of eastern spruce budworm (Lepidoptera: Tortricidae). *Can. Entomol.* **109,** 487–496.

Harvey, G. T. (1983). Environmental and genetic effects on mean egg weight in spruce budworm (Lepidoptera: Tortricidae). *Can. Entomol.* **115**, 1109–1117.

Harvey, G. T. (1985). Egg weight as a factor in the overwintering survival of spruce budworm (Lepidoptera: Tortricidae) Larvae. *Can. Entomol.* **117**, 1451–1461.

Hastings, A., and Caswell, H. (1979). Role of environmental variability in the evolution of life-history strategies. *Proc. Natl. Acad. Sci. U.S.A.* **9**, 4700–4703.

Haukioja, E. (1980). On the role of plant defenses in the fluctuation of herbivore populations. *Oikos* **35**, 202–213.

Haukioja, E., and Hakala, T. (1975). Herbivore cycles and periodic outbreaks. Formulation of a general hypothesis. *Rept. Kevo Subarctic Res. Station*, **12**, 1–9.

Haukioja, E., and Neuvonen, S. (1985). The relationship between male size and reproductive potential in *Epirrita autumnata* (Lep., Geometridae). *Ecol. Entomol.* **10**, 267–270.

Haukioja, E., and Neuvonen, S. (1987). Insect population dynamics and induction of plant resistance: The testing of hypotheses. *In* "Insect Outbreaks" (P. Barbosa and J. C. Schultz, eds.), pp. 411–432. Academic Press, New York.

Haukioja, E., Neuvonen, S., Hanhimaki, S., and Niemala, P. (1988). The autumnal moth in Fennoscandia. *In* "Dynamics of Forest Insect Populations: Patterns, Causes, Implications" (A. A. Berryman, ed.), pp. 163–178. Plenum, New York.

Henrich, S., and Travis, J. (1988). Genetic variation in reproductive traits in a population of *Heterandria formosa* (Pisces: Poeciliidae). *J. Evol. Biol.* **1**, 275–280.

Hoffman, A. A., and Harshman, L. G. (1985). Male effects on fecundity in *Drosophila melanogaster*. *Evolution* **39**, 638–644.

Holliday, N. J. (1977). Population ecology of the winter moth (*Operophtera brumata*) on apple in relation to larval dispersal and time of budburst. *J. Appl. Ecol.* **14**, 803–814.

Huffaker, C. B., and Messenger, P. S. (eds.) (1976). "Theory and Practice of Biological Control." Academic Press, New York.

Hunter, M. D. (1987). Opposing effects of spring defoliation on late season oak caterpillars. *Ecol. Entomol.* **12**, 373–382.

Hunter, M. D. (1990). Differential susceptibility to variable plant phenology and its role in competition between two insect herbivores on oak. *Ecol. Entomol.* **15**, 401–408.

Hunter, M. D., and Willmer, P. G. (1989). The potential for interspecific competition between two abundant defoliators on oak: Leaf damage and habitat quality. *Ecol. Entomol.* **14**, 267–277.

Hunter, M. D., Watt, A. D., and Docherty, M. (1991). Outbreaks of the winter moth on Sitka spruce in Scotland are not influenced by nutrient deficiencies of trees, tree budburst, or pupal predation. *Oecologia* **86**, 62–69.

Janssen, G. M., De Jong, G., Joose, E.N.G., and Scharloo, W. (1988). A negative maternal effect in springtails. *Evolution* **42**, 828–834.

Jeker, T. B. (1981). Durch Insektenfrass induzierte, resistenzahnliche Phanomene bei Pflanzen. Wechselwirkungen zwischen Grauerle, *Alnus incana* (L.) und den Erlenblattkafern *Agelastica alni* L. und *Melasoma aenea* L. sowie zwischen stumpflattrigem Ampfer, *Rumex obtusifolius* L. und Ampferblattkafer, Gastrophysa viridula Deg. *Dissertation No. 6895*, Eidgem. Technische Hochschule, Zurich.

Karban, R. (1987). Environmental conditions affecting the strength of induced resistance against mites in cotton. *Oecologia* **73**, 414–419.

Kimmins, J. P. (1971). Variations in the foliar amino acid composition of flowering and nonflowering balsam fir [*Abies balsamea* (L.) Mill.] and white spruce [*Picea glauca* (Moench) Voss] in relation to outbreaks of spruce budworm (*Choristoneura fumiferana* (Clem.)]. *Can. J. Zool.* **49**, 1005–1011.

Krebs, C. J., and Myers, J. H. (1974). Population cycles in small mammals. *Adv. Ecol. Res.* **8**, 267–399.

Lacey, E. P., Real, L., Antonovics, J., and Heckel, D. G. (1983). Variance models in the study of life histories. *Am. Nat.* **122**, 114–131.

Lance, D. R., Elkinton, J. S., and Schwalbe, C. P. (1986). Feeding rhythms of gypsy moth larvae: Effect of food quality during outbreaks. *Ecology* **67,** 1650–1654.

Leonard, D. E. (1970). Intrinsic factors causing qualitative changes in populations of *Porthetria dispar* (Lepidoptera: Lymantriidae). *Can. Entomol.* **102,** 239–249.

Levins, R. (1968). "Evolution in Changing Environments: Some Theoretical Explorations." Princeton University Press, New Jersey.

Lomnicki, A. (1988). "Population Ecology of Individuals." Princeton University Press, New Jersey.

Ludwig, D., Jones, D. D., and Holling, C. S. (1978). Qualitative analysis of insect outbreak systems: The spruce budworm and forest. *J. Anim. Ecol.* **47,** 315–332.

Martinat, P. J. (1987). The role of climatic variation and weather in forest insect outbreaks. *In* "Insect Outbreaks" (P. Barbosa and J. C. Schultz, eds.), pp. 241–268. Academic Press, New York.

Mason, R. R., and Overton, W. S. (1983). Predicting size and change in nonoutbreak populations of the Douglas-fir tussock moth (Lepidoptera: Lymantriidae). *Environ. Entomol.* **12,** 799–803.

Mason, R. R., and Wickman, B. E. (1988). The Douglas-fir tussock moth in the Interior Pacific Northwest. *In* "Dynamics of Forest Insect Populations: Patterns, Causes, Implications" (A. A. Berryman, ed.), pp. 179–210. Plenum, New York.

Mattson, W. J., and Haack, R. A. (1987). The role of drought stress in provoking outbreaks of phytophagous insects. *In* "Insect Outbreaks" (P. Barbosa and J. C. Schultz, eds.), pp. 365–410. Academic Press, New York.

Mattson, W. J., Simmons, G. A., and Witter, J. A. (1988). The spruce budworm in eastern North America. *In* "Dynamics of Forest Insect Populations: Patterns, Causes, Implications" (A. A. Berryman, ed.), pp. 310–330. Plenum Press, New York.

May, R. M. (1975). Biological populations obeying difference equations: Stable points, stable cycles and chaos. *J. Theor. Biol.* **49,** 511–524.

May, R. M., Conway, G. R., Hassell, M. P., and Southwood, T. R. E. (1974). Time-delays, density-dependence and single-species oscillations. *J. Anim. Ecol.* **43,** 747–770.

McCambridge, W. F., and Knight, F. B. (1972). Factors affecting spruce beetles during a small outbreak. *Ecology* **53,** 830–839.

McClure, M. S. (1988). The armored scales of hemlock. *In* "Dynamics of Forest Insect Populations: Patterns, Causes, Implications" (A. A. Berryman, ed.), pp. 45–66. Plenum, New York.

Milsum, J. H. (1968). "Positive Feedback—A General Systems Approach to Positive/Negative Feedback and Mutual Causality." Pergamon, New York.

Mitter, C., and Schneider, J. C. (1987). Genetic change and insect outbreaks. *In* "Insect Outbreaks" (P. Barbosa and J. C. Schultz, eds.), pp. 505–532. Academic Press, New York.

Montgomery, M. E. (1990). Role of site and insect variables in forecasting defoliation by the gypsy moth. *In* "Population Dynamics of Forest Insects" (A. D. Watt, S. R. Leather, M. D. Hunter, and N. A. Kidd, eds.), pp. 73–84. Intercept, Andover, Hampshire, UK.

Montgomery, M. M., and Wallner, W. E. (1988). The gypsy moth: A westward migrant. *In* "Dynamics of Forest Insect Populations: Patterns, Causes, Implications" (A. A. Berryman, ed.), pp. 353–376. Plenum, New York.

Moran, N., and Hamilton, W. D. (1980). Low nutritive quality as defense against herbivores. *J. Theor. Biol.* **86,** 247–254.

Morris, R. F. (1963). The dynamics of epidemic spruce budworm populations. *Mem. Entomol. Soc. Can.* **31,** 1–332.

Morris, R. F. (1967). Influence of parental food quality on the survival of *Hyphantria cunea*. *Can. Entomol.* **99,** 24–33.

Morris, R. F., Cheshire, W. F., Miller, C. A., and Mott, D. G. (1958). The numerical response of avian and mammalian predators during a gradation of the spruce budworm. *Ecology* **39,** 487–494.

Mousseau, T. A., and Dingle, H. (1991). Maternal effects in insect life histories. *Annu. Rev. Entomol.* **36,** 511–534.

Mousseau, T. A., and Roff, D. A. (1989). Adaptation to seasonality in a cricket: Patterns of phenotypic and genotypic variation in body size and diapause expression along a cline in season length. *Evolution* **43,** 1483–1496.

Myers, J. H. (1990). Population cycles of western tent caterpillar: Experimental introductions and synchrony of fluctuations. *Ecology* **71,** 986–995.

Myers, J. H., and Williams, K. S. (1984). Does tent caterpillar attack reduce the food quality of red alder foliage? *Oecologia* **62,** 74–79.

Nolte, D. J. (1974). The gregarization of locusts. *Biol. Rev. Cambridge Philos. Soc.* **49,** 1–14.

Pashley, D. P. (1988). Quantitative genetics, development, and physiological adaptation in host strains of fall armyworm. *Evolution* **42,** 93–102.

Peters, T. M., and Barbosa, P. (1977). Influence of population density on size, fecundity, and developmental rate of insects in culture. *Annu. Rev. Entomol.* **22,** 431–450.

Pimentel, D. (1961). On a genetic feed-back mechanism regulating populations of herbivores, parasites and predators. *Am. Nat.* **95,** 65–79.

Price, P. W. (1987). The role of natural enemies in insect populations. *In* "Insect Outbreaks" (P. Barbosa and J. C. Schultz, eds.), pp. 287–313. Academic Press, New York.

Price, P. W., Bouton, C. E., Gross, P., McPheron, B. A., Thompson, J. N., and Weis, A. E. (1980). Interactions among three tropic levels: Influence of plants on interactions between insect herbivores and natural enemies. *Annu. Rev. Ecol. Syst.* **11,** 41–65.

Raffa, K. F. (1988). The mountain pine beetle in western North America. *In* "Dynamics of Forest Insect Populations: Patterns, Causes, Implications" (A. A. Berryman, ed.), pp. 505–530. Plenum, New York.

Raffa, K. F., and Berryman, A. A. (1983). The role of host-plant resistance in the colonization behavior and ecology of bark beetles. *Ecol. Monogr.* **53,** 27–49.

Ramachandran, R. (1987). Influence of host-plants on the wind dispersal and the survival of an Australian geometrid caterpillar. *Entomol. Exp. Appl.* **44,** 289–294.

Rankin, M. A., and Singer, M. C. (1984). Insect movement: Mechanism and effects. *In* "Ecological Entomology" (C. B. Huffaker and R. L. Rabb, eds.), pp. 185–216. Wiley Interscience, New York.

Rausher, M. D. (1984). Tradeoffs in performance on different hosts: Evidence from within and between site variation in the beetle *Delayala guttata. Evolution* **38,** 582–595.

Redfearn, A., and Pimm, S. L. (1987). Insect outbreaks and community structure. *In* "Insect Outbreaks" (P. Barbosa and J. C. Schultz, eds.), pp. 99–134. Academic Press, New York.

Rhoades, D. F. (1979). Evolution of plant chemical defense against herbivores. *In* "Herbivores: Their Interaction with Secondary Plant Metabolites" (G. A. Rosenthal and D. H. Janzen, eds.), pp. 4–55. Academic Press, New York.

Rhoades, D. F. (1983). Herbivore population dynamics and plant chemistry. *In* "Variable Plants and Herbivores in Natural and Managed Systems" (R. F. Denno and M. S. McClure, eds.), pp. 155–220. Academic Press, New York.

Richerson, J. V., Cameron, E. A., White, D. E., and Walsh, M. (1978). Egg parameters as a measure of population quality of the gypsy moth, *Lymantria dispar. Ann. Entomol. Soc. Amer.* **71,** 60–64.

Risch, S. J. (1987). Agricultural ecology and insect outbreaks. *In* "Insect Outbreaks" (P. Barbosa and J. C. Schultz, eds.), pp. 217–240. Academic Press, New York.

Rossiter, M. C. (1987). Genetic and phenotypic variation in diet breadth in a generalist herbivore. *Evol. Ecol.* **1,** 272–282.

Rossiter, M. C. (1991a). Environmentally based maternal effects: A hidden force in population dynamics? *Oecologia* **87,** 288–294.

Rossiter, M. C. (1991b). Maternal effects generate variation in life history: Consequences of egg weight plasticity in the gypsy moth. *Functional Ecology* **5,** 386–393.

Rossiter, M. C., Schultz, J. C., and Baldwin, I. T. (1988). Relationships among defoliation, red oak phenolics, and gypsy moth growth and reproduction. *Ecology* **69,** 267–277.

Rossiter, M. C., Yendol, W. G., and Dubois, N. R. (1990). Resistance to *Bacillus thuringiensis* in gypsy moth (Lepidoptera: Lymantriidae): Genetic and environmental causes. *J. Econ. Entomol.* **86,** 2211–2218.

Royama, T. (1977). Population persistence and density dependence. *Ecol. Monogr.* **47,** 1–35.

Schroeder, L. A. (1986). Changes in tree leaf quality and growth performance of lepidopteran larvae. *Ecology* **67,** 1628–1636.

Schultz, J. C., Nothnagle, P. J., and Baldwin, I. T. (1982). Seasonal and individual variation in leaf quality of two northern hardwood species. *Amer. J. Bot.* **69,** 753–759.

Schultz, T. W. (1980). "Investing in People: The Economics of Population Quality." University of California Press, Berkeley, California.

Skaller, F. M. (1985). Patterns in the distribution of gypsy moth (*Lymantria dispar*) (Lepidoptera: Lymantriidae) egg masses over an 11-year-period population cycle. *Environ. Entomol.* **14,** 106–117.

Smith-Gill, S. J. (1983). Developmental plasticity: Developmental conversion *versus* phenotypic modulation. *Am. Zool.* **23,** 47–55.

Stenseth, N. C. (1981). On Chitty's theory for fluctuating populations. The importance of genetic polymorphism in the generation of regular density cycles. *J. Theor. Biol.* **90,** 9–36.

Turchin, P. (1990). Rarity of density dependence or population regulation with lags? *Nature* **344,** 660–663.

Uvarov, B. P. (1961). Quantity and quality in insect popualtions. *Proc. R. Entomol. Soc. Lond.* **25,** 52–59.

Valentine, H. T. (1983). The influence of herbivory on the net rate of increase of gypsy moth abundance: A modeling analysis. *In* "Proc. Forest Defoliator–Host Interactions: A Comparison Between Gypsy Moth and Spruce Budworms", pp. 105–111. U.S. Dept. Agric. For. Serv. Gen. Tech. Rep. NE-85, Washington, D.C.

Via, S. (1984). The quantitative genetics of polyphagy in an insect herbivore. I. Genotype–environment interaction in larval performance on different host plant species. *Evolution* **38,** 881–895.

Wainhouse, D., and Gate, I. M. (1988). The Beech Scale. *In* "Dynamics of Forest Insect Populations: Patterns, Causes, Implications" (A. A. Berryman, ed.), pp. 67–86. Plenum, New York.

Wallace, B. (1982). Phenotypic variation with respect to fitness: The basis for rank–order selection. *Biol. J. Linn. Soc.* **17,** 269–274.

Watt, K. E. F. (1963). The analysis of the survival of large larvae in the unsprayed area. *Mem. Entomol. Soc. Can.* **31,** 52–63.

Wellington, W. G. (1957). Individual differences as a factor in population dynamics: The development of a problem. *Can. J. Zool.* **35,** 293–323.

Wellington, W. G. (1960). Qualitative changes in natural populations during changes in abundance. *Can. J. Zool.* **38,** 289–314.

Wellington, W. G. (1964). Qualitative changes in populations in unstable environments. *Can. Entomol.* **96,** 436–451.

Wellington, W. G. (1965). Some maternal influences on progeny quality in the western tent caterpillar, *Malacosoma pluviale* (Dyar). *Can. Entomol.* **97,** 1–14.

Wellington, W. G. (1977). Returning the insect to insect ecology: Some consequences for pest management. *Environ. Entomol.* **6,** 1–8.

Wellington, W. G., Cameron, P. J., Thompson, W. A., Vertinsky, and Lansberg, A. S. (1975). A stochastic model for assessing the effects of external and internal heterogeneity on an insect population. *Res. Popul. Ecol.* **17,** 1–28.

White, T.C.R. (1978). The importance of a relative shortage of food in animal ecology. *Oecologia* **33,** 71–86.

Williams, K. S., and Myers, J. H. (1984). Previous herbivore attack on red alder may improve food quality for fall webworm larvae. *Oecologia* **63,** 166–170.

Zera, A. J., Innes, D. I., and Saks, M. E. (1983). Genetic and environmental determinants of wing polymorphism in the waterstrider, *Limnoporus canaliculatus*. *Evolution* **37:** 513–522.

3

Small-Mammal Herbivores in a Patchy Environment: Individual Strategies and Population Responses

Richard S. Ostfeld

Institute of Ecosystem Studies
Mary Flagler Cary Arboretum
New York Botanical Garden
Millbook, New York

I. Introduction

A. Some Natural History Notes

Some of the smallest-bodied of the mammalian herbivores are the arvicolid (microtine) rodents (the voles, lemmings, and their allies.) Most members of this Holarctic group consume largely graminoids and herbaceous dicotyledons and live in open habitats, such as meadows and tundra. Arvicolids as a group show many characteristics typical of r-selected species (MacArthur and Wilson, 1967): they are small (usually < 80 g), they develop very rapidly, reach sexual maturity early, have many offspring per litter and frequent litters per breeding season, and have a short lifespan (Hasler, 1975; Keller, 1985). As a result of their postpartum estrus and mating, most female arvicolids gestate one litter while they are lactating the previous one. They tend to have higher mass-specific metabolic rates than expected for a mammal of their size (McNab, 1980), yet arvicolids do not store appreciable amounts of fat. To support this intensely metabolically demanding lifestyle, voles and lemmings must consume a great deal of food, and they are selective of plant species and specific parts of plants that maximize energy and nutrient intake (Batzli, 1985). They also detect and avoid plants containing toxic chemicals (Berger *et al.*, 1977).

Arvicolid rodents are also a preferred prey of many vertebrate predators including hawks, owls, snakes, and carnivorous mammals (Korpimäki, 1985; Pearson, 1985). Because the morphological and behavioral defenses of these rodents are not well developed, arvicolids must rely on crypsis to avoid predation. Particularly useful for escaping predation by raptors and some carnivorous mammals is dense vegetative cover near the ground (Birney *et al.*, 1977).

Herbivorous rodents thus rely on plants for two critical resources: food and cover. Even slight differences in nutritional value or vegetative cover between two different sites may translate into large differences in the fitness of individuals inhabiting them. We should therefore expect these animals to be quite sensitive to habitat patchiness.

B. Scope of This Chapter

In this chapter, I shall explore how patchiness, defined as discrete spatial variation in the composition of plant communities, influences small-mammal herbivores at the levels of the individual and population. In the remainder of this section, I shall define two spatial scales of patchiness, *resource patchiness* and *habitat patchiness*, which have their primary effects on the consumer *individual* and *population*, respectively.

In Section II, I shall explore individual-level responses to resource patchiness in the context of hypotheses relating the abundance and distribution of key resources to the behavioral tactics of animals (e.g., Rubenstein and Wrangham, 1986). I shall rely on this theoretical literature to elaborate hypotheses that apply specifically to small-mammal herbivores. Small-mammal herbivores show remarkable variability in territoriality and mating systems that I believe stems from variation in plant patchiness. Both phenotypic and genotypic changes in social behavior will be discussed.

In Section III, population-level responses to habitat patchiness will be considered. Population ecologists have been developing and testing hypotheses to explain the dynamics of small-mammal herbivore populations through time (e.g., Elton, 1942; Chitty, 1967; Taitt and Krebs, 1985; Heske *et al.*, 1988; Lidicker, 1988). Some of these hypotheses incorporate assumptions about habitat patchiness and the efficacy of dispersal between patch types in regulating population size. But the types of interactions between animals in different patch types, and the parameters of habitat patchiness relevant to dispersal and population regulation, are under debate. I shall explore the relationships among plant patchiness, dispersal, gene flow, and population dynamics. This section will culminate with a conceptual model of population dynamics in patchy landscapes in which patches differ in size, quality, and phenology.

Finally, the interactions between individual and population level responses to plant patchiness will be discussed in Section IV. In particular, I shall consider the ways in which social organization (arguably produced by individual responses to patchiness) influences population responses to patchiness. I shall describe a case study based on my own work on California voles and provide a verbal model predicting the effects of different social systems on population dynamics in patchy environments.

Although there is some evidence that small-mammal herbivores can change the floristic composition of grassland habitats (e.g., Batzli and Pitelka, 1971; Cockburn and Lidicker, 1983), their influence on plant patchiness has not been explored adequately. Thus, despite the likelihood that resource distribution is important to *both* directions of the animal–plant interaction, I shall consider only the effects of plant distribution on animals. However, the need for studies of the impact of small-mammal herbivores on plant distribution should be stressed.

C. Resource Patchiness and Habitat Patchiness

Some ecologists recently have emphasized the importance of recognizing that ecological processes occurring at different scales may be fundamentally different in character and outcome (Wiens, 1989). Processes taking place at larger spatial scales may or may not be explicable in terms of those occurring at smaller scales. Patchiness relevant to systems studied by ecologists can occur at a range of scales from the microscopic to the global or even celestial. Each scale of patchiness is likely to have its most important effects at a particular level of biological organization. As this essay explores implications of patchiness for individuals and populations of small mammals, it is useful to distinguish two spatial scales relevant to these levels of organization.

Patches of some plant or plants that are exploited as food or cover, and that are equal to or smaller than the average size of an individual home range, can be referred to as *resource patches*. This term underscores the potential for a single animal to exploit an entire patch as a resource (usually food), and potentially to exclude others from the same patch. For instance, in a meadow, edaphic, historic, or other effects often lead to the formation of small patches in which certain species of plants are abundant, and surrounded by areas where they are scarce (Fig. 1). Patches smaller than about 100 to 200 m^2 (a typical home-range size for meadow-dwelling arvicolids) are likely to have their most direct effects on the behavioral tactics of individuals.

Resource patches are nested within *habitat patches*, which are defined as distinct plant community types, each of which is larger than an individual home range. Because groups of individuals will experience the same habitat patch, and other groups will occupy different patches, the most direct effect of habitat patchiness will be on groups or subpopulations. When a habitat patch is surrounded by unsuitable habitat, an entire population may be limited to a single patch. More often, however, a population will occupy an area composed of several habitat patches. The arrangement of habitat patches in any given area will have its most direct effect on the animal

(A) HABITAT
     PATCHINESS

(B) RESOURCE
     PATCHINESS

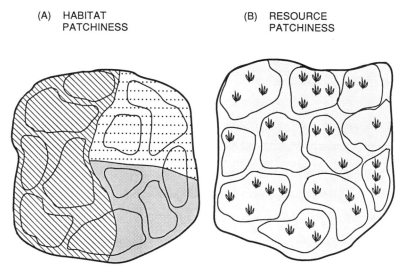

Figure 1   Schematic representation of habitat patchiness (A) and resource patchiness (B) as they pertain to small-mammal herbivores. Each diagram represents a portion of a landscape (heavy border) within which there are several individual territories (light border). Different shadings represent habitat patches, and small symbols indicate resource patches. In (A), animal territories usually occur in a single patch type, but there are two cases of a territory encompassing portions of two patch types. For simplicity, habitat patches are homogeneous (i.e., devoid of resource patches). In (B) only one habitat patch is represented, within which there are numerous resource patches.

population. There may be individuals that experience qualitatively different habitats because their home ranges encompass transition zones between habitat patches, rather than being entirely within one patch type (Fig. 1).

Of course, patchiness occurs along a continuum of patch-sizes, and it will not always be easy to pigeonhole patches according to the above dichotomy. This is particularly evident for patches near the home-range-size criterion that separates the categories. The categorization nevertheless is useful for conceptualizing individual versus population responses to a patchy environment.

II. Individuals and Resource Patchiness

A. Resource Patchiness and Territoriality

Variation in social organization among arvicolid species is dramatic. In some species, adult females defend territories, but males do not, and mating tends to be promiscuous for both sexes. In others, adult males but not females are territorial, and mating is polygynous. In yet others, mating is monogamous, and a joint territory is defended (see Heske and Ostfeld, 1990; Ostfeld, 1990, for reviews). Recently, I used a comparative approach to try to understand possible ecological causes of this social diversity (Ostfeld, 1985). I found that social organization appeared to be causally related to diet, such that arvicolid species that rely largely on forbs, or fruits, seeds, and fungi tend to exhibit female-territoriality and promiscuous mating, whereas those that consume mostly graminoids show male-territoriality and polygyny (Table 1; see also Cockburn, 1988). I did not find a relationship between diet and social organization in the two known monogamous species.

I interpreted this pattern as a result of animal–plant interactions. Although I could find insufficient data to evaluate it, I have observed that forbs (for grassland-dwelling arvicolids) and fruits, seeds, and fungi (for woodland arvicolids) tend to occur in resource patches, whereas graminoids (in grasslands) have a more homogeneous distribution at the scale of vole home ranges (Ostfeld, 1985). Female arvicolids, which are highly dependent on obtaining nutritious resources to support reproduction, can be expected to defend a territory when a clumped distribution of food plants lowers the costs of defense, but not when a homogeneous resource renders food undefendable (see Davies and Houston, 1984). Thus, for species whose diet consists mainly of forbs, fruits, seeds, or fungi, females are expected to be territorial, but for those consuming mainly grasses and sedges, females are expected to be nonterritorial. When female arvicolids are aggregated as a result of being nonterritorial, they may compose a defendable resource for males, whose fitness is highly dependent on access to mates. This leads to polygynous mating, in which single males have more or less exclusive access to several females (Heske and Ostfeld, 1990). But

Table 1 Social System and Diet of Species of Arvicolid Rodent

Species	Diet	Source	Territorial sex	Source	Consistent with Ostfeld (1985) Females	Consistent with Ostfeld (1985) Males
Arvicola terrestris	Grass	Holisová (1975, 1976)	Female	Jeppsson (1990)	No	Yes
Clethrionomys gapperi	Seed, fruit	Dyke (1971) Vickery (1979)	Female	Bondrup-Nielsen (1986) Mihok (1979)	Yes	Yes
C. glareolus	Seed, fruit	Hansson (1971a, 1979)	Female	Bujalska (1973, 1990) Mazurkiewicz (1971)	Yes	Yes
C. rufocanus	Seed, fruit, forb	Hansson (1985) Viitala (1977)	Female	Kawata (1985) Viitala (1977)	Yes	Yes
C. rutilus	Seed, fruit	West (1982)	Female	Burns (1981) Viitala & Hoffmeyer (1985)	Yes	Yes
Microtus agrestis	Grass	Godfrey (1953) Hansson (1971b)	Male	Myllymäki (1977a)	Yes	Yes
M. arvalis	Forb, grass	Yu *et al.* (1980) Holisová (1975)	Female[a]	Boyce & Boyce (1988)	?	?
M. breweri	Grass	Rothstein & Tamarin (1977)	Female	Zwicker (1989)	No	Yes
M. californicus	Grass	Batzli & Pitelka (1971)	Male	Heske (1987) Ostfeld (1986)	Yes	Yes
M. ochrogaster	Forb	Cole & Batzli (1979) Fleharty & Olson (1969)	Female, Male[b]	Getz *et al.* (1987)	Yes	No
M. oeconomus	Grass	Batzli & Jung (1980)	Male	Tast (1966) Viitala (1980)	Yes	Yes
M. pennsylvanicus	Forb	Lindroth & Batzli (1984) Neal *et al.* (1973)	Female	Madison (1980) Ostfeld *et al.* (1988) Webster & Brooks (1981)	Yes	Yes
M. richardsoni	Forb	Anderson *et al.* (1976)	Female	Ludwig (1984)	Yes	Yes
M. xanthognathus	Horsetail[c]	Wolff & Lidicker (1980)	Male	Wolff & Lidicker (1980)	Yes	Yes

[a] Females sometimes nest communally.
[b] Males sometimes overlap each other.
[c] Considered grass-like in distribution (Ostfeld, 1985).

when females are uniformly distributed as a result of being territorial, males may be forced to assume a more mobile, searching strategy for obtaining mates, and will therefore be nonterritorial. In this latter system, pair bonds are not established, and mating is promiscuous. Nonterritorial males may still be able to monopolize multiple mates, especially if females enter estrus asynchronously (Ims, 1988a).

Factors other than plant patchiness have been proposed to influence territorial systems in small rodents. These factors include defense against infanticide (Wolff, 1989), defense of burrow sites (Jeppsson, 1990), and the temporal distribution of mates (Ims, 1987) (reviewed by Ostfeld, 1990). Resource patchiness may also interact with these other factors to influence behavior.

B. Changes in Social Behavior: Phenotypic Variation in Space and Time

Despite the general conformance of arvicolid species to particular territorial and mating systems, recent studies have shown that considerable intraspecific variability exists (Madison, 1990; Ostfeld and Klosterman, 1990). This variability occurs among subspecies (Heske and Ostfeld, 1990), among populations (Madison, 1990; Jeppsson, 1990), and within individuals (Ims, 1988b; Ostfeld and Klosterman, 1990). Interpopulation comparisons have revealed striking differences in social organization within several species, including *Clethrionomys glareolus*, *Microtus agrestis*, and *Arvicola terrestris* (Andrzejewski and Mazurkiewicz, 1976; Myllymäki,1977a; Viitala, 1977; Jeppsson, 1990). In each of these species, social organization varies with habitat type and the distribution of food (e.g., *M. agrestis*) or nest sites (*A. terrestris*). However, other species, such as *M. pennsylvanicus* and *M. californicus*, seem to conserve their social organization despite differences in habitat and in population density (Madison, 1980; Webster and Brooks, 1981; Ostfeld *et al.*, 1985,1988; Ostfeld and Klosterman, 1986; Heske, 1987). Perhaps further studies will show that even these latter species are responsive to habitat variation.

Many observations reveal a capacity of small mammals to exhibit plasticity in behavioral strategy. Short-term manipulations of food abundance and changes in population density have resulted in modification of the pattern of spacing, and perhaps territoriality, of individuals (Ims, 1988b; Ostfeld,1986; Ylonen *et al.*, 1988). In addition, the agonistic behavior of voles, as measured by rates of wounding in free-ranging individuals and by staged encounters, has been shown to increase with increasing population density (Krebs, 1970; Christian, 1971; Rose and Gaines, 1976; Reich *et al.*, 1982). Finally, seasonal changes in behavior, coincident with transitions between breeding and nonbreeding seasons and with temperature fluctuations, are well known (Madison, 1990; Madison and McShea, 1987; West and Dublin, 1984).

Despite the demonstration of variation in behavioral patterns in small-mammal herbivores, the underlying causes of this variation remain poorly understood. The degree to which behavioral variation is genetically based is not known in most cases. In a few instances, however, it is clear that individuals are phenotypically plastic. Changes in the aggressive behavior of female voles throughout their reproductive cycles are clearly associated with circulating hormones (Seabloom, 1985). Transitions between territorial and affiliative behavior with the change of seasons is also clearly a phenotypic response (Madison and McShea, 1987; Wolff and Lidicker, 1980). However, most other instances of behavioral variation, at both the population and individual levels, cannot be ascribed with confidence to either a proximate or ultimate cause. This is particularly true of responses to resource distribution. To date, the distribution of food and mates has not been manipulated with a systematic design, and no long-term experiments have been carried out. Systematic, long-term experiments will be necessary to determine the degree of phenotypic plasticity in social behavior of small mammals, and to identify the proximate causes of behavioral plasticity.

C. Changes in Social Behavior: Genotypic Variation in Space and Time

An influential model of population cycles in arvicolids requires that agonistic behavior, dispersal, and reproductive effort have a strong genetic component (Chitty, 1967; Krebs, 1978). According to the Chitty–Krebs model, cyclic changes in animal density are caused by oscillating natural selection for and against aggressive morphotypes with low reproductive output. At low density, selection favors docile individuals with a high reproductive rate. As density increases and space and resources become more scarce, agonistic interactions become increasingly important, and selection favors aggressive individuals who devote less energy to reproduction. These aggressive individuals cause the emigration (dispersal), or death of docile ones until the population consists predominantly of aggressive morphotypes. At this point, a population crash results from social interference and low reproductive output. Then the cycle begins anew.

For this model to be plausible, a high heritability of aggressive behavior and reproductive traits must co-occur with strong (oscillating) natural selection to produce highly aggressive voles during high densities, and docile ones during the low density phase of the cycle. However, neither high heritability, strong selection, nor the predicted timing of behavioral traits has been demonstrated conclusively in arvicolids (Anderson, 1975, Boonstra and Boag, 1987; Boag and Boonstra, 1988; Krebs, 1970; Mihok, 1981), and the original proponents of the model now seem to have abandoned it (Chitty, 1987; Krebs, 1985).

Nevertheless, certain genotypes (represented by electrophoretic markers) have been shown to vary in frequency over the course of a cycle (reviewed by

Gaines, 1985); thus, it remains possible that selection acts on some traits in a density-dependent manner. However, genetic drift and gene flow may be more important than selection in causing these changes in genotype frequency. Dispersing voles are, in some cases, a genotypically nonrandom subset of the population (reviewed by Gaines and Johnson, 1987), indicating that the genetic structure of populations is influenced by dispersal and gene flow. Genetic drift is likely to be important in any population that goes through repeated bottlenecks (Falconer, 1981), which may be the case for many arvicolids.

The spatial distribution of arvicolid resources may be critical to the distribution of genotypes. Unfortunately, most small-mammal ecologists have focused their attention on temporal, rather than spatial, variation in genetic structure, and the interaction between temporal and spatial variation has been almost entirely neglected. One important exception is the work of Bowen (1982; Bowen and Koford, 1987), who studied allozyme variation in a cyclic population of *M. californicus* in a patchy meadow. During a period of low density, this population was divided into small subpopulations inhabiting particular patch types. The genetic differentiation of these subpopulations (demes), as determined by Wright's F_{ST} (Wright, 1965), was high, especially considering that they were separated by only 50–200 m. As density increased, genetic differentiation decreased, probably as a result of gene flow between refugia and into the interstices, and an increase in the effective population size, N_e. After the population crashed, genetic differentiation of demes was rapidly reestablished and remained high throughout the low-density phase (Fig. 2).

A high degree of genetic differentiation among demes was discovered in another small-mammal herbivore, the pocket gopher, *Thomomys bottae* (Patton and Feder, 1981; Daly and Patton, 1986). In this species, adults of both sexes are strongly philopatric and territorial, and successful immigration into intact populations appears to be rare. N_e in several disjunct fields ranged from 12 to 26, and was probably maintained at such low levels by a strongly female-biased adult sex ratio, and the active repulsion of potential immigrants (Patton and Feder, 1981). Areas from which pocket gophers had been experimentally removed were rapidly recolonized, resulting in the erosion of F_{ST} among demes. However, genetic differentiation among demes was reestablished after a single generation due to the social restructuring of the population (Patton and Feder, 1981; Lidicker and Patton, 1987).

Pocket gopher populations are stable compared to most arvicolids, and thus, patches of vacant habitat may not be as regular an occurrence for the former. Nevertheless, both examples indicate the potential importance of gene flow (or its absence) among habitat patches for the genetic structuring of small-mammal populations. Further research should concentrate on the degree of spatial concordance between habitat and genetic discontinuities.

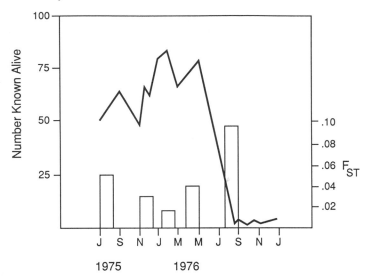

Figure 2   The relationship between population density (number known alive, solid line) and genetic differentiation between demes (F_{ST}, open bars) of the California vole, *Microtus californicus*. Data from Bowen (1982) and Bowen and Koford (1987).

III. Populations and Habitat Patchiness

A. Habitat Types

Small-mammal ecologists have long recognized that habitat patchiness influences herbivorous rodents (Elton, 1942; Naumov, 1972), and there have been several recent attempts to classify habitat types according to their suitability for herbivore populations (reviewed by Cockburn, 1988). Despite their varying terminologies, all of these classifications define optimal habitat as that which supports stable breeding populations of herbivores. Different types of suboptimal habitats are recognized by being favorable for only short periods, or unpredictably in time. Nearly all current classifications fail to recognize that suboptimal habitats may be chronically unfavorable, never supporting the same rates of reproduction per capita that optimal habitats do. One exception is the *source-sink* scheme of Pulliam (1988), but this model makes the unrealistic assumption that survival rate is as high in suboptimal (sink) habitat as in optimal (source) habitat. I shall argue that assumptions concerning the relative rates of natality, mortality, immigration, and emigration in different patch types may strongly influence the expected population dynamics.

How can different patch types be recognized? Taitt and Krebs (1985) contended that defining optimal habitat on the basis of the distribution of animals, and not the vegetation, is circular. They seem to suggest that herbivorous mammals may be distributed across habitat types as a result of chance events, and that this distribution may not reflect habitat quality. In reality, demographic performance is the best indicator of habitat quality, if the differences in performance among habitat types are constant through time. A mechanistic explanation for differences in demographic performance based on some aspects of the vegetation is helpful, but not necessary.

In most studies of the effects of habitat patchiness on small-mammal populations, patches are initially recognized on the basis of vegetative characteristics (e.g., Hansson, 1977; Cole and Batzli, 1979; Cockburn and Lidicker, 1983; Bondrup-Nielson, 1987; Ostfeld and Klosterman,1986; Dickman and Doncaster, 1987), and animal demography is then sampled in each patch type. Moreover, the quality of patches has been assayed independent of the animal population *in situ*. For instance, several studies have shown that patch quality, as indicated by population parameters, is associated with food quality or quantity (Batzli and Jung, 1980; Cole and Batzli, 1979; Krohne, 1980; Ostfeld and Klosterman, 1986; Kincaid and Cameron, 1985; Dickman and Doncaster, 1989). Others have demonstrated an important role for the density of vegetation cover (Birney *et al.*, 1977; Southern, 1979; Dickman and Doncaster, 1987; Adler and Wilson, 1989). Finally, most of these studies showed that differences across patch types in performance of small mammals were repeatable and stable through time, reducing the likelihood that chance events were important.

Van Horne (1983) and Pulliam (1988) warned against using animal density as the only indicator of population performance in different patches. In many cases, animal density may be high in patches of poor quality, for example, if suboptimal patches are regularly invaded by emigrants from nearby high-quality patches, or if suboptimal patches are sampled only in the brief period during which they are favorable, or if suboptimal patches are occupied by transients who fail to form a social system that limits population size. In addition, animal density may be temporarily low in patches of high quality due to stochastic events and isolation from a source of immigrants. Therefore, a combination of demographic variables, including density, age structure, reproductive rate, and ratio of emigrants to immigrants should be used to define herbivore performance in habitat patches (Van Horne, 1983; Cockburn and Lidicker, 1983; Ostfeld and Klosterman, 1986; Pulliam, 1988). A selective review of herbivorous small-mammal species known to respond to habitat patchiness, and the demographic variables that respond, are shown in Table 2. Clearly, small-mammal populations often inhabit patchy landscapes, and their demography is affected by patch quality.

B. Seasonal, Annual, and Multiannual Dynamics in Patchy Environments

Many studies have shown that the pattern of habitat occupancy (or habitat selection) of small-mammal herbivores changes on a seasonal, annual, or multiannual basis (Table 3). A clear example of seasonal dynamics is the

Table 2 Discoveries of Differences in Demographic Features within Populations of Small-Mammal Herbivores

Species	Higher quality habitat characterized by	Demographic feature(s) varying with habitat quality	Source
Clethrionomys gapperi	Higher density of deciduous tree cover	Density, persistence, recruitment, transiency, density	Bondrup-Nielson (1987)
C. glareolus	Higher density of vegetation cover	Density	Dickman & Doncaster (1987)
C. glareolus	More continuous forest; forest type	Emigration; probability of maturing	Gliwicz (1989)
C. glareolus	Higher density of ground cover	Density	Southern (1979)
C. glareolus	Higher soil moisture	Density, survivorship	Bock (1972)
Microtus agrestis	Higher density of vegetation cover	Density	Dickman & Doncaster (1987)
M. agrestis	Higher density of ground cover; certain grass & forb species	Density	Hansson (1977)
M. californicus	Certain grass & forb species; greater seasonal constancy	Density, adult sex ratio, age structure	Cockburn & Lidicker (1983)
M. californicus	Higher food quality; certain grass species	Density, sex ratio, persistence, recruitment	Ostfeld & Klosterman (1986)
M. californicus	Certain grass & forb species; greater seasonal constancy	Density, sex ratio, persistence, recruitment	Ostfeld *et al.* (1985)
M. longicaudus	Higher log, shrub, and herb cover	Density, survivorship	Van Horne (1982)
M. ochrogaster	Higher food quality; percent cover of alfalfa	Density, length of breeding season, litter size, growth rates of individuals	Cole & Batzli (1979) Getz *et al.* (1979)
M. pennsylvanicus	Higher density of grass cover (along a gradient)	Density, survival rate	Adler & Wilson (1989)
Sigmodon hispidus	Type of herbaceous plant cover	Density, sex ratio, age structure	Kincaid & Cameron (1985)

Table 3   Small-Mammal Herbivores That Change Their Pattern of Habitat Occupancy through Time[a]

Species	Temporal scale	Associated with Increasing Population Density?	Source
Clethrionomys glareolus	Multiannual	Yes	Bock (1972) Mazurkiewicz (1981) Naumov (1972)
C. glareolus	Annual	Yes	Wallin (1973)
Lemmus lemmus	Seasonal, multiannual	Yes	Kalcla *et al.* (1961, 1971)
L. obensis	Seasonal	No	Batzli (1975)
L. sibiricus (= *L. trimucronatus*)	Seasonal	No	Batzli *et al.* (1983)
L. sibiricus	Multiannual	Yes	Pitelka (1973)
Microtus agrestis	Multiannual	Yes	Hansson (1977)
M. brandti	Multiannual	Yes	Naumov (1972)
M. californicus	Multiannual	Yes	Bowen (1982) Lidicker (1975)
M. longicaudius	Multiannual	Yes	Van Horne (1982)
M. montanus	Multiannual	Yes	Randall & Johnson (1979)
M. oeconomus	Seasonal	No	Tast (1966)
M. pennsylvanicus	Annual	Yes	Grant (1971)

[a] Based on a review of studies in which small-mammal populations expand and shrink the range of habitat types occupied seasonally (> once per year), annually (once per year), or multiannually (once every few years). Whether or not an increase in number of habitat types occupied is associated with increasing population density is also shown.

Norwegian lemming (*Lemmus lemmus*), which winters in alpine habitats under good snow cover, emigrates when these habitats are flooded in spring to willow thickets and peatlands, and concentrates summer activities in moist areas with high productivity of mosses and graminoids (Batzli, 1975). Seasonal habitat shifts in this and other lemming and vole species appear to be caused by marked seasonal changes in the suitability of habitat types, and seem to be unrelated to seasonal changes in population density (Batzli *et al.*, 1983; Elton, 1942; Tast, 1966).

In contrast, changes in habitat use between years, and over a multiannual cycle, are strongly related to changes in herbivore density. The typical pattern is for small-mammal herbivores to be restricted to optimal habitat patches during years of low density and to begin occupying suboptimal habitats as density increases. During high-density years, the distribution of herbivores among patches is even, or nearly so (e.g., Bock, 1972; Naumov, 1972; Pitelka, 1973; Wallin, 1973; Randall and Johnson, 1979; Wolff, 1980; Lidicker, 1985; Table 3). Unfortunately, it is not yet clear whether changing patterns of habitat occupancy are primarily a cause or an effect of population fluctuations.

C. Models of Interactions between Patchiness and Population Dynamics

After the discovery that enclosing a vole population causes a pathologically rapid increase in vole numbers, followed by overexploitation of food and a precipitous decline (the *fence effect*) (e.g., Lousch, 1956; Krebs *et al.*, 1969), small-mammal ecologists became more interested in the role of emigration (dispersal) in regulating population size. Unfortunately, dispersal has rarely been studied in the context of habitat patchiness. Instead, dispersal is usually examined by contriving a situation in which *dispersers* can be distinguished from *residents* in order to see whether members of these two classes are demographically or genetically different (Gaines and McClenaghan, 1980). Such contrivances include removal grids, in which all small mammals are removed from a study plot, and any immigrant from adjacent areas is classified as a disperser; exit tubes, which are openings in otherwise vole-proof fences through which dispersers but not residents travel; and various kinds of mown, trenched, or mounded turf over which only a disperser would traverse (Gaines and Johnson, 1987). Aside from the high potential for inaccurately classifying individuals using these methods (Gaines and McClenaghan, 1980), this approach has contributed little insight into the dynamics of dispersal between patch types and its role in regulating numbers. Any role of habitat patchiness in population dynamics will depend on the dynamics of dispersal of individuals between patch types.

Several mathematical, graphical, and verbal models contend that habitat patchiness plays an important role in population dynamics. The two models of Rosenzweig and Abramsky (1980) seem to be alone in predicting that habitat patchiness will *destabilize* population dynamics and potentially lead to multiannual cycles. Both their Predation Hypothesis and Phenological Hypothesis are based on coevolutionary relationships between voles and their food plants. The Phenological Hypothesis, which Rosenzweig and Abramsky (1980) deem more plausible, contends that selection will favor voles that respond appropriately to phenological changes in plant chemicals that activate or inhibit reproduction (Berger *et al.*, 1977, 1981; Sanders *et al.*, 1981). Because different habitat patches will have phenological differences in production of these chemicals, Rosenzweig and Abramsky expect selection for tight local adaptation of voles to their resources. Locally adapted voles will not negatively affect their food plants, because they will breed only at times when plants can support population growth, and in the absence of immigration, the system will be stable. However, immigrants from other patches, differing in their sensitivity to plant chemicals, will subvert local adaptation, breed at inappropriate times, have a negative effect on the food plants, and cause a crash. After the crash, selection for the appropriate level of sensitivity will again result in local coadaptation. Consequently, multiannual cycles should occur only in vole populations that inhabit a patchy environment (Rosenzweig and Abramsky, 1980).

Genotypic differences among vole populations found in different habitat patches at low density (Bowen, 1982; Dobrowolska, 1981) would lend credence to the Phenological Hypothesis if they occur as a result of strong selection for local adaptation to food plants. However, no theoretical or empirical studies have been conducted to determine the strength of selection, degree of genetic isolation, and the number of generations necessary to achieve such tight local adaptation to food sources. All these variables may have to be unrealistically high for the model to apply to herbivorous rodents. As indicated above, genetic drift may play a major role in the genetic differentiation of vole populations. In addition, Tamarin *et al.* (1984) have demonstrated that vole populations may cycle even when immigration is prevented by fences, which indicates that the disruption of local adaptation is not a necessary condition. Finally, one of the most stable microtine populations studied to date (Ostfeld, 1988) inhabited a distinctly patchy environment (Ostfeld and Klosterman, 1986).

Other models of population responses to patchy environments are nearly unanimous in concluding that patchiness should be stabilizing (Roff, 1975; Hassell, 1980; Łomnicki, 1980, 1982; Stenseth, 1980, 1983; Wolff, 1980). The essential role of habitat patchiness is to facilitate density-dependent emigration, which can be very effective in regulating population size (Lidicker, 1975; Łomnicki, 1980; Stenseth, 1983). Emigration is envisioned as preventing overcrowding in high-quality habitat patches, and homogenizing population density among patches. However, in order for emigration to be important in stabilizing population dynamics, it must decrease the density of the local population *and* result in higher mortality of emigrants than residents (Łomnicki, 1982). There is ample evidence that both conditions are met (summarized by Lidicker, 1975, 1985; Gaines and McClenaghan, 1980).

Given the mortality and other risks associated with emigration, Anderson (1988) has questioned whether emigration ever evolves as an adaptive strategy. He contends that its prevalence is due to the advantage to residents of forcing the emigration of their conspecifics. However, Łomnicki (1982) has pointed out that individuals differ substantially in their ability to garner limited resources. Those that often lose in intraspecific competition for these resources may have higher fitness if they emigrate and colonize patches having lower herbivore density, even if these patches have fewer resources. In addition, Wolff *et al.* (1991) point out that inclusive fitness of emigrants must be considered; the emigrants as well as the relatives that they leave behind may benefit by outbreeding, which is facilitated by dispersal. Lidicker (1975) and Stenseth (1983) argue that individuals emigrating before the habitat is saturated (*presaturation dispersers* of Lidicker, *adaptive dispersers* of Stenseth) are more likely than those emigrating under crowded conditions (*saturation dispersers* or *nonadaptive dispersers*) both to have nonzero fitness and to affect population dynamics.

To date, field tests of the role of habitat patchiness in the population dynamics of small-mammal herbivores have not been conducted. Such tests would require either comparative or experimental studies at large temporal and spatial scales. Large areas differing naturally in their degree of patchiness (but otherwise very similar), or homogeneous areas, some of which are experimentally made patchy, would have to be monitored over several years.

D. How Patchy Is Patchy?

Patchy environment and *habitat heterogeneity* are usually ill defined in the literature on rodent population dynamics. There are many ways in which patchy environments may vary, and different types of patchiness may affect population dynamics in different ways. Patches may be small or large, semi-isolated or connected, marginally or vastly different in quality, but there has been surprisingly little exploration of the population-level effects of these kinds of landscape variation (but see Fahrig and Merriam, 1985). Buechner (1989) and Stamps *et al.* (1987) describe models of the effects of patch size and edge permeability on the rate of emigration from source habitats, but have not explored the dynamics of emigration in fluctuating populations, or the role of emigration in regulating population size.

Recently, Lidicker (1988) described one important way in which patchy landscapes may differ: the ratio of optimal to marginal patch area, or ROMPA. According to Lidicker (1988; p. 229), vole populations in a landscape with a high ROMPA will be unlikely to undergo cycles because "so much optimal habitat is available that harsh season populations are not reduced sufficiently to prevent complete recovery of the population during the following breeding season." In contrast, recovery may take several years when only small numbers survive harsh seasons in small refugia (low ROMPA), and cycles will result. Gaines *et al.* (in press) disagree. They argue that when marginal habitat is abundant relative to optimal habitat (low ROMPA), vole populations will be stabilized (noncyclic) because dispersal into marginal habitats will prevent irruptions in optimal patches. When marginal habitat is scarce (high ROMPA), they argue, dispersal sinks will fill rapidly, further emigration from optimal patches will be curtailed [causing *frustrated dispersal* (Lidicker, 1975)], and the population may exceed carrying capacity and crash (i.e., cycle). Dispersal from optimal to marginal patches will be suppressed when high density in marginal patches causes social repulsion of potential immigrants [the *social fence* of Hestbeck (1982)].

I believe these different perspectives are each a part of an emerging description of the importance of patch differences to metapopulation dynamics. Our view of optimal versus suboptimal (or marginal) habitats has been too restrictive, in that it has not included two other issues: the relative difference in quality between optimal and suboptimal habitats, and whether suboptimal habitat is always or only seasonally poorer than optimal habitat.

E. A Conceptual Model of Population Dynamics When Patches Differ in Size, Quality, and Phenology

If suboptimal habitat is much poorer in quality than optimal habitat, such that mortality vastly exceeds natality, the population in suboptimal habitat should never achieve densities at which it would prevent emigration from optimal patches. Suboptimal habitat of quality this poor will, therefore, tend to stabilize population density in optimal habitat, and in the meta-population, regardless of its size. If mortality only slightly exceeds natality in suboptimal habitat, these patches are much more likely to achieve densities that would eventually frustrate emigration from the optimal patch. In this case, the size of the suboptimal patches should determine the time during which emigration will be effective in regulating density of the optimal patch(es). When optimal habitat is scarce relative to suboptimal (low ROMPA), it may take several years for density in suboptimal habitat to be high enough to suppress emigration, and multiannual cycles may be more likely. When optimal habitat is abundant relative to suboptimal (high ROMPA), emigration may be suppressed every year leading to annual fluctuations (Table 4). The length of time necessary for populations in suboptimal patches to increase in density to the point that they inhibit

Table 4 Predictions of the Effects of the Juxtaposition of Habitat Patches of Different Sizes and Qualities on Population Dynamics[a]

	Patch size	
Patch quality	Optimal (O) ≤ suboptimal (S) (High ROMPA)	Optimal < < suboptimal (Low ROMPA)
Optimal > suboptimal (*in situ* population growth possible in suboptimal)	S fills up quickly Frustrated emigration likely Dispersal sink temporary O can temporarily exceed carrying capacity (K) Dynamics unstable Crashes frequent Annual cycles likely	S fills up slowly but eventually Delay in frustrated emigration Dispersal sink temporary O can periodically exceed K Dynamics unstable Crashes infrequent Multiannual cycles likely
Optimal > > suboptimal (*in situ* population growth not possible in suboptimal)	S fills slowly if at all Frustrated emigration unlikely Dispersal sink usually effective O regulated below K by emigration Dynamics stable Potential for irregular outbreaks	S never fills No frustrated emigration Dispersal sink always effective O regulated below K by emigration Dynamics stable

[a] ROMPA is an acronym for the Ratio of Optimal to Marginal Patch Area (Lidicker, 1988). It is assumed that the area of optimal patches never exceeds that of marginal patches. The rationales for these predictions are described in the text.

emigration from adjacent optimal patches will depend not only on relative size of patches, but also on other factors including reproductive rate, emigration rate, and length of breeding season in both optimal and suboptimal habitat. Table 4 gives a series of predictions regarding population dynamics in optimal and suboptimal patches, depending on relative size and quality of patch types.

Whether suboptimal patches are seasonally or consistently poorer than optimal patches will also be important. If they are seasonally poorer, then the timing of their change in quality will be critical. If suboptimal patches increase in quality when populations in optimal habitats are growing (i.e., the phenologies of the two habitat types are in phase), then the suboptimal patch may be able, at least temporarily, to absorb many immigrants and provide an effective dispersal sink regardless of ROMPA. However, if quality of suboptimal patches collapses when density in optimal patches is still increasing (i.e., they are out of phase), then dispersal is likely to be curtailed and an irruption and crash are likely, again with little effect of ROMPA.

This model links the spatial and temporal distribution of resources with an index of the quality of these resources, to predict the overriding form of population dynamics in different habitats. The model could be tested with field experiments. One approach would be to create suboptimal patches in a homogeneous meadow by mowing, herbiciding, or replacing existing vegetation with a toxic, poorly digestible, or low-stature plant. Patch quality could be degraded to two levels, assayed by animal survivorship and natality, and suboptimal patches of two (or more) sizes created. Phenology could be altered, for instance, by replacing perennial with annual vegetation. Population dynamics could then be followed over time. Another approach is to create dispersal sinks (suboptimal patches) by removing small mammals from patches of designated sizes. Rate and timing of removal would mimic patches of different quality and phenology.

IV. Interactions between Resource Patchiness and Habitat Patchiness

A. A Case Study with *Microtus californicus*

1. Habitat Patchiness, Seasonality, and Cycles

Two populations of the California vole (*M. californicus*) recently have been studied to determine the importance of habitat patchiness in population dynamics. One occupied a coastal meadow at the Bodega Marine Laboratory in Sonoma County, California (Bodega population), and the other occurred in an inland meadow at the Russell Reservation in Contra Costa County (Russell population). The Bodega population was studied intensively for 1 year, but censused at least annually for 4 years (Ostfeld and

Klosterman, 1986), and the Russell population was monitored at least monthly for six years (Bowen, 1982; Cockburn and Lidicker, 1983; Heske *et al.*, 1984; Ostfeld *et al.*, 1985).

In both localities, optimal and suboptimal patches were identified. Optimal patches were characterized by vole populations with higher average density, longer persistence (survivorship), and higher per capita rate of juvenile recruitment. Despite a similar patchy mosaic, and low ROMPA, at both study sites, the two populations had starkly contrasting population dynamics. The Russell population underwent typical multiannual cycles with a 3- to 4-year period between peaks, but the Bodega population was extremely stable. The reasons for this difference have not been confirmed, but there is evidence that the relative quality of marginal patches and seasonal changes in their quality may be important.

The small, optimal patches at Bodega Bay were surrounded by a large expanse (hundreds of hectares) of dunes covered with the beach grass *Ammophila arenaria*, which has been shown to be a poor-quality food for California voles (Ostfeld and Klosterman, 1986). In addition, vole burrows in the sandy substrate of this habitat type were subject to frequent collapse (Ostfeld, personal observations, 1983). As a result, this patch type supported a relatively sparse vole population with high mortality and low reproductive output. The low quality of the beach grass patch remained stable seasonally and from year to year. The Bodega landscape clearly had both a low ROMPA and a large difference between the quality of optimal and suboptimal patches. Perhaps emigration was consistently effective in regulating population size. The Bodega population appears to be a prototype of the lower right quadrant of Table 4.

The Russell population occurred in an isolated 8-hectare meadow. Suboptimal habitat at Russell Reservation was annual grassland, which varied seasonally in quality and consisted of grasses superior in nutritional quality to the beach grass at Bodega Marine Laboratory (Batzli and Pitelka, 1971). In fact, the annual grassland was clearly of poorer quality than the perennial grass patches (optimal habitat) only during the summer dry season (Cockburn and Lidicker, 1983). Therefore, a low ROMPA, relatively little difference in quality between optimal and suboptimal patches, and the fact that the capacity of the suboptimal patches to absorb emigrants from the optimal patches plummeted seasonally, may have destabilized this population. The Russell population would represent the upper right panel of Table 4.

In neither locality was there any evidence that voles affected the patch structure of their habitat. However, because voles are selective grazers (Batzli, 1985), and at high density are capable of degrading their habitat (Krebs *et al.*, 1969), the role of small-mammal herbivores in altering plant resource distribution and quality should be explored further.

2. Sex-Specific Influences on Population Structure

Males and females showed strikingly different responses to habitat patchiness at both Bodega Bay and Russell Reservation. At both localities, females (but not males) tended to aggregate in optimal patches, causing adult sex ratios to be female-biased in these patches. Sex ratios in suboptimal patches tended to be even (Fig. 3). Females (but not males) had a higher probability of persisting in optimal than in suboptimal patches (Cockburn and Lidicker, 1983; Ostfeld *et al.*, 1985; Ostfeld and Klosterman, 1986; Fig. 3). At Russell Reservation, the persistence rates of females (but not of males) varied in response to the vegetation growing season, whereas the persistence rates of males (but not females) varied in response to population density (Ostfeld *et al.*, 1985; Table 5).

The responsiveness of female *M. californicus* to both spatial and temporal variation in habitat quality, and their lack of response to population density, were interpreted as arising from rather weak social regulation of density, which allowed aggregation and persistence in optimal patches. In contrast, males were relatively unresponsive to spatial and temporal variation in habitat quality, but had reduced fitness at high density. Their lack of response to patchiness and seasonality was interpreted as a result of strong social regulation of density, such that increasing social interference led to a decline in persistence (Ostfeld *et al.*, 1985; Ostfeld and Klosterman, 1986). These interpretations have been supported by radiotelemetry studies, which have revealed that male California voles are strongly territorial and polygynous, and females nonterritorial, or only weakly so (Ostfeld, 1986).

3. A Role for Low Resource Patchiness

The patches at Bodega Bay and Russell Reservation were larger than the home ranges of individual voles, therefore I consider them habitat patches. Within patches, there was no obvious subdivision, although this has not been measured directly in either locality. Therefore, despite a high degree of habitat patchiness, the level of resource patchiness appeared to be low. As explained above, female arvicolids living in areas with low resource patchiness are expected not to defend territories. And indeed, female California voles lacked strong territoriality, which allowed them to aggregate. This aggregation is seen as a direct response to habitat (but not resource) patchiness. In a situation in which females are territorial (e.g., in *M. pennsylvanicus*), they would be expected to be less responsive to habitat patchiness because of social interference. This expectation is being tested.

As described previously, a clumped distribution of females in populations of California voles (Ostfeld and Klosterman, 1986; Ostfeld, 1986), could lead to male territoriality (Ostfeld, 1985), and consequently for males to be more uniformly distributed across habitat patches. This expectation was confirmed at both Bodega Bay and Russell Reservation. When males are nonterritorial (e.g., in species with territorial females, such as *M. penn-*

(A)

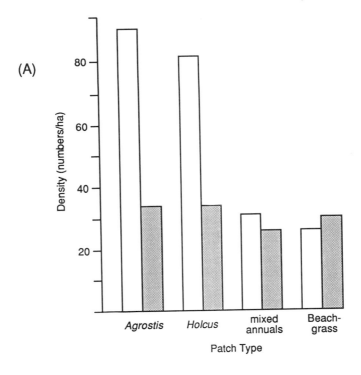

(B)

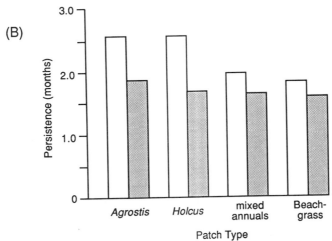

Figure 3   Density (A) and persistence (B) of California voles in four different patch types at Bodega Bay, California. Females are represented by open bars and males, by shaded bars. Data from Ostfeld and Klosterman (1986).

Table 5   Mean Probability (P) That Individual
California Voles, Once Captured, Would be
Trapped Again at a Subsequent
Trapping Session[a]

Conditions		Males		Females	
		P	N	P	N
Season	Wet	0.28	418	0.39	897
		NS		**	
	Dry	0.23	387	0.27	473
Density	Low	0.36	165	0.35	248
		**		NS	
	High	0.23	641	0.34	1122

[a] From Ostfeld *et al.*, (1985). *P* is often considered a measure of survivorship, although, strictly speaking, it measures *persistence*, since disappearances are attributable to emigration as well as death. *N*, number of individuals in each sample;
** = $p < 0.01$; NS, not significant by G-test of independence.

sylvanicus), they would *not necessarily* be expected to aggregate in optimal habitat patches, since a critical resource for males is estrous females, which, due to territoriality, may be uniformly distributed across habitat patches. Thus, the degree of resource patchiness may strongly influence a species' response to habitat patchiness by influencing the territorial or associative behavior of individuals.

B. Predicting the Effects of Social System on Population Dynamics in a Patchy Environment

To my knowledge, it has never been demonstrated that the reproductive success of a female vole was limited by access to mates. Because females seem to have no difficulty finding mates even at low population density, the population natality rate should be largely independent of male density (Stenseth and Framstad, 1980). In addition, in some arvicolid species, females, but not males, regulate the breeding density of the population (Boonstra, 1977; Redfield *et al.*, 1978; Boonstra and Rodd, 1983). Therefore, in predicting the effects of social behavior on population dynamics, it is probably appropriate to focus on females. An important role of males, however, should not be dismissed.

Envision a population of herbivorous mammals that occupies an area having high resource patchiness as well as high habitat patchiness. Assume that the population enters the breeding season at low density and occupies only optimal habitat. As argued above, the resource patchiness should elicit territorial behavior among females and result in unequal access of individuals to food and shelter. Territorial behavior functions to prevent aggregation in high-quality habitat patches (e.g., Saitoh, 1985). As the breeding season progresses and density increases, the rate of increase in optimal habitat patches should approach zero long before saturation density as females evict their daughters and prevent immigration of unrelated females (e.g., Boonstra and Rodd, 1983; Boonstra *et al.*, 1987). Territorial behavior will tend to prevent overexploitation of resources. Suboptimal patches will be colonized relatively early in order of their quality, and soon the distribution of females across habitat patches will become relatively uniform. Despite supporting low reproductive output, suboptimal patches may achieve moderate density owing to steady immigration from optimal patches. If suboptimal patches become saturated, further emigration from optimal patches may be curtailed, giving rise to frustrated dispersal. However, territorial females may be effective in preventing frustrated dispersers from breeding (Bujalska, 1973), and causing further depletion of optimal patches. After a seasonal decline in habitat suitability, optimal patches should recover quickly because they were unlikely to have been overexploited in the prior year. Thus, territoriality by females will tend to stabilize the subpopulation in optimal patches, but, by inducing dispersal, shorten the length of time that suboptimal habitat is an effective dispersal sink. I believe this type of social system would be more likely to give rise to annual than to multiannual cycles.

Now imagine another population of herbivorous mammals in an area with low resource patchiness, but an identical degree of habitat patchiness (i.e., habitat patches are internally homogeneous). Again, assume a low-density population inhabiting mainly optimal patches at the start of the breeding season. Female territoriality is expected to be relaxed, and as a consequence, population density in the optimal patches will increase quickly, while fewer emigrants are produced than if females were territorial. Substantial emigration may occur only after resources in the optimal patches have begun to be depleted. The distribution of females among habitat patches is expected to be uneven throughout most of the breeding season, giving rise to a consistently greater impact on the optimal patches. Suboptimal patches should not become saturated in most years, but when they do, they will inhibit emigration from the already deteriorating optimal patches. Frustrated dispersers will remain in optimal patches, breed, and help precipitate a crash by contributing to resource overexploitation. Thus, lack of territoriality by females destabilizes population dynamics in optimal patches (by allowing rapid growth and intense resource exploitation), but lengthens the time that suboptimal patches may be effective dispersal sinks. Because of overexploitation, immediate recovery of resources is not likely, and multiannual cycles may result.

The above scenarios are not intended as comprehensive models of vole cycles. The scenarios are intended to point out ways in which social organization and habitat patchiness may interact to make multiannual cycles more or less likely. I have not specified relative size, quality, or phenologies of different patch types because that level of complexity far exceeds the logic of simple verbal models.

A first step in testing whether small mammals having different social systems respond differently to habitat patchiness is to construct mathematical models or computer simulations in which sex-specific demography and patch parameters (e.g., size, quality) are made explicit, and population responses predicted. The logical consistency of the conceptual model could then be evaluated. Empirical tests will probably require parallel experiments with species that typically employ different social systems. Responses of socially divergent small mammals to the experimental creation of patchy landscapes, with suboptimal patches of different sizes, qualities, and phenologies, could be used to evaluate the model.

V. Conclusions

Plant patchiness, when it occurs on a small scale (resource patchiness), may influence behavioral strategies of individuals, and, when it occurs on a large scale (habitat patchiness), may affect population dynamics. Evidence is presented that the pattern of sex-specific territoriality in small-mammal herbivores is related to the degree of resource patchiness. However, social behavior of these animals is not fixed, and both phenotypic and genotypic variation in response to resources and population density is observed. The relative importance of natural selection, genetic drift, gene flow, and phenotypic plasticity in giving rise to spatial and temporal variation in behavioral characters has yet to be discovered. A critical issue is whether selection is strong enough to cause local adaptation of small mammals to particular habitat patches. A related problem is whether discontinuities in genetic structure and habitat structure are concordant or discordant. Dispersal of individuals among habitat patches will be of fundamental importance to the patterns discovered.

By affecting social organization, resource patchiness may influence the way a population responds to habitat patchiness. The role of social behavior *per se* in the population dynamics of arvicolid rodents remains enigmatic. There is no clear relationship between mode of social organization and magnitude of population fluctuations (Table 6), and it is increasingly evident that social behavior plays a limited role in vole cycles (Heske *et al.*, 1988; Hansson and Henttonen, 1988; Tamarin *et al.*, 1990). However, I argue that social behavior may interact with habitat patchiness to influence population dynamics. Female territoriality will generally be a stabilizing influence (Sten-

Table 6 Territorial Systems and the Tendency of Selected
Vole Populations to Fluctuate[a]

Species	Territorial Sex	S
Clethrionomys gapperi	Female	0.27
C. glareolus	Female	0.22, 0.47, 0.52
C. rufocanus	Female	0.68
C. rutilus	Female	0.35
Microtus breweri	Female	0.16
M. pennsylvanicus	Female	0.16, 0.24
M. agrestis	Male	0.30, 0.53, 0.62
M. californicus	Male	0.09, 0.89
M. oeconomus	Male	0.75
M. ochrogaster	Female–male pair	0.23, 0.82

[a] Temporal dynamics are measured by

$$S = \sqrt{\frac{\Sigma(\log N_i - \overline{\log N_i})^2}{n - 1}}$$

where N_i = density obtained by annual census, and n = the number of censuses (Lewonthin, 1966). S has been shown to discriminate cyclic ($S > 0.50$) from noncyclic ($S < 0.50$) populations (Henttonen *et al.* 1985). S-values were obtained from Ostfeld (1988); territorial systems are as indicated in Table 3.1. Some species show geographic variation in S. When more than one entry appears for a given species, a range of S- values is given to represent interpopulational variation.

seth, 1985), but by accelerating emigration from optimal patches, it may reduce the length of time that suboptimal patches are effective dispersal sinks. Thus, female territoriality may, in fact, destabilize population dynamics in a patchy landscape.

More realistic models of the population consequences of habitat patchiness should incorporate variation in patch size and connectedness, relative patch quality, and seasonal changes in patch quality. Empiricists will then be kept busy for years to come.

Acknowledgments

I am very grateful to Lorrie Klosterman for collaborations and for her extensive comments on a draft of the manuscript; Bill Lidicker and Andrew Cockburn for influencing my thinking on patchiness; Mike Gaines for sending and discussing an unpublished manuscript; and Mark Hunter for inviting me to prepare this paper. Financial support was received from the Mary Flagler Cary Charitable Trust. This is a contribution to the program of the Institute of Ecosystem Studies.

References

Adler, G. H., and Wilson, M. L. (1989). Demography of the meadow vole along a simple habitat gradient. *Can. J. Zool.* **67**, 772–774.

Anderson, J. H. (1975). Phenotypic correlations among relatives and variability in reproductive performance in populations of the vole *Microtus townsendii.* Unpublished Ph.D. dissertation, University of British Columbia, Vancouver, B.C., Canada.

Anderson, P. K. (1988). "Rodent Dispersal: A Resident Fitness Hypothesis." Am. Soc. Mammal. Spec. Publ. no. 9. Provo, Utah.

Anderson, P. K., Whitney, P. H., and Huang, J. P. (1976). *Arvicola richardsoni:* Ecology and biochemical polymorphism in the front ranges of southern Alberta. *Acta Theriol.* **21**, 425–468.

Andrzejewski, R., and Mazurkiewicz, M. (1976). Abundance of food supply and the size of the bank vole's home range. *Acta Theriol.* **21**, 237–253.

Batzli, G. O. (1975). The role of small mammals in arctic ecosystems. *In* "Small Mammals: Their Productivity and Population Dynamics" (F. B. Golley, K. Petrusewicz, and L. Ryszkowski, eds.), pp. 243–268. Cambridge University Press, London.

Batzli, G. O. (1985). Nutrition. *In* "Biology of New World *Microtus*" (R. H. Tamarin, ed.), pp. 779–811. Am. Soc. Mammalogists Spec. Pub. No. 8, Provo, Utah.

Batzli, G. O., and Jung, H. G. (1980). Nutritional ecology of microtine rodents: Resource utilization near Atkasook, Alaska. *Arctic Alpine Res.* **12**, 483–499.

Batzli, G. O., and Pitelka, F. A. (1971). Condition and diet of cycling populations of the California vole, *Microtus californicus. J. Mammal.* **52**, 141–163.

Batzli, G. O., Pitelka, F. A., and Cameron, G. N. (1983). Habitat use by lemmings near Barrow, Alaska. *Hol. Ecol.* **6**, 255-262.

Berger, P. J., Sanders, E. H., Gardner, P. D., and Negus, N. C. (1977). Phenolic plant compounds functioning as reproductive inhibitors in *Microtus montanus. Science* **195**, 575–577.

Berger, P. J., Negus, N. C., Sanders, E. H., and Gardner, P. D. (1981). Chemical triggering of reproduction in *Microtus montanus. Science* **214**, 69–70.

Birney, E. C., Grant, W. E., and Baird, D. D. (1977). Importance of vegetative cover to cycles of *Microtus* cycles. *Ecology* **60**, 349–361.

Boag, P. T., and Boonstra, R. (1988). Quantitative genetics of life history traits in meadow voles (*Microtus pennsylvanicus*). *In* "Evolution of Life Histories of Mammals" (M. S. Boyce, ed.), pp. 149–168. Yale University Press, New Haven, Connecticut.

Bock, E. (1972). Use of forest associations by bank vole populations. *Acta Theriol.* **17**, 203–219.

Bondrup-Nielson, S. (1986). Investigation of spacing behaviour of *Clethrionomys gapperi* by experimentation. *J. Anim. Ecol.* **55**, 269–279.

Bondrup-Nielson, S. (1987). Demography of *Clethrionomys gapperi* in different habitats. *Can. J. Zool.* **65**, 277–283.

Boonstra, R. (1977). Effects of conspecifics on survival during population declines in *Microtus townsendii. J. Anim. Ecol.* **46**, 835–851.

Boonstra, R., and Boag, P. T. (1987). A test of the Chitty hypothesis: Inheritance of life history traits in meadow voles. *Evolution* **41**, 929–947.

Boonstra, R., Krebs, C. J., Gaines, M. S., Johnson, M. L., and Craine, I. T. M. (1987). Natal philopatry and breeding systems in voles (*Microtus* spp.). *J. Anim. Ecol.* **56**, 655–673.

Boonstra, R. and Rodd, F. H. (1983). Regulation of breeding density in *Microtus pennsylvanicus. J. Anim. Ecol.* **52**, 757–780.

Bowen, B. S. (1982). Temporal dynamics of microgeographic structure of genetic variation in *Microtus californicus. J. Mammal.* **63**, 625–638.

Bowen, B. S., and Koford, R. R. (1987). Dispersal, population size, and genetic structure of *Microtus californicus:* Empirical findings and computer simulation. *In* "Mammalian Dispersal

Patterns" (D. Chepko-Sade and Z. T. Halpin, eds.), pp. 180–189. University of Chicago Press, Chicago, Illinois.

Boyce, C. C. K., and Boyce, J. L., III. (1988). Population biology of *Microtus arvalis*. I. Lifetime reproductive success of solitary and grouped breeding females. *J. Anim. Ecol.* **57,** 711–722.

Buechner, M. (1989). Are small-scale landscape features important factors for field studies of small mammal dispersal sinks? *Landscape Ecol.* **2,** 191–199.

Bujalska, G. (1973). The role of spacing behavior among females in the regulation of reproduction in the bank vole. *J. Reprod. Fert. (Suppl.)* **19,** 465–474.

Bujalska, G. (1990). Social system of the bank vole, *Clethrionomys glareolus*. *In* "Social Systems and Population Cycles in Voles" (R. H. Tamarin, R. S. Ostfeld, S. R. Pugh, and G. Bujalska, eds.), pp. 155–168. Birkhäuser-Verlag, Basel, Switzerland.

Burns, G. R. (1981). Population dynamics of island populations of *Clethrionomys rutilus*. *Can. J. Zool.* **59,** 2115–2122.

Chitty, D. (1967). The natural selection of self-regulatory behaviour in animal populations. *Proc. Ecol. Soc. Aust.* **2,** 51–78.

Chitty, D. (1987). Social and local environments of the vole *Microtus townsendii*. *Can. J. Zool.* **65,** 2555–2566.

Christian, J. J. (1971). Population density and reproductive efficiency. *Biol. Reprod.* **4,** 248–294.

Cockburn, A. (1988). "Social Behaviour in Fluctuating Populations." Croom Helm, London.

Cockburn, A., and Lidicker, W. Z., Jr., (1983). Microhabitat heterogeneity and population ecology of an herbivorous rodent, *Microtus californicus*. *Oecologia (Berl.)* **59,** 167–177.

Cole, F. R., and Batzli, G. O. (1979). Nutrition and population dynamics of the prairie vole, *Microtus ochrogaster*, in central Illinois. *J. Anim. Ecol.* **48,** 455–470.

Daly, J. C., and Patton, J. L. (1986). Growth, reproduction, and sexual dimorphism in *Thomomys bottae* pocket gophers. *J. Mammal.* **67,** 256–265.

Davies, N. B., and Houston, A. I. (1984). Territory economics. *In* "Behavioural Ecology: An Evolutionary Approach." (J. R. Krebs, and N. B. Davies, eds), pp. 148–169. Blackwell, Oxford, England.

Dickman, C. R., and Doncaster, C. P. (1987). The ecology of small mammals in urban habitats. I. Populations in a patchy environment. *J. Anim. Ecol.* **56,** 629–640.

Dickman, C. R., and Doncaster, C. P. (1989). The ecology of small mammals in urban habitats. II. Demography and dispersal. *J. Anim. Ecol.* **58,** 119–127.

Dobrowolska, A. (1981). Serum transferrin polymorphism in the common vole, *Microtus arvalis* (Pall. 1779). *Bull. l'Acad. Pol. Sci. ser. sci. biol.* **29,** 149–154.

Dyke, G. R. (1971). "Food and Cover of Fluctuating Populations of Northern Cricetids." Unpublished Ph.D. dissertation, University of Alberta, Edmonton, Canada.

Elton, C. S. (1942). "Voles, Mice and Lemmings." Clarendon Press, Oxford, England.

Fahrig, L. and Merriam, G. (1985). Habitat patch connectivity and population survival. *Ecology* **66,** 1762–1768.

Falconer, D. S. (1981). "Introduction to Quantitative Genetics." Longman, London, England.

Fleharty, E. D., and Olson, L. E. (1969). Summer food habits of *Microtus ochrogaster* and *Sigmodon hispidus*. *J. Mammal.* **50,** 475–486.

Gaines, M. S. (1985). Genetics. *In* "Biology of New World *Microtus*" (R. H. Tamarin, ed.), pp. 845–883. Am. Soc. Mammalogists Spec. Pub. no. 8, Provo, Utah.

Gaines, M. S., and Johnson, M. L. (1987). Phenotypic and genotypic mechanisms for dispersal in *Microtus* populations and the role of dispersal in population regulation. *In* "Mammalian Dispersal Patterns" (D. Chepko-Sade, and Z. T. Halpin, eds.), pp. 162–179. University of Chicago Press, Chicago.

Gaines, M. S., and McClenaghan, L. R., Jr. (1980). Dispersal in small mammals. *Annu. Rev. Ecol. Syst.* **11,** 163–196.

Gaines, M. S., McClenaghan, L. R., Jr., and Rose, R. K. (1978). Temporal patterns of allozymic variation in fluctuating populations of *Microtus ochrogaster*. *Evolution* **32,** 723–739.

Gaines, M. S., Stenseth, N. C., Johnson, M. L., Ims, R. A., and Bondrup-Nielson, S. (1991). A response to solving the enigma of population cycles with a multifactorial perspective. *J. Mammal*, in press.

Getz, L. L., Hofmann, J. E., and Carter, C. S. (1987). Mating system and population fluctuations of the prairie vole, *Microtus ochrogaster*. *Am. Zool.* **27**, 909–920.

Getz, L. L., Verner, L., Cole, F. R., Hofmann, J. E., and Avalos, D. E. (1979). Comparisons of population demography of *Microtus ochrogaster* and *M. pennsylvanicus*. *Acta Theriol.* **24**, 319–349.

Gliwicz, J. (1989). Individuals and populations of the bank vole in optimal, suboptimal, and insular habitats. *J. Anim. Ecol.* **58**, 237–247.

Godfrey, G. K. (1953). The food of *Microtus agrestis hirtus* (Bellamy, 1839) in Wytham, Berkshire. *Saugetierkundliche Mitteilungen* **1**, 148–151.

Grant, P. R. (1971). The habitat preference of *Microtus pennsylvanicus*, and its relevance to the distribution of this species on islands. *J. Mammal.* **52**, 351–361.

Hansson, L. (1971a). Small rodent food, feeding, and population dynamics. A comparison between granivorous and herbivorous species in Scandinavia. *Oikos* **22**, 183–198.

Hansson, L. (1971b). Habitat, food, and population dynamics of the field vole *Microtus agrestis* (L.) in south Sweden. *Viltrevy (Stockh.)* **8**, 267–378.

Hansson, L. (1977). Spatial dynamics of field voles *Microtus agrestis* in heterogeneous landscapes. *Oikos* **29**, 539–544.

Hansson, L. (1979). Condition and diet in relation to habitat in bank voles *Clethrionomys glareolus:* Population or community approach? *Oikos* **33**, 55–63.

Hansson, L. (1985). *Clethrionomys* food: Generic, specific and regional characteristics. *Ann. Zool. Fennici* **22**, 319–328.

Hansson, L., and Henttonen, H. (1988). Rodent dynamics as community processes. *Trends Ecol. Evol.* **3**, 195–200.

Hasler, J. F. (1975). A review of reproduction and sexual maturation in the microtine rodents. *Biologist* **57**, 52–86.

Hassell, M. P. (1980). Some consequences of habitat heterogeneity for population dynamics. *Oikos* **35**, 150– 160.

Henttonen, H., McGuire, A. D., and Hansson, L. (1985). Comparisons of amplitudes and frequencies (spectral analyses) of density variations in long-term data sets of *Clethrionomys* species. *Ann. Zool. Fennici* **22**, 221–227.

Heske, E. J. (1987). Spatial structuring and dispersal in a high-density population of the California vole *Microtus californicus*. *Hol. Ecol.* **10**, 137–148.

Heske, E. J., and Ostfeld, R. S. (1990). Sexual dimorphism in size, relative size of testes, and mating systems in North American voles. *J. Mammal.* **71**, 510–519.

Heske, E. J., Ostfeld, R. S., and Lidicker, W. Z., Jr. (1984). Competitive interactions between *Microtus californicus* and *Reithrodontomys megalotis* during two peaks of *Microtus* abundance. *J. Mammal.* **65**, 271–280.

Heske, E. J., Ostfeld, R. S., and Lidicker, W. Z., Jr. (1988). Does social behavior drive vole cycles? An evaluation of competing models as they pertain to California voles. *Can. J. Zool.* **66**, 1153–1159.

Hestbeck, J. B. (1982). Population regulation of cyclic small mammals: The social fence hypothesis. *Oikos* **39**, 157–163.

Holisová, V. (1975). The foods eaten by rodents in reed swamps of Nesyt fishpond. *Zool. Listy* **24**, 223–237.

Holisová, V. (1976). The food eaten by the water vole (*Arvicola terrestris*) in gardens. *Zool. Listy* **25**, 193– 208.

Ims, R. A. (1987). Male spacing systems in microtine rodents. *Am. Nat.* **139**, 475–484.

Ims, R. A. (1988a). The potential for sexual selection in males: Effect of sex ratio and spatiotemporal distribution of receptive females. *Evol. Ecol.* **2**, 338–352.

Ims, R. A. (1988b). Spatial clumping of sexually receptive females induces space sharing among male voles. *Nature* **335**, 541–543.

Jeppsson, B. (1990). Effects of density and resources on the social system of water voles. *In* "Social Systems and Population Cycles in Voles" (R. H. Tamarin, R. S. Ostfeld, S. R. Pugh, and G. Bujalska, eds.), pp. 168–184. Birkhauser-Verlag, Basel.

Kalela, O., Koponen, T., Lind, E. A., Skaren, V., and Tast, J. (1961). Seasonal change of habitat in the Norwegian lemming, *Lemmus lemmus* (L.). *Ann. Acad. Sci Fenn., Ser A, IV. Biologica* **55**, 1–72.

Kalela, O., Kilpelainen, L., Koponen, T., and Tast, J. (1971). Seasonal differences in habitats of the Norwegian lemming, *Lemmus lemmus* (L.), in 1959 and 1960 at Kilpisjärvi, Finnish Lapland. *Ann. Acad. Sci. Fenn., Ser. A., IV. Biologica* **178**, 1–22.

Kawata, M. (1985). Mating system and reproductive success in a spring population of the red-backed vole, *Clethrionomys rufocanus bedfordiae*. *Oikos* **45**, 181–190.

Keller, B. J. (1985). Reproductive patterns. *In* "Biology of New World *Microtus*" (R. H. Tamarin, ed.), pp. 725–778. Am. Soc. Mammalogists Spec. Pub. No. 8, Provo, Utah.

Kincaid, W. B., and Cameron, G. N. (1985). Interactions of cottonrats with a patchy environment: Dietary responses and habitat selection. *Ecology* **66**, 1769–1783.

Korpimäki, E. (1985). Rapid tracking of microtine populations by their avian predators: Possible evidence for stabilizing predation. *Oikos* **45**, 281–284.

Krebs, C. J. (1970). *Microtus* population biology: Behavioral changes associated with the population cycle in *M. ochrogaster* and *M. pennsylvanicus. Ecology* **51**, 34–52.

Krebs, C. J. (1978). A review of the Chitty hypothesis of population regulation. *Can. J. Zool.* **56**, 2463–2480.

Krebs, C. J. (1985). Do changes in spacing behaviour drive population cycles in small mammals? *In* "Behavioural Ecology: Ecological Causes of Adaptive Behaviour" (R. M. Sibly, and R. H. Smith, eds.), pp. 295–312. Blackwell, Oxford, England.

Krebs, C. J., Keller, B. L., and Tamarin, R. H. (1969). *Microtus* population biology: Demographic changes in fluctuating populations of *Microtus ochrogaster* and *M. pennsylvanicus* in southern Indiana. *Ecology* **50**, 587–607.

Krohne, D. T. (1980). Intraspecific litter size variation in *Microtus californicus*. II. Variation between populations. *Evolution* **34**, 1174–1182.

Lewontin, R. C. (1966). On the measurement of relative variability. *Syst. Zool.* **15**, 141–142.

Lidicker, W. Z., Jr. (1975). The role of dispersal in the demography of small mammals. *In* "Small Mammals: Their Productivity and Population Dynamics" (F. B. Golley, K. Petrusewicz, and L. Ryszkowski, eds.), pp. 103–128. Cambridge University Press, London.

Lidicker, W. Z., Jr. (1985). Population structuring as a factor in understanding microtine cycles. *Acta Zool. Fennica* **173**, 23–27.

Lidicker, W. Z., Jr. (1988). Solving the enigma of microtine "cycles." *J. Mammal.* **69**, 225–235.

Lidicker, W. Z., Jr., and Patton, J. L. (1987). Patterns of dispersal and genetic structure in populations of small rodents. *In* "Mammalian Dispersal Patterns" (D. Chepko-Sade, and Z. T. Halpin, eds.), pp. 144–161. University of Chicago Press.

Lindroth, R. L., and Batzli, G. O. (1984). Food habits of the meadow vole (*Microtus pennsylvanicus*) in bluegrass and prairie habitats. *J. Mammal.* **65**, 600–606.

Łomnicki, A. (1980). Regulation of population density due to individual differences and patchy environment. *Oikos* **35**, 185–193.

Łomnicki, A. (1982). Individual heterogeneity and population regulation. *In* "Current Problems in Sociobiology" (King's College Sociobiology Group, eds.), pp. 153–167. Cambridge University Press, London.

Lousch, C. D. (1956). Adrenocortical activity in relation to the density and dynamics of three confined populations of *Microtus pennsylvanicus. Ecology* **37**, 701–713.

Ludwig, D. (1984). *Microtus richardsoni* microhabitat and life history. *In* "Winter Ecology of Small Mammals" (J. F. Merritt, ed.), pp. 319–331. Special Publication No. 10, Carnegie Museum of Natural History, Pittsburgh, Pennsylvania.

MacArthur, R. H., and Wilson, E. O. (1967). "The Theory of Island Biogeography." Princeton University Press, Princeton, New Jersey.

Madison, D. M. (1980). Space use and social structure in meadow voles, *Microtus pennsylvanicus*. *Behav. Ecol. Sociobiol.* **7**, 65–71.

Madison, D. M. (1990). Social organizational modes in models of microtine cycles. *In* "Social Systems and Population Cycles in Voles" (R. H. Tamarin, R. S. Ostfeld, S. R. Pugh, and G. Bujalska, eds.), pp. 25–34. Birkhäuser-Verlag, Basel, Switzerland.

Madison, D. M., and McShea, W. J. (1987). Seasonal changes in reproductive tolerance, spacing, and social organization in meadow voles: A microtine model. *Am. Zool.* **27**, 899–908.

Mazurkiewicz, M. (1971). Shape, size, and distribution of home ranges of *Clethrionomys glareolus* (Schreber, 1780). *Acta Theriol.* **16**, 23–60.

Mazurkiewicz, M. (1981). Spatial organization of a bank vole population in years of small or large numbers. *Acta Theriol.* **26**, 31–45.

McNab, B. K. (1980). Food habits, energetics, and the population biology of mammals. *Am. Nat.* **116**, 106–124.

Mihok, S. (1979). Behavioral structure and demography of subarctic *Clethrionomys gapperi* and *Peromyscus maniculatus*. *Can. J. Zool.* **57**, 1520–1535.

Mihok, S. (1981). Chitty's hypothesis and behaviour in subarctic red-backed voles *Clethrionomys gapperi*. *Oikos* **36**, 281–295.

Myllymäki, A. (1977a). Intraspecific competition and home range dynamics in the field vole *Microtus agrestis*. *Oikos* **29**, 553–569.

Myllymäki, A. (1977b). Demographic mechanisms in the fluctuating populations of the field vole *Microtus agrestis*. *Oikos* **29**, 468–493.

Naumov, N. P. (1972). "The Ecology of Animals." University of Illinois Press, Urbana, Illinois.

Neal, B. R., Pulkinen, D. A., and Owen, B. D. (1973). A comparison of faecal and stomach contents analysis in the meadow vole (*Microtus pennsylvanicus*). *Can. J. Zool.* **51**, 715–721.

Ostfeld, R. S. (1985). Limiting resources and territoriality in microtine rodents. *Am. Nat.* **126**, 1–15.

Ostfeld, R. S. (1986). Territoriality and mating system of California voles. *J. Anim. Ecol.* **55**, 691–706.

Ostfeld, R. S. (1988). Fluctuations and constancy in populations of small rodents. *Am. Nat.* **131**, 445–452.

Ostfeld, R. S. (1990). The ecology of territoriality in small mammals. *Trends Ecol. Evol.* **5**, 411–415.

Ostfeld, R. S., and Klosterman, L. L. (1986). Demographic substructure in a California vole population inhabiting a patchy environment. *J. Mammal.* **67**, 693–704.

Ostfeld, R. S., and Klosterman, L. L. (1990). Microtine social systems, adaptation, and the comparative method. *In* "Social Systems and Population Cycles in Voles" (R. H. Tamarin, R. S. Ostfeld, S. R. Pugh, and G. Bujalska, eds.), pp. 35–44. Birkhauser-Verlag, Basel, Switzerland.

Ostfeld, R. S., Lidicker, W. Z., Jr., and Heske, E. J. (1985). The relationship between habitat heterogeneity, space use, and demography in a population of California voles. *Oikos* **45**, 433–442.

Ostfeld, R. S., Pugh, S. R., Seamon, J. O., and Tamarin, R. H. (1988). Space use and reproductive success in a population of meadow voles. *J. Anim. Ecol.* **57**, 385–394.

Patton, J. L., and Feder, J. H. (1981). Microspatial genetic heterogeneity in pocket gophers: Non-random breeding and drift. *Evolution* **35**, 912–920.

Pearson, O. P. (1985). Predation. *In* "Biology of New World *Microtus*" (R. H. Tamarin, ed.), pp. 535–566. Am. Soc. Mammal. Spec. Pub. No. 8, Provo, Utah.

Pitelka, F. A. (1973). Cyclic pattern in lemming populations near Barrow, Alaska. *In* "Alaskan Arctic Tundra" (M. E. Britton, ed.), pp. 199–216. Arctic Institute of North America Technical Paper no.25, Edmonton, Canada.

Pulliam, H. R. (1988). Sources, sinks, and population regulation. *Am. Nat.* **132**, 652–661.

Randall, J. A., and Johnson, R. E. (1979). Population densities and habitat occupancy by *Microtus longicaudus* and *M. montanus*. *J. Mammal.* **60**, 217–219.

Redfield, J. A., Taitt, M. J., and Krebs, C. J. (1978). Experimental alteration of sex ratios in populations of *Microtus townsendii,* a field vole. *Can. J. Zool.* **56,** 17–27.

Reich, L. M., Wood, K. M., Rothstein, B. E., and Tamarin, R. H. (1982). Aggressive behavior of male *Microtus breweri* and its demographic implications. *Anim. Behav.* **30,** 117–122.

Roff, D. A. (1975). Population stability and the evolution of dispersal in a heterogeneous environment. *Oecologia(Berlin)* **19,** 217–237.

Rose, R. K., and Gaines, M. S. (1976). Levels of aggression in fluctuating populations of the prairie vole, *Microtus ochrogaster,* in eastern Kansas. *J. Mammal.* **57,** 43–57.

Rose, R. K., and Gaines, M. S. (1981). Relationships of genotype, reproduction, and wounding in Kansas prairie voles. *In* "Mammalian Population Genetics", (M. H. Smith, and J. Joule, eds.), pp. 161–179. University of Georgia Press, Athens, Georgia.

Rosenzweig, M. L., and Abramsky, Z. (1980). Microtine cycles: the role of habitat heterogeneity. *Oikos* **34,** 141–146.

Rothstein, B. E., and Tamarin, R. H. (1977). Feeding behavior of the insular beach vole, *Microtus breweri. J. Mammal.* **58,** 84–85.

Rubenstein, D. I., and Wrangham, R. W. (eds.) (1986). "Ecological Aspects of Social Evolution: Birds and Mammals." pp. 1–551. Princeton University Press, Princeton, New Jersey.

Saitoh, T. (1985). Practical definition of territory and its application to the spatial distribution of voles. *J. Ethol. (Kyoto)* **3,** 143–149.

Sanders, E. H., Gardner, P. D., Berger, P. J., and Negus, N. C. (1981). 6-Methoxybenzoxazolinone: A plant derivative that stimulates reproduction in *Microtus montanus. Science* **214,** 67–69.

Seabloom, R. W. (1985). Endocrinology. *In* "Biology of New World *Microtus*" (R. H. Tamarin, ed.), pp. 685–724. Am. Soc. Mammalogists Spec. Pub. no. 8, Provo, Utah.

Southern, H. N. (1979). Population processes in small mammals. *In* "Ecology of Small Mammals" (D. M. Stoddart, ed.), pp. 63–101. Chapman and Hall, London.

Stamps, J. A., Buechner, M., and Krishnan, V. V. (1987). The effects of edge permeability and habitat geometry on emigration from patches of habitat. *Am. Nat.* **129,** 533–552.

Stenseth, N. C. (1980). Spatial heterogeneity and population stability: Some evolutionary consequences. *Oikos* **35,** 165–184.

Stenseth, N. C. (1983). Causes and consequences of dispersal in small mammals. *In* "The Ecology of Animal Movement" (I. Swingland, and P. Greenwood, eds.), pp. 63–101. Oxford University Press, Oxford, England.

Stenseth, N. C. (1985). Models of bank vole and wood mouse populations. *In* "The Ecology of Woodland Rodents: Bank Voles and Wood Mice" (J. R. Flowerdew, J. Gurnell, and J. H. W. Gipps, eds.), *Symp. Zool. Soc. Lond.* No. 55, 339–376, Oxford Science Publ., Oxford, England.

Stenseth, N. C., and Framstad, E. (1980). Reproductive effort and optimal reproductive rates in small rodents. *Oikos* **34,** 23–34.

Taitt, M. J., and Krebs, C. J. (1985). Population dynamics and cycles. *In* "Biology of New World *Microtus*" (R. H. Tamarin, ed.), pp. 567–620. Am. Soc. Mammal. Spec. Pub. no. 8, Provo, Utah.

Tamarin, R. H., Reich, L. M., and Moyer, C. A. (1984). Meadow vole cycles within fences. *Can. J. Zool.* **62,** 1796–1804.

Tamarin, R. H., Ostfeld, R. S., Pugh, S. R., and Bujalska, G. (eds.) (1990). "Social Systems and Population Cycles in Voles" Birkhäuser-Verlag, Basel, Switzerland.

Tast, J. (1966). The root vole, *Microtus oeconomus* (Pallas), as an inhabitant of seasonally flooded land. *Ann. Zool. Fennici* **3,** 127–171.

Van Horne, B. (1982). Demography of the longtail vole *Microtus longicaudus* in seral stages of coastal coniferous forest, southeast Alaska. *Can. J. Zool.* **60,** 1690–1709.

Van Horne, B. (1983). Density as a misleading indicator of habitat quality. *J. Wildl. Man.* **47,** 893–901.

Vickery, W. L. (1979). Food consumption and preferences in wild populations of *Clethrionomys gapperi* and *Napeozapus insignis. Can. J. Zool.* **57,** 1536–1542.

Viitala, J. (1977). Social organization in cyclic subarctic populations of the voles *Clethrionomys rufocanus* (Sund.) and *Microtus agrestis* (L.). *Ann. Zool. Fennici* **14**, 53–93.

Viitala, J., and Hoffmeyer, I. (1985). Social organization in *Clethrionomys* compared with *Microtus* and *Apodemus:* Social odours, chemistry and biological effects. *Ann. Zool. Fennici* **22**, 359–371.

Wallin, L. (1973). Relative estimates of small-mammal populations in relation to the spatial pattern of trappability. *Oikos* **24**, 282–286.

Webster, A. B., and Brooks, R. J. (1981). Social behavior of *Microtus pennsylvanicus* in relation to seasonal changes in demography. *J. Mammal.* **62**, 738–751.

West, S. D. (1982). Dynamics of colonization and abundance in central Alaskan populations of the northern red-backed vole, *Clethrionomys rutilus. J. Mammal.* **63**, 128–143.

West, S. D., and Dublin, H. T. (1984). Behavioral strategies of small mammals under winter conditions: Solitary or social? *In* "Winter Ecology of Small Mammals" (J. F. Merritt, ed), pp. 293–299. Carnegie Mus. Nat. Hist, Spec. Publ. no. 10, Pittsburgh, Pennsylvania.

Wiens, J. A. (1989). Spatial scaling in ecology. *Funct. Ecol.* **3**, 385–397.

Wolff, J. O. (1980). The role of habitat patchiness in the population dynamics of snowshoe hares. *Ecology* **50**, 111–130.

Wolff, J. O. (1989). Behavior. *In* "Advances in the Study of *Peromyscus* (Rodentia)" (G. L. Kirkland, Jr. and L. N. Layne, eds), pp. 271–191. Texas Tech University Press, Lubbock, Texas.

Wolff, J. O., and Lidicker, W. Z., Jr. (1980). Population ecology of the taiga vole, *Microtus xanthognathus*, in interior Alaska. *Can. J. Zool.* **58**, 1800–1812.

Wolff, J. O., Lidicker, W. Z., Jr., Chesser, R. K., and Smith,. M. H. (1991). Review of Anderson, P. K., Dispersal in rodents: A resident fitness hypothesis. *J. Mammal.* **72**, in press.

Wright, S. (1965). The interpretation of population structure by F-statistics with special regard to systems of mating. *Evolution* **19**, 395–420.

Ylönen, H., Kojola, T., and Viitala, J. (1988). Changing female spacing behaviour and demography in an enclosed breeding population of *Clethrionomys glareolus. Hol. Ecol.* **11**, 286–292.

Yu, O., Verge, Y., and Gounot, M. (1980). Modele d'interaction entre campagnols *Microtus arvalis* et praire permanente. *Rev. Ecol. (Terre Vie)* **34**, 373–426.

Zwicker, K. (1989). Home range and spatial organization of the beach vole, *Microtus breweri. Behav. Ecol. Sociobiol.* **25**, 161–170.

4

Plant Genotype: A Variable Factor in Insect–Plant Interactions

Arthur E. Weis and Diane R. Campbell
Department of Ecology and Evolutionary Biology
University of California, Irvine
Irvine, California

I. Introduction: Plant Genotype and Insect Resources

In the bottom-up view of ecology explored in this volume, plant genes have a special role. Genes expressed at the producer level of a food chain mediate the transformation of mineral nutrients and energy from light from the abiotic environment into a form that can be used by species at higher trophic levels. Since the plant genotype encodes the resource *transformation rules,* genetic variance within a plant population can be a source of resource diversity that influences consumer population size and community structure.

Insects, be they herbivores or pollinators, do not respond directly to plant genes, but rather to plant phenotypes, and phenotypes within a plant population can vary for reasons having to do with genetics, environment, or both. In the simplest case, the population variance observed in a phenotypic trait is the sum of the variation caused by genetic differences among individuals, and that caused by differences in the environments in which individuals develop (Falconer, 1981). Genetic contributions to phenotypic variance can also change with the environment.

Because insects respond to plant phenotypes, and because phenotypic differences are not always, nor necessarily often, caused by genetic factors, why should ecologists be interested in genotypic variation? Over the short ecological time scale, a genetic cause for plant phenotypic variation may be inconsequential for most insects. However, natural selection imposed by herbivores and pollinators can lead to evolutionary changes in plant gene frequencies. Plant population genetic changes can, in turn, alter the rate and pattern of resource flow to higher trophic levels. The role of pollinators as plant gamete vectors can even play a role in plant speciation, which can increase the number of channels for resources to flow to higher trophic levels.

In this chapter, we examine some of the effects that within-population genetic variation in plants are known to have on insect consumers. There are several important differences between herbivores and pollinators, not the least of which is the fact that the former are usually plant antagonists, while the later are usually plant mutualists. Thus, plants are under selection to avoid one and attract the other. Further, many insect herbivores spend most of their developmental period on the plant, and perhaps on only one individual. On the other hand, pollinators briefly visit many plant individuals. If the plant is self-incompatible or vulnerable to inbreeding depression, it is in the genetic interest of the plant to encourage the pollinator to keep moving (Waser and Price, 1983). For these reasons, genetic variation in plants is likely to have different effects on these two types of consumers, and we will deal with each in turn. However, we will conclude the chapter by exploring an important similarity in the selective regimes these two types of consumers place on their plants. For both pollination and herbivory, the impact on the fitness of a plant genotype depends on its probability of encounter with these insects, and on its subsequent reaction to such encounters. We will point out that because the probability of plant–insect encounter can change with insect population density, the intensity, and under some circumstances the direction, of natural selection that pollinators and herbivores place on plant characters can fluctuate with the wax and wane of insect populations.

II. Ecological Consequences of Plant Genetic Variation for Herbivores

A. Plant Genetic Variation and Herbivore Demography

Like many environmental factors, plant genotype can influence the capacity of a plant individual to support growth and reproduction of consumers. Plant properties can influence the ecological efficiency of herbivores, particularly through nutritional and defensive modes, and thus contribute to the absolute limit on herbivore population size. Quite simply, a population of plants with genotypes that code for nutritious tissues, unprotected by chemical or mechanical barriers, will probably support a larger herbivore population than an equal-sized plant population composed of genotypes that code for nutrient-poor or highly defended tissues, or a plant population composed of any mixture of the two genotypes. Although it has been debated whether or not herbivore population size is often limited by carnivore attack (Hairston *et al.*, 1960; Erlich and Birch, 1967; Crawley, 1983), plant quality, which of necessity is mediated by genotype, may keep herbivore population growth rates in the range at which natural enemies can be an effective agent of population regulation (Lawton and McNeil, 1979; Price *et al.*, 1980). What is the evidence that plant genotype is a contributing factor to herbivore population size?

Different plant species represent genetic variants at a particularly coarse level. Thus, the differential effects of plant species on herbivore population dynamics serve as the model for exploring the effects of within-species differences in suitability to herbivores. Feeding experiments have frequently shown strong differences in herbivore vital statistics when they are fed on alternative host plants. The plant properties implicated include nutrient content, especially nitrogen and water (Scriber and Slansky, 1981), concentration of defensive compounds (e.g., Berenbaum and Feeny, 1981), and plant growth phenology (e.g., Mitter *et al.*, 1979). Carnivore attack rates on herbivores can also vary with host plant species. One might expect within-population genetic differences in host plant chemistry, morphology, or phenology to have similar effects on herbivore populations.

A note of caution in assessing the effects of plant genotype on herbivore populations has been urged by Karban (1991). As he points out, demonstrating that hosts of different genotype vary in their suitability for herbivore growth does not prove that those hosts also differ in effect on herbivore population size. Physiological experiments can indicate only the potential for ecological phenomena, and their results should be interpreted as such. Ecological experiments on the proper temporal and spatial scales are required to put plant genotype into perspective. To date, relatively few studies in natural populations have addressed the demographic consequences of host plant genotypic variation for herbivores, but interest in this area is growing (see Fritz and Simms, 1991).

Among the best-studied systems that have revealed plant genotypic effects on herbivore vital statistics is the interaction between the tall goldenrod *Solidago altissima* (Asteraceae) and its gall-inducing herbivore, *Eurosta solidaginis* (Diptera: Tephritidae). This host plant is a herbaceous perennial that spreads by rhizomes to form clones that frequently grow to several hundred stems; the aboveground parts are deciduous, and new stems grow from the rhizomes each spring. The *Eurosta* female punctures the terminal bud and injects an egg with her ovipositor, leaving a visible scar. When the egg hatches, the insect larva induces a spheroid swelling on the plant stem (Uhler, 1951; Weis *et al.*, 1989), in which it is supplied with all its nutritional needs; the insect totally depends on the plant to complete its life cycle. Goldenrod is well suited for studying plant genotypic effects on insects because it can be easily cloned from rhizome cuttings; the effects of genes can be separated from environment through common garden experiments. The self-incompatible mating system also facilitates hand crosses to study transmission of traits from parent to offspring.

In the field, it has been shown that gall infestation rates on goldenrod clones are consistent from year to year (McCrea and Abrahamson, 1987; Maddox and Root, 1990). In a 3-year study of 100 different clones, McCrea and Abrahamson (1987) showed that the rank order correlation among years for the number of galls per stem varied from 0.50 to 0.64. The consistency in infestation was not because of low dispersal by the insect— even when surviving insects were removed from clones at the end of the season, during the next spring incoming colonists would infest them at similar levels. Two mechanisms contributed to differential infestation; both the fraction of stems that were punctured (i.e., oviposition was attempted), and the fraction of punctured stems that formed galls differed among clones in the field. Since field observations such as these cannot distinguish between genetic influences on infestation rate and very localized environmental effects, a common garden experiment was performed using 13 replicates each of the 15 most heavily and 15 most lightly infested clones; these were denoted as most susceptible and most resistant, respectively. In the common garden, the number of galls per stem, the proportion of stems punctured, and the proportion of punctured stems to form galls was higher on the susceptible clones than on the resistant clones; the proportion of stems punctured in the most susceptible clone was nearly 90%, but less than 40% on the most resistant (Anderson *et al.*, 1989). These data indicate that females are more likely to attempt oviposition on some clones than on others, and that these preferred clones are more likely to produce galls. Further work by Anderson *et al.* (1989) suggested that the resistant clones may show a hypersensitive response; that is, the cells surrounding the larva in a newly forming gall die suddenly, leaving the insect without food. These experiments show that plant properties can have a very strong effect on gallmaker survivorship.

Beyond these direct effects on gallmaker demography, genetic variation in *S. altissima* strongly affects *Eurosta*'s vulnerability to parasitoids and predators. Gall size is an important determinant of insect survivorship (Weis and Abrahamson, 1986; Abrahamson *et al.*, 1989). Gallmakers in small galls (those with diameters less than ca. 20 mm) are vulnerable to attack by the parasitoid wasp *Eurytoma gigantea* (Hymenoptera: Eurytomidae), which penetrates the gall wall with its ovipositor to lay its egg within the gall's central chamber. Large galls (greater than ca. 21 mm) are too thick for the wasp's ovipositor to reach the center (Weis *et al.*, 1985), and so *Eurosta* larvae are invulnerable to attack when in large galls. Although from the gallmaker's perspective, bigger is better, biggest may not be best. Downy woodpeckers frequently peck open galls and eat the gallmaker during winter months when food is scarce; this visually hunting predator is much more likely to attack large galls than small ones (Weis and Abrahamson, 1986; Abrahamson *et al.*, 1989). Although insect genotype has an influence on gall size (Weis and Abrahamson, 1986; Weis and Gorman, 1990), repeated experiments have shown that around 20% of the variance in gall diameter is explained by plant genotype.

The goldenrod–gallmaker system shows that plant genotype influences the likelihood of success in a series of events through the life cycle of the gallmaker, *viz.*, whether an oviposition is attempted, the likelihood that an attempt actually results in an oviposition, the survivorship of the newly hatched larva, and the likelihood that the mature larva escapes natural enemies. Although the potential for a plant genetic influence on *Eurosta* population size at the end of a generation is clear, it is unknown how much of that potential can be realized. Do the within-generation effects of plant genotypic variation on gallmaker survival have an across-generation effect on gallmaker population density? Weather may influence overwintering survival of the fly and could have environmental effects on plant quality that outweigh genotypic effects. This question could be resolved with large-scale experimentation, whereby gallmaker density is measured on plant populations of controlled genotype, consisting of all resistant, all susceptible, and mixtures of the two kinds of plants.

Plant genotype is known to influence insect population growth for several species of aphids (Homoptera: Aphididae) (Moran, 1981; Service, 1984;

Whitham, 1983). Plant influences on aphid population growth can be readily assessed during the parthenogenetic phase of their lifecycle, which may last several months during the growing season. In a study of *Uroleucon rudbeckiae* (Homoptera; Aphididae) on clonal replicates of black-eyed susan (*Rudbeckia laciniata*; Asteraceae), Service (1984) showed that variance in aphid colony growth rate could be explained by differences in host plant genotype. However, aphid genotype also influenced colony growth rate in that the effect of plant genotype on aphid colony growth varied with colony genotype. Thus, the size of the local population, i.e., the number of aphids per plant stem, depended on genetically variable plant properties and on synergistic effects that arose from specific plant–insect genotype combinations. As a result, the size attained by the entire population (the sum of aphids in all colonies) at the end of the season will depend, in part, on the frequency distribution of plant and insect properties, and therefore, on the frequency distribution of plant and aphid genes.

Genetic substructuring of the herbivore population may occur when variance in herbivore performance is affected by a *plant genotype–insect genotype* interaction effect. Low vagility may result in a herbivore population structure in which the herbivores that infest each plant are genetically related. Karban (1989a) reported a case in which local herbivore demes are adapted to specific plant genotypes. He studied *Apterothrips secticornis* (Thysanoptera; Thripidae), an insect of low vagility and short generation time, on clones of the perennial herb *Erigeron glaucus* (Asteraceae). Three field plants were cloned and grown in a common garden, and infested with the thrips from each field clone in a cross-classified fashion. In each case, the thrips grew best on the clones taken from their field plant. Thus, low gene flow, due to low vagility, can lead to *genotype matching* in plant–insect interactions.

Variation in the insect's host-choice behavior can lead to a similar genetic structuring of herbivore populations. One might expect evolution of host-choice behavior by ovipositing adult insects to be driven by subsequent offspring performance. If this is a strong selective force on adult behavior, one would expect host choice and offspring performance to be positively correlated. Studies of choice between alternative host plant species does not always show this type of preference–performance correlation (see Thompson, 1988a), although such patterns of insect preference among genotypes within a plant species are less well documented. A curious type of correspondence between choice and performance has been found by Ng (1988) in *Euphydryas editha*, (Lepidoptera; Nymphalidae), a butterfly that feeds on *Pedicularis semibarbata* (Scrophulariaceae) in its larval stages. When ovipositing, some females are discriminating, preferring some plant genotypes over the rest. Other females are much less choosy. Offspring of the discriminating females grow better on the chosen plant genotypes than on the rest, whereas growth in the offspring of the nondiscriminators is more even across plant genotypes. It would be interesting to know what prevents recombination from breaking up this correlation between discrimination and performance.

Given that genetic structuring of the insect population can arise in response to plant genetic variation, what will be the effect on insect population dynamics? Genetic structuring of the insect population will increase population growth rates over those in the absence of structuring, although maximal population size may not be effected. Imagine an extreme case in which there are two plant genotypes, **A** and **B**, and two insect genotypes, **a** and **b**, and that insect **a** does well on plant **A**, but poorly on **B**, and that the reciprocal relationship applies to **b**. Assume also that plants are very long-lived, and that the insects do not exert an appreciable effect on plant population size. When hosts are occupied at random in each generation, the concordant **A–a** and **B–b** combinations will add many individuals to the population, but the discordant **A–b** and **B–a** combinations will not. The insect population size will depend on the frequencies of the alternative genotypes in both the plant and insect populations. If **A** is the more common plant genotype, and **b** the more common insect genotype, most insects will be on plants to which they are unsuited, and so population size will be low. As natural selection increases the frequency of **a,** the insect population will grow to a large size, but the population growth rate will be slow until the **a** genotype becomes common. On the other hand, if there is substructuring of the insect population because either (1) the insect chooses the plant to which its offspring is suited, or (2) following random colonization, subpopulations build up on each plant over several generations, then the insect population will grow at a faster rate, than by random recolonization of plant genotypes every generation. Either of these latter mechanisms will promote *genotype matching* between plant and herbivore, and thus, most insects will be on the plants on which they do best.

B. Plant Genetic Variation and Parasitoid Demography

Plant properties are known to have strong effects on the foraging success of many natural enemies of herbivores (Price *et al.*, 1980; Hare, 1991). What are the effects of plant genetic variation on population size of species at higher trophic levels? Some indications of these effects can be seen in two plant–gallmaker–parasitoid systems. Price (1988) examined the numerical response of parasitoids to the gallmaker *Euura lasiolepis* (Hymenoptera: Tenthredinidae) on arroyo willow, *Salix lasiolepis* (Salicaceae). The plant varies genetically in attractiveness to gall-inducing sawflies (Fritz and Price, 1988). If parasitoids had typical numerical responses, higher rates of parasitoid attack would be expected in willows with high gall densities. However, parasitism rates were highest on the plants with intermediate numbers of galls. The preferred plants grow larger galls, and thus, the sawfly is probably more protected from parasitism on these. Any positive density-dependent response of the parasitoid was countered by a negative correlation between gallmaker density and gallmaker susceptibility. Plant genetic variation can act to lower parasitoid population size in this case; host plants differ in refuge quality, and sawflies aggregate on the most protective plants, so that much of the parasitoid's resource base (the sawflies) is unavailable. But in other cases, plant genetic variance could have the opposite effect on parasitoid population size; if herbivores are attracted to the best plants and thus, grow into superior hosts, parasitoid reproductive success could be enhanced.

The herbivore's behavioral response to genetic variation in plant quality may either reduce or increase the density of resources available to the higher trophic level. But in addition to density effects, the frequency *per se* of high- and low-quality plants can influence the ability of parasitoids to discover suitable hosts. This may be the case with goldenrod, its stem galler, and the galler's parasitoid. In laboratory experiments, Weist *et al.* (1985) showed that the parasitoid is unable to distinguish small galls, which are vulnerable to attack, from large ones, which are not. Because the wasp spends time investigating galls too large to penetrate, it was suspected that large galls act as *false targets*, which distract the parasitoid and thereby reduce its rate of successful attack. In an experiment performed to test the false-target effect, small galls were offered to the parasitoid at a constant density of eight per cage; in these same cages the density of large, invulnerable galls was set at either zero or eight. The addition of the large galls so distracted the parasitoids that the number of successful attacks in these cages was only 60% as great as the number in the cages without false targets (Weis *et al.*, 1985). In natural populations composed predominantly of plant genotypes producing large galls, rates of successful parasitoid attack would be limited not only by the low density of vulnerable hosts, but also by the high frequency with which distracting, invulnerable hosts are encountered (Weis and Kapelinski, manuscript in preparation).

C. Plant Genetic Variation and Herbivore Community Structure

The species richness of the herbivore community associated with a plant will depend on a variety of factors, as many of the other chapters in this volume will attest. Individual variation among hosts, including variation due to genetic causes, can influence insect community structure by several means (Fritz, 1991).

The key question in assessing the effect of plant genotype on herbivore community structure is whether insect species are similar or different in their preference and performance on the various plant genotypes (Fritz, 1991; Fritz and Price, 1988; Karban, 1989b; Maddox and Root, 1987, 1990). To illustrate, suppose that two insect species responded identically to variation in the same host plant character, such that both species aggregated on the plants expressing a particular character state. On these preferred plants, the local density of insects could be high enough that density-dependent interspecific processes could be strong, even if herbivore density averaged over all plants were low. Conversely, if the two herbivore species had opposite responses to the plant character, such that plants of alternative character states supported a single species of herbivore, intraspecific processes would dominate; each herbivore species along with its preferred plant genotype would constitute a quasi-independent community. It is interesting to note that if the role of the plant in mediating herbivore abundance were unknown, totally opposite conclusions could be drawn from patterns of herbivore species co-occurrence. The high frequency of co-occurrence when herbivores prefer the same plants could be misinterpreted as facilitation, whereas low co-occurrence when preferences differ could be taken as evidence of localized competitive exclusion (Fritz, 1991; see also Holt, 1977).

In one of the most extensive empirical studies of plant genetic influences on herbivore community composition, Maddox and Root (1990) found all three of the possible patterns of herbivore co-occurrence, i.e., positive, negative, and no association. They studied abundances of 17 insect species, mostly specialists, on goldenrod (*S. altissima*) in an experimental garden. Plant genotype was varied by using individuals from 18 maternal half-sibships. [A maternal half-sibship consists of plants grown from seed produced by a single plant. When mating is random, the seeds from a single plant will be sired by many different pollen donors. If so, the seeds within the half-sibship are, on average, identical by descent from the seed parent at one quarter of all loci. Thus, half-sibships are partial replicates of the seed parent's genotype (Falconer, 1981)].

Insects were censused regularly through the season so that each plant half-sibship could be assigned an infestation score for each of the 17 herbivore species. Correlation analysis of these infestation scores thus revealed patterns of herbivore co-occurrence that could be explained by plant genetic variation. Of the 186 pairwise genetic correlations, 36 were significantly different from zero. Of the significant correlations, 27 were positive. One of the more surprising findings of the study was that correlated herbivores were not predictably of the same guild (Maddox and Root, 1990). For instance, the abundance of the stem galler *Eurosta solidaginis* was positively correlated with the abundance of the three other galling species, but the abundance of the leaf-chewing *Trirhabda* sp. was uncorrelated with that of the three other leaf chewers. Cluster analysis on the genetic correlations revealed that groups of species tended to co-occur in what Maddox and Root called herbivore suites. One suite, for instance, included one of the four leaf chewers, one of the five sucking insects, two of the three leaf miners, and the four gallmakers. A plant genotype infested by one of these species tended to be infested by the other seven. The abundance of insects of this suite on a plant genotype was uncorrelated with the abundance of any of the nine species in the other suites. The specific plant genes and characters that mediated these patterns of co-occurrence are unknown, but invite further investigation. One caution that should be raised with this study is that herbivore abundances on the half-sibships were not measured independently. It is possible that interactions among the herbivores on the plants could have influenced the observed patterns of co-occurrence and thus obscured the underlying influence of plant genotype (Maddox and Root, 1990).

Can plant genetic correlations in infestation levels influence competition intensity among herbivores? The gallmaking sawflies of arroyo willow, *S. lasieolepis*, may present an instance whereby preferences for the same host genotypes can bring herbivores into competition. Four species of sawflies induce galls on this willow; one forms its gall on the developing shoot and the others on the leaves or leaf petioles. A common garden experiment has established that infestation by the herbivores varies with plant genotype, and that infestation levels by the four species are for the most part positively genetically correlated (Fritz and Price, 1988). Asymmetrical competition is possible in this community since the shoot galler retards stem elongation, and thereby reduces the number of leaves available to leaf and petiole gallers. In field experiments (Fritz *et al.*, 1986), high infestations by the shoot galler caused a slight reduction in the infestations of some of the other species. The intensity of competition varies among willow clones (Fritz, 1990).

When herbivore species have opposite responses to variation in a plant character, the potential for competition can be reduced. Such may be the case with two insect seed predators of cocklebur, *Xanthium strumarium* (Asteraceae). One seed predator more easily attacks burrs (fruits) with short spines while the other attacks burrs with long spines. Direct competition may be very weak in such a case, since infestation levels of the seed predators differ among fields according to mean spine length—differences in host plant requirements prevent interaction. Ironically, if herbivory causes a major decrease in plant fitness, then over evolutionary time two species such as these may facilitate one another. By exerting opposing selection on capsule morphology, the plant is prevented from evolving complete resistance to either (Hare and Futuyma, 1978).

To summarize, genetic variation in plant characters that attract insects, or that influence insect survival and performance, may lend apparent structure to herbivore communities even when the species are noninteractive. When individualistic responses by the herbivore species associated with a plant are positively or negatively genetically correlated, the structure of the herbivore assemblage on an individual plant may be predicted by its genotype alone. However, interactions among the herbivores, such as competition, can be amplified by concentrating herbivore species on preferred plants' genotypes.

III. Variable Attack Rates and Natural Selection on Plant Defense

In the previous section we documented the effect of genetically variable plant characters on the size of herbivore populations and the structure of herbivore component communities in ecological time. As we stated at the outset, genotypic differences are not the only source of phenotypic variation among individual plants. Variation in the external environment may cause plant phenotypic variation as great as that caused by genetic variation, and perhaps environmental effects are greater on average. Why then should ecologists care if the effects of host plant variability on herbivores are rooted in genetics? The obvious answer is that genetically variable characters are those that can evolve. Even if the influence of plant genetics on herbivore populations and communities is weak at any one time, the cumulative selective effects of differential herbivory on genetically variable plant traits can change the characteristics of the host plant in ways that profoundly alter herbivore ecology. In keeping with the theme of this volume, it can be said that by changing plant gene frequencies, differential herbivory can change the *transformation rules* whereby energy and mineral nutrients become available to consumers.

Herbivores are potential agents of natural selection through their adverse effects on the growth, survival, and reproduction of plants. Some evidence that herbivory is a selective force is indicated in cases in which geographic differences in herbivore abundance are correlated with among-population differences in plant defense levels (Janzen, 1975; Dirzo, 1984; McNaughton and Tarrants, 1983), although other environmental differences among sites may contribute to these differences.

A. Consumption and Selection

Herbivory has the general effect of depressing components of plant fitness such as survival, growth, or reproductive success (reviewed by Marquis, 1991). Two types of characters could be favored by the herbivore selection pressure. First are characters that decrease the probability that insect herbivores will find, settle, and feed on the plant. Plant varieties or genotypes that are less infested or suffer less damage are refered to as resistant (Painter, 1951; McCrea and Abrahamson, 1987; Marquis, 1990; Maddox and Root, 1990; Rausher and Simms, 1989; Simms and Rausher, 1989; Simms, 1991). Plant production of secondary chemicals that repel herbivores (or failure to produce secondary chemicals that attract them) is a commonly proposed mechanism of resistance. Likewise, mechanical barriers such as trichomes, thorns, and leaf toughness can discourage feeding. These are resistance mechanisms since they reduce herbivore consumption of plant tissue. The second variety of defenses include those that operate after the damage has been done. Defenses acting at this point have been called tolerance, and consist of physiological adjustments that allow the plant to increase growth to compensate for tissues lost to herbivores (McNaughton, 1983; Belsky, 1986; Paige and Whitham, 1987).

Genetically based differences in plant resistance have been observed in a number of plant species to a variety of insect herbivores. The studies of McCrea and Abrahamson (1987) and Maddox and Root (1990) on goldenrod, and of Fritz and Price (1988) on willow described in a previous section, amply show that plant genotypes differ in infestation levels. Berenbaum *et al.* (1986) showed that concentrations and proportions of secondary chemicals that act as resistance mechanisms in wild parsnip (*Pastinaca sativa*: Umbelliferae) also vary with plant genotype. Does genetic variation in infestation or chemistry translate to genetic variation in damage?

Genotypic differences in damage levels were found for a tropical shrub *Piper arieianum* (Piperaceae) by Marquis (1990) during a 3.5-year study. This plant species was attacked by 95 insect species at the Costa Rica study site; herbivory levels could be assigned to 15 taxonomic groups based on their characteristic damage patterns. Since insect herbivory is known to depress fitness in this shrub (Marquis 1984), genotypic differences in damage would indicate that herbivores select on resistance traits. Marquis found that genotypes differed significantly in damage levels in some plots and at some dates, but plant genotype never accounted for more than 21% of the variance in damage. The individual herbivore taxa did not respond identically to plant genotype, so that no one genotype seemed resistant to all insects. Thus selection in this system is probably episodic, and it probably acts on a variety of resistance mechanisms simultaneously.

The *Piper* case shows that to understand the rate at which defensive traits can evolve, one must study the frequency and intensity of the selection episodes, an ecological and physiological problem. Fluctuations in herbivore numbers can cause the intensity of selection to wax and wane, and the relationship between selection intensity and herbivore density need not be monotonic: when herbivores are rare, infestations on all plants may be too low to depress fitness (no selection), and when herbivores are at extreme outbreak levels, all plants could be so damaged that none has any reproductive success (also no selection). If mean tolerance levels in the plant population are high enough to buffer fitness against intermediate herbivore damage levels, the selective advantage of resistance could be negligible.

B. Constraints on Selection for Increased Defense

The rate of defense evolution will also depend on the degree to which the expression of defensive characters is independent of other important plant functions, a developmental and genetic problem. Despite a continued selection pressure from a particular herbivore, evolution of increased defense can be constrained by counteracting selection forces. It should be expected, for instance, that in some cases a trait that defends against one herbivore permits or encourages attack by another. Secondary chemicals produced by members of the squash family (Curcubitaceae) are repellent to some leaf-feeding beetles, but act as feeding stimulants to others (Carrol and Hoffman, 1980). Similarly, the opposite effects of spine length on the two cocklebur seed predators may prevent this defense from evolving to completely exclude either one (Hare and Futuyma, 1978). The rate of evolution of traits with opposite effects on different herbivores will depend critically on the relative strength of selection by the two enemies.

Many authors have suggested that levels of plant defense are under an evolutionary constraint determined by the balance between the benefits and costs of defensive characters (McKey, 1974; Feeny, 1976; Mooney and Gulmon, 1982; Coley *et al.*, 1985; Berenbaum *et al.*, 1986; Simms and Rausher, 1987; reviewed by Simms, 1991). The physiological cost associated with the development of defensive characters can place a drain on plant growth and reproduction that exceeds the benefit of reduced herbivore damage. Theoretically, plant populations should evolve to a point at which the defensive ability of the average individual is at the maximal difference between the cost and benefit of the defensive trait (Simms and Rausher, 1987). Under a broad range of conditions, therefore, plant populations at evolutionary equilibrium should maintain an intermediate level of defense. If the defensive character state is determined by alternative alleles at a single locus, a high cost to defense could maintain genetic polymorphism. Simms (1991) reviewed studies on both cultivated and natural populations that have tested the cost–benefit balance hypothesis; she found that costs are not ubiquitous. Here we will recount work on two systems in which within-population genetic variation in defense is known, and tests for costs have been performed. Demonstration of a balance between cost and benefit in the within-population case would present the strongest support for the hypothesis.

Cyanogenesis, the release of cyanide from damaged tissue, is a reputed defense against herbivory found in plant species from many taxa. It has been particularly well studied in the herbaceous perennial, *Trifolium repens* (Fabaceae), in which its expression is controlled by two unlinked loci (Nass, 1972). One locus (Ac/ac) controls cyanogenic glycoside synthesis, and the other locus (Li/li) controls production of an enzyme that hydrolyzes the glycoside. When these two components are mixed in damaged tissue, cyanide is released. Cyanogenesis is polymorphic in many *Trifolium* populations, a pattern which has been attributed to a balance between the cost of cyanogenesis and the benefit of reduced herbivore damage. Costs due to autotoxicity may be incurred under stressful environmental conditions; both frost (Daday, 1965) and drought stress (Foulds and Grime, 1972) can lead to breakdown of vacuolar membranes that normally keep the glycoside and the enzyme compartmentalized. If auto-toxicity exacts a cost, one would expect cyanogenic plants to make up for the cost by the benefit gained from reduced herbivory. Dirzo and Harper (1982) compared the growth and reproduction of cloned replicates of both cyanogenic and acyanogenic *Trifolium* genotypes exposed to natural levels of herbivory. Rather than showing a benefit to cyanogenesis, their results indicated that acyanogenic plants grew faster. It seems then that cyanogenesis in this population is not only costly, but is also ineffective as a defense. These results leave open the question as to what maintains balance in the polymorphism (Simms, 1991).

Field studies by Simms and Rausher (1987,1989) used multivariate statistical analysis of selection (Lande and Arnold, 1983) to examine the cost of resistance to herbivory in the morning glory *Ipomoea purpurea* (Convolvulaceae). A variation of the half-sib method was used to replicate parental genotypes. They measured the proportion of tissue damaged by four different herbivores as well as lifetime seed set in this annual species, and then performed multiple regression of relative seed set on proportion damaged. When relative fitness (measured through some fitness component like seed set) is regressed over some quantitative measure of phenotype (like proportion damaged), the coefficients of the regression equation can be interpreted as selection gradients, that is, a measure of the intensity of natural selection (Lande and Arnold, 1983). Directional selection intensity is estimated by the coefficients of the linear multiple regression of fitness on phenotype. A positive linear regression coefficient means that selection favors increase in the trait, while a negative coefficient means selection is for reduction. Stabilizing and disruptive selection can be estimated by the quadratic regression coefficients in a polynomial regression. Negative quadratic coefficients can indicate stabilizing selection (intermediate phenotypes have highest fitness), while positive quadratic coefficients can indicate disruptive selection (intermediate phenotypes have lowest fitness). Simms and Rausher argued (1987) that if there is a cost to resistance, stabilizing selection on resistance should be observed. Their reasoning is that the most-defended plants will have low damage levels, but the costs of the defense will keep fitness low. On the other hand, undefended plants will pay no cost, but herbivore damage will keep fitness low. If this is the case, then the polynomial regression of fitness on damage level should peak at an intermediate damage level. The selection analysis was performed two ways, first on the phenotypic measures of damage and seed set from 2800 individual plants, and on the genetic measures (breeding values) of the 30 half-sibships to which they belonged. When phenotypic damage scores by the four herbivores were considered (statistically holding damage by all other herbivores constant), only corn earworm damage had a significant stabilizing selection gradient on seed set (Rausher and Simms, 1989). No significant stabilizing selection gradients were found when genetic scores of seed set were regressed over the genetic scores for damage. However, significant directional selection gradients indicated that selection favors increased resistance

(Rausher and Simms, 1989; Simms and Rausher, 1989). In a recent reanalysis of this experiment, Simms (1990) showed there may be a cost of resistance to certain combinations of the four types of herbivores that attack morning glory.

If so few cases indicate a cost to defense, does that mean that defense can evolve free of cost constraints? Simms (1991) argues that this is not necessarily the case. In the initial evolutionary response to selection by herbivores, the benefit gained through a defense mechanism when insects attack may come at a cost to fitness when insects do not attack. However, continued selection may ameliorate the cost of defense. An analogous situation was seen in the evolution of pesticide resistance in the sheep blow fly, which was attributed to a single allele (McKinzie *et al.,* 1982). Strains newly selected for resistance reverted to susceptibility if switched from a pesticide-rich to a pesticide-free medium, thus indicating a cost to the resistance gene. However, after continued selection, modifier alleles at other loci increased in frequency, which negated the cost (McKinzie *et al.,* 1982). Similar mechanisms could be at work in plant defenses against herbivory.

IV. Plant Genetic Variation and Animal Pollinators

A. Plant Variation and Its Effect on Populations of Pollinators

A fundamentally different kind of resource offered by plants is that of a reward (e.g., nectar, pollen, oils and resins, aromatic substances) that attracts animal pollinators. How do pollinators respond to genetic variation in plants? Are these responses similar to those we have described for herbivores?

Behavioral specialization of pollinators in response to plant variation is well known and commonly takes the form of floral preference on the basis of reward rates. For example, bumblebees and honeybees choose higher-rewarding flowers species over lower-rewarding ones (Heinrich, 1979a; Seeley, 1985), and these choices can be manipulated by experimentally enriching the sugar content of low-ranking blossoms (Heinrich, 1976). Even when the pollinator population as a whole exhibits no preference, individual pollinators may show a second form of specialization in which they remain constant to a single flower species for at least part of a foraging bout, thereby reducing handling time (Waser 1986). These choices among flower species represent responses to genetic variation at the coarsest level. Many pollinators also make choices among plants *within* a population. Bumblebees and hummingbirds visit high-nectar-producing plants over low-rewarding ones in the same population (Mitchell and Waser, 1991; Real and Rathcke, 1991; Chapter 5) and also respond to other floral traits, such as color patterns, that influence the time required to extract a reward (Waser and Price, 1983 and 1985). Honeybees make similar choices in experiments with artificial flowers (Waddington and Holden, 1979). For most natural populations we do not know the extent to which these differences in nectar production rate or standing crop are heritable, but in at least some crop plants, nectar volume is partly heritable and responds to artificial selection under greenhouse conditions (Teuber and Barnes 1979; Teuber *et al.*, 1983).

In choosing flower species to visit, pollinators may respond to the level of variation in rewards as well as the mean value. Bumblebees and wasps exhibit risk-averse behavior, choosing floral types with lower variance in nectar rewards (Real, 1981; Waddington *et al.*, 1981). So, intraspecific genetic variation could discourage pollinators from visiting a particular plant species. Pollinators can also choose among individual plants on the basis of the degree of variation across flowers in reward rates. However, in the one case in which the relative variances of nectar volumes for individuals were examined in detail, they fluctuated greatly over time (Real and Rathcke, 1988), making consistent choices on the basis of heritable variation unlikely (Chapter 5).

Although behavioral responses to plant variation are well studied, much less is known about the extent to which these foraging choices influence the population dynamics of pollinators. For social bees, most investigators have assumed a direct relationship between net rates of energy return and colony reproductive output because of the central importance of sugar and pollen to colony economics (Heinrich, 1979b; Pyke, 1978; Seeley, 1985). Indeed, the total weight of undisturbed honeybee hives falls during periods of inclement weather when foraging is reduced (Seeley, 1985). However, reproductive output of natural colonies not managed for honey production and especially population dynamics have been largely ignored. For pollinators other than social bees, even less is known about population regulation. Migrant rufous hummingbirds behave according to foraging models for energy maximizers, in which reproductive success is limited by net energy intake (Hixon and Carpenter, 1988), and suffer little predation (Miller and Gass, 1985), but their individual reproductive success cannot be followed over a lifetime.

Variation among plant species can contribute to species diversity of competing pollinators by allowing resource partitioning on the basis of flower type (Feinsinger, 1978; Feinsinger and Colwell, 1978; Inouye, 1978; Pyke, 1982; Ranta, 1982; Bowers, 1985; but see Chapter 11). Stable partitioning on the basis of genetic variants *within* a plant population appears, however, to be rare. This represents a major difference with the rich communities of sedentary insect herbivores that may develop across different genotypes of a single plant species. It is commonplace for a plant species to receive visits from many different pollinator species, and those pollinators can differ in preference, but few pollinator populations are dependent on a single plant species (Feinsinger, 1983; Schemske, 1983). Moreover, in plant populations that are serviced by two types of pollinators with sharply contrasting preferences, floral form is likely to be under disruptive selection (Galen *et al.*, 1987). *Complete* partitioning in this case would lead to reproductive isolation and speciation, because of the special role of pollinators as gamete vectors. The highly specific relationships between figs and fig wasps (Agaonidae), which both pollinate the flowers and develop in the seeds, illustrate the potential for this process. Introduction of a small group of figs with a highly unusual flowering phenology is likely to lead to rapid speciation of both fig and wasp (Ramirez, 1970; Kiester *et al.*, 1984).

In some cases two or more pollinator species partition a single plant species on the basis of differences that are not at the level of individual plants and are not likely to be genetic. The stingless bee *Trigona fuscipennis* forages in large groups and restricts its visit to large, dense clumps of the shrub *Cassia*, while its congener *T. fulviventris* forages individually and visits isolated plants (Johnson and Hubbell, 1975). In a second example, hummingbirds visit the outermost flowers of jewelweed and lower their nectar below the level accessible to bumblebees, while bumblebees visit mostly the inner flowers (Laverty and Plowright, 1985). In these cases, the partitioning should not generate reproductive isolation and plant speciation.

B. Variation in Pollination and the Evolution of Floral Form

Genetic variation among individual plants is subject to strong selection imposed by animal pollinators. Indeed, broad associations between floral features and types of animal pollinators (Grant and Grant, 1965; Proctor and Yeo, 1972) support the idea that floral diversity is largely the result of pollinator-mediated selection.

To demonstrate pollinator-mediated selection requires relating fitness to floral character expression. This requirement is complicated by production of hermaphroditic flowers in about 70% of all angiosperm species (Yampolsky and Yampolsky, 1922). In such plants, selection can arise from an effect on male fitness (number of seeds sired) or female fitness (number of seeds produced). Both male and female reproduction can be divided into several lifecycle stages: (1) flower production, during which the two sexes remain together, (2) pollination, including export and receipt of pollen from compatible plants and (3) post-pollination events, including fertilization and predispersal seed predation (Fig. 1). Pollinator-mediated selection requires first that a plant trait influence pollen export or receipt. To produce net selection, that variation in pollination success must then result in variation in the total number of seeds sired or produced. For female function, the latter step requires that the total seed production of a plant, or quality of those seeds, be pollen-limited. Although pollen-limitation has been observed in many populations (e.g., Bierzychudek, 1981; Campbell, 1985; Galen, 1985; Hainsworth *et al.*, 1985), sexual selection theory suggests that seed production will more often be limited by nutritive resources, making selection of floral traits primarily through male function (Charnov,

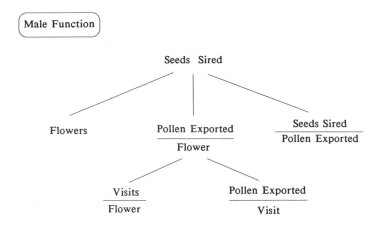

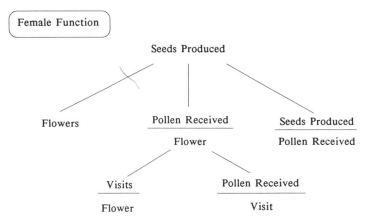

Figure 1 Multiplicative components of reproductive success in a hermaphroditic angiosperm. The first level of each hierarchy divides total male or female reproductive success into components corresponding to flower production, pollination, and postpollination stages of the life cycle. The second level divides pollination success into visit rate and effectiveness of pollen transfer per visit.

1979; Willson, 1979; Sutherland and Delph, 1984; Bell, 1985; Stanton *et al.*, 1986).

Pollination success itself depends on the rate of visits by pollinators and on the quantity of pollen exported or received during a visit (Fig. 1). In addition, the quality of pollen, including the distance it has traveled between mates, can strongly influence plant fitness. Waser (1983) reviews the myriad ways that color, nectar rewards, and flower morphology influence pollination success.

In relatively few cases do we have quantitative measurements of effects of floral traits on plant fitness components. Data on male fitness are especially scarce, despite the postulated importance of male function to floral evolution. Estimates of fitness components are badly needed since they allow comparisons of the relative importance of different forms of selection, such as male versus female function and pollinators versus other selective agents.

Fitness effects of single-locus traits in natural plant populations have been measured for flower color. In wild radish, *Raphanus raphanistrum* (Brassicaceae), petal color is controlled by a single locus (Kay, 1978). Yellow-flowered individuals are visited preferentially by most insects and father 72% of seeds (Stanton *et al.*, 1986). Fruit production does not differ between the two morphs, so selection appears to be entirely through male function in this case, and averaging over male and female function, relative fitnesses of yellow and white morphs are 1.00 and 0.69. In larkspur, *Delphinium nelsonii* (Ranunculaceae), hummingbirds and bumblebees visit blue flowers preferentially over rare albino mutants because they take longer to orient at and extract nectar in the absence of the contrasting nectar guides on blue flowers (Waser and Price, 1981, 1983, 1985). In natural populations, blue-flowered plants set more seed, so, in contrast to wild radish, flower color is selected through female function. Male fitness was not estimated, but assuming a similar difference in male function, albinos have a relative fitness of 0.62.

Selection on continuous traits can be estimated by finding selection differentials or selection gradients using the fitness regression approach outlined in previous sections. A selection differential gives the within-generation shift in the mean value of a trait induced by both direct selection on that trait and indirect selection of correlated characters. As shown by Arnold and Wade (1984), selection differentials can be estimated separately for different stages of the life cycle. With multiplicative fitness components as shown in Figure 1, selection differentials for the different stages sum to the total shift in the mean trait value. This methodology has been used to estimate selection of floral traits in two different species, *Ipomopsis aggregata* and *Polemonium viscosum*, in the Polemoniaceae. This plant family is of special interest because it provides a classic example of associations between floral morphology and pollinator type (Grant and Grant, 1965).

Red-flowered *Ipomopsis aggregata* are pollinated predominantly by hummingbirds. Individual plants with wide corolla tubes are more successful at exporting pollen per flower (Campbell, 1989) judging from the use of dyes to estimate pollen movement. Corolla width has no detectable effect on import of pollen. Instead, position of the stigma and proportion of time spent by the protandrous flowers in the female phase (proportion pistillate) influence pollen receipt, while having little or no effect on pollen export

(Campbell, 1989). Because seed production in these populations is limited by low levels of pollen receipt (Hainsworth *et al.,* 1985; Campbell, 1991), these effects on pollen receipt result in some net selection through female function as measured by total number of viable seeds produced (Campbell, 1991). Preliminary results from half-sibship analysis suggest that stigma position and proportion pistillate have high heritabilities under field conditions. Thus, in this species, selection through both male and female functions has the potential to shape floral traits. Although intensities of these two kinds of selection (measured by the absolute value of the standardized selection differential) are similar when averaged over traits, different traits affect male versus female reproduction.

In *Polemonium viscosum* corolla width influences both male and female reproductive functions. In high-elevation populations, total receipt of outcross pollen and seed production are higher for wider flowers (Galen and Newport, 1988; Galen, 1989), as is pollen removal (Galen and Stanton, 1989). In this system corolla width strongly influences the rate of visitation by bumblebees (Galen, 1989). In *I. aggregata,* corolla width also correlates with visit rate of the hummingbird pollinators, apparently because wider flowers secrete nectar at a higher rate, but the relationship is very weak (Campbell *et al.,* 1991). The primary mechanism of selection in this case, instead, appears to involve differences in pollen export on a per-visit basis. Indeed the standardized selection differential for pollen exported per visit in experiments with captive hummingbirds is 0.42, compared to a selection differential of only 0.23 for visits per flower (Campbell *et al.,* 1991). The contrast between the two plant species suggests that traits that function in attraction and influence primarily visit rate are likely to affect both male and female pollination success, while traits that affect per-visit efficiencies may influence pollen export and import in different ways.

Evolutionary responses of floral traits to pollinator-mediated selection are subject to several kinds of constraints: (1) the possibility of opposing effects on male and female fitness, (2) genetic and phenotypic correlations between traits (Lande and Arnold, 1983), and (3) costs incurred by negative effects on fitness during other stages of the life cycle. The latter two constraints can be generated by the combined influences of herbivores and pollinators of a single plant species. For example, in morning glory (*Ipomoea purpurea*), the genetic loci that determine flower color have pleiotropic effects on stem color, which in turn might influence herbivore attack (Schoen *et al.* 1984). In addition, a floral trait that is attractive to pollinators might itself draw seed predators. Oviposition rates by dipteran seed predators (Hylemya: Anthomyiidae) depend on some features of floral display in *Ipomopsis aggregata* (A. Brody, personal communication, 1991). The female flies lay single eggs on the inside of the calyx of a flower or floral bud, and the developing larva typically consumes all of the seeds in the fruit. Because some floral traits affect the proportion of fruits destroyed as well as rates of seed set per pollen grain received, the net patterns of selection through female function differ from those expected from selection during pollination alone (Campbell, 1991). For example, selection during pollination for proportion of time in the pistillate phase is strongly directional, but based on total number of viable seeds produced by a plant, selection is stabilizing (Campbell, 1991).

Different types of pollinators to the same species may favor different floral morphologies. In *Polemonium viscosum*, plants growing on alpine tundra have sweet-smelling flowers visited primarily by large-bodied queen bumblebees, *Bombus kirbyellus*, while plants growing at lower elevations where ants are attracted to sweet flowers are skunky-scented and attract smaller-bodied flies (Diptera, Muscidae and Anthomyidae) (Galen *et al.,* 1987). These two kinds of pollinators prefer large flowers and narrow, relatively short flowers, respectively. In populations visited by bumblebees, selection clearly favors wider, longer flowers (Galen, 1989). Differences in selection on morphology thus reinforce the differences between the two scent morphs.

Such floral variation may allow the support of two different types of insect populations. In this case, much of the variation is between populations so that in at least some populations, selection on morphological traits is directional (Galen 1989). Such differences in selection provide a scenario for allopatric speciation of plants in response to different pollinator types, each of which is associated with plants on a different adaptive peak. Different pollinators of variants *within* a single plant population can also result in disruptive selection, as Galen *et al.* (1987) suggest for mixed populations of *P. viscosum.* This disruptive selection could then lead to sympatric speciation (Rice, 1987), although the conditions required are controversial (Felsenstein, 1982; Diehl and Bush, 1989). Regardless of the mechanism of speciation, increased diversity of flower species increases the diversity of resources available to pollinators and may thereby support a greater number of pollinator species.

The potential role of pollinators in plant speciation is illustrated by the *Ipomopsis aggregata* (Polemoniaceae) species complex, which is thought to provide examples of incipient speciation (Grant, 1981). The classic picture of speciation in this family relies on adaptation to different pollinators to generate reproductive isolation (Grant and Grant, 1965). Crosses between all taxa in the complex produce F_1 hybrids, suggesting little postzygotic isolation, although crosses involving *I. arizonica* show reduced fertility (D. Wilken, personal communication, 1991). In some areas red-flowered, broad-tubed *I. aggregata* undergoes extensive hybridization with white-flowered, narrow-tubed *I. tenuituba*, while in other locales, the two species grow in close proximity without hybridizing (Grant and Wilken, 1988; Wolf and Soltis, 1991). It is unclear whether areas of hybridization represent primary intergradation or zones of secondary intergradation with breakdown of ethological isolating mechanisms following habitat disturbance. Under the first hypothesis, if divergence is mediated by adaptation to hummingbird versus hawkmoth pollinators, ongoing selection on floral traits such as corolla width should be disruptive in hybrid populations. This situation also presents an excellent opportunity to examine forces that maintain and limit hybrid zones (Barton and Hewitt, 1985; Hewitt, 1989) by asking whether areas with and without hybridization differ in gene flow mediated by animal pollinators, or in selection.

Hybrid zones have the potential to support a greater diversity of plant consumers. For phytophagous insects, there is strong support for this hypothesis. Cottonwoods in areas of hybridization suffer higher rates of herbivore attack (Whitham, 1989). Whitham suggests that these areas act as sinks for herbivore populations. How plant hybridization influences pollinator species diversity is less clear. In some areas hawkmoths prefer to visit white flowers while hummingbirds prefer red flowers of *I. aggregata* (Paige and Whitham, 1985), suggesting that a hybrid zone could maintain both pollinators. But hybridization implies some pollen transfer between species by a single pollinator. Indeed Elam and Linhart (1988) detected no ethological isolation between color morphs in mixed populations, suggesting the possibility of competition between hummingbirds and hawkmoths in plant hybrid zones. In addition, the floral visitors include migrating rufous hummingbirds that travel thousands of kilometers over the course of the summer. Their population dynamics are probably little affected by plant composition at a single 1- to 2-week stopping point during migration. Plant hybrid zones may be likely to affect dynamics of consumer populations only when those consumers are relatively sedentary.

V. Points of Contact in the Study of Selection on Plants by Herbivores and Pollinators

As we mentioned in our introduction, there are some fundamental differences between herbivores and pollinators; the former are antagonists that frequently reside on the plant, while the latter are mutualists that make brief contact with the plant during plant reproductive bouts. Even though their effects on plant reproductive success tend in opposite directions, there is structural similarity in the selective regimes that herbivores and pollinators impose on the plant.

For both pollinators and herbivores, the insect's effect on plant reproductive success is initiated with discovery and settlement, which make traits influencing the likelihood of encounter key determinants of plant fitness. However, the precise selective consequences once an encounter is made depend on a chain of physiological and developmental events; for pollination these include the process of pollen transfer, fertilization, seed maturation, and dispersal, while for herbivores, these include physiological responses to wounding (including induction of defenses and compensatory growth) and impairment of subsequent growth and reproduction. It is quite possible that the fitness consequences of individual variation in the plant's frequency of encounter with insects, as determined by its attractiveness, will depend on individual variation in the operation of these chains of postencounter events. For instance, a plant producing a floral display that attracts twice as many pollinators may gain nothing in seed production if resources for filling seeds are limiting (although reproductive success through male function may be enhanced), and by the same token, increasing allocation to seeds may yield no increase in fitness if pollinators are not attracted. Similar fitness relationships between plant traits expressed before and after herbivore feeding are discussed below. Because pre- and postencounter plant traits can draw on a common resource pool, an improvement in one set of characters may be possible only at a cost to the other.

The structural similarity of the pollinator- and herbivore-induced selective regimes extends to the critical role that insect population dynamics plays in each. Taking an extreme case, a plant that is highly attractive to herbivores may have equal fitness to a plant that is highly repellent if there are no herbivores in the habitat. Additionally, pollinators and herbivores are active foragers, and so the degree of preference for one plant phenotype over the other need not be fixed. Conflicting stimuli or search image formation (Rausher, 1978) can change the degree or direction of preference as the ratios of the different phenotypes change.

Having outlined these points of contact in the evolutionary ecology of plant defense and plant pollination, we feel that there are several important issues that empiricists undertaking studies of natural selection in these interactions should consider:

1. the interaction between pre- and postencounter characters on plant fitness;
2. the potential wax and wane of selective pressures on plants caused by fluctuations in insect population density; and
3. changes in the intensity and direction of selection caused by frequency-dependent attack on genotypes of differential resistance or attractiveness.

Here we present a brief sketch of an analytical framework that can serve to tie these three issues together. This framework relies on the developmental genetic concept of the reaction norm (Schmalhausen, 1949; Weis and Gorman, 1990; Garvilets and Scheiner, *in press*; de Jong, 1990). To understand the reaction norm concept, consider the view that genotypes do not code for a single phenotype, but rather code for a range of phenotypes. The particular phenotype expressed by an individual of a given genotype will depend on the environment. A trait such as plant height may be under genetic control, but the height achieved by any one plant is influenced not only by genes, but also by the availability of water, nutrients, and light. Of the many genotypes in a population, some consistently yield tall plants, and others consistently yield small plants. Some are able to increase size dramatically when given extra resources, and others are unable to capitalize on such opportunities. Not only can there be genetic variation in height, but genetic variation in the reaction of height to resource availability. One would expect selection to favor genotypes with reaction norms that give the fittest phenotypes within the expected range of environments (de Jong, 1990; Weis and Gorman, 1990; Weis, 1991). Thompson (1988b) introduced the concept of the interaction norm, whereby the fitness consequences of a species interaction can shift with environmental variation; in this spirit we look at fitness consequences to a plant as the density of an interacting animal varies.

A. Defense Reaction Norms

In the case of plant defense, which we will discuss first, resistance and tolerance can be evaluated as reaction norms, and we suggest that selection will favor the reaction norms that give the most defended phenotypes across the expected range of herbivore population densities. As we discuss defense reaction norms here, we will assume that herbivore population size is the only environmental factor that varies (cf. Maddox and Cappuccino, 1986).

The resistance level expressed by a plant genotype depends on how many herbivores are available to attack. Here we define resistance as the inverse of susceptibility, which is the number of herbivore individuals per unit plant or the unit damage per unit plant. For clarity of discussion we will deal with susceptibility directly. The level of susceptibility actually expressed is of necessity tied to herbivore population size—all plants express zero damage in habitats without herbivores, but can be expected to show increased damage with increased herbivore population size (Fig. 2A). The algebraic function relating expected damage to herbivore density is the susceptibility reaction norm. Reaction norm differences among genotypes can be tied to underlying resistance mechanisms such as secondary chemistry. In the simplest possible conditions, the fittest susceptibility reaction norm is the one with the lowest slope, i.e., low damage regardless of herbivore population size. However, the relationship between susceptibility and fitness may not be so simple; low susceptibility may have a high fitness cost, or, high susceptibility may be inconsequential if the plant can increase growth rate to compensate for tissue loss. The fitness consequences of a genotype's susceptibility reaction norm depends on its tolerance reaction norm.

Figure 2   Susceptibility, tolerance, and fitness reaction norms for three hypothetical plant genotypes. This example illustrates the consequences of positive correlation between susceptibility and tolerance (a negative correlation between resistance and tolerance). (A) The damage inflicted by herbivory increases with herbivore abundance; genotype **x** has low susceptibility (high resistance), and so little damage is inflicted even when herbivores are abundant in the

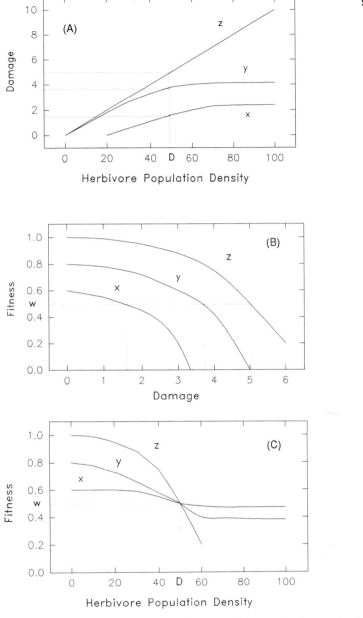

habitat, but genotype **z** is highly susceptible (low resistance) and is severely damaged at high herbivore density. (B) The fitness consequences of damage for the three hypothetical genotypes; although genotype **x** is unlikely to be damaged, when it is, it suffers a large loss in fitness, but genotype **z,** which is susceptible to attack, is able to withstand more damage. (C) When the two reaction norms are combines, the relationship of fitness to herbivore population density is revealed; genotype **z** has the greatest relative fitness at low herbivore densities because of its superior tolerance, but at high herbivore densities, **x** is superior because of its superior resistance.

The tolerance reaction norm is the algebraic function that relates expected fitness to herbivore damage (Fig. 2B). On average, one would expect that fitness of a genotype would decline with increasing herbivore damage (but see Belksy, 1986; Paige and Whitham, 1987). The expected fitness at zero damage (the *y* intercept of the function) indicates a basal fitness, while the slope and curvature of the function indicate the sensitivity of fitness to damage. Genotypes with a high intercept and shallow slope in the tolerance reaction norm would have high fitness regardless of herbivore intensity. Tolerance reaction norms may be difficult to measure. It would require that each plant genotype examined be replicated a number of times, and that the number of herbivore individuals (and hence herbivore damage) be experimentally manipulated.

Susceptibility determines the developmental environment (Weis, 1991) in which tolerance is expressed, and tolerance determines the selectional environment for susceptibility. To illustrate how the two defense components interact, consider the following thought experiment in a hypothetical plant population attacked by a herbivore of constant population density. Suppose that all genotypes experience 10% defoliation, but that a new allele, which reduces defoliation to 5%, enters the population. Will that allele spread? It depends on the level of tolerance; if all genotypes can tolerate 12% defoliation without loss of fitness, then the new allele is selectively neutral. Now suppose that a new tolerance allele, which allows plants to withstand 15% defoliation, enters the population. If plants never lose more than 10% of their foliage, this second new allele is also neutral. Thus the selective value of change in one defense component depends on the level of the other. However, herbivore population density is not constant in real habitats, as assumed in this thought experiment, so the selection intensity on the two novel alleles may change as herbivore populations grow and decline.

To analyze the dynamics of selection on susceptibility and tolerance, the relationship between plant fitness and herbivore density must be established. The susceptibility and tolerance reaction norms can be combined to construct a third reaction norm. By substituting the function that describes susceptibility (damage versus herbivore density) into the damage term of the tolerance function (fitness versus damage), one gets the expected fitness of the genotype as a function of herbivore density. The importance of the *fitness–herbivore density* function (when constructed from independent empirical measures of susceptibility and tolerance reaction norms on randomly drawn, replicated genotypes) is that it can be used to calculate the relative contributions of resistance and tolerance to expected fitness at any given herbivore population density (Fig. 2C). By so doing, it is possible to determine how selection on each may wax and wane with fluctuations in herbivore population size.

As suggested, the relative fitness of genotypes can change with herbivore density. Figure 2C shows the combined susceptibility and tolerance reaction norms for three genotypes. In this hypothetical case, there is a cost to resistance; i.e., the least susceptible genotype does poorest when herbivores are absent, while the most susceptible does best under this condition. However, there is an intermediate herbivore population size at which all three genotypes have equal fitness. When population levels are below this point, the genotype with higher tolerance will have greater fitness, but when density rises above this point, the genotype with the greater resistance will have the advantage. With temporal fluctuation in herbivore density, this pattern of reaction norms can lead to reversals in the direction of selection if the generation time of the plant is short compared to the periodicity of the fluctuations in herbivore abundance. When herbivore density varies spatially, the net direction of selection will not change; the expected fitness of each genotype averaged over the actual range of herbivore densities can be calculated by integrating the fitness–herbivore density reaction norm over the frequency distribution of herbivore abundances, using a method suggested by Weis and Gorman (1990).

Up to this point, we have considered the situation in which a genotype's susceptibility to attack was independent of the susceptibilities of all other genotypes, or in other words, selection was not frequency dependent. Dolinger *et al.* (1973) have suggested that frequency-dependent selection by herbivores may maintain within-species diversity in plant secondary chemistry. It is possible that an undefended plant individual will suffer less damage when it grows in the midst of resistant plants than when grown only with other equally susceptible plants. This could occur, for instance, if the defended plants produce a repellent that stimulates ovipositing females to leave the habitat. This type of *defense by association* has been suggested as a mechanism that can explain changes in herbivore abundance with different mixtures of plant species (Atsatt and O'Dowd, 1976). But, by the same token, a single susceptible individual in the midst of many resistant plants may flag its presence to ovipositing females and thus suffer greater damage than if surrounded by other susceptible plants. Kinsman (1982), used two clones of evening primrose (*Oenothera biennis*; Onagraceae), and found that colonization of the *susceptible* and *resistant* clones by herbivores was the same whether grown alone or in mixture. Although her evidence was negative, this type of question needs to be examined in several systems and at several spatial scales before we can make any general statements about frequency-dependent herbivore attack and its consequences for selection on defense.

The reaction norm approach will be of use in this regard, since frequency-dependent attack will be reflected in the shape and slope of the susceptibility reaction norm. Consider Figure 3. This graph depicts the hypothetical relationship of damage to herbivore population density for a poorly defended genotype at three different genotypic frequencies. When the susceptible plant is rare (and the defended plant, abundant), the potential selection against susceptibility can decrease for two reasons: first, if most

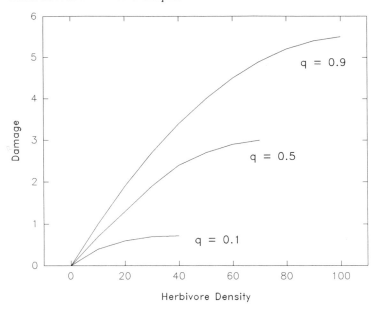

Figure 3 Susceptibility reaction norms for a genotype as genotypic frequencies change. When the susceptible genotype is abundant, herbivore populations can be large, so much damage can be inflicted, but when the susceptible genotype is rare, damage inflicted can be less because (1) high density of resistant plants limits herbivore density, and (2) frequent encounters with the resistant plants can discourage foraging herbivores, causing them to migrate from the habitat.

plants are well defended, the population size of the herbivore will tend to be small, and thus fewer herbivores are available to colonize the susceptible plant; second, if the other plants deter settlement or oviposition, herbivores will leave the habitat before they discover the susceptible genotype, as suggested previously. This form of frequency dependence is asymmetrical in that the fitness of the susceptible morph changes as a function of its frequency, but the fitness of the resistant morph may be constant (e.g., if it is never eaten). We hope questions on frequency dependence will be incorporated into future studies on natural selection of defense.

B. Reaction Norms in Pollination Systems

The link between plant fitness and pollinator population density can also be drawn by considering the components of successful plant reproduction as a chain of reaction norms. Viewed in terms of plant female function, pollinator abundance is a factor in the developmental environment that influences visitation, which then is a factor in the developmental environment for pollen receipt, and so on until ovules are successfully fertilized and seeds matured. By multiplying through, the relationship between fitness and pollinator emerges. For simplicity of discussion we will collapse this chain of reaction norms into two links: the relationship of stigmatic pollen load to pollinator abundance and the relationship of seed production to pollen load.

Plant genotypes may differ in their ability to attract available pollinators and thus can vary in mean pollen load per stigma. Figure 4A illustrates two

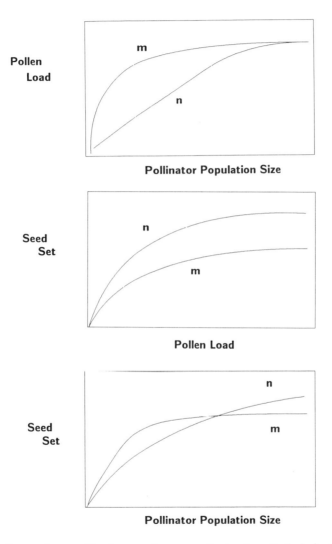

Figure 4 Pre- and postpollination reaction norms for two hypothetical plant genotypes, reflecting female reproductive success. (A) Genotype **m** is the more attractive, and the stigma achieves a saturating load even when pollinators are scarce, while genotype **n** is less attractive, and will be saturated only when pollinators are abundant. (B) Genotype **m** is unable to produce as many seeds as **n** because of its greater allocation to attraction. (C) At low pollinator densities, the greater investment in attraction by genotype **m** pays off, but at high pollinator densities, genotype **n** is superior because it gets as much pollen and allocates more to seed production.

attractiveness reaction norms representing two different relationships between pollinator population density and pollen load. The very attractive plant may have its stigmatic surface saturated even if pollinators are scarce, while less-attractive plants reach saturation only when pollinators are unusually abundant. Assuming for the moment that the plants are identical during the postpollination phase, it becomes apparent that the fitness differences between these two genotypes can vary with pollinator number. They have equal fitness when pollinator densities are extreme, but a large fitness difference when pollinators are at intermediate densities. Selection intensity on the traits determining attractiveness and efficiency of pollen transfer would vary accordingly with pollinator abundance (Campbell, 1989; Schemske and Horvitz, 1988).

Not all pollen placed on the stigma fertilizes an ovule, and not all fertilized ovules develop into mature seeds. Thus a variety of plant characters will influence the relationship of pollen load to seed set. Genotypic variation in these processes accounts for variation in the reaction norm relating fitness (through female function) to pollen load (Fig. 4B). Generally, fitness increases at a decelerating rate with pollen load (Silander and Primack, 1978; Snow, 1982, 1986; Campbell, 1986; Kohn and Waser, 1985; Waser and Price, 1991), although a fitness decline at high loads is possible (Young, 1988). The relationship of fitness to pollinator density is established by substituting the attractiveness reaction norm function (pollen load versus pollinator abundance) into the pollen load term of the postpollination reaction norm (fitness versus pollen load).

Just as the cost of resistance can be reflected in the tolerance reaction norm, so too the cost of attractiveness can be reflected in the postpollination reaction norm. If attractive floral displays and rewards come at the cost of energy and nutrients available for seed maturation (Lovett-Doust and Harper, 1980; Ashman and Baker, 1991; Pyke 1991) or at the cost of increased seed predation (A. Brody, personal communication, 1991), then the upper limit on female success in an attractive plant may be less than that in an unattractive one. If this is the case, the relative fitnesses of attractive and unattractive plants may reverse at some intermediate pollinator population density (Fig. 4C). For this reason, the average pollinator population size, as well as the variation in size, can determine whether or not the trade-off between attraction and postpollination contributions to fitness becomes an important constraint in floral evolution. If pollinator populations are consistently low, the cost of attraction will be paid, but not if it is consistently high.

The selective value of an attraction trait may be frequency dependent. Pollination biologists have made more progress in evaluating the strength of frequency dependence than have those investigating herbivory. Floral choice and visitation rate can depend on relative frequencies of plant species as well as per flower rewards (Thomson, 1978; Heinrich, 1979a). Pollination of many nectarless species, especially orchids, has been attributed to Batesian mimicry, a process that involves frequency-dependent visit rate (Wiens, 1978; Brown and Kodric-Brown, 1979). In addition, it has been recognized for some time that pollen receipt should increase with frequency of a plant species relative to competing species when there is interspecific pollen movement (Levin and Anderson, 1970). However, less is known about pollinator responses to different frequencies of floral morphs within a plant species. Frequency dependence, in this case, will typically favor the morph in the majority. Waser and Price (1981) tested whether bumblebees switched visitation rates to white mutants of the montane perennial *Delphinium nelsonii* when present in different ratios to the more common blue form. Epperson and Clegg (1987) performed a similar experiment with white and colored morphs of *Ipomoea purpurea*. Only in the second case were white morphs undervisited when rare. However, it is unclear if the size of the floral arrays were of the appropriate scale to induce changes in pollinator behavior. Large-scale field experiments may be the surest way to determine whether frequency-dependent selection operates on attraction.

C. Closing Thoughts

The common experience of ecologists is that the environment a species occupies can change dramatically from year to year and place to place. Among evolutionary biologists, especially theoretical population geneticists, the study of selection in varying environments has a long and venerable history. We present this application of the reaction norm concept as one way to examine the relationship between population ecology and the evolutionary biology of species interactions. At the same time, the reaction norm approach underlines the structural similarities between the evolution of pollinator–plant and herbivore–plant interactions. The necessary field experiments to grow plants in natural habitats under varying densities of hebivores or pollinators are daunting. Nonetheless, these kinds of experiments are needed to move forward to a full understanding of the role of insects as selective agents in plant evolution.

Acknowledgments

We wish to thank Mark Hunter, Ellen Simms, and Nick Waser for their helpful comments on the manuscript. Authors were supported by National Science Foundation Grants BSR-8614895 (A.E.W.) and BSR-8996306 (D.R.C).

References

Abrahamson, W. G., Satler, J. F., McCrea, K. D., and Weis, A. E. (1989). Variation in selection pressures on the goldenrod gall fly and the competitive interactions of its natural enemies. *Oecologia* **79**, 15–22.

Anderson, S. S., McCrea, K. D., Abrahamson, W. G., and Hartzel, L. M. (1989). Host genotype choice by the ball gallmaker *Eurosta solidaginis* (Diptera: Tephritidae). *Ecology* **70**, 1048–1054.

Arnold, S. J., and Wade, M. J. (1984). On the measurement of natural and sexual selection: Theory. *Evolution* **38**, 709–719.

Ashman, T., and Baker, I. (1991). Variation in floral sex allocation with time of season and currency, *submitted.*

Atsatt, P. R., and O'Dowd, D. J. (1976). Plant defense guilds. *Science* **193**, 24–29.

Barton, N. H., and Hewitt, G. M. (1985). Analysis of hybrid zones. *Annu. Rev. Ecol. System.* **16**, 113–148.

Bell, G. (1985). On the function of flowers. *Proc. R. Soc. Lond. (Biol.)* **223**, 224–265.

Belsky, A. J. (1986). Does herbivory benefit plants? A review of the evidence. *Am. Nat.* **127**, 870–892.

Berenbaum, M. R., and Fenny, P. (1981). Toxicity of angular furancoumarins to swallowtail butterflies: Escalation in a coevolutionary arms race? *Science* **212**, 927–929.

Berenbaum, M. R., Zangrel, A. R., and Nitao, J. K. (1986). Constraints on chemical and coevolution: Wild parsnips and the parsnip webworm. *Evolution* **40**, 1215–1228.

Bierzychudek, P. (1981). Pollinator limitation of plant reproductive effort. *Am. Nat.* **117**, 838–840.

Bowers, M. A. (1985). Bumblebee colonization, extinction, and reproduction in subalpine meadows in northeastern Utah. *Ecology* **66**, 914–927.

Brown, J. H., and Kodric-Brown, A. (1979). Convergence, competition, and mimicry in a temperate community of hummingbird-pollinated flowers. *Ecology* **60**, 1022–1035.

Campbell, D. R. (1985). Pollinator sharing and seed set of *Stellaria pubera:* Competition for pollination. *Ecology* **66**, 544–553.

Campbell, D. R. (1986). Predicting palnt reproductive success from models of competition for pollination. *Oikos* **47**, 257–266.

Campbell, D. R. (1989). Measurements of selection in a hermaphroditic plant: Variation in male and female pollination success. *Evolution* **43**, 318–334.

Campbell, D. R. (1991). Effects of floral traits on sequential components of fitness in *Ipomopsis aggregata. Am. Nat.*, in press.

Campbell, D. R., Waser, N. M., Price, M. V., Lynch, E. A., and Mitchell, R. J. (1991). Components of phenotypic selection: Pollen export and flower corolla width in *Ipomopsis aggregata. Evolution*, in press.

Carroll, C. R., and Hoffman, C. A. (1980). Chemical feeding deterrent mobilized in response to insect herbivory and counteradaptation by *Epilachna tredecimnota. Science* **209**, 414–416.

Charnov, E. L. (1979). Simultaneous hermaphroditism and sexual selection. *Proc. Nat. Acad. Sci. U.S.A.* **76**, 2480–2484.

Coley, P. D., Bryant, J. P., and Chapin, F. S., III. (1985). Resource availability and plant antiherbivore defense. *Science* **230**, 895–899.

Crawley, M. J. (1983). "Herbivory. The Dynamics of Animal–Plant Interactions." Blackwell Scientific, Oxford, England.

Daday, H. (1965). Gene frequencies in wild populations of *Trifolium repens* L. IV. Mechanisms of natural selection. *Heredity* **20**, 355–365.

de Jong, G. (1990). Quantitative genetics of reaction norms. *J. Evol. Biol.* **3**, 447–468.

Diehl, S. R., and Bush, G. L. (1989). The role of habitat preference in adaptation and speciation. *In* "Speciation and Its Consequences" (D. Otte and J. A. Endler eds.), pp. 345–365. Sinauer, Sunderland, Massachusetts.

Dirzo, R. (1984). Herbivory: A phytocentric overview. *In* "Perspectives in Plant Population Biology" (R. Drizo and J. Sarukham, eds.), pp. 141–165. Sinauer, Sunderland, Massachusetts.

Dirzo, R., and Harper, J. L. (1982). Experimental studies on slug–plant interactions. IV. The performance of cyanogenic and acyanogenic morphs of *Trifolium repens* in the field. *J. Ecol.* **70**, 119–138.

Dolinger, P., Ehrlich, P. R., Fitch, W. L., and Breedlove, D. E. (1973). Alkaloid and predation patterns on Colorado lupine populations. *Oecologia* **13,** 191–204.

Elam, D. R., and Linhart, Y. B. (1988). Pollination and seed production in *Ipomopsis aggregata:* Differences among and within flower color morphs. *Am. J. Botany* **75,** 1262–1274.

Ehrlich, P. R., and Birch, L. C. (1967). The balance of nature and population control. *Am. Nat.* **101,** 97–101.

Epperson, B. K., and Clegg, M. T. (1987). Frequency-dependent variation for outcrossing rate among flower-color morphs of *Ipomoea purpurea. Evolution* **41,** 1302–1311.

Falconer, D. S. (1981). "Introduction to Quantitative Genetics." 2nd Ed. Longman House, Burnt Mill, U.K.

Feeny, P. P. (1976). Plant aparency and chemical defense. *Rec. Adv. Phytochem.* **10,** 1–40.

Feinsinger, P. (1978). Ecological interactions between plants and hummingbirds in a successional tropical community. *Ecol. Monogr.* **43,** 269–287.

Feinsinger, P. (1983). Coevolution and pollination. *In* "Coevolution" (D. J. Futuyma, and M. Slatkin eds.), pp. 282–310. Sinauer, Sunderland, Mass.

Feinsinger, P., and Colwell, R. K. (1978). Community organization among neotropical nectar-feeding birds. *Am. Zool.* **18,** 779–795.

Felsenstein, J. (1982). Skepticism towards Santa Rosalia, or why are there so few kinds of animals? *Evolution* **35,** 124–138.

Foulds, W., and Grime, J. P. (1972). The response of cyanogenic and acyanogenic phenotypes of *Trifolium repens* to soil moisture supply. *Heredity* **28,** 181–187.

Fritz, R. S. (1990). Variable competition coefficients between insect herbivores on genetically variable host plants. *Ecology* **71,** 1208–1211.

Fritz, R. S. (1991). Community structure and species interactions of phytophagous insects on resistant and susceptible host plants. *In* "Ecology and Evolution of Plant Resistance." (R. S. Fritz and E. L. Simms eds.) University of Chicago Press, Chicago, *in press.*

Fritz, R. S., and Price, P. W. (1988). Genetic variation among palnts and insect community structure: Willows and sawflies. *Ecology* **69,** 845–856.

Fritz, R. S., Sacchi, C. F., and Price, P. W. (1986). Competition versus host plant phenotype in species composition: Willow sawflies. *Ecology* **67,** 1608–1618.

Fritz, R. S., and Simms, E. L. (eds.). (1991). "Ecology and Evolution of Plant Resistance." University of Chicago Press, Chicago.

Galen, C. (1985). Regulation of seed-set in *Polemonium viscosum:* Floral scents, pollination, and resources. *Ecology* **66,** 792–797.

Galen, C. (1989). Measuring pollinator-mediated selection on morphometric floral traits: Bumblebees and the alpine sky pilot, *Polemonium viscosum. Evolution* **43,** 882–890.

Galen, C., and Newport, M. E. A. (1988). Pollination quality, seed set, and flower traits in *Polemonium viscosum:* Complementary effects of variation in flower scent and size. *Am. J. Bot.* **75,** 900–905.

Galen, C., and Stanton, M. L. (1989). Bumble bee pollination and floral morphology: Factors influencing pollen dispersal in the alpine sky pilot, *Polemonium viscosum* (Polemoniaceae). *Am. J. Bot.* **76,** 419–426.

Galen, C., Zimmer, K. A., and Newport, M. E. A. (1987). Pollination in floral scent morphs of *Polemonium viscosum:* A mechanism for disruptive selection on flower size. *Evolution* **41,** 599–606.

Gavrilets, S., and Scheiner, S. M. The genetics of phenotypic plasticity. IV. Evolution of reaction norm shape. *J. Evol. Biol., in press.*

Grant, V. (1981). "Plant Speciation," 2nd Ed. Columbia University Press, New York.

Grant, V., and Wilken, D. H. (1988). Natural hybridization between *Ipomopsis aggregata* and *I. tenuituba* (Polemoniaceae). *Botanical Gazette* **149,** 213–221.

Grant, V., and Grant, K. A. (1965). "Flower Pollination in the Phlox Family." Columbia University Press, New York.

Hainsworth, F. R., Wolf, L. L., and Mercier, T. (1985). Pollen limitation in a monocarpic species, *Ipomopsis aggregata. J. Ecol.* **73,** 263–270.

Hairston, N. G., Smith, F. E., and Slobodkin, L. B. (1960). Community structure, population control, and competition. *Am. Nat.* **94,** 421–425.

Hare, D. J. (1991). Effects of plant variation on herbivore–natural enemies interactions. *In* "Ecology and Evolution of Plant Resistance." (R. S. Fritz and E. L. Simms eds.) University of Chicago Press, Chicago, *in press.*

Hare, J. D., and Futuyma, D. J. (1978). Different effects on variation in *Xanthium strumarium* L. (Compositae) on two insect seed predators. *Oecologia* **37,** 109–112.

Heinrich, B. (1976). The foraging specializations of individual bumblebees. *Ecol. Monogr.* **46,** 105–128.

Heinrich, B. (1979a). Majoring and minoring by foraging bumblebees, *Bombus vagans:* An experimental analysis. *Ecology* **60,** 245–255.

Heinrich, B. (1979b). "Bumble-Bee Economics." Harvard University Press, Cambridge, Massachusetts.

Hewitt, G. M. (1989). The subdivision of species by hybrid zones. *In* "Speciation and Its Consequences." (D. Otte and J. A. Endler eds.), pp. 85–110. Sinauer, Sunderland, Massachusetts.

Hixon, M. A., and Carpenter, F. L. (1988). Distinguishing energy maximizers from time minimizers: A comparative study of two hummingbird species. *Am. Zool.* **28,** 913–925.

Holt, R. (1977). Predation, apparent competition, and the structure of prey communities. *Theor. Popul. Biol.* **12,** 197–229.

Inouye, D. W. (1978). Resource partitioning in bumblebees: Experimental studies of foraging behavior. *Ecology* **59,** 672–678.

Janzen, D. H. (1975). Behavior of *Hymenaea courbaril* when its predispersal seed predator is absent. *Science* **189,** 145–147.

Johnson, L. K., and Hubbell, S. P. (1975). Contrasting foraging strategies and coexistence of two bee species on a single resource. *Ecology* **56,** 1398–1406.

Karban, R. (1989a). Fine-scale adaptation of herbivorous thrips to individual host plants. *Nature* **340,** 60–61.

Karban, R. (1989b). Community organization of *Erigeron glaucus* foliovores: Effects of competition, predation, and host plant. *Ecology* **70,** 1028–1039.

Karban, R. (1991). Plant variation: Its effects on populations of herbivorous insects. *In* "Ecology and Evolution of Plant Resistance." (R. S. Fritz and E. L. Simms (eds.), University of Chicago Press, Chicago, *in press.*

Kay, Q. O. N. (1978). The role of preference and assortative pollination in the maintenance of flower colour polymorphism. *In* "The Pollination of Flowers by Insects." (A. J. Richards ed.), pp. 175–190. Academic Press, New York.

Kiester, A. R., Lande, R., and Schemske, D. W. (1984). Models of coevolution and speciation in plants and their pollinators. *Am. Nat.* **124,** 220–243.

Kinsman, S. (1982). Herbivore responses to *Oenothera biennis* (Onagraceae): Effects of host plant's size, genotype, and resistance of conspecific neighbors. Unpublished Ph.D. dissertation, Cornell University, Ithaca, New York.

Kohn, J. R. and Waser, N. M. (1985). The effect of *Delphinium nelsonii* pollen on seed set in *Ipomopsis aggregata,* a competitor for hummingbird pollination. *Am. J. Bot.* **72,** 1144–1148.

Lande, R., and Arnold, S. J. (1983). The measurement of selection on correlated characters. *Evolution* **37,** 1210–1226.

Laverty, T. M. and Plowright, R. C. (1985). Competition between hummingbirds and bumble bees for nectar in flowers of *Impatiens biflora. Oecologia* **66,** 25–32.

Lawton, J. H., and McNeil, S. (1979). Between the devil and the deep blue sea: On the problems of being a herbivore. *In* "Population Dynamics." (R. M. Anderson, B. D. Turner, and L. R. Taylor (eds.), pp. 223–244. Blackwell Scientific, Oxford, England.

Levin, D. A. and Anderson, W. W. (1970). Competition for pollinators between simultaneously flowering species. *Am. Nat.* **104,** 455–467.

Lovett-Doust, J. and Harper, J. L. (1980). The resource cost of gender and maternal support in an andromonoecious umbellifer *Smyrnium olusatrum. New Phytol.* **85,** 251–264.

Maddox, G. D., and Cappuccino, N. (1986). Genetic determination of plant susceptibility to an herbivorous insect depends on environmental context. *Evolution* **40,** 863–866.

Maddox, G. D., and Root, R. B. (1987). Resistance to 16 diverse species of herbivorous insects within a population of goldenrod, *Solidago altissima:* Genetic variation and heritability. *Oecologia* **72,** 8–14.

Maddox, G. D., and Root, R. B. (1990). Structure of the encounter between goldenrod (*Solidago altissima*) and its diverse insect fauna. *Ecology* **71,** 2115–2125.

Marquis, R. J. (1984). Leaf herbivores decrease fitness of a tropical plant. *Science* **226,** 537–539.

Marquis, R. J. (1990). Genotypic variation in leaf damage in *Piper arieianum* (Piperaceae) by a multispecies assemblage of herbivores. *Evolution* **44,** 104–120.

Marquis, R. J. (1991). Selective impact of herbivores. *In* "Ecology and Evolution of Plant Resistance." (R. S. Fritz and E. L. Simms (eds.). University of Chicago Press, Chicago, *in press.*

McCrea, K. D., and Abrahamson, W. G. (1987). Variation in herbivore infestation: Historical vs. genetic factors. *Ecology* **68,** 822–827.

McKey, D. (1974). Adaptive patterns in alkaloid physiology. *Am. Nat.* **108,** 305–320.

McKinzie, J. A., Whitten, M. J., and Adena, M. A. (1982). The effect of genetic background on fitness of diazinon resistance genotypes of the Australian sheep blowfly, *Lucilia cuprina. Heredity* **49,** 1–9.

McNaughton, S. J. (1983). Compensatory growth: A response to herbivory. *Oikos* **40,** 329–336.

McNaughton, S. J., and Tarrants, J. L. (1983). Grass leaf silification: Natural selection for an inducible defense against herbivores. *Proc. Nat. Acad. Sci. U.S.A.* **80,** 790–791.

Miller, R. S., and Gass, C. L. (1985). Survivorship in hummingbirds: Is predation important? *Auk* **102,** 175–178.

Mitchell, R. J., and Waser, N. M. (1991). Adaptive significance of *Ipomopsis aggregata* nectar production: Pollen transfer to and from single flowers, *submitted.*

Mitter, C., Futuyma, D. J., Schneider, J. C., and Hare, J. D. (1979). Genetic variation and host plant relationships in a parthenogenic moth. *Evolution* **33,** 777–790.

Mooney, H. A., and Gulmon, S. L. (1982). Constraints on leaf structure and function in reference to herbivory. *BioSci.* **32,** 198–206.

Moran, N. (1981). Intraspecific variability in herbivore performance and host quality: A field study of *Uroleucon caligatum* (Homoptera: Aphididae) and its *Solidago* hosts (Asteraceae). *Ecol. Entomol.* **6,** 301–306.

Nass, H. G. (1972). Cyanogenesis in *Sorgum bicolor, Sorgum sundanense, Lotus,* and *Trifolium repens*—A review. *Crop Sci.* **12,** 502–506.

Ng, D. (1988). A novel level of interactions in plant–insect systems. *Nature* **344,** 611–613.

Paige, K. N., and Whitham, T. G. (1985). Individual and population shifts in flower color by scarlet gilia: A mechanism for pollinator tracking. *Science* **227,** 315–317.

Paige, K. N., and Whitham, T. G. (1987). Overcompensation in response to mammalian herbivory: The advantage of being eaten. *Am. Nat.* **129,** 407–416.

Painter, R. H. (1951). "Insect Resistance of Crop Plants." Macmillan, New York.

Price, P. W. (1988). Inversely density dependent parasitism: The role of plant refuges for hosts. *J. Anim. Ecol.* **57,** 89–96.

Price, P. W., Bouton, C. E., Gross, P., Mcpheron, B. A., Thompson, J. N., and Weis, A. E. (1980). Interactions among three trophic levels: Influences of plants on interactions between herbivores and natural enemies. *Annu. Rev. Ecol. Syst.* **11,** 41–65.

Proctor, M., and Yeo, P. (1972). "The Pollination of Flowers." Taplinger, New York.

Pyke, G. H. (1978). Optimal foraging. Movement patterns of bumblebees between inflorescences. *Theor. Popul. Biol.* **13,** 72–98.

Pyke, G. H. (1982). Local geographic distributions of bumblebees near Crested Butte, Colorado: Competition and community structure. *Ecology* **63**, 555–573.

Pyke, G. H. (1991). What does it cost a plant to produce floral nectar? *Nature* **350**, 58–59.

Ramirez, W. (1970). Host specificity of fig wasps (Agaonidae). *Evolution* **24**, 680–691.

Ranta, E. (1982). Species structure of North European bumblebee communities. *Oikos* **38**, 202–209.

Rausher, M. D. (1978). Search image for leaf shape in a butterfly. *Science* **200**, 1071–1073.

Rausher, M. D., and Simms, E. L. (1989). Evolution of resistance to herbivory in *Ipomoea purpurea*. I. Attempts to detect selection. *Evolution* **43**, 563–572.

Real, L. A. (1981). Uncertainty and pollinator–plant interactions: The foraging behavior of bumblebees and wasps on artificial flowers. *Ecology* **62**, 20–26.

Real, L. A., and Rathcke, B. J. (1988). Patterns of individual variability in floral resources. *Ecology* **69**, 728–735.

Real, L. A., and Rathcke, B. J. (1991). Individual variation in nectar production and its effect on fitness in *Kalmia latifolia*. *Ecology* **72**, 149–155.

Rice, W. R. (1987). Speciation via habitat specialization: The evolution of reproductive isolation as a correlated character. *Evol. Ecol.* **1**, 301–314.

Schemske, D. (1983). Limits to specialization and coevolution in plant–animal mutualism. In "Coevolution." (M. H. Nitecki, ed.), pp. 67–110. University of Chicago Press, Chicago.

Schemske, D. W., and Horvitz, C. C. (1988). Plant–animal interactions and fruit production in a neotropical herb: A path analysis. *Ecology* **69**, 1128–1137.

Schmalhausen, I. I. (1949). "Factors of Evolution: The Theory of Stabilizing Selection." University of Chicago Press, Chicago.

Schoen, D. J., Giannasi, D. E., Ennos, R. A., and Clegg, M. T. (1984). Stem color and pleitropy on genes determining flower color in the common morning glory. *J. Hered.* **75**, 113–116.

Scriber, J. M., and Slansky, F., Jr. (1981). Nutritional ecology of immature insects. *Annu. Rev. Entomol.* **26**, 183–211.

Seeley, T. D. (1885). "Honeybee Ecology: A Study of Adaptation in Social Life." Princeton University Press, Princeton, New Jersey.

Service, P. (1984). Genotypic interactions in an aphid–host plant relationship: *Uroleucon rudbeckiae* and *Rudbeckia laciniata*. *Oecologia* **61**, 271–276.

Silander, J. A., and Primack, R. B. *1978*. Pollination intensity and seed set in the evening primrose (*Oenothera fruitcosa*). *Am. Mid. Nat.* **100**, 213–216.

Simms, E. L. (1990). Examining selection on the multivariate phenotype: Plant resistance to herbivores. *Evolution* **44**, 1177–1188.

Simms, E. L. (1991). Costs of plant resistance. *In* "Ecology and Evolution of Plant Resistance." (R. S. Fritz and E. L. Simms eds.), University of Chicago Press, Chicago, *in press*.

Simms, E. L., and Rausher, M. D. (1989). Evolution of resistance to herbivory in *Ipomoea purpurea*. II. Natural selection by insects and costs of resistance. *Evolution* **43**, 573–585.

Simms, E. L., and Rausher, M. D. (1987). Costs and benefits of plant resistance to herbivory. *Am. Nat.* **130**, 570–581.

Snow, A. A. (1982). Pollination intensity and potential seed set in *Passiflora vitifolia*. *Oecologia* **55**, 231–237.

Snow, A. A. (1986). Pollination dynamics in *Epilobium canum* (Onagraceae): Consequences for gametophytic selection. *Am. J. Bot.* **73**, 139–151.

Stanton, M. L., Snow, A. A., and Handel, S. N. (1986). Floral evolution: Attractiveness to pollinators increases male fitness. *Science* **232**, 1625–1627.

Sutherland, S., and Delph, L. F. (1984). On the importance of male fitness in plants: Patterns of fruit-set. *Ecology* **65**, 1093–1104.

Teuber, L. R. and Barns, D. K. (1979). Environmental and genetic influences on alfalfa nectar. *Crop Sci.* **19**, 874–878.

Teuber, L. R., Barnes, D. K., and Rincker, C. M. (1983). Effectiveness of selection for nectar volume, receptacle diameter, and seed yield characteristics in alfalfa. *Crop Sci.* **23**, 283–289.

Thompson, J. N. (1988a). Evolutionary ecology of the relationship between oviposition preference and performance of offspring in phytophagous insects. *Entomologia Experimentalis et Applicalis* **48**, 3–14.

Thompson, J. N. (1988b). Variation in interspecific interactions. *Annu. Rev. Ecol. Syst.* **19**, 65–87.

Thomson, J. D. (1978). Effects of stand composition on insect visitation in two-species mixtures of *Hieracium. Am. Mid. Nat.* **100**, 431–440.

Uhler, L. D. (1951). Biology and ecology of the goldenrod gall fly *Eurosta solidaginins* (Fitch). *Cornell University Experiment Station Memoir* **300**, 1–51.

Waddington, K. D., and Holden, L. R. (1979). Optimal foraging: On flower selection by bees. *Am. Nat.* **114**, 179–196.

Waddington, K. D., Allen, T., and Heinrich, B. (1981). Floral preferences of bumblebees (*Bombus edwardsii*) in relation to intermittent versus continuous rewards. *Anim. Behav.* **29**, 779–784.

Waser, N. M. (1983). The adaptive nature of floral traits: Ideas and evidence. *In* "Pollination Biology." (L. Real (ed.), pp. 241–285. Academic Press, Orlando, Florida.

Waser, N. M. (1986). Flower constancy: Definition, cause and measurement. *Am. Nat.* **127**, 593–603.

Waser, N. M., and Price, M. V. (1981). Pollinator choice and stabilizing selection for flower color in *Delphinium nelsonii. Evolution* **35**, 376–390.

Waser, N. M., and Price, M. V. (1983). Pollinator behavior and natural selection for flower colour in *Delphinium nelsonii. Nature* **302**, 422–424.

Waser, N. M., and Price, M. V. (1985). The effect of nectar guides on pollinator preference: Experimental studies with a montane herb. *Oecologia* **67**, 121–126.

Waser, N. M. and Price, M. V. (1991). Outcrossing distance effects in *Delphinium nelsonii:* Pollen loads, pollen tubes, and seed set. *Ecology* **72**, 171–179.

Weis, A. E. (1991). Plant variation and the evolution of phenotypic plasticity in herbivore performance. *In* "Ecology and Evolution of Plant Resistance." (R. F. Fritz and E. L. Simms, eds.) University of Chicago Press, Chicago, *in press.*

Weis, A. E., and Abrahamson, W. G. (1986). Evolution of host–plant manipulation by gall makers: Ecological and genetic factors in the Eurosta-Solidago system. *Am. Nat.* **127**, 681–695.

Weis, A. E., Abrahamson, W. G., and McCrea, K. D. (1985). Host gall size and oviposition success by the parasitoid *Eurytoma gigantea. Ecol. Entomol.* **10**, 341–348.

Weis, A. E., and Kapelinski, A. D. (1991). Density, size, frequency, and enemy attack rates as factors in selection intensity on *Eurosta's* gall size: Hermon Bumpus meets C. S. Holling. Manuscript in preparation.

Weis, A. E., and W. L. Gorman. (1990). Measuring selection on reaction norms: An exploration of the *Eurosta–Solidago* system. *Evolution* **44**, 820–831.

Weis, A. E., Wolfe, C. L., and Gorman, W. L. (1989). Genotpyic variation and integration in histological features of the goldenrod ball gall. *Am. J. Bot.* **76**, 1541–1550.

Whitham, T. G. (1983). Host manipulation of parasites: Within-plant variation as a defense against rapidly evolving pests. *In* "Variable Plants and Herbivores in Natural and Managed Systems." (R. F. Denno and M. S. McClure eds.), pp. 14–41. Academic Press, New York.

Whitham, T. G. (1989). Plant hybrid zones as sinks for pests. *Science* **244**, 1490–1493.

Wiens, D. (1978). Mimicry in plants. *Evol. Biol.* **11**, 365–403.

Willson, M. F. (1979). Sexual selection in plants. *Am. Nat.* **113**, 777–790.

Wolf, P. G., and Soltis, P. S. (1991). Estimates of gene flow among populations, geographic races, and species of perennial plants in the *Ipomopsis aggregata complex,* submitted.

Yampolsky, E., and Yampolsky, H. (1922). Distribution of sex forms in phanerogamic flora. *Bib. Genet.* **3**, 1–62.

Young, H. J. (1988). Differential importance of beetle species pollinating *Dieffenbachia longispatha* (Araceae). *Ecology* **69**, 832–844.

5

Nectar Distributions, Pollinator Behavior, and Plant Reproductive Success

Beverly J. Rathcke
Department of Biology
University of Michigan
Ann Arbor, Michigan

I. Introduction

Nectar distributions of flowers have often been used as model systems for studying the foraging behavior of pollinators. However, the links between nectar production, pollinator behavior, and plant reproductive success have seldom been completed. Understanding these links will be essential to understanding the evolution of nectar distributions and plant–pollinator interactions. Although some nectar traits are matched with pollinator types (Baker and Baker, 1983), virtually nothing is known about whether pollinator behavior acts to fine-tune nectar distributions within plants. Nectar distributions can be defined by the variance as well as the average nectar production rate per flower (NPR). Variation in resources is usually considered to have negative (or at least neutral) effects on animal–plant interactions in most other chapters in this volume. However, within-plant variation in leaf quality has been postulated to have defensive consequences and to benefit plant fitness in plant–herbivore antagonisms (Whitham and Slobodchikof, 1981). Here I hypothesize that within-plant variation in NPR may benefit plant fitness in plant–pollinator mutualisms.

Nectar production is a complex trait in several ways. Nectar can influence pollinator behavior and hence can determine competitive ability for pollinator service, mating possibilities, and gene transmission. Nectar production could also be determined by the resource-allocation strategies of plants. Nectar production can be governed by physiological processes (e.g., photosynthesis, growth, etc.) as well as by morphology (e.g., nectary size). In general, the adaptive value of physiological traits has been much less studied than has that of morphological traits (Endler, 1986), and this is certainly true for floral biology. Several studies have demonstrated that variation in floral morphology and color can determine plant fitness through pollinator behavior (e.g., Clegg and Epperson, 1988; Campbell, 1989; Stanton *et al.*, 1986). Although nectar distributions may often underlie these links (Waser, 1983), the effects of nectar variation on components of plant fitness are virtually unexplored.

In this chapter I examine evidence that nectar distributions affect plant reproductive success, and I explore possible constraints to selection. I first present graphical models of optimal nectar distributions (mean and variance) and then address the following questions: What are the patterns of variation both between and within plants in local populations, and do they reflect heritable or environmental variation? How do pollinators respond to different nectar distributions and, in turn, effect pollen transport and plant reproductive success? What other factors may select for nectar distributions? Finally, I discuss the relatively few studies in which the links between nectar production, pollinator foraging, and plant reproductive success have been measured.

II. Models of Optimal Nectar Allocation

Nectar rewards can influence many aspects of pollinator behavior including visit frequency to plants, the number of flowers probed per plant, probe time in each flower, and movement after leaving a plant. In turn, these pollinator behaviors can effect pollen deposition (female function) and pollen removal and transfer (male function), and pollen transport can influence plant reproductive success (RS), i.e., total seeds produced and total seeds sired. The strength of these links will depend on many specific characteristics of both pollinators and plants (see Zimmerman, 1988 and the following), and I review this evidence in the following sections. Here I present two graphical models of optimal nectar distributions for mean and variance in NPR that may be generally applicable for outcrossing plants with many flowers. Optimal NPR is defined to be the mean (or variance) that confers maximal reproductive success upon an individual plant (see Pyke, 1981).

For outcrossing plants with many flowers, mean NPR is predicted to be under stabilizing selection for an intermediate optimum (Fig. 1A modified from Zimmerman, 1988). The following relationships are hypothesized: Initially, female and male RS increase with higher NPR for two reasons: more pollinators visit and return to plants (but see Pyke, 1981; Zimmerman, 1988) and pollinators visit more flowers within plants. As a result, more outcross pollen is transferred to and from plants, and both female and male RS benefit. However, these benefits reach a maximum and begin to decline as pollinators visit more flowers within plants and transfer increasing amounts of self-pollen. Female RS declines because self-pollen reduces fertilization by outcross pollen or reduces progeny viability or vigor. Male RS declines because self-pollen is a loss for male RS. Female RS declines less steeply than male RS because small amounts of outcross pollen can be sufficient for maximal seed set (Bell, 1985; Stanton *et al.*, 1986).

In the second model, I predict that the decline in plant RS (above the optimal NPR) can be reduced by increasing the variance in NPR within plants (Fig. 1B). As NPR variation increases (for the maximal average NPR shown in Fig. 1A and the same flower number), more flowers have lower NPR than average. Pollinators are more likely to encounter low-rewarding flowers and to leave. A higher proportion of outcross–self-pollen is transferred within plants, and both male and female plant RS increase. Plant RS will eventually asymptote, because there are limits to how much pollen can be transferred and then decline as pollinators avoid visiting plants with very high variance.

Total nectar allocation and total plant reproductive success may be determined by resource allocation (Pyke, 1981) and reallocation (Zimmerman and Pyke, 1988a), and these are discussed later. In these two models, I empha-

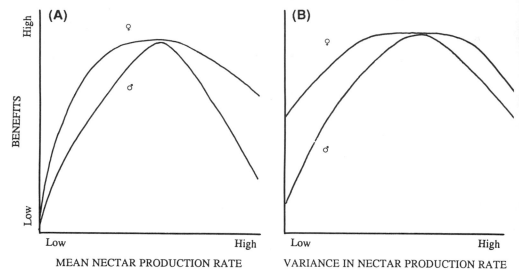

Figure 1 Predicted female and male reproductive success based on the relationships between nectar production rate per flower, pollinator behavior, and pollen transport. (A) As mean NPR increases, both female and male reproductive success initially increase and then decrease. An intermediate optimal NPR is predicted. (B) As the variance in NPR increases (within the maximum nectar allocation shown in (A) among the same number of flowers) the negative effects of high NPR can be decreased or eliminated. At some high level, pollinators may avoid plants with high variance, so plant reproductive success declines.

size the adaptive value of how the total nectar is distributed among the flowers of a plant.

III. Variation in Nectar Rewards

It is meaningless to address the issues of how nectar distributions influence pollinator foraging or how selection might act upon nectar distribution strategies without documenting natural variation in nectar rewards between and within plants.

A. Measurements of Nectar Variation

As a first step, the most appropriate measurements of nectar must be established. For pollinators and plants, the most appropriate single parameter is probably sugar production per flower. Empirical evidence shows that most variation in sugar production is determined by variation in nectar volume; coefficients of variation (CVs) for nectar volume commonly range between 40 and > 100%, whereas CVs for sugar concentration are usually

5– < 25% (Bertsch, 1983; Pleasants, 1983; Southwick, 1983; Corbet and Delfosse, 1984; Marden, 1984b; Devlin and Stephenson, 1985; Zimmerman and Pyke, 1986; Real and Rathcke, 1988). Therefore, I emphasize variation in nectar volume per flower per time in this review, but with the caution that sugar concentration, sugar composition, and other constituents can also influence pollinator behavior and resource allocation within plants.

For pollinators, nectar available as standing crop is determined by depletion as well as by NPR. However, standing crop can give a misleading picture of resource availability if foraging is nonrandom (Possingham, 1990). If pollinators concentrate their foraging in high-rewarding patches, standing crop will underestimate nectar availability in high-rewarding patches and overestimate it in low-rewarding patches. In fact, pollinators commonly show area-restricted foraging and stay longer in high-nectar plants (or areas) than in low-nectar plants (Heinrich, 1979c; Waser and Mitchell, 1990). As expected with nonrandom foraging, distributions of NPR and standing crop often become uncorrelated as pollinators forage (Zimmerman, 1988). Therefore, NPR is probably a better estimator of nectar availability than is standing crop, and I emphasize variation in NPR in this review. These results also suggest that pollinators do respond to nectar rewards and may exert strong selection on NPR (but see Zimmerman, 1988).

B. Nectar Variation between and within Plants

Variation in NPR appears to be the rule both between and within plants. Between-plant variation in average NPR has been found to be significant in virtually every species that has been studied (Pleasants, 1983; Marden, 1984b; Wyatt and Shannon, 1986; Real and Rathcke, 1988; Thomson *et al.*, 1989; Hodges, 1990), the exceptions being some hummingbird-pollinated species with unusually high within-plant variation (Feinsinger, 1978; see following).

Within-plant variation in NPR at any one time is also common and is often very high. CVs (standard deviation × 100/mean) range between 20 and 40% (Pleasants, 1983; Hodges, 1990) and > 100% (Marden, 1984b; Real and Rathcke, 1988). Some tropical species show extreme variation within individuals with many flowers producing no nectar (*blanks*) and a few flowers producing high-nectar rewards (*bonanzas*) (Feinsinger, 1978, 1983). This bonanza–blank pattern is commonly found in species visited by short-billed hummingbirds. In contrast, relatively little variation is found in species visited by long-billed hummingbirds (Feinsinger and Colwell, 1978; Linhart and Feinsinger, 1980). The floral populations of two temperate bee-pollinated herbs exhibited a skewed distribution. Most flowers had little or no nectar, and a few flowers had high nectar, but the contributions of within- and between-plant variation were not separated (Marden, 1984b; Zimmerman and Pyke, 1986).

Within-plant variation can show predictable spatial patterns. Vertical inflorescences often exhibit nectar gradients with either increasing (Best and Bierzychudek, 1982; Hodges, 1981; Waddington and Heinrich, 1979) or decreasing nectar from bottom to top (Devlin and Stephenson, 1985). In *Polemonium*, NPR is correlated between flowers in the same inflorescence (Zimmerman and Pyke, 1986).

Given that within-plant variation in NPR exists, we can ask whether some plants are more variable than others in a local population. This has seldom been tested. CVs were significantly different among individuals of *Ipomopsis aggregata* (Pleasants, 1983) but not among individuals of *Impatiens capensis* (Marden, 1984b) or *Kalmia latifolia* (Real and Rathcke, 1988).

IV. Sources of Variation in Nectar Production

The significant between-plant variation in NPR discussed above provides the material for natural selection, but the sources of variation will determine whether selection can mold nectar distributions. For natural selection to act, individual heritable variation must exist. If NPR is greatly influenced by environmental factors, the impact of selection may be weakened. If NPR is associated with other plant characteristics, these characteristics could serve as cues to pollinators, and the associations could be molded by natural selection if heritable variation exists.

A. Genetic Variation

The extent to which the observed between-individual variation in NPR can be attributed to heritable genetic differences is totally unknown for natural plant populations. However, significant heritable variation in NPR has been demonstrated in several crop species (Pedersen, 1953; Hawkins, 1971; Teuber and Barnes, 1979; Murrell *et al.*, 1982; Teuber *et al.*, 1983). For alfalfa, nectar production was increased significantly in two cycles of recurrent selection and was shown to be predominantly under additive genetic control (Teuber and Barnes, 1979; Teuber *et al.*, 1983). The genetics of NPR are likely to be complex because NPR can be determined by many physiological and morphological factors, including regulation of photosynthetic activity, capacity of the sugar-conducting system, nectary enzyme complement, and nectary size (Pederson, 1953; Shuel, 1955).

Because NPR is commonly influenced by environmental factors (see below), it will be difficult to sort out the genetic component. However, we must consider that this plasticity in response to environment may be adaptive, and this has not been studied for NPR (see Chapter 2 for a similar argument for insect life history traits). It is unknown whether the responses of NPR to environmental changes (reaction norms) show significant between-plant variation within populations or whether they benefit plant fitness. Genetic effects of plants as variable resources for animals are considered further in Chapter 4.

B. Environmental Variation

Many environmental factors have been correlated with nectar standing crop, including air temperature (Shuel, 1955; Corbet, 1990), amount of sunlight (Kenoyer, 1916; Pederson, 1953; Shuel, 1955; Walker *et al.*, 1974; Pleasants, 1983), relative humidity (Bertsch, 1983), soil moisture (Shuel and Shivas, 1953; Huber, 1956; Waser, 1983; Pleasants, 1983; Southwick and Southwick, 1983; Zimmerman, 1983; Corbet and Delfosse, 1984), fertilizer (Ryle, 1954; Shuel, 1955, 1957; Pederson, 1957), and season (Cruden *et al.*, 1983; Pleasants, 1983; Zimmerman and Pyke, 1986) (see also Zimmerman, 1988). Foragers can also influence NPR through nectar removal, which stimulates NPR in some species (Raw, 1953; Cruden *et al.*, 1983) and reduces it in others (although this reduction may reflect damage caused by removal in some cases) (Feinsinger, 1978; McDade and Kinsman, 1980; Cruden *et al.*, 1983; Zimmerman and Pyke, 1988b).

Despite this wealth of description, the extent to which intrapopulational variation in nectar per flower can be attributed to local environmental variation is unknown. Most factors are identified through large-scale correlations, and causality is seldom established. Some factors affect evaporation of nectar rather than production, and these processes are seldom separated (Corbet, 1990, but see Marden, 1984b). Correlations may be especially misleading because many environmental factors can covary and their interaction can determine NPR. For example, temperature is often positively correlated with the amount of sunlight, which usually increases nectar production (Pederson, 1953; Shuel, 1955; Walker *et al.*, 1974; Pleasants, 1983), but it is often negatively correlated with soil moisture, which usually decreases NPR (Shuel and Shivas, 1953; Pleasants, 1983; Southwick and Southwick, 1983; Zimmerman, 1983).

Environmental factors may also indirectly influence NPR through influencing plant growth and resource allocation. Shuel, (1955) suggests that rapid vegetative growth may be able to outcompete nectaries for limited carbohydrates because nitrogen fertilization, which increases vegetative growth, decreased NPR in several cultivars (Shuel, 1955; Ryle, 1954). In contrast, potassium fertilizer increased NPR (Ryle, 1954). If such indirect effects are important, variation in plant condition due to past abiotic or biotic environments could give rise to between–individual variation in NPR, even if current environmental factors were uniform.

Given the possible interactions and indirect effects of environmental factors on NPR, controlled experiments are necessary to determine causality, and these are rare. In one of the few field-manipulation studies, supplemental watering increased nectar production in *Delphinium* during a dry period and not during a wetter period (Zimmerman, 1983). The next step is to separate the effects of the genotype, environment, and genotype–environment interactions on NPR.

C. Associations with Other Plant Characteristics

Between-plant variation in NPR is associated with either gender, plant size, or floral size in some species. In dioecious species, the flowers of males commonly produce more nectar than flowers of females, and females may mimic the higher-rewarding males to gain pollination (Baker, 1976; Bawa, 1980; Willson and Agren, 1989). Larger plants with more flowers often have higher NPR (Pleasants and Chaplin, 1983; Devlin *et al.*, 1987; but see Marden, 1984b; Zimmerman and Pyke, 1986). Individuals with larger flowers often show higher NPR (Plowright, 1981; Murrell *et al.*, 1982; Teuber *et al.*, 1983; Cresswell and Galen, 1991). Whether these associations are maintained by selection from pollinator behavior or plant allocation or by genetic linkages is totally unknown (e.g., Cresswell and Galen, 1991).

Within a plant, NPR may vary with floral age, sexual stage, and position. As flowers age, NPR typically peaks shortly after opening and then declines (Percival, 1946; Cruden *et al.*, 1983; Southwick and Southwick, 1983; Devlin *et al.*, 1987; Rathcke, 1988b). In dichogamous species, flowers change sexual function with age, and NPR often changes as well (Percival, 1946; Fahn, 1949; Cruden, 1976; Feinsinger, 1978; Bawa, 1980; Bullock and Bawa, 1981; Best and Bierzychudek, 1982; Devlin *et al.*, 1987) but not in all species (Willson and Bertin, 1979; Cruden *et al.*, 1983; Pleasants, 1983; Marden, 1984b). Relative position of a flower in an inflorescence can also influence NPR (Percival, 1946; Pleasants, 1983; Devlin *et al.*, 1987). The nectar distributions generated by these associations could be strongly influenced by developmental timing and by genetic and environmental variation.

Spatial patterns of NPR can be formed by age or position effects within individual plants. In vertical inflorescences, flowers often open either from bottom to top or vice versa and produce nectar gradients (Waddington and Heinrich, 1979; Best and Bierzychudek, 1982; Hodges, 1981; Devlin and Stephenson, 1985). In contrast, flowers in the radial inflorescences of *Monarda fistulosa* open continuously and form a spatially unpredictable patchwork of new, high producers (Cruden *et al.*, 1983; Cresswell, 1989).

Changes in flower color after fertilization are associated with the cessation of NPR in some species. These color cues may direct pollinator behavior and be adaptive for plant pollination (Gori, 1989).

Although the associations of NPR with different plant characteristics can explain some variation between and within plants, it is unlikely to explain all variation in NPR. In an unusually complete study of NPR (sugar/flower) in *Lobelia cardinalis,* Devlin *et al.* (1987) found that five factors showed significant effects on NPR (flower age, sexual stage, flower position, flower number, and date), but that 42% of the variance remained unexplained. Despite associations with floral cues, nectar/flower can be highly unpredictable for pollinators.

V. Nectar, Pollinator Behavior, and Potential Effects on Plant Reproductive Success

Given the variation in nectar distributions that I have just described, what does this mean for pollinator behavior and the evolution of nectar distribution patterns? Nectar obtained by pollinators can influence plant choice (visit frequency to plants), behavior within plants (i.e., probe time per flower and number of flowers probed), and behavior after leaving (distance and direction of movement). How do these pollinator behaviors determine pollen deposition (female function) and pollen removal and dispersal (male function)? What are the potential effects on the female and male RS of plants? Here I review evidence for and against the assumptions of the optimal nectar models for NPR (Fig. 1A) and the variance in NPR (Fig. 1B).

The nectar/flower obtained by pollinators will be determined by depletion as well as by NPR (see Variation) and by the morphological match between pollinator and flower (Waser, 1983; Harder, 1985; Chapter 13). Pollen transport can also be determined by the morphologies of pollinators and flowers (Galen and Plowright, 1985; Neff and Simpson, 1990). These relationships will vary with specific pollinators and plants; however, changes in the average or variance in nectar/flower may elicit some similar responses.

A. Visit Frequency to Plants

1. Average Nectar per Flower

Individual plants with higher average NPR are visited more frequently by bumblebees in at least two species (Thomson *et al.*, 1989; Real and Rathcke, 1991). This influence of nectar on visits and returns depends on pollinators recognizing high-rewarding plants or remotely assessing nectar (before visiting) (Zimmerman, 1988), and this is not unusual. Bee pollinators quickly learn to associate nectar rewards and floral cues such as color or size (Heinrich, 1979b; Marden, 1984a; Real, 1981; Cresswell and Galen, 1991) and may also remember the location of plants (Waddington, 1983; Thomson *et al.*, 1989). Some bee pollinators can assess floral rewards remotely by seeing or smelling nectar (Marden, 1984a) or by smelling scent markings left by previous visitors (Frankie and Vinson, 1977; Cameron, 1981). However, pollinators may be unlikely to recognize or remember individuals of small herbaceous species growing in dense mixtures, and visit frequency is likely to be independent of nectar/flower in these plant communities (Pyke, 1981; Pleasants and Zimmerman, 1983; Zimmerman, 1988).

Increased visit frequency should increase the chance that flowers will gain adequate outcross pollen for maximal seed set and increased female RS. It may also increase the number of potential donors/stigma, ensure compatible mates and promote pollen tube competition, both of which can increase progeny quality (Schemske and Pautler, 1984; Vander Kloet and Tosh, 1984; Stanton *et al.*, 1986; Young and Stanton, 1990; but see Bertin, 1988;

Cruzan, 1989). Male RS should also benefit as more pollen is removed and dispersed to more recipients. The existence of floral features for restricted pollen dispensing (Harder and Thomson, 1989; Harder, 1990) support the suggestion that many visits are beneficial for male RS.

2. Within-Plant Variation in Nectar per Flower

Within-plant variation in nectar/flower could either increase or decrease visit frequency depending on whether pollinators exhibit either risk-averse or risk-prone foraging behavior (Real and Caraco, 1986), and whether they recognize or assess within-plant variation before visiting, as discussed previously. In laboratory experiments, bumblebees and wasps preferred artificial flowers (different colors) with constant nectar rewards over those with variable rewards with the same average reward (Real, 1981; Real *et al.*, 1982) and express risk-averse behavior. For such foragers, increased variation could decrease visit frequency and reduce plant RS. In contrast, risk-prone foragers may prefer more variable plants, but this has not been demonstrated. Also, risk-averse foragers will switch their preference to the more-variable flower type if the average nectar is increased sufficiently above that in the constant type (Real *et al.*, 1982). A more variable but higher-rewarding plant could be more attractive than a constant, lower-rewarding plant for both risk-averse and risk-prone foragers.

The influence of within-plant variation in NPR on visit frequency to plants in natural populations was examined in one study, but the results were inconclusive (Real and Rathcke, 1988; 1991). Visit frequency increased as both the average and variance in lifetime NPR increased, so risk-averse behavior was not evident. However, because the variance and average were correlated, their effects on visit frequency could not be distinguished. Some plant species effectively reduce the nectar variation perceived by pollinators by advertising the presence or absence of nectar with color changes (Gori, 1989), but whether they benefit from increased visit frequency over control plants that do not advertise is unknown (see below).

B. Pollinator Behavior within Plants

1. Average Nectar per Flower

Many studies show that higher nectar rewards cause foragers to stay longer within plants (or inflorescences) for two reasons: they increase probe time per flower (Hodges and Wolf, 1981; Zimmerman, 1983; Galen and Plowright, 1985; Neff and Simpson, 1990) and they probe more flowers before leaving (Heinrich, 1979c; Pyke, 1978; Hartling and Plowright, 1978; Pleasants and Zimmerman, 1979; Waddington, 1981; Zimmerman, 1983; Galen and Plowright, 1985; Hodges, 1985; Kato, 1988; Cresswell, 1990; Neff and Simpson, 1990).

Increased probe time within a flower commonly increases pollen deposition to some asymptotic value determined by diminishing returns (Pederson, 1953; Plowright and Hartling, 1981; Thomson and Plowright, 1980; Thomson, 1986; Harder and Thomson, 1989; Zimmerman, 1988; Harder, 1990, but see Young and Stanton, 1990). Increased pollen deposition could benefit female RS depending on the pollen quality (see above). However, if the pollen deposited by the pollinator is from a single plant, this could exclude subsequent pollen, reduce the number of pollen donors, and reduce female RS.

For male RS, increased probe time can increase pollen removal to some asymptote (Galen and Stanton, 1989; Harder, 1990). This could benefit male RS, but the existence of floral features for restricted pollen dispensing (Harder and Thomson, 1989) suggests that many visits are beneficial for pollen success. Increased probe time may be relatively more beneficial if pollinator visits are uncertain than if they are common (Harder and Thomson, 1989; Harder, 1990). Increased probe time can also increase pollen loss (Harder and Thomson, 1989).

By probing more flowers per plant, pollinators often increase pollen deposition per stigma (Galen and Plowright, 1985; Hodges, 1990) but this increase is caused largely by pollinators depositing relatively more self-pollen on stigmas. Increased self-pollen can reduce the quantity or quality of seed per female by clogging stigmas and preventing fertilization by outcross pollen (Waser, 1983), or by reducing the number or fitness of progeny because of inbreeding depression (Charlesworth and Charlesworth, 1987). The ratio of outcross- to self-pollen transferred will depend on pollen carryover, i.e., the amount of initial pollen deposited at each flower visited in sequence (Thomson and Plowright, 1980) . Pollen carryover has been shown to decrease exponentially; most outcross pollen is deposited on the first one or few flowers, and subsequent flowers receive mostly self-pollen (Galen and Plowright, 1985). However, some pollen carryover can last to more than 50 flowers (Thomson and Plowright, 1980), and it is likely to vary with specific plant and pollinator characteristics.

If deposition of self-pollen increases, male RS might be more adversely affected than female RS. The probability of dispersal to compatible stigmas would be reduced, and the pollen of all but the last few flowers might be wasted.

2. Within-Plant Variation in Nectar per Flower

Bumblebees often leave after visiting only one to two low-nectar flowers (Hartling and Plowright, 1978; Hodges, 1985; Cresswell, 1989) or two to four low-nectar inflorescences (Thomson *et al.*, 1982; Cibula and Zimmerman, 1986). If within-plant variation in nectar/flower increases the likelihood that pollinators will encounter low-reward flowers early in their visit, they may leave the plant sooner and probe fewer flowers/plant. On the other hand, increased variation could produce some very high-rewarding flowers, or bonanzas, and cause pollinators to stay longer than average because of intermittent reinforcement (Feinsinger, 1978). These conflicting effects on pollinator behavior within plants will determine the effects on plant RS; their relative effects have not been quantified in relation to within-plant nectar variation.

If pollinators leave plants sooner, they may transfer relatively less self-pollen and increase both female and male RS (Fig. 1B). Some plants effectively reduce within-plant variation in nectar available to pollinators by advertising their nectar and reproductive status with color changes, and pollinators are less likely to probe flowers that can no longer receive or donate pollen (Cruzan et al., 1988; Gori, 1989; Delph and Lively, 1989; Cresswell and Galen, 1991; but see Casper and La Pine, 1984). Such advertising is assumed to benefit the plant; however, it could also increase self-pollen transfer if pollinators probe more flowers within each plant before leaving.

Nectar variation can produce predictable spatial patterns that direct pollinator movement within plants. For example, bumblebees usually move upward on vertical inflorescences with increasing nectar gradients from bottom to top (Waddington and Heinrich, 1979; Best and Bierzychudek, 1982; Waddington, 1983; Hodges, 1985). However, such pollinator manipulation by plants may be constrained by innate pollinator behavior. When nectar gradients are reversed, pollinator movements are not; they usually continue to move upward although they will start higher on the inflorescence (Waddington and Heinrich, 1979; Waddington, 1983; Devlin and Stephenson, 1985).

C. Pollinator Movement between Plants

1. Average Nectar per Flower

As average nectar per flower increases, pollinators often show area-restricted foraging by decreasing their flight distance to the next flower (Gill and Wolf, 1977; Pyke, 1978; Heinrich, 1979c; Zimmerman, 1981a; Waddington, 1983; Galen and Plowright, 1985; Zimmerman and Cook, 1985; Zimmerman, 1988) and by increasing their turning (Heinrich, 1979; Waddington, 1983; Zimmerman, 1988).

Area-restricted foraging should reduce gene flow distance (Levin and Kerster, 1969; Schaal, 1978; Zimmerman, 1988) and may be detrimental to plant RS if neighboring plants are related and show inbreeding depression (Price and Waser, 1979; Waser and Price, 1983).

2. Within-Plant Variation in Nectar per Flower

Increased within-plant variation in nectar/flower could cause increased variation in flight distances and turning angles. Bumblebees tend to make longer, straighter flights after visiting nonrewarding flowers (Marden, 1984a), which could reduce inbreeding effects and increase plant RS. This behavior could increase outbreeding depression, but this is unlikely in reality because most visits are to nearby neighbors and are usually shorter than optimal distances (Waser, 1982; Waser and Price, 1983).

D. Summary

Pollinators (especially bees) respond quite predictably to change in average nectar per flower, and they behave as assumed in the first optimal nectar model (Fig. 1A). However, the effect of nectar on visit frequency is in dispute and may not be applicable to some plant species (Zimmerman, 1988). More data are needed on pollen carryover and how this varies with different pollinator and plant traits. Low pollen carryover is crucial to the predicted decline in plant RS and the generality of the model (Fig. 1A). Pollinator response to within-plant variation, as assumed in the second model (Fig. 1B), is untested for natural plant populations.

VI. Other Selective Factors and Constraints

Selection for optimal NPR and distributions of NPR within plants (within-plant variation) could be modified or weakened by many factors other than the pollinator behaviors discussed in the previous section. Some factors could change the optimum for NPR, and others could weaken or oppose stabilizing selection for an optimum and maintain between-plant variation in NPR. Which factors, if any, will need to be invoked for understanding the evolution of nectar distributions is an open question.

A. Behavior of Floral Visitors

In the optimal-nectar models (Fig. 1A and 1B), nectar/flower is assumed to be the major factor influencing pollinator foraging at flowers. However, a better predictor of foraging may be energy gain/time, and this will depend on handling time (Harder, 1988), and travel time (Hartling and Plowright, 1978; Cibula and Zimmerman, 1986), as well as on nectar obtained per flower. Handling time will vary with the match between pollinator and floral morphology (Harder, 1985), and travel time will vary with distances between flowers and plants. Both may change the predicted optimum for NPR.

Nectar foragers are often more attracted to larger plants with more flowers (Geber, 1985; Primack and Kang, 1989). This could reflect lower travel costs between flowers and select for a lower optimal NPR in larger plants (Pyke, 1981). In fact, NPR is negatively correlated with plant size in several species (Marden, 1984b; Pleasants and Chaplin, 1983; Zimmerman and Pyke, 1986, but see Pyke, 1978; Geber, 1985), but whether this represents some optimal strategy for either pollination or resource allocation remains to be tested.

As plant densities increase, travel costs decrease, and foragers often stay longer and visit more flowers (Hartling and Plowright, 1979; Rathcke, 1984). They also tend to move shorter distances between plants (Levin and Kerster, 1969; Schaal, 1978; Waddington, 1981, 1983). Because plant density commonly varies greatly in time and space, gene flow and selection on NPR could also be highly variable and weaken stabilizing selection for one optimal NPR.

The predictions of optimal foraging also depend upon the currency. Nectar is often the appropriate currency, but not always. Many bees switch between foraging for nectar and pollen, and each behavior may affect pollen transport differently. Small bees collect both pollen and nectar on any one trip, and they seldom show optimal foraging for nectar (Neff and Simpson, 1989).

Responses of foragers to nectar/flower may also vary with the nectar rewards of plant neighbors (Rathcke, 1983; Rathcke, 1988a). Higher-rewarding species (or morphs) could decrease visits to lower-rewarding neighbors (Rathcke, 1988a) and select for increased NPR. This could promote nectar equivalence or an upward-spiraling *rewards* race among species (Waser, 1983). On the other hand, high-rewarding neighbors could increase visits to low-rewarding plants (Rathcke, 1983). For example, honeybees virtually never visit nectarless muskmelons when they are planted alone, but when they are planted among nectar-producing muskmelons, they visit frequently enough to cause nearly normal seed set (Bohn and Mann, 1960). Cheaters that eliminate the cost of NPR could be favored by selection but only when rare. Selection for NPR would be frequency dependent.

If pollinators compete for nectar, less-preferred floral species may be visited (Heinrich, 1979b; Inouye, 1978), and select for lower optimal NPR. Aggression among bees may increase inter-plant movements (Frankie et al., 1976) and reduce the potentially negative effects of high NPR caused by self-pollen transfer so that increased NPR could be optimal.

In natural communities most plants have many potential pollinators (but see Chapter 11 for a comparison of temperate and tropical bee communities). If pollinators respond differently to the same nectar distributions, they may impose a variable selection regime for NPR. Variable selection on NPR could also be promoted by differences in pollination effectiveness and by unpredictable visit frequency over time and space (Horvitz and Schemske, 1990).

Some floral visitors may be nectar parasites that collect nectar but do not effect pollination (McDade and Kinsman, 1980). If pollinator behavior selects for increased NPR, nectar parasites may also increase and push the optimal NPR to lower values (Feinsinger, personal communication, 1991).

Overall, unpredictability in the behavior of different floral visitors could impose limits on the fine-tuning of selection for nectar traits.

B. Plant Size and Resource Allocation

The evolution of nectar distributions will depend on the effects of variation on relative plant fitness. Ultimately, plant fitness depends upon total seed production/plant, and this can depend on plant size and allocation strategies. Larger plants within a population have more resources and often produce and sire more seeds by a combination of greater attraction (see above) and simply because they have more flowers and fruits (Barrett and Eckert, 1990; but see Waller, 1989). As a result, variation in NPR may be overwhelmed by the effects of high variation in seed and pollen production (Primack and Kang, 1989), and selection on nectar traits may be weakened.

Plant fitness will depend on how available resources are allocated to various functions, such as nectar production, seed provisioning, or vegetative growth, and what trade-offs are made (Pyke, 1981; Zimmerman, 1988). Changes in average nectar per flower could depend on either allocation to nectar or allocation to flowers or both. In contrast, changes in the variance in NPR (Fig. 1B) should be relatively cost-free and unconstrained by allocation costs.

The importance of nectar allocation in constraining NPR will depend on the costs. The total cost of nectar production has seldom been estimated, but it can be substantial. Nectar production can account for 30 to 40% of the energy budget of plants (Pederson, 1953; Southwick, 1984), although these estimates are probably high (Zimmerman, 1988). Most perennial plants allocate a total of 8 to 15% of their annual gain to sexual reproduction, of which nectar is only a part (Mooney, 1972; Zimmerman, 1988). In reality, the costs of nectar could be minimal if the currency is considered. Nectar consists largely of carbohydrates (sugars) (Baker and Baker, 1983), and plants are often assumed to have an excess of carbohydrate to the extent that they have been called pathological overproducers of carbohydrates (Crane, 1975). If so, trade-offs may be nonexistent. The opposing argument is that carbohydrates are limited, especially in rapidly growing plants. Nectaries may compete with rapidly growing plant parts for carbohydrates (Shuel, 1955; 1957) or with other plant functions such as seed provisioning (Haig and Westoby, 1988), defense, or food for other mutualists such as mycorrhizae. Trade-offs in water use are unstudied. The importance of trade-offs for nectar allocation is totally hypothetical at this point and demands further study.

VII. Completing the Links: Nectar Production, Pollinator Behavior, and Plant Reproductive Success

To understand natural selection as an evolutionary force, we need to measure lifetime fitness, although this has been attempted for few animals (Endler, 1986) and no plants (Primack and Kang, 1989). Measuring some component of plant RS over one season, much less over a plant lifetime, can be fraught with difficulties, and protocols are just being established. Studies linking nectar production, pollinator behavior, and plant RS are just beginning and are often incomplete. At this early stage, measurements of RS in relation to variation in nectar production are few but provide a valuable start.

A. Measuring Plant Reproductive Success

Most researchers have concentrated on measuring maternal fitness for good reason. Paternal fitness requires the use of genetic markers or fingerprinting (Primack and Kang, 1989) and has not been measured in any study of nectar variation. Given that paternal fitness has been argued to drive the evolution of other floral traits (Stanton *et al.*, 1986), this leaves a major gap in our understanding of nectar distributions. Maternal RS (total seed set) can be limited by pollen and by other resources, and these factors need to be separated to understand how selection might act on nectar distributions. Pollen limitation has commonly been inferred when pollen-augmented flowers show more seeds/flower than do naturally pollinated flowers. Zimmerman and Pyke (1988a) point out that this procedure may be insufficient for detecting pollination limitation. If plants compensate for pollen limitation of early flowers by reallocating resources into more flowers and more seed provisioning, pollen will not necessarily limit total plant seed set. They outline new protocols for establishing whether or not total seed set is pollen- or resource-limited and whether compensation occurs.

Compensation for pollen limitation of flowers has been observed in two perennial herbs. When seed-set of early flowers was pollen limited, plants produced more flowers and more seeds per fruit, so that total seed set was not pollen limited or was reduced less than expected (Zimmerman and Pyke, 1988a; Lawrence, 1991). However, Lawrence (1991) found that small plants could not compensate because they had few available meristems. For these individuals, reproduction appears to be module limited rather than resource limited (Watson, 1984).

Compensation by resource reallocation can buffer total seed set from pollen limitation, but it probably incurs some cost. If so, selection should favor optimal allocation strategies. Haig and Westoby (1988) hypothesize that plants should evolve resource allocation strategies that result in seed set's being limited by both pollen (fertilization) and resources (seed provisioning). When plants are pollen limited, selection should favor increased allocation for pollinator attraction; when plants are seed provisioning limited, selection should favor reduced allocation to pollen attraction and more to provisioning. Allocation to pollinator attraction or seed-provisoning may involve trade-offs with other functions as well. Plasticity that would allow plants to reallocate resources as conditions change during their lifetime may be especially adaptive (Cohen and Dufas, 1990). Whether the plasticity in nectar production that is commonly observed (See Variation) reflects adaptive resource reallocation strategies remains to be tested.

B. Completing the Links: Evidence from Field Studies

1. Average Nectar per Flower

The links between NPR, pollinator behavior, and some component of plant RS have been completed in few studies. One of these studies supports the prediction of an intermediate optimal NPR (Hodges, 1990; see Fig. 1A), and the others are ambiguous. For *Mirabilis multiflora*, Hodges (1990) showed that percentage seed set/plant was maximal at intermediate nectar volumes/flower and was reduced at both lower and higher nectar levels. He also demonstrated that nectar volume was positively correlated with visit frequency and with both pollen deposition and removal. He argues that the decrease in seed set at higher nectar levels is probably caused by increased transfer of self-pollen because the hawkmoth pollinators visited more flowers within plants and because self-pollen can probably block seed-set. In addition, individual differences in NPR were constant over days, seasons, and years, suggesting genetic differences in NPR. These results strongly suggest that stabilizing selection for an intermediate optimum exists. Indeed, the most common NPR value is the optimal level for maximal seed set. However, some between-plant variation remains, perhaps because of complex genetics (Hodges, 1990).

The links between nectar, pollinator visit frequency, and fruit set were also measured in mountain laurel (*Kalmia latifolia*) in Virginia (Real and Rathcke, 1991). Average lifetime NPR of plants was correlated with visit frequency by bumblebees, the only pollinator. In turn, visit frequency was correlated with fruit/flower. However, these results suggest directional selection for increased NPR rather than stabilizing selection for an intermediate optimum.

This lack of support for an intermediate optimum in NPR in *K. latifolia* may not be a strong test because of the limited range of natural nectar variation exhibited. If nectar had been artifically augmented above natural levels, fruit set might have declined as predicted, because inbreeding depression can significantly reduce fruit set (Rathcke, 1988a; Rathcke and Real, unpublished data, 1986). An experiment to do just this was thwarted by a drought that eliminated flower production during one year. On the other hand, many individuals had low NPR, and fruit set was pollination limited, indicating selection for higher NPR. However, nectar production of individuals was not significantly correlated between years, and any directional selection is probably very weak (Real and Rathcke, 1988).

In another study of *K. latifolia* in Rhode Island, low NPR was linked to fewer bumblebee visits/flower, reduced fruit set, and pollen limitation of fruit set (Rathcke, 1988a). In shaded habitats, individuals produced almost no nectar, they were seldom visited by bumblebees, and fruit/flower was pollen limited (Rathcke, 1988a). In contrast, plants in sunny areas produced more nectar, were visited more frequently, and fruit/flower was not pollen limited. Although NPR is linked to fruit set, this nectar variation is probably largely caused by environmental rather than genetic variation and is not under direct selection. However, the phenotypic plasticity in NPR could be adaptive if this nectar reduction represents a trade-off with other plant functions, such as plant defense or vegetative growth. This remains to be tested.

NPR also appears to be under directional selection for increased production in *Delphinium nelsonii* (Zimmerman, 1983). By watering one population, Zimmerman (1983) increased nectar per flower and found that bumblebees visited more flowers, stayed longer per flower, and flew shorter distances than in the unwatered population. In addition, seeds/flower increased in the watered population and seeds/flower were pollen limited only in the unwatered population. Although the increased seed production could also have been caused by increased resources in the watered population, these results suggest that NPR should be under directional selection for increased production.

In a few other studies, nectar has been correlated with seed-set, but the link with pollinator behavior is missing, and causes are ambiguous. Increased nectar concentration was significantly correlated with percentage of fruit set and the number of mature fruit in *Asclepias exaltata* (Wyatt and Shannon, 1986). Exclusion of nectar-robbing ants increased nectar standing crop and seedpod initiation in *Asclepias syriaca;* however, neither total pollinia removal nor insertion was increased (Fritz and Morse, 1981). Other similar studies show no relationship between nectar variation and fruit set (Wood and Wood, 1963; Pyke *et al.*, 1988).

2. Nectar Variation within Plants

The effect of within-plant variation in NPR on plant reproductive success (Fig. 1B) has not yet been rigorously tested. For *Kalmia latifolia,* individuals showed significant differences in within-plant variation in NPR (Real and Rathcke, 1988). However, the variance was significantly correlated with the mean. As a result, their relative effects in causing the correlations between average NPR, visit frequency, and fruit set cannot be distinguished. Within-plant variation may be important because inbreeding depression is significant, and plants commonly have thousands of synchronous flowers (Rathcke, 1988a). Coefficients of variation were not significantly different among individuals and cannot account for the increase in visits or fruit set (Real and Rathcke, 1988, 1991). However, pollinators may respond to absolute variance, and coefficients of variation may not be the best measure of effective variation, but this is untested. For *Mirabilis multiflora,* coefficients of variation were not consistent for individuals within or between seasons and therefore seem unlikely to be under stabilizing selection (Hodges, 1990a).

Spatial nectar gradients in vertical inflorescences are known to influence pollinator behavior and to have great potential for determining plant RS, so I discuss two studies here, although neither maternal nor paternal RS was measured. In foxglove (*Digitalis purpurea*), the lower female flowers have more nectar. Bumblebees always start low and move upward and leave after visiting an average of five flowers (Best and Bierzychudek, 1982). Best and Bierzychudek (1982) argue that this ensures that each pollinator visits at least one male flower, and they conclude that female function is favored over male function because mostly female flowers were visited. In contrast, *Lobelia cardinalis* shows the opposite gradient with the upper, male flowers having more nectar. Hummingbird pollinators usually first visit flowers in midinflorescence and then move upward (Devlin and Stephenson, 1985). Devlin and Stephenson (1985) estimated that female flowers would have fewer visits than male flowers during their lifetimes, but that these few visits would be sufficient for seed-set so that male function is favored by the nectar gradient. Whether conflicts between female and male function are important in the evolution of nectar distributions needs to be explored.

VIII. Conclusions

The links between nectar rewards, pollinator behavior, and plant reproductive success are many, but few have been empirically revealed. As a consequence, the models of resource allocation have been subjected to few tests (Fig. 1A and 1B). However, the prediction of an intermediate optimum for NPR is supported in one recent study (Hodges, 1990).

Other studies are inconclusive but suggest that NPR is under directional selection for increased production rather than stabilizing selection for an intermediate value. If this be true, why has not selection favored increased NPR in *Kalmia latifolia*, which is commonly pollen limited because it is a poor competitor for bumblebees (Rathcke, 1988a; 1988b)? If no negative effects of increased NPR ensue, why has it not entered a "rewards race" with other higher-rewarding species? Alternatively, it seems that NPR would be constrained by foragers or allocation. Increasing NPR above the natural levels seems likely to increase self-pollen transfer and reduce plant RS by high inbreeding depression and pollen wastage. Nectar allocation may also be limited by trade-offs with other functions, such as seed provisioning or defense. This needs to be tested in general. In this latter case, nectar may be optimal in the context of whole-plant fitness but not necessarily for pollination.

The finding that nectar variation can affect seed-set (female function) as well as pollen removal (male function) contrasts with studies on other floral traits, such as color or corolla size, where male function appears to be the driving selective force because female function is always maximized (Clegg and Epperson, 1988; Stanton *et al.*, 1986; Murcia, 1990, but see Campbell, 1989). This difference may reflect the different roles of floral advertisements and rewards (Hodges, 1990). Advertisements attract pollinators and affect visit frequency. Rewards can also influence pollinator behavior after visiting and determine pollen deposition, carryover, and removal, which, in turn, may more strongly affect both female and male function than does attraction alone.

Although rewards may influence both male and female function, effects may be greater on male function if female function is more easily maximized (Figs. 1A and 1B). If so, male function may be more important for fine-tuning nectar distributions, and conflicting selection pressures may arise from male and female functions. Female and male flowers commonly produce different amounts of nectar; whether these differences represent different nectar optima needs to be explored.

The effects of within-plant variation in NPR have not been fully linked with behavior, pollen transport, and plant reproductive success in any natural population. However, within-plant variation is common and is known to affect pollinator behavior. Whether it has evolved for this purpose is an open question.

The significant between-plant variation in NPR found in most plant populations suggests that selection is not strongly stabilizing to some optimum as predicted in the model (Fig. 1A). Alternatively, variation may be maintained by variable selection for different optima or by selection for optimal resource-allocation strategies. NPR is commonly highly plastic; this plasticity could be optimal. The possible links to other plant processes such as growth suggest that total plant responses will need to be examined to understand the evolution of NPR traits. Distinguishing the relative contributions of the many complex factors that may contribute to the evolution of nectar distributions will be difficult and challenging.

References

Baker, H. G. (1976). "Mistake" pollination as a reproductive system with special reference to the Caricaceae. *In* "Tropical Trees: Variation, Breeding and Conservation" (J. Burley and B. T. Styles, eds.), pp. 161–169. Academic Press, New York.

Baker, H. B., and Baker, I. (1983). Floral nectar sugar constituents in relation to pollinator type. *In* "Handbook of Experimental Pollination Biology" (C.E. Jones and R. J. Little, eds.), pp. 117–141. Van Nostrand-Reinhold, New York.

Barrett, S. C. H., and Eckert, C. G. (1990). Current issues in plant reproductive ecology. *Isr. J. Bot.* **39**, 5–12.

Bawa, K. S. (1980). Mimicry of male by female flowers and intrasexual competition for pollinators in *Jacartia dolichaula* (D. Smith) Woodson (Caricaceae). *Evolution* **34**, 467–474.

Bell, G. (1985). On the function of flowers. *Proc. R. Soc. Lond.* **224**, 223–265.

Bell, G. (1986). The evolution of empty flowers. *J. Theor. Biol.* **118**, 253–258.

Bertin, R. I. (1988). Paternity in plants. *In* "Plant Reproductive Ecology: Patterns and Strategies." (J. Lovett Doust and L. Lovett Doust, eds.), pp. 30–59. Oxford University Press, New York.

Bertsch, A. (1983). Nectar production of *Epilobium angustifolium* L. at different air humidities: Nectar sugar in individual flowers and the optimal foraging theory. *Oecologia* **59,** 40–48.

Best, L. S., and Bierzychudek, P. (1982). Pollinator foraging on foxglove (*Digitalis purpurea*): A test of a new model. *Evolution* **36,** 70–79.

Bierzychudek, P. (1981). Pollinator limitation of plant reproductive effort. *Am. Nat.* **117,** 838–840.

Bohn, G. W., and Mann, L. K. (1960). Nectarless, a yield-reducing mutant character in the muskmelon. *Am. Soc. Hort. Proc.* **76,** 455–459.

Bullock, S. H., and Bawa, K. S. (1981). Sexual dimorphism and the annual flowering pattern of *Jacartia dolichauda* (D. Smith) Woodson (Caricaceae) in a Costa Rican rain forest. *Ecology* **62,** 143–150.

Cameron, S. A. (1981). Chemical signals in bumblebee foraging. *Behav. Ecol. Sociobiol.* **9,** 257–260.

Campbell, D. R. (1989). Measurements of selection in a hermaphroditic plant: Variation in male and female pollination success. *Evolution* **43,** 318–334.

Carpenter, F. L. (1976). Plant–pollinator interactions in Hawaii: Pollination energetics of *Metrosideros collina* Myrtaceae. *Ecology* **57,** 1125–1144.

Caspar, B. B., and LaPine, T. R. (1984). Changes in corolla color and other floral characteristics in *Cryptantha humilis* (Boraginaceae): Cues to discourage pollinators? *Evolution* **38,** 128–141.

Charlesworth, D., and Charlesworth, B. (1987). Inbreeding depression and its evolutionary consequences. *Annu. Rev. Ecol. Syst.* **18,** 237–268.

Cibula, D. A., and Zimmerman, M. (1986). The effect of plant density on departure decisions: Testing the marginal value theorem using bumblebees and *Delphinium nelsonii. Oikos* **43,** 154–158.

Clegg, M. T., and Epperson, B. K. (1988). Natural selection of flower color polymorphisms in morning glory populations. *In* "Plant Evolutionary Biology" (L.D. Gottlieb and S. K. Jain, eds.), pp. 255–274. Chapman and Hall, New York.

Cohen, D., and Dufas, R. (1990). The optimal number of female flowers and fruits-to-flower ratio in plants under pollination and resource limitation. *Am. Nat.* **135,** 218–241.

Corbet, S. A. (1990). Pollination and the weather. *Isr. J. Bot.* **39,** 13–30.

Corbet, S. A., and Delfosse, E. S. (1984). Honeybees and the nectar of *Echium plantagineum* L. in southeastern Australia. *Austr. J. Ecol.* **9,** 125–139.

Crane, E. (ed.) (1975). "Honey: A Comprehensive Survey." Heinemann and Bee Research Association, London, England.

Cresswell, J. E. (1989). "Optimal Foraging Theory Applied to Bumblebees Gathering Nectar from Wild Bergamont." Ph.D. thesis, University of Michigan, Ann Arbor, Michigan.

Cresswell, J. E. (1990). How and why do nectar-foraging bumblebees initiate movements between inflorescences of wild Bergamot *Monarda fistulosa* (Lamiaceae)? *Oecologia* **82,** 450–460.

Cresswell, J. E., and Galen, C. (1991). Frequency-dependent selection and adaptive surfaces for floral character combinations: The pollination of *Polemonium viscosum. Am. Nat.,* in press.

Cruden, R. W. (1976). Intraspecific variation in pollen–ovule ratios and nectar secretion— preliminary evidence for ecotypic adaptation. *Ann. Missouri Bot. Gard.* **63,** 277–289.

Cruden, R. W., Hermann-Parker, S. M., and Peterson, S. (1983). Patterns of nectar production and plant–pollinator coevolution. *In* "Biology of Nectaries" (T.S. Elias and B. Bentley, eds.), pp. 80–125. Columbia University Press, New York.

Cruzan, M. B. (1989). Pollen tube attrition in *Erythronium grandiflorum. Am. J. Bot.* **76,** 562–570.

Cruzan, M. B., Neal, P. R., and Wilson, M. F. (1988). Floral display in *Phlox incisa:* Consequences for male and female reproductive success. *Evolution* **42,** 505–515.

Delph, L. F., and Lively, C. M. (1989). The evolution of floral color change: Pollinator attraction versus physiological constraints in *Fuchsia excorticata. Evolution* **43,** 1252–1262.

Devlin, B., and Stephenson, A. G. (1985). Sex differential floral duration, nectar secretion, and pollinator foraging in a protandrous species. *Am. J. Bot.* **62,** 303–310.

Devlin, B., and Stephenson, A. G. (1987). Sexual variations among plants of a perfect-flowered species. *Am. Nat.* **130,** 199–218.

Devlin, B., Horton, J. B., and Stephenson, A. B. (1987). Patterns of nectar production of *Lobelia cardinalis. Am. Midl. Nat.* **117,** 289–295.

Doust, J. L. (1989). Plant reproductive strategies and resource allocation. *TREE* **4,** 230–234.

Endler, J. A. (1986). "Natural Selection in the Wild." Princeton University Press, Princeton, New Jersey.

Fahn, A. (1949). Studies in the ecology of nectar secretion. *Palest. J. Bot. Jerus. Ser.* **1B,** 207–224.

Feinsinger, P. (1978). Ecological interactions betwen plants and hummingbirds in a successional tropical community. *Ecol. Monogr.* **48,** 269–287.

Feinsinger, P. (1983). Variable nectar secretion in a *Heliconia* species pollinated by hermit hummingbirds. *Biotropica* **15,** 48–52.

Feinsinger, P., and Colwell, R. K. (1978). Community organization among neotropical nectar-feeding birds. *Am. Zool.* **18,** 779–795.

Frankie, G. W., and Vinson, S. B. (1977). Scent marking of passion flowers in Texas by females of *Xylocopa virginica texana* (Hymenoptera: Anthophoridae). *J. Kansas Entomol. Soc.* **50,** 613–635.

Frankie, G. W., Opler, P. A., and Bawa, K. S. (1976). Foraging behavior of solitary bees: Implications for outcrossing of a neotropical forest tree species. *J. Ecol.* **64,** 1049–1057.

Fritz, R. S., and Morse, D. H. (1981). Nectar production of *Asclepias syriaca* by ants: Effects on nectar levels, pollinia insertion, pollinaria removal, and pod production. *Oecologia* **50,** 316–319.

Galen, C., and Plowright, R. C. (1985). The effects of nectar level and flower development on pollen carry-over in inflorescences of fireweed (*Epilobium angustifolium*) (Onagraceae). *Can. J. Bot.* **63,** 488–491.

Galen, C., and Stanton, M. L. (1989). Bumblebee pollination and floral morphology: Factors influencing pollen dispersal in the alpine sky pilot, *Polemonium viscosum* (Polemoniaceae). *Am. J. Bot.* **76,** 419–426.

Geber, M. A. (1985). The relationship of plant size to self-pollination in *Mertensia ciliata. Ecology* **66,** 762–771.

Gill, F. B., and Wolf, L. L. (1977). Nonrandom foraging by sunbirds in a patchy environment. *Ecology* **58,** 1284–1296.

Gori, D. F. (1989). Floral color change in *Lupinus argenteus* (Fabaceae): Why should plants advertise the location of unrewarding flowers to pollinators? *Evolution* **43,** 870–881.

Haig, D., and Westoby, M. (1988). On limits to seed production. *Am. Nat.* **131,** 757–759.

Harder, L. (1985). Morphology as a predictor of flower choice by bumble bees. Ecology **66,** 198–210.

Harder, L. D. (1990). Pollen removal by bumble bees and its implications for pollen dispersal. Ecology **71,** 1110–1125.

Harder, L. D., and Real, L. (1987). Why are bumblebees risk averse? *Ecology* **68,** 1104–1108.

Harder, L. D., and Thomson, J. D. (1989). Evolutionary options for maximizing pollen dispersal of animal-pollinated plants. *Am. Nat.* **133,** 323–344.

Hartling, L. K., and Plowright, R. C. (1978). Foraging by bumble bees on patches of artificial flowers: A laboratory study. *Can. J. Zool.* **57,** 1866–1870.

Hawkins, R. P. (1971). Selection for height of nectar in the corolla tube of English singlecut red clover. *J. Agric. Sci.* **77,** 348–350.

Heinrich, B. (1976). The foraging specialization of individual bumblebees. *Ecology* **57,** 874–889.

Heinrich, B. (1979a). "Majoring" and "minoring" by foraging bumblebees, *Bombus vagans:* An experimental analysis. *Ecology* **60,** 245–255.

Heinrich, B. (1979b). "Bumblebee Economics." Harvard University Press, Cambridge, Massachusetts.

Heinrich, B. (1979c). Resource heterogeneity and patterns of movement in foraging bumblebees. *Oecologia* **40,** 234–245.

Hodges, D. M. (1981). Optimal foraging in bumblebees: Hunting by expectation. *Anim. Behav.* **29,** 1166–1171.

Hodges, C. M. (1985). Bumble bee foraging: The threshold departure rule. *Ecology* **66,** 179–187.

Hodges, C. M., and Wolf, L. L. (1981). Optimal foraging in bumblebees: Why is nectar left behind in flowers? *Behav. Ecol. Sociobiol.* **9,** 41–44.

Hodges, S. A. (1990). "The Roles of Nectar Production, Hawkmoth Behavior and Pollen Movement in Natural Selection for Nectar Production in *Mirabilis multiflora.*" Ph.D. thesis, University of California, Berkeley.

Horvitz, C. C., and Schemske, D. W. (1990). Spatiotemporal variation in insect mutualists of a neotropical herb. *Ecology* **71,** 1085–1097.

Huber, H. (1956). Die Abhangigkeit der Nektarsekretion von Temperatur, Luft, und Bodenfeuchingkeit. *Planta* **48,** 47–98.

Inouye, D. (1978). Resource partitioning in bumblebees: Experimental studies of foraging behavior. *Ecology* **59,** 672–678.

Kato, M. (1988). Bumblebee visits to *Impatiens* spp.: Pattern and efficiency. *Oecologia* **76,** 364–370.

Kenoyer, L. A. (1916). Environmental influences on nectar secretion. *Bot. Gaz.* **63,** 249–265.

Lawrence, W. S. (1991). Resource allocation, pollination limitation, and compensatory response: Plant size–dependent reproductive patterns in *Physalis longifolia. Am. Nat.*, in press.

Levin, D. A., and Kerster, H. W. (1969). Density-dependent gene dispersal in *Liatris. Am. Nat.* **103,** 61–74.

Marden, J. H. (1984a). Remote perception of floral nectar by bumblebees. *Oecologia* **64,** 232–240.

Marden, J. H. (1984b). Intrapopulation variation in nectar secretion in *Impatiens capensis. Oecologia* **63,** 418–422.

McDade, L. A., and Kinsman, S. (1980). The impact of floral parasitism in two neotropical hummingbird-pollinated plant species. *Evolution* **34,** 944–958.

Mooney, H. A. (1972). The carbon balance of plants. *Annu. Rev. Ecol. Syst.* **3,** 315–346.

Murcia, C. 1990. Effect of floral morphology and temperature on pollen receipt and removal in *Ipomoea trichocarpa. Ecology* **71,** 1098–1109.

Murrell, D. C., Tomes, D. T., and Shuel, R. W. (1982). Inheritance of nectar production in birdsfoot trefoil. *Can. J. Plant Sci.* **62,** 101–105.

Neff, J. L., and Simpson, B. B. (1990). The roles of phenology and reward structure in the pollination biology of wild sunflower (*Helianthus annuus* L. Asteraceae). *Isr. J. Bot.* **39,** 197–216.

Newport, M. E. A. (1989). A test for proximity-dependent outcrossing in the alpine skypilot, *Polemonium viscosum. Evolution* **43,** 1110–1113.

Opler, P. A., and Bawa, K. S. (1978). Sex ratios in tropical forest trees. *Evolution* **32,** 812–821.

Pankiw, P., and Bolton, J. L. (1965). Characteristics of alfalfa flowers and their effects on seed production. *Can. J. Plant Sci.* **45,** 333–342.

Pedersen, M. W. (1953). Seed production in alfalfa as related to nectar production and honeybee visitation. *Bot. Gaz.* **115,** 129–138.

Percival, M. S. (1946). Observations on the flowering and nectar secretion of *Rubus fruticosus. New Phytol.* **45,** 111–123.

Pleasants, J. M. (1983). Nectar production patterns in *Ipomopsis aggregata* (Polemoniaceae). *Am. J. Bot.* **70,** 1468–1475.

Pleasants, J. M. and Chaplin, S. J. (1983). Nectar production rates in *Asclepias quadrifolia*: Causes and consequences of individual variation. *Oecologia* **59,** 232–238.

Pleasants, J. M., and Zimmerman, M. (1979). Patchiness in the dispersion of nectar resources: Evidence for hot and cold spots. *Oecologia* **41,** 283–288.

Pleasants, J. M., and Zimmerman, M. (1983). The distribution of standing crop of nectar: What does it really tell us? *Oecologia* **57,** 412–414.

Plowright, R. C. (1981). Nectar production in the boreal forest lily *Clintonia borealis. Can. J. Bot.* **59,** 156–160.

Plowright, R. C., and Hartling, L. K. (1981). Red clover pollination by bumble bees: A study of the dynamics of a plant–pollinator relationship. *J. Appl. Ecol.* **18,** 639–647.

Possingham, H. P. (1990). The distribution and abundance of resources encountered by a forager. *Am. Nat.* **133,** 42–60.

Price, M. V., and Waser, N. M. (1979). Pollen dispersal and optimal out-crossing in *Delphinium nelsonii. Nature* **277,** 294–296.

Price, M. V., and Waser, N. M. (1982). Experimental studies of pollen carryover: Hummingbirds and *Ipomopsis aggregata. Oecologia* **54,** 353–358.

Primack, R. B., and Kang, H. (1989). Measuring fitness and natural selection in wild plant populations. *Annu. Rev. Ecol. Syst.* **20,** 367–396.

Pyke, G. (1978). Optimal foraging: Movement patterns of bumblebees between inflorescences. *Theor. Popul. Biol.* **13,** 72–98.

Pyke, G. H. (1981). Optimal nectar production in a hummingbird-pollinated plant. *Theor. Popul. Biol.* **20,** 326–343.

Pyke, G. H., Day, L. P., and Wale, K. A. (1988). Pollination ecology of Christmas Bells (*Blandfordia nobilis* Sm.): Effects of adding artificial nectar on pollen removal and seed-set. *Aust. J. Ecol.* **13,** 279–284.

Rathcke, B. (1983). Competition and facilitation among plants for pollination. *In* "Pollination Biology" (L. Real, ed.), pp. 305–329. Academic Press, New York.

Rathcke, B. (1988a). Interactions for pollination among coflowering shrubs. *Ecology* **69,** 446–457.

Rathcke, B. (1988b). Flowering phenologies in a shrub community: Competition and constraints. *J. Ecol.* **76,** 975–994.

Rathcke, B., and Real, L. Interpopulational variation in the breeding system of mountain laurel (*Kalmia latifolia*). (submitted).

Raw, G. R. (1953). The effect on nectar secretion of removing nectar from flowers. *Bee World* **34,** 23–25.

Real, L. (1981). Uncertainty and pollinator–plant interactions: The foraging behavior of bees and wasps on artificial flowers. *Ecology* **62,** 20–26.

Real, L., and Caraco, T. (1986). Risk and foraging in stochastic environments: Theory and evidence. *Annu. Rev. Ecol. Syst.* **17,** 371–390.

Real, L., Ott, J., and Silverfine, E. (1982). On the trade-off between the mean and the variance in foraging: Effect of spatial distribution and color preference. *Ecology* **63,** 1617–1623.

Real, L., and Rathcke, B. J. (1988). Patterns of individual variability in floral resources. *Ecology* **69,** 728–735.

Real, L., and Rathcke, B. (1991). Individual variation in nectar production and its effects on fitness in *Kalmia latifolia. Ecology* **72,** 149–155.

Ryle, M. (1954). The influence of nitrogen, phosphate, and potash on the secretion of nectar. *J. Agric. Sci.* **44,** 400–407.

Schaal, B. A. (1978). Density-dependent foraging on *Liatris pynostachya. Evolution* **32,** 452–454.

Schemske, D. W., and Pautler, L. P. (1984). The effects of pollen composition on fitness components in a neotropical herb. *Oecologia* **62,** 31–36.

Schemske, D. W., and Horvitz, C. C. (1989). Temporal variation in selection on a floral character. *Evolution* **43,** 461–465.

Seeley, T. D. (1985). "Honeybee Ecology." Princeton University Press, Princeton, New Jersey.

Shuel, R. W. (1955). Nectar secretion in relation to nitrogen supply, nutritional status, and growth of the plant. *Can. J. Agr. Sci.* **35,** 124–138.

Shuel, R. W. (1957). Some aspects of the relation between nectar secretion and nitrogen, phosphorus, and potassium nutrition. *Can. J. Plant Sci.* **37,** 220–236.

Shuel, R. W., and Shivas, J. A. (1953). The influence of soil physical condition during the flowering period on nectar production in snapdragon. *Plant Physiol.* **28**, 645–651.

Snow, A. A., and Whigham, D. F. (1989). Costs of flower and fruit production in *Tipularia discolor* (Orchidaceae). *Ecology* **70**, 1286–1293.

Southwick, E. E. (1983). Nectar biology and nectar feeders of common milkweed *Asclepias syriaca* L. *Bull. Torrey Bot. Club* **110**, 324–334.

Southwick, A. K., and Southwick, E. E. (1983). Aging effect on nectar production in two clones of *Asclepias syriaca. Oecologia* **56**, 121–125.

Southwick, E. E. (1982). Lucky hit nectar rewards and energetics of plant and pollinators. *Comp. Physiol. Ecol.* **7**, 41–44.

Southwick, E. E. (1984). Photosynthate allocation to floral nectar: A neglected energy investment. *Ecology* **65**, 1775–1779.

Southwick, E. E., Loper, G. M., and Sadwick, S. E. (1981). Nectar production, composition, energetics, and pollinator attractiveness in spring flowers of western New York. *Am. J. Bot.* **68**, 994–1102.

Stanton, M. L., Snow, A. A., and Handel, S. N. (1986). Floral evolution: Attractiveness to pollinators influence male fitness. *Science* **232**, 1625–1627.

Teuber, L. R., and Barnes, D. K. (1979). Environmental and genetic influences on alfalfa nectar. *Crop Sci.* **19**, 874–878.

Teuber, L. R., Barnes, D. K., and Rincker, C. M. (1983). Effectiveness of selection for nectar volume, receptacle diameter, and seed yield characteristics in alfalfa. *Crop Sci.* **23**, 283–289.

Thomson, J. D. (1986). Pollen transport and deposition by bumble bees in *Erythronium*: Influence of floral nectar and bee grooming. *J. Ecol.* **74**, 329–342.

Thomson, J. D., and Plowright, R. C. (1980). Pollen carryover, nectar rewards, and pollinator behavior with special reference to *Diervilla lonicera. Oecologia* **46**, 68–74.

Thomson, J. D., and Thomson, B. A. (1989). Dispersal of *Erythronium grandiflorum* pollen by bumblebees: Implications for gene flow and reproductive success. *Evolution* **43**, 657–661.

Thomson, J. D., Maddison, W. P., and Plowright, R. C. (1982). Behavior of bumble bee pollinators of *Aralia hispida* Vent. (Araliaceae). *Oecologia* **54**, 326–336.

Thomson, J. D., McKenna, M. A., and Cruzan, M. B. (1989). Temporal patterns of nectar and pollen production in *Aralia hispida:* Implications for reproductive succcess. *Ecology* **70**, 1061–1068.

Travis, J. (1989). The role of optimizing selection in natural populations. *Annu. Rev. Ecol. Syst.* **20**, 279–296.

Vander Kloet, S. P., and Tosh, D. (1984). Effects of pollen donors on seed production, seed weight, germination, and seedling vigor in *Vaccinium corymbosum* L. *Am. Midl. Nat.* **112**, 392–396.

Waddington, K. D. (1981). Factors influencing pollen flow in bumblebee-pollinated *Delphinium virescens. Oikos* **37**, 153–159.

Waddington, K. D. (1983). Foraging behavior of pollinators. *In* "Pollination Biology" (L. Real, ed.), pp. 213–239. Academic Press, New York.

Waddington, K. D., and Heinrich, B. (1979). The foraging movements of bumblebees on vertical inflorescences: An experimental analysis. *J. Comp. Physiol.* **134**, 113–117.

Waddington, K. D., Allen, T., and Heinrich, B. (1981). Floral preferences of bumblebees (*Bombus edwardsii*) in relation to intermittent versus continuous rewards. *Anim. Behav.* **29**, 779–284.

Walker, A. K., Barnes, D. K., and Furgata, B. (1974). Genetic and environmental effects on quantity and quality of alfalfa nectar. *Crop Sci.* **14**, 235–238.

Waller, D. M. (1989). Plant morphology and reproduction. *In* "Plant Reproductive Ecology: Patterns and Strategies" (J. Lovett Doust and L. Lovett Doust, eds.) pp. 203–227. Oxford University Press, New York.

Wardlaw, I. (1968). The control and pattern of movement of carboyhydrates in plants. *Bot. Rev.* **34**, 79–105.

Waser, N. M. (1982). A comparison of distances flown by different visitors to flowers of the same species. *Oecologia* **55,** 251–257.

Waser, N. M. (1983). The adaptive nature of floral traits: Ideas and evidence. *In* "Pollination Biology" (L. Real, ed.), pp. 242–286. Academic Press, New York.

Waser, N. M., and Mitchell, R. J. (1990). Nectar standing crops in *Delphinium nelsonii* flowers: Spatial autocorrelation among plants? *Ecology* **71,** 116–123.

Waser, N. M., and Price, M. V. (1983). Optimal and actual outcrossing in plants. *In* "Handbook of Experimental Pollination Biology" (C. E. Jones and R. J. Little, eds.), pp. 341–359. Van Nostrand-Reinhold, New York.

Watson, M. A. (1984). Developmental constraints: Effects on population growth and patterns of resource allocation in a clonal plant. *Am. Nat.* **123,** 411–426.

Whitham, T. G., and Slobodchikof, C. N. (1981). Evolution by individuals, plant herbivore interactions, and mosaics of genetic variability: The adaptive significance of somatic mutations in plants. *Oecologia* **49,** 287–292.

Willson, M. F., and Agren, J. (1989). Differential floral rewards and pollination by deceit in unisexual flowers. *Oikos* **55,** 23–29.

Wood, G. W., and Wood, F. A. (1963). Nectar production and its relation to fruitset in the lowbush blueberry. *Can. J. Bot.* **41,** 1675–1679.

Wyatt, R., and Shannon, T. R. (1986). Nectar production and pollination of *Asclepias exaltata*. *Syst. Bot.* **11,** 326–334.

Young, H. J., and Stanton, M. L. (1990). Influences of floral variation on pollen removal and seed production in wild radish. *Ecology* **71,** 536–542.

Zimmerman, M. (1981a). Optimal foraging, plant density, and the marginal value theorem. *Oecologia* **49,** 148–153.

Zimmerman, M. (1981b). Patchiness in the dispersion of nectar resources: Probable causes. *Oecologia* **49,** 154–157.

Zimmerman, M. (1983). Plant reproduction and optimal foraging: Experimental nectar manipulations in *Delphinium nelsonii*. *Oikos* **41,** 57–63.

Zimmerman, M. (1984). Reproduction in *Polemonium*: A five-year study of seed production and implications for competition for pollinator service. *Oikos* **42,** 225–228.

Zimmerman, M. (1987). Reproduction in *Polemonium*: Factors influencing outbreeding potential. *Oecologia* **72,** 624–632.

Zimmerman, M. (1988). Nectar production, flowering phenology, and strategies for pollination. *In* "Plant Reproductive Ecology: Patterns and Strategies" (J. Lovett Doust and L. Lovett Doust, eds.), pp. 157–178. Oxford University Press, New York.

Zimmerman, M., and Cook, C. W. (1985). Pollinator foraging, experimental nectar robbing, and plant fitness in *Impatiens capensis*. *Am. Midl. Nat.* **113,** 84–91.

Zimmerman, M., and Pyke, G. H. (1986). Reproduction in *Polemonium*: Patterns and implications of floral nectar production and standing crops. *Am. J. Bot.* **73,** 1405–1415.

Zimmerman, M., and Pyke, G. H. (1988a). Reproduction in *Polemonium*: Assessing the factors limiting seed set. *Am. Nat.* **131,** 723–738.

Zimmerman, M., and Pyke, G. H. (1988b). Experimental manipulations of *Polemonium foliosissimum*: Effects on subsequent nectar production, seed production, and growth. *J. Ecol.* **76,** 777–789.

6

Plant Resources as the Mechanistic Basis for Insect Herbivore Population Dynamics

Peter W. Price
Department of Biological Sciences
Northern Arizona University
Flagstaff, Arizona

I. Bottom-Up and Top-Down Effects

Assessing the relative importance of food, competition, predation, parasitism, and abiotic factors in the population dynamics of a species, its habitat utilization, and its role in the community, provides a central theme in ecology. Whether effects on dynamics work principally from below through food quality and quantity or through natural enemies has been long debated (e.g., Hairston *et al.,* 1960; Murdoch, 1966; Ehrlich and Birch, 1967). In limnological systems this bottom-up versus top-down view of regulation in trophic webs has received focused attention, and generalizations are available. Lake productivity is driven from below by nutrient input, turnover time of the water, and vertical mixing, and these factors account for about half of the variation measured in lake productivity (Carpenter and Kitchell, 1987, 1988; Schindler, 1978; Schindler *et al.,* 1978). The other half of the variation is probably accounted for by top-down effects, or trophic cascades of influence, with predators acting as important regulators of production in the trophic levels below (Carpenter *et al.,* 1985; Carpenter and Kitchell, 1988). "Trophic cascades and physicochemical factors act at different time scales to determine the productivity of lakes. Nutrient loading and water retention time set the long-term potential productivity of a lake, while interannual variability around that potential derives from species interactions and food web effects on nutrient cycling" (Carpenter and Kitchell, 1988, p. 764).

In this chapter I concentrate on insect herbivores, excluding large mammalian herbivores, and on north temperate landscapes on which most work has been conducted. Many of the generalizations addressed, I believe, apply to other landscapes at other latitudes, but development of supporting arguments will be presented elsewhere.

In terrestrial systems the bottom-up versus top-down debate has yet to reach a balanced view on relative importance, although in the plant-herbivore literature, calls for comprehensive study of both plant effects and enemy effects should be a focus of research energy in the future (Barbosa, 1988; Bernays and Graham, 1988; Courtney, 1988; Fox, 1988; Janzen, 1988; Jermy, 1988; Rausher, 1988; Schultz, 1988; Thompson, 1988). Emphasis on one factor as the major regulator of host plant specificity, or population dynamics, is generally scorned as oversimplification, but general principles on when, why, and how bottom-up and top-down factors play a role in plant–herbivore interactions remain elusive. Even hypotheses that predict pattern in bottom-up and top-down effects are poorly developed, and the temporal and spatial scales of influences need more attention.

Indeed, we have much to learn from the limnologists in terms of identifying major abiotic and biotic forces acting on species and communities, the temporal scales on which they act, and the community-wide experimental approaches they have used to understand pattern. In spite of the many major differences between lake systems, built up from small phytoplankton to large generalized predators, and terrestrial systems such as those based on large plants and small insect herbivores, there are compelling reasons for extrapolating the view expressed by Carpenter and Kitchell, cited above, to plant and insect herbivore relationships. First, the primary producers in these systems depend on abiotic factors such as sunlight and nutrients, and they form the basis of the food web. Therefore, quantity and quality of primary production is likely to affect all upper trophic levels. Secondly, 100% of primary consumers depend on primary producers for food, making the ecological and evolutionary ties between food and feeder inevitably very strong, and perhaps of overriding importance. Third, the second trophic level does not depend on natural enemies but, given adequate primary production, plant feeders can reach population sizes exploitable by predators and other enemies, which may then have a major impact on herbivore populations. The net result of these relationships is that the primary producers must determine the long-term carrying capacity for herbivore populations, and deviations in population size below this capacity may well be influenced profoundly on a shorter time scale by natural enemies of herbivores, as well as by abiotic factors.

Following this logic, evidence is mustered in this chapter to support the view that plants as resources, coupled with insect behavior, determine the carrying capacity for insect herbivore populations. This evidence applies particularly where disturbance and plant-succession phenomena have been strong forces on the evolution of three-trophic-level systems. Abiotic forces such as water and nutrients for plants influence profoundly this carrying capacity. Superimposed on these relationships are the effects of natural enemies, which can be weak to strong, depending partly on pattern and partly on idiosyncratic phenomena.

II. Plants Set the Carrying Capacity for Insect Herbivore Populations

A. A Landscape Perspective

A major pattern providing a template in which insect herbivores and their enemies must forage for food is a patchy landscape composed of vegetation in different stages of ecological succession. In primeval north temperate landscapes the scale of the patches was very different from the present, with small patches of early succession developing after local disturbance, and large tracts of more or less mature forest (e.g., Loucks, 1970; Pickett and White, 1985; Shugart, 1987). Temporal scales also differed because early successional sites would change rapidly from weedy colonizers, to perennial herbs, shrubs, colonizing trees, and ultimately the long-lived dominant trees of a mature vegetation. Even within woody plant species, spatial and temporal scales would change because regeneration and rapid growth would occur in relatively small patches after disturbance, and ultimately these patches would revert to the slower growth of the mature forest trees (e.g., Shugart and West, 1981; Franklin, *et al.*, 1987).

These vegetational patterns provide the basis for major differences in carrying capacity for insect herbivores exploiting different plant species, and the same plant species in different stages of succession (Fig. 1). As used in this chapter, the carrying capacity, K, is conceived of as the population size per insect species that can be supported by the plant biomass available to this species in any stage of succession. It is most usefully conceived of as the population of reproductive adults supported, such that influences of plant biomass on both natality and survivorship in the population dictate K. This carrying capacity, say for leaf chewers, will be relatively low in early succession because plants are small, and patches are relatively small. The relatively low biomass of foliage sets a low limit on population size for insect herbivores from generation to generation. If such patches have a high diversity of weedy species and herbivore species are specialized to one or a few plant species (e.g., Feeny, 1976; Rhoades, 1979), the carrying capacity will be set at an even lower level defined by available plant biomass. As woody plants enter succession, edible biomass increases, both in biomass per unit area, and in the area occupied as successional stages converge toward large trees over large areas. For leaf-feeding insect herbivores capable of maturing by feeding generally within and between tree species, the leaf

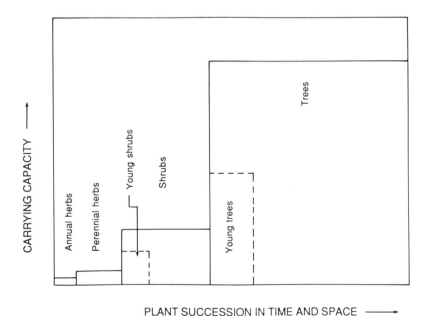

Figure 1   A schematic view of change in carrying capacity for insect herbivores provided by plants in different stages of ecological succession. The succession scale relates to time and space because small patches of weedy species are ephemeral, and large stands of trees cover extensive areas.

biomass in an extensive mature forest is, relatively speaking, very high per unit area, and per tract of this vegetation over the landscape. For any generation of such herbivores, K is very high whether or not the population is close to or distant from K. However, many insect herbivores specialize on very vigorous plant growth seen in young shrubs and trees (Price *et al.*, 1990), and their distribution and abundance is likely to be limited to patchy sites following weedy stages of succession in which woody species are colonizing. Here carrying capacity for herbivores will be low relative to mature forest because plants are smaller, patches are relatively small, and for specialists on vigorous plant growth, rapidly growing modules are likely to be in short supply relative to equivalent modules in general (Price *et al.*, 1990). Hence, young shrubs and trees are designated in Figure 1 as providing lower carrying capacity for a shorter time than mature vegetation of a similar kind. All these ecological constraints of carrying capacity through plant succession imposed from generation to generation on insect herbivores must also impose longer-term evolutionary strictures on the evolution of herbivore life history traits, as discussed later in this chapter.

The evolutionary responses of insect herbivores to plant succession should reflect the same degree of variability that succession imposes on their environment. Herbivores utilizing weedy annuals must colonize small patches rapidly, breed, and disperse to new patches. The carrying capacity is low and ephemeral. In some ways weediness of the host plant requires that its herbivores adopt a similarly *weedy strategy*. At the other end of succession, the scales of time, space, and plant size are orders of a magnitude larger. Mature forest tends to be stable over decades and centuries, large tracts of land support similar forest trees, and plant biomass and food potential are very high. This stable community provides a high carrying capacity, and insect herbivores are likely to evolve a more sedentary lifestyle, characteristic of their host plant population (Barbosa *et al.*, 1989; Roff, 1990), and dramatically different from the weedy strategy.

Within woody host plant species, vegetational dynamics also generate very different arenas for foraging insect herbivores. Patches of woody regeneration after disturbance are relatively small and progress rapidly toward maturity within decades. Insects adapted to exploit rapidly growing plants with relatively large modules, such as new shoots and leaves, must be effective colonizers of these patches and will be constrained by a relatively low and confined carrying capacity (Fig. 1). Insects adapted to exploit mature forest trees need not be strong colonizers, for their habitat and food plants will remain over large areas for centuries.

Superimposed on this very general view of differences in carrying capacity for insect herbivore populations is the quantitative and qualitative variation of specific resources provided by a plant population or community. Simple quantitative variation in resource supply from year to year in host plant populations can be decisive in defining herbivore population size. This tracking close to carrying capacity is well documented for the red pine cone beetle, which is dependent largely on cone production (Mattson, 1980), and the thistle-feeding lady beetle, which utilizes thistle shoots (Ohgushi and Sawada, 1985) (Table 1). For shoot borers and gallers, plant module size is frequently limiting, such that modules of large size or of high quality for other reasons actually set the carrying capacity qualitatively, and well below the quantitative availability of modules (Table 1, Examples 3–7). Other plant age effects restrict utilization by the poplar leaf beetle to the younger age classes (Example 8), while *Dendroctonus* bark beetles are constrained by the availability of mature and stressed pines (Example 9). Among acridid

Table 1   Plant Resource Quantity and/or Quality in Regulation of Distribution and Abundance of Insect Herbivores.[a]

Herbivore species or group	Plant factor	Source
Conophthorus resinosae Red pine cone beetle	Cone availability	Mattson (1980)
Henosepilachna niponica Thistle-feeding lady beetle	Number of thistle shoots per unit area	Ohgushi and Sawada (1985)
Euura lasiolepis Arroyo willow shoot galler	Age and vigor	Craig *et al.* (1986, 1989) Preszler and Price (1988), Price and Clancy (1986a), Price (1988)
Euura mucronata Sallow willow bud galler	Age and vigor	Price *et al.* (1987a,b)
Euura exiguae Coyote willow shoot galler	Age and vigor	Price (1989)
Pemphigus betae Poplar leaf galler	Age and genotype	Whitham (1978, 1989) Kearsley and Whitham (1989)
Dioryctria albovitella Pinyon pine cone & shoot moth	Genotype and water stress	Whitham and Mopper (1985)
Chrysomela confluens Poplar leaf beetle	Age	Kearsley and Whitham (1989)
Dendroctonus bark beetles	Age and stress	Berryman (1982)
Acridid grasshoppers	Food quantity and quality	Dempster (1963), White (1976), Rainey (1982)

[a] Species are listed in the order discussed in the text. English names denote the kind of insect involved but are not necessarily formally accepted common names.

grasshoppers in dry localities, both quantity and quality of forage define population size (Example 10).

All these examples and the sources cited in Table 1 show that plants as resources can have an overriding influence on population sizes and distributions of insect herbivores (see also Coley, 1983; Coley *et al.*, 1985). Top-down effects from natural enemies are weak. This point is documented in Examples 1–5, 9, 10 (Table 1), and in examples 6–8, personal discussions have reinforced the implicit exclusion of carnivores as driving forces in the systems. "Red pine cone beetles . . . suffer negligible mortality from parasites or predators" (Mattson, 1980, p. 390). "Parasites and predators apparently have little if any significant effect" (White, 1976, p. 119). Berryman (1982) stressed the essential role of the plant–herbivore interaction in bark beetle population dynamics, with natural enemies not changing "the general qualitative properties of the system" (p. 313).

B. Carrying Capacity and Latent Population Dynamics

Plants may play such a central role in setting the carrying capacity for insect herbivores that populations are kept in a stable state over several generations. Carrying capacity may be viewed generally as the mean population supportable by plant biomass across a plant population, but it is also instructive to examine available biomass per plant as it varies within a plant population. Certain plants in a population are favorable to a herbivore species and others are not, and these differential carrying capacities influencing a population may be relatively stable over a decade or more. Stable but limiting plant-imposed carrying capacity may result in latent population dynamics characteristic of many specialized endophytic species that utilize rapidly developing plant parts in spring and early summer (Table 2). A common feature among endophytes such as gallers and shoot borers is that the ovipositing female selects the site at which the larva will feed. Larval performance will then have strong selective feedback on female preference of oviposition site toward optimal module utilization. An optimal module may be a large leaf or a rapidly growing shoot, but females are likely to evolve a capacity to detect optimal sites for larval survival (e.g., Craig *et al.*, 1989). Another related trait among these kinds of herbivores is the relatively short length of the egg stage, which is less than 30 days in the examples provided. This would be expected for herbivores laying into and utilizing rapidly developing plant parts (Table 2). For gallers, attack usually precedes differentiation of a meristem, so gall differentiation can proceed independently of normal module development. For shoot and cone borers, attack may precede strong lignification, providing more edible tissue for larvae.

Population regulation close to and at carrying capacity probably results from rapid negative feedback through female competition for oviposition

Table 2 Insect Herbivore Species with Latent Population Dynamics[a]

Herbivore species	Female site and time of oviposition	Time from egg-laying to larval feeding	Feeding site of first instar larva and time of feeding	Limiting resource	Source
Euura lasiolepis and *E. exiguae* Willow shoot gallers	Through very young leaf petioles into stem in May to June	20 days	Within gall at oviposition site	Long, vigorous shoots	Price and Craig (1984) Craig *et al.* 1986, 1989; Price 1989
Euura mucronata Willow bud galler	Into very young axillary buds in June	20 days	Within gall at oviposition site	Large buds on long shoots	Price *et al.* 1987a,b
Diplolepis fusiformans and *D. spinosa* Rose shoot gallers	Into rapidly growing shoots	?	Within gall at oviposition site	Long, vigorous shoots	Caouette & Price 1989
Pemphigus betae Poplar leaf galler	Within gall early in leaf development	0 days (viviparous)	Within gall at oviposition site	Large leaves	Whitham 1978, 1980, 1989
Daktulosphaira vitifoliae Grape phylloxera	Within gall on young leaves	20–28 days	Within gall at oviposition site	Leaves on rapidly growing shoots	Kimberling *et al.* 1990
Dioryctria albovitella Pinyon pine cone and shoot moth	Into young shoots	7–28 days	Within shoot from oviposition site	Large, vigorous shoots	Whitham & Mopper 1985; Furniss & Carolin 1977
Conophthorus resinosae Red pine cone beetle	Into young cones in June and July	?	Within cone from oviposition site	Cones	Mattson 1980 Baker 1972
Eucosma gloriola White-pine shoot borer	On needle sheaths on new shoots in May and June	?	Within new shoot	Large shoots, particularly terminals	Baker 1972
Pissodes strobi White-pine weevil	In bark of leading shoots in April & May	7–10 days	Within leading shoot	Large, vigorous leading shoots	Baker 1972
Rhynchites betulae Birch leafrolling weevil	In large young rolled leaves on long shoots	?	In leaf rolls formed by female	Large leaves	Personal observ. Price *et al.* 1990

[a] The causes of strong limitation of larval resources, and the coupling between female oviposition site and larval feeding site. The first five examples are galling species and the last five are other forms of endophytes.

sites (Price *et al.*, 1990). This is well understood in the willow–galling sawfly system described in Section IV, A.

Among the examples listed in Table 2 are several species considered to be forest pests (Examples 6–9). In the primeval forests in which they evolved and adapted, they were probably generally rare, restricted to small patches of young vigorous plants regenerated after disturbance. They may have persisted in such sites for about a decade, at relatively stable densities, until tree vigor declined into a more mature growth form (cf. Price *et al.*, 1990). The pest status of such herbivores probably increased as human harvesting and management generated more even-aged and extensive stands of young host plants. Of the so-called pests, it is notable that they do not kill trees by general defoliation, but attack leading shoots, thus reducing shoot dominance, or kill seeds, commodities essential to effective forestry practice. Relatively low populations can become pests when prime resources such as leading shoots and cones are targets of herbivore attack.

Not surprisingly, abiotic environmental variation changes the plant–herbivore relationship in many cases, acting to change the carrying capacity for insect populations (Table 3). Abiotic factors commonly involve rainfall, or soil drainage, temperature, and soil nutrient status. Such factors can be of central importance to the insect–host plant interaction, as is water availability to willows and, as a consequence, the dynamics of the shoot galler, *Euura lasiolepis* (Price and Clancy, 1986b; Preszler and Price, 1988; Price *et al.*, 1990). Qualitative variation in the same individual plant can result in 50% survival from egg to cocooned larva in one year and 0.5% survival in an adjacent year; or a 100-fold difference in survival resulting from differences in winter precipitation (Price and Clancy, 1986b). In other cases the abiotic effect via plant quality may be less important, as in the case of temperature effects on the fall webworm. Temperature has strong direct effects on development of larvae and pupae, and weaker effects on changing the relative phenology of host plant and herbivore such that larvae feed on older, poorer-quality leaves in a cool season compared to a warmer season (Morris, 1967, 1969).

Such abiotic environmental variation affects species with latent population dynamics and those with eruptive dynamics. In fact, the interface between the host plant and its variable abiotic environment can represent an evolutionary force acting on insect dispersal strategies and dietary breadth (Hunter, 1990). In Table 3, Examples 1–7 are known to be eruptive species, and 8–10 are more latent. In all cases, except for the fall webworm, effects of abiotic factors on plant quality play a central role in the dynamics of the herbivore, defining where populations will be high and low (Examples 1, 2, 4–10 in Table 3), and when they will be high and low (Examples 1, 2, 4, 5, 7, 8, 10 in Table 3). For example, *Eucalyptus* foliage feeders reach high densities on trees under which sheep rest and seek shelter, and leave copious dung, which fertilizes the soil and the trees. This results in higher nutrient

Table 3 Abiotic Factors Influence Plant Quality for Herbivores, Influencing Distribution and Abundance

Herbivore species	Abiotic factor	Effect on plant	Effect on herbivore	Source
Schistocerca gregaria Desert locust	High rainfall	Increased biomass	More food of better quality	Rainey (1982)
Dendroctonus Bark beetles	Low rainfall	Stress	Reduced plant resistance	Berryman (1982)
Hyphantria cunea Fall webworm	Cool temperatures	Delayed growth in spring	Reduce overlap with young leaves	Morris (1967, 1969)
Selidosema suavis Looper caterpillar on pine	Hardpan and poor drainage	Stress	Higher early larval survival	White (1974)
Choristoneura fumiferana Spruce budworm	Warm summers	Increased flowering	Increased survival	Greenbank (1963)
Neodiprion swainei Swaine jack-pine sawfly	Poor, sandy soil, dense stands	Stress	Unknown	McLeod (1970)
Neodiprion sertifer European pine sawfly	Drought on infertile soils	Stress	Unknown	Larsson and Tenow (1984)
Eucalyptus Foliage feeders	High nitrogen in soil	High nitrogen in leaves	Better nutrition	Landsberg and Wylie 1988, Landsberg and Ohmart (1989)
Dioryctria albovitella Pinyon-pine cone and shoot moth	Dry cinder soils	Stress	Reduced plant resistance	Whitham and Mopper (1985)
Euura Shoot and bud gallers	High water supply to roots	Increased plant vigor	Higher quality modules and increased survival	Price *et al.* (1990)

content of leaves for herbivores, and nitrogen passes from below to above threshold requirements for chrysomelids and other herbivores. Outside sheep paddocks, trees without supplemented nutrients set a very low carrying capacity for insect herbivores, while adjacent sites with sheep set such a high carrying capacity that trees may be killed by insect attack (Landsberg and Wylie, 1988; Landsberg and Ohmart, 1989; Beckmann, 1989).

Thus, carrying capacity for insect populations set by plants is the result of a complex interplay of the detailed requirements of individual herbivores, and abiotic variables affecting the availability of these requirements. The world is green principally because most green foliage cannot be effectively utilized by herbivores to achieve high performance and survival. We have not needed to invoke the effects of natural enemies up to this point in the development of the thesis. In most places, for most of the time, plants provide low carrying capacities for insect herbivores, and constrain populations from reaching high and very damaging levels that weaken and kill plants.

C. Carrying Capacity in Eruptive Species

A minority of species of insect herbivore are known to defoliate and kill trees. It is clear that some species can escape the strictures of plant-imposed regulation and reach epidemic numbers limited only by a very high carrying capacity composed of practically all foliage in a forest. What are the mechanistic processes involved with such epidemic population dynamics? Many hypotheses have been proposed (see Price, 1984; Barbosa and Schultz, 1987; Watt *et al.*, 1990), but my thesis argues that the plant–herbivore interaction is still central (Price *et al.*, 1990).

A common theme in life histories of eruptive species among insect herbivores is that tight linkage between female oviposition site and larval feeding site is lost (Table 4, cf. Table 3). The capacity of females to select with precision the site of larval feeding is disrupted, meaning that optimal sites for high larval performance cannot be assessed by females. In fact, females may become very unspecific in oviposition, laying eggs on bark, or in soil, well away from larval feeding sites (Examples 2–7, 9, 10). In other cases females lay eggs on one type of foliage and larvae feed on another age class, as in spruce budworm and Swaine jack-pine sawfly. Part of the syndrome of these eruptive species is the exophytic leaf-chewing habit of larvae, the frequent, but not inevitable, long duration of the egg stage, capacity to feed on several host species, and a greater-than-expected frequency of impaired flight capacity (Table 4, see also Barbosa, *et al.* 1989). Well-known eruptive species are principally forest tree feeders. One case in Table 4 is a little-studied eruptive species, *Disonycha pluriligata*, on a shrubby willow.

If females are unable to evaluate plant quality relevant to larval feeding sites, and first-instar larvae are extremely inefficient foragers in terms of selecting among plants and among plant parts, then the most viable strategy

Table 4 Eruptive Insect Herbivore Species and the Causes of Loss of Tight Linkage between Female Oviposition Site and Larval Feeding Site[a]

Herbivore species	Female site and time of oviposition, and flight	Time from egg-laying to larval feeding	Feeding site of first instar larva and time of feeding	Source
Choristoneura fumiferana Spruce budworm	On maturing leaves in June and July. Lay 50% of eggs before flying.	10 months	Best sites in buds and staminate cones in April and May of following year, often after dispersing passively on silken thread.	Morris (1963) Mattson *et al.* (1988)
Operophtera brumata Winter moth	On bark and lichens in November and December. Female flightless.	4 months	In buds in early April	Varley *et al.* (1973)
Porthetria dispar Gypsy moth	On tree trunks and limbs, stumps and stones, in July and August. Female flightless.	9 months	On young leaves in May often after dispersing passively on silken thread.	Baker (1972)
Malacosoma disstria Forest tent caterpillar	On bark of twigs in summer. Flight normal.	9 months	Expanding buds in spring.	Baker (1972)
Alsophilia pometaria Fall cankerworm	On twigs and branches in November and December. Female flightless.	5 months	On young leaves in late April and May.	Baker (1972)
Lambdina punctata Western oak inch worm	In leaf litter, base of tree, on associated plants and under bark scales from August to October.	8 months	On developing buds and leaves in May and June.	Furniss & Carolin (1977)

Paleacrita vernata Spring cankerworm	In bark crevices and under scales on trunk and branches in March and April. Female wingless.	1–2 months	On young leaves in early May.	Baker (1972)
Neodiprion swainei Swaine jack-pine sawfly	Into young needles in June and July. Female proovigenic and a poor flier.	28 days	On mature needles grown in previous years 1 to 3.	Tripp (1965); Ghent & Wallace (1958) Personal observ.
Disonycha pluriligata Willow flea beetle	In soil at base of host plant in June. Flight normal.	10 days	On young leaves in June and July.	Dodge and Price (1991a,b).
Didymuria violescens. *Eucalyptus* stick insect	One forest floor, dropped from canopy. From December to March (summer). Female flightless.	6–18 months	Youngest leaves on terminal shoots in spring (September to October).	Readshaw (1965)

[a] All examples are North Temperate except last example, from Australia.

is probably to evolve with a generalized capacity, enabling survival on almost any foliage quality. Then larvae can consume all foliage in a forest canopy, numbers of larvae per unit area can become very high, and trees can be killed by repeated defoliation. The carrying capacity for such herbivores is defined only by the total biomass of foliage in the canopy. Negative feedback resulting in population regulation results from larval competition for food and ultimately mass starvation of larvae. The world really is green for these unusual herbivores. As an example, the spruce budworm can feed and mature on current foliage or foliage developed in previous years (Blais, 1953; Miller, 1957), although fecundity reduction occurs on older foliage. Nevertheless, all leaves on balsam fir contribute to the carrying capacity for budworm, and trees can be killed. Apparently, the high carrying capacity imposed by plants on eruptive insect herbivore species is infrequently ex- ploited, or our forests would be much less green than they normally are. The understanding of why this is remains a major challenge.

How this syndrome associated with eruptive species evolved is unre- solved. Broadly adapted larvae feeding on abundant leaf resources in for- ests could select for unspecific oviposition, reduced selection on dispersal ability, and allow for long egg-stage duration through many of the winter months. Conversely, basic life history traits may cause dissociation of female oviposition and larval feeding, resulting in strong selection on larvae for general utilization of leaf resources. For example, within the genus *Chori- stoneura,* all the species known to me have the basic life-history pattern of oviposition onto maturing foliage. A nonfeeding, first-instar larva spins a hibernaculum, molts to the second instar, and overwinters. Only in the spring does feeding commence on the youngest of foliage in buds and male cones, and this may follow two episodes of passive dispersal on silken threads after hatching and after leaving the hibernaculum (Mattson *et al.,* 1988; Baker, 1972; Furniss and Carolin, 1977). A characteristic common to a genus suggests an ancient pattern that sets a phylogenetic constraint on the ecology of all members (Price *et al.,* 1990). A case can be made that such an unusual life history among lepidopterans was forced by females emerg- ing at the time of maturing foliage when it was maladaptive for small larvae to establish and feed on tough needles. It would be hard to envisage feed- back on female phenology working from larval feeding capabilities in the spring. We do know that the host-utilization patterns of larvae are fre- quently less constrained than ovipositing female preferences (e.g., Kogan, 1977), so female behavior is more constrained against evolutionary flexibil- ity than larval abilities.

If this argument is admissible, then the *phylogenetic constraint* of female oviposition on maturing foliage forces an *adaptive syndrome* of characters in larvae including general utilization of foliage and dissociation of oviposition site and larval feeding site. One of the *emergent properties* of these evolved characters is a very high carrying capacity for populations defined by all the foliage in a mature balsam fir forest (Price *et al.*, 1990). When the "grip" of a tight link between oviposition site and feeding site is lost, it appears that the system becomes permissive of eruptive population dynamics. The potential exists for a population to fully exploit a carrying capacity set only by the amount of foliage on a tree and in a forest, resulting in tree defoliation and perhaps ultimate death.

Other factors may override this potential much of the time or all of the time. The point is that the pure interaction between the plant host, oviposition behavior of females, and larval feeding capacity, in cases such as those listed in Table 4, produces a very permissive system relative to population increase. Sooner or later, environmental conditions combine to permit a population to exploit fully the high carrying capacity, and an eruption or epidemic will occur. The proximate factors allowing population eruptions and declines are no doubt complex and frequently debated (e.g., Price, 1984; Myers, 1988; Barbosa and Schultz, 1987), but they do not detract from the fundamental logic of ultimate regulation of carrying capacity being set by the interplay of plant resources, female insect oviposition behavior, and the capabilities of larvae to exploit plant food.

This perspective leaves a major question unresolved. Why are most insect herbivores noneruptive, especially those that are external leaf feeders in north temperate forests? I suspect that in many of these cases, factors beyond the plant–herbivore interaction are involved, including abiotic forces such as weather and its various components, and the role of natural enemies. Thus, populations are kept well below the carrying capacity defined by the vegetation. I also suspect that in many other cases the plants as resources are more constraining than we realize and more detailed studies are warranted to unravel the details of this interaction.

III. Bottom-Up Effects on Natural Enemies

A. A Landscape Perspective

Vegetation type exerts continuing influence up the trophic system to the natural enemies of insect herbivores. As vegetational succession proceeds from herbs to shrubs to trees, so the number of parasitoids per host insect increases (Askew, 1980; Hawkins and Lawton, 1987; Hawkins, 1988). For example, on average there is a doubling of species richness of parasitoids on lepidopteran hosts from herbaceous plants with 4 to 5 species per host to over 10 per host on trees (Hawkins, 1988). The mechanisms resulting in this pattern have not been resolved and are probably multiple, but one factor commonly considered to be important is the increasing stability and continuity of habitats as succession proceeds (Askew and Shaw, 1986; Hawkins, 1988).

A contributing factor is no doubt the increasing prevalence of generalist parasitoids in later stages of succession (Hawkins *et al.*, 1990). As plants in succession change defensive strategies from diverse toxins in early succession to more general defenses like tannins, so the herbivores can become more general in feeding across plant species (Feeny, 1976; Rhoades, 1979). Using Lepidoptera in North America and the British Isles, Futuyma (1976) supported this generalization with empirical data. It seems that the parasitoids can also become more general in late succession because of the richer herbivore fauna and the smaller array of toxins sequestered by hosts.

B. Patterns of Impact by Natural Enemies

These trends imposed by vegetation on the third trophic level also translate into the dynamics of interaction between parasitoids and their host insects. Species of insect herbivore that support richer parasitoid communities are also affected more, with greater mortality incurred (Price and Pschorn-Walcher, 1988; Gross and Price, 1988). This pattern is reinforced by successes in applied biological control of insect herbivores when the probability of success increases as the number of parasitoid species per host increases, even when similar numbers of parasitoid species are introduced per host (Hawkins and Gross, 1991). The result of this differential success in relation to vegetation type was noted by Varley (1959) many years ago when, by his estimates, 33% of biocontrol attempts had been successful in stable habitats like orchards and forests, but only 5% of attempts were successful in disturbed vegetation such as field and garden crops, more similar to early vegetational succession.

These broad patterns of influence up the trophic system, and down again, all suggest that during early stages of plant succession, parasitoids, and perhaps natural enemies in general, are likely to have relatively weak effects on the population dynamics of insect herbivores, and these effects will increase as the vegetation matures and long-term stable habitats over large areas are established.

C. Correlation or Causation?

It is possible that natural enemies simply respond to an increasing food resource as insect herbivore populations become larger and more widespread with vegetational succession, but have little or no real effect on dynamic patterns imposed largely by the plant–herbivore interaction. That plants commonly dictate the efficacy of natural enemies is well documented (Bergman and Tingey, 1979; Price *et al.*, 1980; Price and Clancy, 1986a; Price, 1988; Gross and Price, 1988). The alternative hypothesis is that natural enemies are frequently the most important factor in the population dynamics of insect herbivores. "A preponderance of evidence supports the view that natural enemies are a principal force in keeping populations of forest Lepidoptera at low densities" wrote Mason (1987, p. 50).

There is no doubt that natural enemies can exert control so strong that potential population sizes of herbivores imposed by plant carrying capacity are never or rarely reached. This is a perspective established early in the field of insect herbivore population dynamics and regulation (e.g., Howard, 1897; Lotka, 1924; De Bach, 1964), and authoritative endorsement persists (Hassell, 1978; Lawton and McNeill, 1979; May, 1981; Anderson, 1981; Hassell and May, 1989). In the future it may be possible to partition arenas of interacting environments, plants, and herbivores into permissive and nonpermissive systems for effective regulation by natural enemies.

Some examples of strong impact by natural enemies are available from natural systems in which experiments have been conducted, and from applied biological control of insect herbivore pests, which emulate the strong experimental approach of studying dynamics with and without natural enemies (Table 5). However, in the cases of biological control, caution must be exercised in interpreting results because the experiments do not control for changes in abiotic environment, plant species and condition, and plant dispersion, given that pests needing control normally appear in managed systems. However, once pest status is established, for whatever reason, then the efficacy of natural enemies is clear enough in many cases.

I find the cases of the mountain ash sawfly and the cassava mealybug particularly convincing examples of the overriding importance of natural enemies in their regulation in native localities. In natural settings the species are, respectively, rare enough to present difficulties in their study, or so rare as to go totally unnoticed! When accidentally imported to new continents, both species became epidemic and damaging to their host plants, and both were very effectively regulated after the introduction of a small number of natural enemies from the native locality.

But the case of the mountain ash sawfly raises a perplexing question of why some species of insect herbivores are so effectively regulated by natural enemies, and others are not. This sawfly is a tenthredinid typical of this large family, in which eruptive population dynamics in native habitats is uncommon. Very local population increase and defoliation may occur in tenthredinids (e.g., Price, 1970, Baker, 1972), but widespread damage showing complete utilization of the plant-imposed carrying capacity is extremely rare in natural settings. But the related diprionid sawflies, a relatively small family compared to tenthredinids, demonstrate a remarkable capacity for eruptive population dynamics, with the genera *Diprion* and *Neodiprion* especially prone to outbreaks and exploitation at high carrying capacities. Even within the family there are epidemic species and rare species (Hanski and Otronen, 1985; Hanski, 1987), and the essential differences in their epidemiology are elusive (Hanski and Otronen, 1985). So why is the strong regulation by natural enemies seen in *Pristiphora geniculata* commonly circumvented by the related diprionid sawflies?

On the other hand, the case of the cassava mealybug seems to conform to a broad phylogenetic pattern. The sedentary homopteran herbivores are particularly prone to strong biological control with many examples of excellent regulation (Sweetman, 1936; De Bach, 1964; Askew, 1971).

The search for pattern in three-trophic-level interactions involving plants, insect herbivores and natural enemies is clouded by important processes in evolutionary time. Naturally, under heavy attack by enemies, herbivore populations are likely to evolve with better escape mechanisms or defenses. Natural enemies may well evolve with counterploys, so the evolving relationship may at times favor the herbivore and at others it may favor the enemy. A balanced relationship may prevail in certain habitats and not in others. This shifting of players over an evolutionary landscape imposes constraints on the development of general patterns, and we may expect to see related herbivore species in different phases of their evolutionary relationship with natural enemies.

A very interesting example concerns *Pieris* butterflies studied by Ohsaki and Sato (1990). In temporary habits *P. rapae* escapes parasitoid attack by rapid colonization of new habitats and departure before discovery. But in more stable habitats, escape is not possible, and this species lacks adequate defenses with consequent heavy impact by the parasitoid *Apanteles glomeratus*. But two congeners live in stable habitats and escape parasitism in different ways. *Pieris melete* has the physiological ability to encapsulate eggs of *A. glomeratus* and is little affected. *Pieris napi* usually attacks the low rock cresses *Arabis gemmifera* and *A. flagellosa*, frequently hidden under taller plants, and the parasitoid does not find larvae on these hosts even though they would act as suitable hosts. Thus, within a single genus we see a diverse array of mechanisms reducing an enemy's efficacy, and even within a species, very different impact in temporary and stable habitats. The role of the plants in the herbivore–enemy interaction is very strong.

Another case illustrates how closely related insect herbivore species can have their parasitoid richness and impact radically altered by simple differences in host plant traits. Comparison of a leaf miner, *Tildenia inconspicuella*, on horsenettle, *Solanum carolinense,* and its congener, *T. georgei,* on groundcherry, *Physalis heterophylla,* demonstrated the importance of the type of plant trichome in the herbivore–enemy interaction (Gross and Price, 1988). Horsenettle has stellate trichomes that constrain reentry of larvae if they leave the mine. Consequently, larvae have evolved the behavior of constructing one large mine without an exit hole. As a result the larvae are vulnerable to parasitoids, which are numerous (nine species) and inflict heavy impact of about 40% mortality. The behavior appears to be derived and specifically adapted to the limitations imposed by *Solanum* trichomes. The more-common condition is illustrated by the *Physalis* leafminer on a host with soft and supple simple trichomes. A larva can leave a mine and start another, frequently shifting mine location and maintaining an exit for each mine. As a result, when attacked by parasitoids, the larvae commonly escape and drop on a silken thread. The number of parasitoids adapted to exploiting such a mobile leafminer is small (four species), and impact is low at about 14% mortality. Simple trichome differences of host plants affect evolved behaviors of herbivores and the composition and impact of the parasitoid community on each.

These examples illustrate well the idiosyncratic nature of evolutionary links between plants, herbivores, and enemies. But they also illustrate the importance of bottom-up effects that reach the third trophic level, both in

Table 5 Natural Enemies Have Overriding Impact on Abundance of Insect Herbivores[a]

Herbivore species	Natural enemy effects and kind of evidence	Source
Pristophora geniculata Mountain ash sawfly	Rare in native Europe where parasitoid attack is very high. Common in Canada as an exotic until biological control reassociated webworm with its parasitoids	Quednau (1984, 1990) Eichorn and Pschorn-Walcher (1978)
Phenacoccus manihoti Cassava mealybug	Rare in native South America where parasitoids are very effective. Introduced pest in Africa until parasitoids were reassociated	Nadel and van Alphen (1987); van Alphen *et al.* (1989)
Icerya purchasi Cottony-cushion scale	Serious exotic pest in California, became uncommon after introduction of predatory coccinellid, *Rodolia cardinalis,* from native location of herbivore	Doutt (1964)
Aonidiella aurantii California red scale	An exotic pest became well controlled after introduction of parasitoids alien to the host	De Bach and Sundby (1963); De Bach *et al.* (1971)
Operophtera brumata Winter moth	Population of an exotic pest became well regulated after reassociation with natural parasitoids	Embree (1966)
Pardia tripunctana Rose bud tortricid	Strong density dependent pupal predation, experimentally tested	Bauer (1985)
Notocelia roborana Rose leaf tortricid	Strong density dependent pupal predation, experimentally tested	Bauer (1985)
Orgya pseudotsugata Douglas fir tussock moth	Experiments excluding predators demonstrated strong effects	Mason (1987)
Labidomera clivicollis Milkweed leaf beetle	Field experiments showed predation on young larvae was critical	Eickwort (1977)
Cameraria sp. Oak leaf-mining moth	Caging experiments demonstrated strong natural enemy effects	Faeth and Simberloff (1981a,b)

[a] The first five cases involve biological control, and the last five examples are from natural systems.

habitat type as in the example of *Pieris* species, and in plant morphology illustrated by the *Tildenia* leaf miners. Here the causal relationships between herbivore and natural enemy impact are well understood, and an evolutionary perspective is essential in their understanding.

More studies of this nature will help decipher the puzzle of whether pattern or idiosyncrasy prevails in natural herbivore–enemy interactions: whether natural enemies are constrained by the resources they exploit, or whether they are commonly capable of constraining their resources. Several detailed studies of herbivore–natural enemy relationships have shown how constraining host plant and herbivore resources are (Price *et al.*, 1980; Price and Clancy, 1986a; Price, 1988; Gross and Price, 1988; Ohsaki and Sato, 1990). At the same time, good examples of regulation by natural enemies are available (e.g., Table 5).

The existing evidence does point in the direction of increasing importance of natural enemies in stable environments provided by late stages in ecological succession of vegetation. Plant population dynamics and defenses impose opportunities and constraints on the herbivores and their natural enemies. The extent to which such openings are exploited and such strictures impose serious limitations is dictated by an evolutionary dynamic that inevitably passes up the trophic system. How much impact passes down the trophic system, and how commonly, needs much more careful and methodical research.

IV. Cascading Effects of Plants through Trophic Webs

As remarked earlier, in lake ecosystems, trophic cascades or top-down effects are based on very small primary producers and build up to relatively large top predators. In pelagic systems there is little or no physical protection, so the cascading effects of top predators can be strong (Carpenter and Kitchell, 1987, 1988; McQueen *et al.*, 1989). In terrestrial systems based on autotrophic plants, the base of the food web is composed of large organisms, and these are fed upon by much smaller insect herbivores. Thus the plant can offer both food and protection to the herbivore, and the herbivore can evolve to maximize its refuge from enemies and minimize their impact (*e.g.* Bernays and Graham, 1988; Price *et al.*, 1987c; Price and Pschorn-Walcher, 1988). This constitutes a major difference between pelagic and terrestrial systems involving insect herbivores. The result in many cases will be that the bottom-up effects in terrestrial systems are likely to be even stronger, and will pass right through the trophic levels, with relatively little feedback from top-down effects.

Thus, instead of seeing trophic cascades down the system as in pelagic habitats, a more fundamental reality will be the cascading effects of the plant up the trophic system through the paths of energy flow. Even abiotic factors may become driving variables by influencing plant production and quality for herbivores, with further impact up the food web.

An example of the cascading effects of the plant hosts is summarized here to illustrate what may be more commonly seen in nature than is now appreciated. When the bottom-up effects are understood in many systems, we will be in a better position to evaluate the role of natural enemies, whether they play an active or passive role in insect herbivore population dynamics, and to what extent either role is influenced from below. This example is probably not representative of all plant, insect herbivore, and natural enemy systems, but it may well capture essential mechanisms for many endophytic species that specialize on vigorous plants or plant parts (*e.g.* Table 2 illustrates the kinds of herbivores: gallers, shoot and cone borers, and some leaf-rollers).

The example concerns the arroyo willow, *Salix lasiolepis*, the shoot-galling sawfly, *Euura lasiolepis*, and its major natural enemies. Much of the research over the past 10 years on this system is reviewed by Price *et al.* (1990). The cascading effects of plant condition involve several scales in time and space and affect both the herbivore and its natural enemies (Fig. 2). Briefer studies on related species of *Euura* conform to the same general patterns. This case study will illustrate the importance of

1. interaction of abiotic factors with plants determine the carrying capacity for insect herbivores;
2. carrying capacity that dictates herbivore population change with regulation imposed by competition for oviposition sites; and
3. natural enemies' negligible impact on population regulation imposed by the plant-insect interaction.

A. Plant Effects on Herbivores

A longer-term time scale of decades involves disturbance and patchy distribution of willow regeneration. Willows colonize mineral soil left by winter runoff (Sacchi and Price, 1991). After establishment, young ramets grow rapidly, producing long shoots. Females show preference for ovipositing in long shoots, and larvae survive better in such shoots (Craig *et al.*, 1989). Therefore, in patches of vigorous willow growth, sawfly populations build to high densities defined by the availability of vigorous shoots. High-density patches are relatively uncommon and are set in a matrix of willows too old to support large populations (Craig *et al.*, 1988a; Sacchi *et al.*, 1988; Price *et al.*, 1990). As these patches age, they become less favorable to sawflies, and populations decline to background levels. In persistently favorable wetter environments, sawflies may maintain a high carrying capacity by causing dieback of attacked shoots and more vigorous, juvenile, and proximal regrowth; a resource regulation effect (Craig *et al.*, 1986).

On a shorter time scale of year-to-year variation, winter precipitation

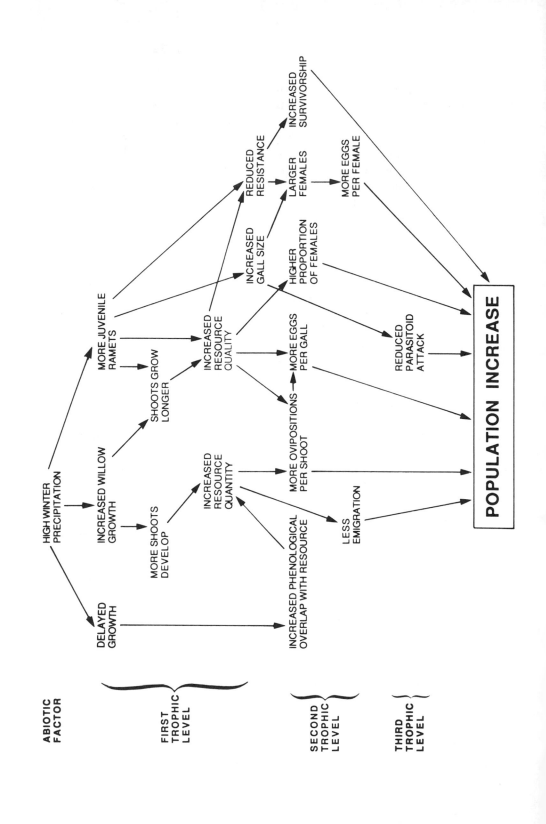

influences willow growth, involving the number of shoots initiated, the rate of shoot growth and ultimate length, and the number of vegetative buds developed for the next growing season (Price and Clancy, 1986b). In cool, moist springs, phenological overlap between shoot growth and sawfly emergence is high, effectively increasing the carrying capacity of each willow clone. After wet winters, and especially after a series of years with high precipitation, sawfly densities can increase dramatically even in relatively dry sites. A series of relatively dry winters has cascading detrimental effects on willow growth and herbivore performance (Fig. 2).

Negative feedback on population size resulting in population regulation is rapid and strong in this system. As a population increases toward the carrying capacity set by shoots of high quality, females begin to compete for oviposition sites (Craig *et al.*, 1990a), and avoid oviposition scars made by previous females on a shoot (Craig *et al.*, 1988b). Females withold eggs rather than deposit them in poor-quality shoots (Preszler and Price, 1988), and presumably emigrate in search of less-crowded sites. Thus competition among females for oviposition sites in vigorous shoots provides rapid negative feedback and population regulation imposed by the interaction of plant resources and female ovipositional preference and competition. Such competition among females may be much more common than generally thought (Price *et al.*, 1990) and may provide an explanation for the weak competition among larval herbivores in many systems (Lawton and Strong, 1981). In the willow–sawfly system, larval competition is weak because of spacing of eggs by female adults (Craig *et al.*, 1988b; Craig *et al.*, 1989).

Spatial scales of variation in plant quality are reflected in the succession of disturbance, willow regeneration, growth and maturity, and in the patchy nature of water availability. Permanent springs are rare in northern Arizona but support vigorous willows and high sawfly populations (Price *et al.*, 1990). Along the temporary streams that run in some years during snowmelt, pockets that hold water longer tend to have larger sawfly populations.

The detailed effects of plant quality on herbivore populations are many and do portray a cascade of widening influence, as abiotic conditions change plant growth (Fig. 2). A recent series of relatively dry winters beginning in the 1986 to 1987 season illustrates the decline of populations (Fig. 3) and the factors involved (Fig. 4).

The real pattern of change involved declining survivorship during eclosion from the egg with reduced establishment of feeding first instar larvae in the gall. This pattern was seen within years when wet and dry sites are

Figure 2   The cascading effects of winter precipitation on plant growth and quality, and consequences for a galling herbivore and its natural enemies.

compared and persists each year, but in a drying spell over years, survivorship of early larvae declines in both wet and dry sites. Usually, mortality is slight after establishment in the gall, but chickadee predation on the 1987 generation during the winter was heavy and concentrated in wet sites. However, the general patterns and behavior of the population were still consistent with the major effects being determined by the plant–herbivore interaction. Chickadee predation accelerated the decline but did not change the major dynamic pattern. Other natural enemies, which act between Stages 2 and 3 (Fig. 4), have little effect on the general patterns of population differences and changes.

B. Plant Effects on Natural Enemies

The plant provides protection from enemies by forming a gall around the egg and larva, which is initiated by the ovipositing female sawfly. On vigorous shoots galls become relatively large, larvae are better concealed from small parasitoids with short ovipositors, and a refuge from attack develops (Price and Clancy, 1986a; Price, 1988). Thus, where plants are vigorous and sawfly populations are high, parasitism by small parasitoids is low. Where plants are poor-quality hosts and galls are relatively small, such parasitoids can attack a much higher proportion of the galls. In space, across clones, each year there is a negatively density-dependent relationship between sawfly density and parasitoid density imposed by plant quality and water availability. This pattern is stable over many years, indicating the weak impact of parasitoids at the centers of sawfly density, and the way in which plant quality dictates access by natural enemies. Small parasitoids are very passive in the system.

Even the larger ichneumonid parasitoid, *Lathrostizus euurae*, is heavily constrained by the plant from attacking a high proportion of sawfly larvae. In this case, gall size is of little consequence, but increasing gall toughness as the gall develops imposes serious limitations. The window of vulnerability of larvae to this parasitoid is small because eggs are not utilized, and galls prevent attack usually by the third larval instar (Craig *et al.*, 1990b). An important component of the plant's influence on the third trophic level was variation in rates of gall toughening among willow clones. The more rapid the toughening, the shorter the window of vulnerability and the lower the percentage of parasitism, such that adjacent clones varied in attack from 13 to 29%. Again the efficacy of this parasitoid is regulated by plant traits, such that it does not respond in a density-dependent way to its host density.

The other major natural enemies of *Euura* have very unpredictable impact. Grasshoppers eat galls and kill larvae, and chickadees feed on larvae in cocoons during the winter. In harsh winters, chickadees may attack in a density-dependent way but without changing the general pattern of population distribution. In milder years, attack is minimal (Fig. 3).

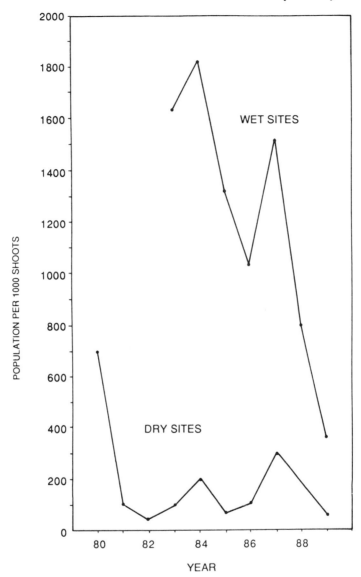

Figure 3 Population change of the shoot-galling sawfly, *Euura lasiolepis,* over a decade. Each generation starts in May and June and ends at the same time in the following year. Thus year 80 gives the population estimate for the generation initiated in 1980. Note the large differences in population size in wet and dry sites, the largely synchronous change in population in these sites, the increase in population resulting from the 1986 generation, and decline since then during a series of drier-than-normal winters.

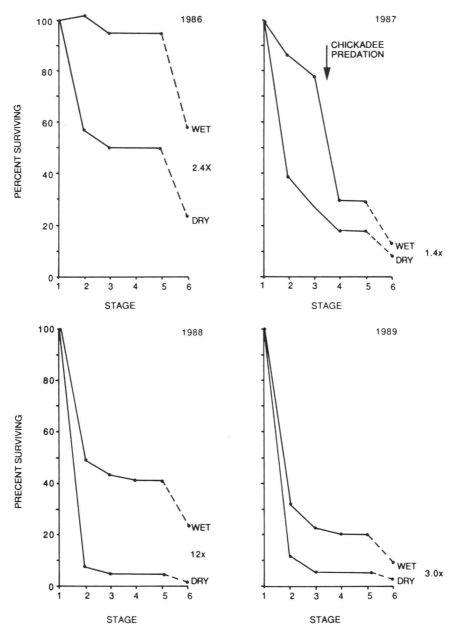

Figure 4 Survivorship curves for four generations of *Euura lasiolepis*, including the 1986 generation that resulted in a population increase, and three subsequent years that resulted in a decline. Wet and dry sites are separated and are represented by three willow clones and 12 willow clones respectively. The stages in development are (1) egg in gall; (2) established first instar larva; (3) late larvae; (4) larva in cocoon in fall; (5) emerging adults; (6) number of females. The number of times the wet site has higher numbers of females produced than dry sites is indicated on the right (e.g., 2.4 X in the 1986 generation).

All natural enemies considered, they do not change the basic pattern of population distribution defined by the plant–herbivore interaction. Caging experiments with and without parasitoids, testing for three sawfly generations, showed no effect of parasitoids on sawfly density (Woodman and Price, 1991). The parasitoids are opportunists, attacking where and when the plant permits. *Euura* or its gall provides only a minor food item for the predators that behave and fluctuate largely independent of sawfly populations. The constant feature of sawfly population dynamics is the bottom-up effect exerted by abiotic variation and its impact on resource quantity and quality. Many of the species listed in Tables 1–3 may well be subjected principally to such resource-based regulation, much of the time. However, more emphasis on experimental testing of the role of natural enemies is needed before patterns can be identified.

V. General Hypotheses

The following working hypotheses summarize the content of this chapter and the evidence presented. They must be regarded as very tentative, but they do present patterns that are testable, especially when coupled with the hypothetical mechanistic relationships discussed in the body of this chapter. They relate especially to primeval north temperate forests.

The hypotheses do emphasize the anticipated stronger effects of resource quantity and quality compared to top-down effects. This reflects the discussion in Section IV, pointing out that plants can provide food and protection for small insect herbivores, making such terrestrial systems significantly different from pelagic trophic webs.

The hypotheses differ from many in plant–herbivore interactions because they are conditional rather than absolute. The condition of the vegetation is variable, with variable impact up the trophic system. This gradient approach accepts that stressed plants can favor some herbivores, but does not accept the universality of this effect. It accepts that natural enemies can play an important role in the regulation of insect herbivore populations, while under certain conditions they will be ineffective. Predictions on patterns and processes in nature help to order our thinking and organization of research, and will ultimately result in a synthetic theory on insect herbivore population dynamics.

1. *The Plant Succession and Carrying Capacity Hypothesis* For insect herbivores the ultimate regulator of insect density is set by the carrying capacity (K) of the host plant species. This K differs by many orders of magnitude as ecological succession proceeds from weedy herbs, to perennial herbs, shrubs, and trees. Biomass production and standing crop available to herbivores is low in early succession and high in late succession, imposing strict

limits on the magnitude of change in population size through which populations can fluctuate.

2. *The Plant Patchiness Hypothesis* Superimposed on this pattern of plant succession and K is the disturbance–recovery regime in natural landscapes, with relatively small areas of disturbed ground reverting to early succession in a large matrix of late successional vegetation. Herbivore species adapting to utilize different plant species on the successional gradient must therefore evolve different syndromes for exploiting the different dispersion patterns of host plants. For early successional herbivores, the directed search for hosts must be effective, feeding must be rapid, and dispersal, frequent. For late successional herbivores, searching for hosts is a minor problem, long-term feeding may evolve, and dispersal can be minimized or undirected because the herbivore is surrounded by edible tree species.

3. *The Plant Qualitative Variation Hypothesis* K defined by plants can be strongly modified by plant quality. Herbivore species favored by vigorous plant growth are likely to concentrate in patches of disturbance where young plants are available. Species favored by mature trees, or senescent growth, or indifferent to plant quality, will be concentrated in extensive stands of mature forest. In woody plant succession many endophytic herbivores such as gallers and shoot borers will concentrate on patches of vigorous regeneration involving shrubs and trees. The exophytic leaf-chewing species are likely to be more common on older plants closer to the mature end of plant life histories.

4. *The Successional Pattern of Natural Enemy Impact Hypothesis* Although bottom-up effects of resources on insect herbivore populations are very strong, natural enemies can impose effective regulation before plant resources become limiting. The impact of natural enemies probably increases with ecological succession, as host insects become more predictable and environments, more stable. The extent to which natural enemies impose regulation on insect populations, before resource effects become important for the herbivore, must be studied in more detail before patterns emerge. Because large plants in late succession provide more deeply concealed resources for some herbivores, this hypothesis probably relates more to exophytic herbivores like leaf chewers than to endophytes like wood borers.

5. *The Idiosyncratic Plant–Herbivore–Enemy Interaction Hypothesis* Three-trophic-level effects can be strong, but current knowledge suggests that idiosyncratic factors play a role even with closely related herbivore species. Adaptive syndromes therefore differ within phylogenies, such as in the *Pieris* and *Tildenia* examples given in Section III,C, and weaken general patterns. The relative importance of bottom-up and top-down effects may therefore differ among related species. Some species may become eruptive and others, latent, and a full understanding of these differences will require detailed study of these three-trophic-level systems.

The study of insect population dynamics based on the template of ecological succession provides a natural ordering of many variables, which should aid in the discovery of pattern, the development of predictions, and the erection of new hypotheses. Insect populations have been studied commonly in isolation as an end in itself, and there has been relatively little emphasis on the discovery of patterns or synthesis. Notable exceptions include the broad overviews developed by Southwood (1975, 1977) and Southwood and Comins (1976), and syntheses in Barbosa and Schultz (1987).

The search for pattern and synthesis must acknowledge that universal truths about nature are probably not available at the level of populations, communities, and ecosystems. Therefore, discovery of pattern must weigh the balance between conforming examples and exceptions. It is common to reject a pattern based on a small number of exceptions, without perceiving the real balance of evidence. Whatever conclusions emerge from a careful evaluation of examples, the search for pattern and synthesis is inevitably creative, and advances the science. Should these hypotheses prove to be wrong, then alternatives will emerge during the research process, and ultimately real patterns in nature will be discovered. This progress will involve both the discovery of pattern in dynamical types of insect herbivores, and the mechanistic processes involved.

Acknowledgments

The research on *Euura* species and three-trophic-level interactions discussed in this chapter were supported by grants from the U. S. National Science Foundation (DEB-7816152, DEB-8021754, BSR-83144594, BSR-8705302, BSR-8715090 and BSR-9020317) and Organized Research Grants from Northern Arizona University. I am grateful to Professor Shoichi Kawano, and the organizing committee for the International Congress of Ecology 1990, held in Yokohama, Japan, for originally suggesting the topic of this book and this chapter. My students have continued to provide me with a unique and invaluable education, although they cannot be held responsible for deficiencies in this chapter. I must express my sincere appreciation for penetrating criticism of an earlier draft by Cathy Bach, Brad Hawkins, Mark Hunter, Jill Landsberg, and Mark Rausher. Stimulating debate influencing issues in this chapter was nurtured by the ambience provided by Henry Africa in Yokohama.

References

Alphen, van J. J. M., Neuenschwander, P., van Dijken, M. J., Hammond, W. N. O., and Herren, H. R. (1989). Insect invasions: The case of the cassava mealybug and its natural enemies evaluated. *Entomologist* **108**, 38–55.
Anderson, R. M. (1981). Population ecology of infectious disease agents. *In* "Theoretical Ecology: Principles and Applications" (R. M. May, ed.), pp. 318–355. Sinauer, Sunderland, Massachusetts.

Askew, R. R. (1971). "Parasitic Insects." American Elsevier, New York.

Askew, R. R. (1980). The diversity of insect communities in leaf-mines and plant galls. *J. Anim. Ecol.* **49**, 817–829.

Askew, R. R., and Shaw, M. R. (1986). Parasitoid communities: Their size, structure, and development. *In* "Insect Parasitoids" (J. Waage and D. Greathead, eds.), pp. 225–264. Academic Press, London.

Baker, W. L. (1972). Eastern forest insects. U.S. Dept. Agr. For. Serv. Misc. Pub. 1175, Washington, D.C.

Barbosa, P. (1988). Some thoughts on "the evolution of host range." *Ecology* **69**, 912–915.

Barbosa, P., Krischik, V., and Lance, D. (1989). Life-history traits of forest-inhabiting flightless Lepidoptera. *Am. Midl. Nat.* **122**, 262–274.

Barbosa, P., and Schultz, J. C. (eds.) (1987). "Insect Outbreaks." Academic Press, New York.

Bauer, G. (1985). Population ecology of *Pardia tripunctana* Schiff. and *Notocelia roborana* Den. and Schiff. (Lepidoptera, Tortricidae)—An example of "equilibrium species." *Oecologia* **65**, 437–441.

Beckmann, R. (1989). Rural dieback: Restoring a balance. *Ecos* **62**, 8–15.

Bergman, J. M., and Tingey, W. M. (1979). Aspects of interaction between plant genotypes and biological control. *Bull. Entomol. Soc. Am.* **25**, 275–279.

Bernays, E., and Graham, M. (1988). On the evolution of host specificity in phytophagous arthropods. *Ecology* **69**, 886–892.

Berryman, A. A. (1982). Population dynamics of bark beetles. *In* "Bark Beetles in North American Conifers: A System for the Study of Evolutionary Biology" (J. B. Mitton and K. B. Sturgeon, eds.), pp. 264–314. University of Texas Press, Austin, Texas.

Blais, J. R. (1953). Effects of the destruction of the current year's foliage of balsam fir on the fecundity and habits of flight of the spruce budworm. *Can. Entomol.* **85**, 446–448.

Caouette, M. R., and Price, P. W. (1989). Growth of Arizona rose and attack and establishment of gall wasps, *Diplolepis fusiformans* (Ashmead) and *D. spinosa* (Ashmead) (Hymenoptera: Cynipidae). *Environ. Entomol.* **18**, 822–828.

Carpenter, S. R., and Kitchell, J. F. (1987). The temporal scale of variance in lake productivity. *Am. Nat.* **129**, 417–433.

Carpenter, S. R., and Kitchell, J. F. (1988). Consumer control of lake productivity. *BioScience* **38**, 764–769.

Carpenter, S. R., Kitchell, J. F., and Hodgson, J. R. (1985). Cascading trophic interactions and lake productivity. *BioScience* **35**, 634–639.

Coley, P. D. (1983). Herbivory and defensive characteristics of tree species in a lowland tropical forest. *Ecol. Monogr.* **53**, 209–233.

Coley, P. D., Bryant, J. P., and Chapin, F. S. (1985). Resource availability and plant antiherbivore defense. *Science* **230**, 895–899.

Courtney, S. (1988). If it's not coevolution, it must be predation? *Ecology* **69**, 910–911.

Craig, T. P., Price, P. W., and Itami, J. K. (1986). Resource regulation by a stem-galling sawfly on the arroyo willow. *Ecology* **67**, 419–425.

Craig, T. P., Price, P. W., Clancy, K. M., Waring, G. M., and Sacchi, C. F. (1988a). Forces preventing coevolution in the three-trophic-level system: Willow, a gall-forming herbivore, and parasitoid. *In* "Chemical Mediation of Coevolution" (K. Spencer, ed.), pp. 57–80. Academic Press, New York.

Craig, T. P., Itami, J. K., and Price, P. W. (1988b). Plant wound compounds from oviposition scars used as oviposition deterrents by a stem-galling sawfly. *J. Insect Behav.* **1**, 343–356.

Craig, T. P., Itami, J. K., and Price, P. W. (1989). A strong relationship between oviposition preference and larval performance in a shoot-galling sawfly. *Ecology* **70**, 1691–1699.

Craig, T. P., Itami, J. K., and Price, P. W. (1990a). Intraspecific competition and facilitation by a shoot-galling sawfly. *J. Anim. Ecol.* **59**, 147–159.

Craig, T. P., Itami, J. K., and Price, P. W. (1990b). The window of vulnerability of a shoot-galling sawfly to attack by a parasitoid. *Ecology* **71,** 1471–1482.

De Bach, P. (1964). "Biological Control of Insect Pests and Weeds." Reinhold, New York.

De Bach, P., and Sundby, R. A. (1963). Competitive displacement between ecological homologues. *Hilgardia* **35,** 105–166.

De Bach, P., Rosen, D., and Kennett, C. E. (1971). Biological control of coccids by introduced natural enemies. *In* "Biological Control" (C. B. Huffaker, ed.), pp. 165–194. Plenum Press, New York.

Dempster, J. P. (1963). The population dynamics of grasshoppers and locusts. *Biol. Rev.* **38,** 490–529.

Dodge, K. L., and Price, P. W. (1991a). Life history of *Disonycha pluriligata* (Coleoptera: Chrysomelidae), and host plant relationships with *Salix exigua* (Salicaceae). *Ann. Entomol. Soc. Am.* **84,** 248–254.

Dodge, K. L., and Price, P. W. (1991b). Eruptive vs. noneruptive species: A comparative study of host plant use by a sawfly, *Euura exiguae* (Hymenoptera: Tenthredinidae) and a leaf beetle, *Disonycha pluriligata* (Coleoptera: Chrysomelidae). *Environ. Entomol.* **20,** 1129–1133.

Doutt, R. L. (1964). The historical development of biological control. *In* "Biological Control of Insect Pests and Weeds" (P. De Bach, ed.), pp. 21–42. Reinhold, New York.

Ehrlich, P. R., and Birch, L. C. (1967). The "balance of nature" and "population control." *Am. Nat.* **101,** 97–107.

Eichhorn, O., and Pschorn-Walcher, H. (1978). Biologie und Parasiten der Ebereschen-Blattwespe, *Pristiphora geniculata* Htg. (Hym: Tenthredinidae). *Z.Angew. Entomol.* **85,** 154–167.

Eickwort, K. R. (1977). Population dynamics of a relatively rare species of milkweed beetle (*Labidomera*). *Ecology* **58,** 527–538.

Embree, D. G. (1966). The role of introduced parasites in the control of the winter moth in Nova Scotia. *Can. Entomol.* **98,** 1159–1168.

Faeth, S. H., and Simberloff, D. (1981a). Population regulation of a leaf-mining insect, *Cameraria* sp. nov., at increased field densities. *Ecology* **62,** 620–624.

Faeth, S. H., and Simberloff, D. (1981b). Experimental isolation of oak host plants: Effects on mortality, survivorship, and abundances of leaf-mining insects. *Ecology* **62,** 625–635.

Feeny, P. (1976). Plant apparency and chemical defense. In "Biochemical Interaction between Plants and Insects" (J. W. Wallace and R. L. Mansell, eds.), pp. 1–40. Plenum Press, New York.

Fox, L. R. (1988). Diffuse coevolution within complex communities. *Ecology* **69,** 906–907.

Franklin, J. F., Shugart, H. H., and Harmon, M. E. (1987). Tree death as an ecological process. *BioScience* **37,** 550–556.

Furniss, R. L., and Carolin, V. M. (1977). "Western Forest Insects." U. S. Dept. Agr. For. Serv. Misc. Pub. 1339, Washington, D.C.

Futuyma, D. J. (1976). Food plant specialization and environmental predictability in Lepidoptera. *Am. Nat.* **110,** 285–292.

Ghent, A. W., and Wallace, D. R. (1958). Oviposition behavior of the Swaine jack-pine sawfly. *For. Sci.* **4,** 264–272.

Greenbank, D. O. (1963). The development of the outbreak; Climate and the spruce budworm; Staminate flowers and the spruce budworm; Host species and the spruce budworm. *In* "The Dynamics of Epidemic Spruce Budworm Populations" (R. F. Morris, ed.). *Mem. Entomol. Soc. Can.* **31,** 19–23, 174–180, 202–218, 219–223.

Gross, P., and Price, P. W. (1988). Plant influences on parasitism of two leafminers: A test of enemy-free space. *Ecology* **69,** 1506–1516.

Hairston, N. G., Smith, F. E., and Slobodkin, L. B. (1960). Community structure, population control, and competition. *Am. Nat.* **94,** 421–425.

Hanski, I. (1987). Pine sawfly population dynamics: Patterns, processes, problems. *Oikos* **50**, 327–335.

Hanski, I., and Otronen, M. (1985). Food quality–induced variance in larval performance: Comparison between rare and common pine-feeding sawflies (Diprionidae). *Oikos* **44**, 165–174.

Hassell, M. P. (1978). "The Dynamics of Arthropod Predator–Prey Systems." Princeton University Press, Princeton, New Jersey.

Hassell, M. P., and May, R. M. (1989). The population biology of host–parasite and host–parasitoid associations. *In* "Perspectives in Ecological Theory" (J. Roughgarden, R. M. May, and S. A. Levin, eds.), pp. 319–347. Princeton University Press, Princeton, New Jersey.

Hawkins, B. A. (1988). Species diversity in the third and fourth trophic levels: Patterns and mechanisms. *J. Anim. Ecol.* **57**, 137–162.

Hawkins, B. A., and Gross, P. (1991). Species richness and population limitation in insect parasitoid–host systems. *Am. Nat.*, in press.

Hawkins, B. A., and Lawton, J. H. (1987). Species richness for the parasitoids of British phytophagous insects. *Nature* **326**, 788–790.

Hawkins, B. A., Askew, R. R., and Shaw, M. R. (1990). Influences of host feeding-niche and food plant type on generalist and specialist parasitoids. *Ecol. Entomol.* **15**, 275–280.

Howard, L. O. (1897). A study of insect parasitism. *U.S. Dept. Agr. Bur. Entomol. Tech. Ser.* **5**, 1–57.

Hunter, M. D. (1990). Differential susceptibility to variable plant phenology and its role in competition between two insect herbivores on oak. *Ecol. Entomol.* **15**, 401–408.

Janzen, D. H. (1988). On the broadening of insect–plant research. *Ecology* **69**, 905.

Jermy, T. (1988). Can predation lead to narrow food specialization in phytophagous insects? *Ecology* **69**, 902–904.

Kearsley, M. J. C., and Whitham, T. G. (1989). Developmental changes in resistance to herbivory: Implications for individuals and populations. *Ecology* **70**, 422–434.

Kimberling, D. N., Scott, E. R., and Price, P. W. (1990). Testing a new hypothesis: Plant vigor and phylloxera distribution on wild grape in Arizona. *Oecologia* **84**, 1–8.

Kogan, M. (1977). The role of chemical factors in insect/plant relationships. *Proc. 15th Int. Cong. Entomol., Washington, D.C, 1976.* pp. 211–227.

Landsberg, J., and Ohmart, C. P. (1989). Levels of insect defoliation in forests: Patterns and concepts. *Trends Ecol. Evol.* **4**, 96–100.

Landsberg, J., and Wylie, F. R. (1988). Dieback of rural trees in Australia. *Geojournal* **17**, 231–237.

Larsson, S., and Tenow, O. (1984). Areal distribution of a *Neodiprion sertifer* (Hym., Diprionidae) outbreak on Scots pine as related to stand condition. *Holarctic Ecol.* **7**, 81–90.

Lawton, J. H., and McNeill, S. (1979). Between the devil and the deep blue sea: On the problem of being a herbivore. *In* "Population Dynamics" (R. M. Anderson, B. D. Turner, and L. R. Taylor, eds.), pp. 223–244. Symp. Brit. Ecol. Soc. No. 20. Oxford, England.

Lawton, J. H., and Strong, D. R. (1981). Community patterns and competition in folivorous insects. *Am. Nat.* **118**, 317–338.

Lotka, A. J. (1924). "Elements of Physical Biology." Williams & Wilkins, Baltimore, Maryland.

Loucks, O. L. (1970). Evolution of diversity, efficiency, and community stability. *Am. Zool.* **10**, 17–25.

Mason, R. R. (1987). Nonoutbreak species of forest Lepidoptera. *In* "Insect Outbreaks" (P. Barbosa and J. C. Schultz, eds.), pp. 31–57. Academic Press, New York.

Mattson, W. J. (1980). Cone resources and the ecology of the red pine cone beetle, *Conophthorus resinosae* (Coleoptera: Scolytidae). *Ann. Entomol. Soc. Amer.* **73**, 390–396.

Mattson, W. J., Simmons, G. A., and Witter, J. A. (1988). The spruce budworm in eastern North America. *In* "Dynamics of Forest Insect Populations" (A. A. Berryman, ed.), pp. 309–330. Plenum, New York.

May, R. M. (1981). Models for two interacting populations. *In* "Theoretical Ecology: Principles and Applications" (R. M. May, ed.), pp. 78–104. Sinauer, Sunderland, Massachusetts.

McLeod, J. M. (1970). The epidemiology of the Swaine jack-pine sawfly, *Neodiprion swainei* Midd. *For. Chron.* **46**, 126–133.

McQueen, D. J., Johannes, M. R. S., Post, J. R., Stewart, T. J., and Lean, D. R. S. (1989). Bottom-up and top-down impacts on freshwater pelagic community structure. *Ecol. Monogr.* **59**, 289–309.

Miller, C. A. (1957). A technique for estimating the fecundity of natural populations of the spruce budworm. *Can. J. Zool.* **35**, 1–13.

Morris, R. F. (ed.) (1963). The dynamics of epidemic spruce budworm populations. *Mem. Entomol. Soc. Can.* **31**, 1–332.

Morris, R. F. (1967). Influence of parental food quality on the survival of *Hyphantria cunea. Can. Entomol.* **99**, 24–33.

Morris, R. F. (1969). Approaches to the study of population dynamics. *In* "Forest Insect Population Dynamics" (W. E. Waters, ed.), pp. 9–28. U.S. Dept. Agr. For. Serv. Res. Paper NE–125, Washington, D.C.

Murdoch, W. W. (1966). Community structure, population control, and competition—a critique. *Am. Nat.* **100**, 219–226.

Myers, J. H. (1988). Can a general hypothesis explain population cycles of forest Lepidoptera? *Adv. Ecol. Res.* **18**, 179–242.

Nadel, H., and van Alphen, J. J. M. (1987). The role of host and host–plant odours in the attraction of a parasitoid, *Epidinocarsis lopezi,* to the habitat of its host, the cassava mealybug, *Phenacoccus manihoti. Entomol. Exp. Appl.* **45**, 181–186.

Ohgushi, T., and Sawada, H. (1985). Population equilibrium with respect to available food resource and its behavioural basis in the herbivorous lady beetle, *Henosepilachna niponica. J. Anim. Ecol.* **54**, 781–796.

Ohsaki, N., and Sato, Y. (1990). Avoidance mechanisms of three *Pieris* butterfly species against the parasitoid wasp *Apanteles glomeratus. Ecol. Entomol.* **15**, 169–176.

Pickett, S. T. A., and White, P. S. (eds.) (1985). "The Ecology of Natural Disturbance and Patch Dynamics." Academic Press, Orlando, Florida.

Preszler, R. W., and Price, P. W. (1988). Host quality and sawfly populations: A new approach to life table analysis. *Ecology* **69**, 2012–2020.

Price, P. W. (1970). A loosestrife sawfly, *Monostegia abdominalis* (Fabricius) (Hymenoptera: Tenthredinidae). *Can. Entomol.* **102**, 491–495.

Price, P. W. (1984). "Insect Ecology." 2nd Ed. Wiley, New York.

Price, P. W. (1988). Inversely density-dependent parasitism: The role of plant refuges for hosts. *J. Anim. Ecol.* **57**, 89–96.

Price, P. W. (1989). Clonal development of coyote willow, *Salix exigua* (Salicaceae), and attack by the shoot-galling sawfly, *Euura exiguae* (Hymenoptera: Tenthredinidae). *Environ. Entomol.* **18**, 61–68.

Price, P. W., and Clancy, K. M. (1986a). Interactions among three trophic levels: Gall size and parasitoid attack. *Ecology* **67**, 1593–1600.

Price, P. W., and Clancy, K. M. (1986b). Multiple effects of precipitation on *Salix lasiolepis* and populations of the stem-galling sawfly, *Euura lasiolepis. Ecol. Res.* **1**, 1–14.

Price, P. W., and Craig, T. P. (1984). Life history, phenology, and survivorship of a stem-galling sawfly, *Euura lasiolepis* (Hymenoptera: Tenthredinidae), on the arroyo willow, *Salix lasiolepis,* in northern Arizona. *Ann. Entomol. Soc. Amer.* **77**, 712–719.

Price, P. W., and Pschorn-Walcher, H. (1988). Are galling insects better protected against parasitoids than exposed feeders?: A test using tenthredinid sawflies. *Ecol. Entomol.* **13**, 195–205.

Price, P. W., Bouton, C. E., Gross, P., McPheron, B. A., Thompson, J. N., and Weis, A. E. (1980). Interactions among three trophic levels: Influence of plants on interactions between insect herbivores and natural enemies. *Annu. Rev. Ecol. Syst.* **11**, 41–65.

Price, P. W., Roininen, H., and Tahvanainen, J. (1987a). Plant age and attack by the bud galler, *Euura mucronata. Oecologia* **73**, 334–337.

Price, P. W., Roininen, H., and Tahvanainen, J. (1987b). Why does the bud-galling sawfly, *Euura mucronata*, attack long shoots? *Oecologia* **74**, 1–6.

Price, P. W., Fernandes, G. W., and Waring, G. L. (1987c). Adaptive nature of insect galls. *Environ. Entomol.* **16**, 15–24.

Price, P. W., Cobb, N., Craig, T. P., Fernandes, G. W., Itami, J. K., Mopper, S., and Preszler, R. W. (1990). Insect herbivore population dynamics on trees and shrubs: New approaches relevant to latent and eruptive species and life table development. *In* "Insect-Plant Interactions" (E. A. Bernays, ed.), Vol. 2, pp. 1–38. CRC Press, Boca Raton, Florida.

Quednau, F. W. (1984). *Pristiphora geniculata* (Htg.), mountain-ash sawfly (Hymenoptera: Tenthredinidae). *In* "Biological Control Programmes Against Insects and Weeds in Canada, 1969–1980." (J. S. Keller and M. A. Hulme, eds.), p. 381–385. Commonwealth Agricultural Bureaux, Farnham, Surrey, England.

Quednau, F. W. (1990). Introduction, permanent establishment, and dispersal in Eastern Canada of *Olesicampe geniculatae* Quednau and Lim (Hymenoptera: Ichneumonidae), an important biological control agent of the mountain ash sawfly, *Pristiphora geniculata* (Hartig) (Hymenoptera: Tenthredinidae). *Can. Entomol.* **122**, 921–934.

Rainey, R. C. (1982). Putting insects on the map: Spatial heterogeneity and the dynamics of insect populations. *Antenna* **6**, 162–169.

Rausher, M. D. (1988). Is coevolution dead? *Ecology* **69**, 898–901.

Readshaw, J. L. (1965). A theory of phasmatid outbreak release. *Austr. J. Zool.* **13**, 475–490.

Rhoades, D. F. (1979). Evolution of plant chemical defense against herbivores. *In* "Herbivores: Their Interaction with Secondary Plant Metabolites." (G. A. Rosenthal and D. H. Janzen, eds.), pp. 3–54. Academic Press, New York.

Roff, D. A. (1990). The evolution of flightlessness in insects. *Ecol. Monogr.* **60**, 389–421.

Sacchi, C. F., and Price, P. W. (1991). The relative role of abiotic and biotic factors in seedling demography of the arroyo willow, *Salix lasiolepis. Amer. J. Bot.*, in press.

Sacchi, C. F., Price, P. W., Craig, T. P., and Itami, J. K. (1988). Impact of shoot galler attack on sexual reproduction in the arroyo willow. *Ecology* **69**, 2021–2030.

Schindler, D. W. (1978). Factors regulating phytoplankton production and standing crop in the world's fresh waters. *Limnol. Oceanogr.* **23**, 478–486.

Schindler, D. W., Fee, E. J., and Ruszczynski, T. (1978). Phosphorus input and its consequences for phytoplankton standing crop and production in the experimental lakes area and in similar lakes. *J. Fish. Res. Bd. Can.* **35**, 190–196.

Schultz, J. C. (1988). Many factors influence the evolution of herbivore diets, but plant chemistry is central. *Ecology* **69**, 896–897.

Shugart, H. H. (1987). Dynamic ecosystem consequences of tree birth and death patterns. *BioScience* **37**, 596–602.

Shugart, H. H., and West, D. C. (1981). Long-term dynamics of forest ecosystems. *Am. Sci.* **69**, 647–652.

Southwood, T. R. E. (1975). The dynamics of insect populations. *In* "Insects, Science, and Society" (D. Pimentel, ed.), pp. 151–199. Academic Press, New York.

Southwood, T. R. E. (1977). The relevance of population dynamics theory to pest status. *In* "Origins of Pest, Parasite, Disease and Weed Problems." (J. M. Cherrett and G. R. Sagar, eds.), pp. 35–54. Symp. Brit. Ecol. Soc. No. 18. Oxford, England.

Southwood, T. R. E., and Comins, H. N. (1976). A synoptic population model. *J. Anim. Ecol.* **45**, 949–965.

Sweetman, H. L. (1936). "The Biological Control of Insects." Comstock, Ithaca, New York.

Thompson, J. N. (1988). Coevolution and alternative hypotheses on insect/plant interactions. *Ecology* **69,** 893–95.

Tripp, H. A. (1965). The development of *Neodiprion swainei* Middleton (Hymenoptera: Diprionidae) in the Province of Quebec. *Can. Entomol.* **97,** 92–107.

Varley, G. C. (1959). The biological control of agricultural pests. *J. Roy. Soc. Arts* **107,** 475–490.

Varley, G. C., Gradwell, G. R., and Hassell, M. P. (1973). "Insect Population Ecology." Blackwell Scientific Publications, Oxford, England.

Watt, A. D., Leather, S. R., Hunter, M. D., and Kidd, N. A. C. (eds.) (1990). "Population Dynamics of Forest Insects." Intercept, Andover, Hampshire, England.

White, T. C. R. (1974). A hypothesis to explain outbreaks of looper caterpillars, with special reference to populations of *Selidosema suavis* in a plantation of *Pinus radiata* in New Zealand. *Oecologia* **16,** 279–301.

White, T. C. R. (1976). Weather, food, and plagues of locusts. *Oecologia* **22,** 119–134.

Whitham, T. G. (1978). Habitat selection by *Pemphigus* aphids in response to resource limitation and competition. *Ecology* **59,** 1164–1176.

Whitham, T. G. (1980). The theory of habitat selection: Examined and extended using *Pemphigus* aphids. *Amer. Nat.* **115,** 449–466.

Whitham, T. G. (1987). Evolution of territoriality by herbivores in response to host plant defenses. *Am. Zool.* **27,** 359–369.

Whitham, T. G. (1989). Plant hybrid zones as sinks for pests. *Science* **244,** 1490–1493.

Whitham, T. G., and Mopper, S. (1985). Chronic herbivory: Impacts on architecture and sex expression of pinyon pine. *Science* **228,** 1089–1091.

Woodman, R. L., and Price, P. W. (1991). Insect parasitoids and gall communities on willow: A critical test of the enemy impact hypothesis. *Ecology,* in press.

7

Factoring Natural Enemies into Plant Tissue Availability to Herbivores

Jack C. Schultz

Department of Entomology
Pennsylvania State University
University Park, Pennsylvania

I. Introduction

A failure to discriminate between resource abundance and resource availability has handicapped the development of powerful generalizations in population biology and community ecology for a long time. For example, despite decades of detailed study, the abundance and degree of exploitation of food resources by insectivorous birds remain largely unknown (Holmes and Schultz, 1988). Similarly, actual host plant use by forest Lepidoptera, even pest species, is poorly known.

Certainly this problem persists because of the difficulty of perceiving resource use by consumers. The prey of most insectivorous birds are too small to be seen or identified in the field, are too rare to be sampled readily, and may be difficult or impossible to identify to genus or species. Rarity, variation, vagility, and short lifespan of insects constrain our attempts to compile exhaustive host plant records for most of them. Even if we had adequate descriptions of resource use, this might not necessarily contribute much to understanding the processes (e.g., population regulation, frequency dependent predation, etc.) central to forming ecological generalizations. These processes are usually context sensitive (e.g., vary with resource density or aggregation), so that mere records of resource use, taken out of context, may not be useful.

All scientists must decide how to allocate research time and resources. A dichotomy frequently exists between formulating generalizations and focusing on mechanisms and processes. Sometimes characterized as *generalist* versus *reductionist* approaches, individuals seem to focus on one approach to the exclusion of the other. Reasons for this are diverse, and probably include constraints on time, personality traits, interest, and training. Whatever the causes, this dichotomy in approach uncouples generalizations from details of process and mechanism.

This uncoupling is responsible for our inability to construct effective generalizations involving resource exploitation. Those who attempt to construct theory and generalizations are rarely the same scientists who have a detailed understanding of actual resource use, and vice versa. More important, these two groups may not even read each other's work. In this chapter I attempt to identify ways in which we may recouple these equally important, complementary approaches to understanding populations and communities.

I focus on resource use by herbivorous insects, and attempt to point out that we have too long ignored processes that compose critical aspects of resource use by this ecologically important group. Specifically, I emphasize the importance of variation in plant tissue food quality to a herbivore's risk of mortality due to natural enemies. Natural enemies include vertebrate and invertebrate predators, parasitoids, and pathogens. I argue that ignoring these risks defeats our attempts to define and understand plant tissues as resources for insects, leading to what I term obese generalizations, which are too broad for the support they have in facts. I compare and contrast the constraints placed on resource availability for herbivores by food quality and mortality risks, suggest ways of integrating them, and provide a few examples in which integration seems particularly promising for generalization. My conclusion is that, at least for the present, sweeping generalizations such as *bottom-up* or *top-down* determination of herbivore numbers are difficult to support. Better and more widespread integration of mechanistic understanding into these conceptual approaches is greatly needed.

II. Apparent versus Available Plant Resources

A. Tissue Variation as Food

It is relatively easy to measure the standing crop or biomass of plant tissues that could in theory be consumed by insects (*e.g.*, Wiegert and Owen, 1971). Sometimes it even appears as though an insect herbivore actually consumes most or all of that biomass, as in the case of irruptive pest outbreaks. The great majority of the time, however, the biomass of plant tissues consumed is minute compared with the plant biomass apparently available and/or numbers of insects exploiting it. Given this, it is difficult to support the argument that herbivorous insect populations are usually limited by food abundance (Hairston *et al.*, 1960; Strong *et al.*, 1984).

However, *apparent* resources (e.g., biomass) are not the same thing as truly *available* resources, or the amount of plant tissue actually useful for growth and reproduction (Ehrlich and Birch, 1967). Our understanding of plant tissue characteristics, including nutrient quality and allelochemistry, has progressed to the point at which it seems clear that all plant tissue is not equally available to be transformed into insect tissue (Chapter 8).

Many authors (see Denno and McClure, 1983) have argued that variation in food quality makes some tissues *unavailable* and reduces the total availability of plants as food. Proponents of this view encounter two logical difficulties. First, because insects have frequently overcome chemical and nutritional barriers to host plant use evolutionarily, one must explain how variation in tissue quality produces lasting reductions in actual exploitation. Despite much speculation and theorizing on the subject, we still do not know what biochemical, physiological, or genetic factors prevent an insect species from exploiting varying or divergent plant tissues. For example, larvae of the gypsy moth (*Lymantria dispar* L.; Lymantriidae) can feed successfully on previous years' needles of white pine (*Pinus strobis* L.) but not on the current year's growth (Schultz, 1983a). These tissues differ quantitatively in a variety of nutritional and chemical traits, but it remains unclear why there is an absolute barrier to feeding on the young needles. In other cases (e.g., Berenbaum, 1978) such barriers are easier to identify.

Second, many, if not most, insect species fail to consume all of the available biomass within a tissue type, even when variation in tissue quality has not been detected. It is tempting to assert that undiscovered variation exists and is actually protecting subsets of a tissue class from exploitation (e.g., Schultz, 1983a,b). Sometimes undetected or short-term, environmentally caused variation does reduce host plant availability, as in the case of damage-induced changes in leaf traits (Schultz, 1988a; Chapter 10). Abiotic factors, such as temperature, light, or soil quality, may also create a shifting mosaic of tissue qualities. Other authors in this volume (Chapters 6, 8) have reviewed the roles of phenological variation (and variation in phenological variation) in denying access to plant tissues. But these plant-centered views all lead us on a never-ending quest for the secret, key plant variable that erects ecological and evolutionary barriers to consumption. They are exclusively phytocentric, or "bottom up" approaches; I suggest that they have utility only in the very few cases in which plant biomass actually determines availability to herbivores (Chapter 6). These situations are likely to be peculiar.

B. Other Influences on Availability

Factors other than food-quality variation often may deny access to plant tissues. Indeed, some of these other modes are easier to understand and quantify as barriers to plant tissue use than is variation in food quality. For example, some fraction of chemically suitable plant tissues may always be located in microenvironments that are physically or physiologically unsuitable for certain insects. These impacts may be difficult to separate from microenvironmentally related variation in food quality.

My focus here is on tissue- or situation-specific variation in risk of mortality from natural enemies. This risk denies access to some plant tissues and reduces overall consumption in three different ways. First, access to high-risk plant tissues may be denied because selection for avoiding risk has produced stereotyped feeding specialization in the herbivore. Second, selection may favor risk-sensitive foraging behavior, in which the decision to feed on a plant species or tissue is contingent on anticipated risk of mortality. These consequences are often important in the evolution of preferences by herbivores for particular plant species, but they are probably common evolutionary causes of tissue specialization as well (Brower, 1958; Schultz, 1983a,b; Bernays and Graham, 1988). Third, there may have been no evolutionary accommodation in feeding behavior of mortality risk; herbivores may continue feeding on high-risk plants or tissues. In this case, access to plant resources is denied at the population level, because exploiting high-risk tissues increases mortality rates and reduces insect population growth.

There are potential demographic effects of all three modes of plant resource denial. Most obviously, denying access to certain plants or tissues reduces overall food availability to individual herbivores, restraining population size and influencing resource utilization by coexisting species. Some herbivores may choose less-valuable food when risk is anticipated, in the sense of lowering their potential fecundities by feeding on suboptimal diets. When there is a temporal (e.g., seasonal) component to risk, the narrowed *window of opportunity* for exploiting plant resources may restrict herbivore populations to one or few generations per year (see Chapter 14). Finally, the actual reduction of herbivore numbers when individuals fail to avoid risky feeding situations has a direct impact on population size and growth. By these means, the interaction between plant traits and natural enemy risks places an additional constraint on the ultimate herbivore population size: on the environment's carrying capacity for the herbivore.

Brower (1958) was among the first to formalize the view that predation risk may deny herbivores access to entire plant populations and/or species. Pointing out that insects bearing specific protective resemblance to particular plant species suffer increased, even intolerable, risk on plant species they don't resemble, he suggested that this effect could exclude the latter species from a diet which is suitable as food. In this view, predation risk *denies access* to suitable host plants, and reduces the overall food abundance for the herbivore. Schultz (1980) argued that the apparent rapid adaptation to, and high degree of specialization on, creosote bush (*Larrea divaricata;* Zygophyllaceae) by herbivorous insects in the Sonoran Desert has been driven by selection pressure from visual predators. Evidence suggests that the unique morphology of this plant combined with predation pressure restricts cryptic insects to it even if they could feed on other plant species in the same habitats (Schultz *et al.*, 1977).

Access to some food is denied to gypsy moth larvae by a different natural enemy: disease. Although larvae attain greater weights and considerably greater fecundity as adults when reared on aspen (*Populus* spp.) than when reared on oak (*Quercus* spp.; Rossiter, 1987; Schultz unpublished data, 1988), they are 20–50 times more susceptible to the gypsy moth nuclear polyhedrosis virus (LdNPV) when feeding on aspen (Keating and Yendol, 1987). Hence, aspen represents high food value in terms of fecundity but low value with respect to mortality risk (Schultz and Keating, 1991). From a behavioral standpoint, both tree species are readily available; larvae rarely show a clear feeding preference for one over the other. However, from the standpoint of population growth, aspen is much less available, because virus-caused mortality reduces population size, balancing or counteracting gains made on the basis of fecundity. Hence we observe episodes of rapid population growth and equally rapid collapse in aspen stands (J. Witter, personal communication, 1990). The true contribution of aspen biomass to gypsy moth population size—and hence aspen's availability—can be understood only if one considers the action of this natural enemy.

The same principle can be applied to variable risk within host plant species or populations. Recently, Zalucki *et al.* (1990) and Oyeyele and Zalucki (1990) have shown that monarch butterfly females oviposit preferentially on milkweed plant individuals and species having intermediate cardenolide contents. Both groups found that fecundity is reduced among larvae developing on high-cardenolide plants, and a long line of research has shown that predation risk increases for individuals developing on low-cardenolide plants. Hence, there is a tradeoff between growth/

fecundity and predation risk. Plant individuals and species with high allelo-chemical concentrations are, in effect, unavailable because they reduce fecundity, and those with low concentrations are unavailable because of increased predation risk. Insect population size is determined, at least in part, by the number of plants having intermediate cardenolide content. The mechanisms involved are both food utilization for growth and minimizing predation risk. One needs to understand both to be able to calculate true food availability for this insect; without the predation mechanism, only half of the story would become apparent.

In the same way that variation among host plant species can determine the efficacy of the gypsy moth virus, variation in leaf traits within northern red oak (*Q. rubra* L.) can make leaves of trees within and between stands differentially available. Both leaf chemistry and the amount of virus re-quired to kill a given proportion of gypsy moth larvae varied among eight stands in central Pennsylvania to an extent that the virus could be 2–4 times more effective in one stand than another (Schultz *et al.*, 1990). Hence, red oak tree leaves were functionally only half as available to support population growth in some stands. The impact of the same range of leaf quality varia-tion on insect growth and fecundity would be comparatively minor (Schultz *et al.*, 1990). Similar effects can be generated within stands (among, or even within trees) by induced responses to differential defoliations (Hunter and Schultz, unpublished data, 1991). The availability of red oak leaves within and between trees and stands may be determined in large part by the interaction between leaf quality and the impact of a natural enemy. In this particular case, the herbivore alters the shape of this interaction by stimulat-ing change in leaf quality (see Fig. 6).

There are numerous published examples of within-plant heterogeneity interacting with natural enemies to deny access by herbivores to host-plant tissues (Denno and McClure, 1983; Jefferies and Lawton, 1984). Whitham (1981, 1983) found that much of the contribution of cottonwood leaves to the growth and size of galling aphid populations is a function of leaf size and its impact on fecundity. However, the actual impact on aphid populations is altered by density-dependent mortality due to predation, which is more intense on large leaves. An accurate portrayal of leaf tissue availability to aphid populations necessarily involves mortality due to natural enemies.

Stamp and Bowers (1990) showed elegantly that harassment by vespid wasps denies access by buckmoth larvae (*Hemileuca lucina;* Saturniidae) to the most nutritious leaves on their host plant (*Spirea latifolia;* Rosaceae) for significant portions of the day. The result is reduced growth rates, and presumed decreased fecundity and increased mortality. As a complicating feature, the impact of this access denial is also influenced by temperature; caterpillars are restricted to cooler microhabitats by the wasps, which fur-ther slows development. Furthermore, the impact of leaf allelochemicals on larval growth is apparently temperature dependent (Stamp, 1990 and personal communication, 1990), so the influence of plant tissue quality variation on the insect is mediated by the activity of natural enemies and abiotic factors.

In a similar interaction of hostplant, enemy, and abiotic influences, Read *et al.* (1970) described a case in which microhabitat conditions determined the ability of a parasitoid to locate and exploit its host. As a consequence, the interaction of these factors determined plant tissue consumption and population density of the herbivore. In this case, resource availability to the herbivore was defined by the interaction between abiotic conditions and natural enemy effectiveness.

Many more examples of resource denial by natural enemies and interactions between plant traits and natural enemy effectiveness could be listed (e.g., see Schultz, 1983a,b; Barbosa and LeTourneau, 1990; Barbosa *et al.*, 1986). Three conclusions emerge from these observations. First, depending on the system, either the host plant or the natural enemy may have the greater impact on resource availability. In some cases, the plant's quality plays a dominant role, and inability to use a tissue to support growth constitutes a barrier to exploitation. In other cases, a tissue may be suitable or even superior as food, but access is denied by the action or risk of natural enemy mortality. In many cases, these two factors interact intimately and in potentially complicated ways. Because either the plant or the enemy, some interaction between them, and/or concatenated interactions with abiotic factors or additional enemies may constrain access to plant resources, it is difficult to generalize about top-down or bottom-up regulation of insect herbivore populations.

Second, examples support the view that there are three general modes of access denial by enemies. When selection from enemies has favored restricted resource use, as either evolved host plant specialization or behavioral avoidance, access to resources is denied to the individual, and by extension to the population. These forms of denial may have both spatial and temporal (e.g., seasonal) components; temporal constraints may reduce the number of generations an insect population may experience per unit time. If actual mortality occurs when a particular plant or plant tissue is consumed, the resource is unavailable for supporting population growth or maintenance because reproductive individuals die. In this sense, access to the resource is denied to the herbivore population. In all of these modes, it is the action of natural enemies interacting with variable plant traits that constrains or limits resource availability.

Third, the action of plant tissue variation on herbivores may involve a multiplicity of mechanisms, resulting in both direct and indirect effects acting via predators, parasitoids, and pathogens. It is difficult to draw conclusions about the availability of plant resources or their interaction with herbivores without elucidating these mechanisms. It is apparent in reviewing the interactions between plant tissue variation and natural enemies that it is difficult to conclude that insect herbivores usually are influenced solely, or even primarily, by the host plant.

III. Modeling Modes of Resource Availability

A. Fish Models and Herbivore Systems

Risk-constrained foraging has received considerable attention in recent years, mainly in studies of vertebrate predators (e.g., Gilliam and Fraser, 1987). Little attempt has been made to do the same with herbivores, especially insects. My goals in this section are twofold: first, I wish to suggest briefly how successful models of risk-constrained foraging by predaceous fish might be applied to herbivore systems. Specifically, I will use these modeling approaches to argue that the effects of natural enemies on herbivore growth and fecundity can be as dramatic as can those of diet quality. Second, I will present simple, graphical approaches to depicting the possible plant—herbivore—enemy interactions as a way of illustrating the diversity of potential outcomes. These approaches are likely to raise more questions than they answer, which, in large part, is the point of undertaking them.

Fish respond to both relative predation risk and habitat profitability in terms of feeding when choosing habitats in which to feed (Werner *et al.*, 1983a). Bluegill sunfish (*Lepomis macrochirus*) respond to changes in resource levels by modifying their selection of food particle size (Werner and Hall, 1974; Mittelbach, 1981; Werner *et al.*, 1983a) and shift habitats as resource levels change (Werner *et al.*, 1983a). The presence of predators in otherwise preferred habitats causes susceptible size classes of sunfish to alter habitat choice and accept a relatively lower return in food value (Werner *et al.*, 1983b). The result is decreased growth by the sunfish experiencing predation risk.

The insight brought to this system by Werner, Hall, and colleagues is that predation risk can be interchanged with reduced growth, bringing two otherwise separate currencies into congruence. Their ability to calculate at least relative growth losses as a function of risk allows one to integrate variation in resource availability as food with variation in resource availability due to risk. Werner *et al.* (1983b) derived a model to calculate survivorship from bluegill size s_1 to size s_2:

$$l_p(s_1,s_2) = [l_{np}(s_1,s_2)]^{c/c_g}$$

where $l_p(s_1,s_2)$ = survivorship in the presence of a predator, and $l_{np}(s_1,s_2)$ = survivorship in the absence of the predator. For all prey (bluegill here) sizes the presence of a predator multiplies daily mortality rate by a factor c, and the growth rate by a factor c_g ($c_g = 1$ indicates no effect). Werner *et al.* (1983b) point out that " . . . halving the growth rate (setting $c_g = 0.5$, holding $c = 1$) has the same effect on survivorship through a size-interval as doubling the daily mortality rate. . . . " Because the decrease in growth rate due to predation has the effect of delaying the entry of fish into the reproductive population, they conclude that " . . . a predator's "indirect" effect on the prey, inducing a lowered growth rate . . . might have greater effects on population demography than the "direct" effect of raising daily mortality rates . . . " (Werner *et al.*, 1983b, p.1546).

There must be many analogous cases involving insect herbivores. One of the most obvious involves the scheduling of feeding by insects susceptible to visually orienting predators (e.g., birds). In temperate forests (and many other habitats), a large fraction of Lepidoptera species feed exclusively at night, often migrating daily between daytime resting sites and nighttime feeding sites (Schultz, 1983b). During spring, this behavior constrains feeding to 8 hours or less per 24-hr period. This risk-constrained feeding time, plus the fact that nighttime temperatures are considerably lower than those during the day, must reduce insect growth rates dramatically. In the terms of Werner *et al.* (1983b) model, c_g must be reduced by at least two thirds, the equivalent of increasing daily mortality rates by 150%.

Slowed growth arising from risk avoidance could reduce fecundity directly if the timing of maturation is fixed or if suboptimal diets are selected, or it may translate into greater mortality due to extended lifespan. These interactions may restrict the lifespan of some herbivores to a fraction of the potential growth season and limit the number of generations to one per year. This is an important constraint on overall population growth.

To determine the impact of risk-avoidance constraints on herbivore populations, we would need to know the degree to which avoiding high-risk diets reduces individual fecundity, either because of time spent not feeding or because of poor diet quality. I am not aware of any quantitative data available to answer this question. In our gypsy moth system, individuals would suffer a 30–50% reduction in fecundity by feeding on oak instead of aspen (Rossiter, 1987); this could be a fecundity cost of avoiding aspen where the risk of disease mortality can be 20 times what it is on oak. However, the gypsy moth does not seem to avoid suitable food because of risk (as far as we know). Instead, there is a direct demographic effect via mortality. Whatever the mechanism, it seems likely that the importance of the host plant to gypsy moth populations may be greater as a mediator of natural enemy effects than as an influence on growth and fecundity (Schultz *et al.*, 1990; Schultz and Keating, 1991).

The Argentine grasshopper, *Astroma riojanum* (Proscopiidae), may represent a case in which feeding behavior is shaped by predation risk. Female grasshoppers select intermediate-aged leaves from pairs of leaves arrayed linearly along stems of its only host plant, *Larrea cuneifolia* (Zygophyllaceae) (Schultz, 1977; Schultz *et al.*, 1977). *Larrea* leaves are coated with a phenolic resin that slows the growth of this grasshopper, and which decreases in concentration with leaf age (Rhoades, 1978). By choosing leaves almost exclusively from the third-youngest age class (Schultz, 1977; Schultz *et al.*,

1977), this insect consumes intermediate levels of resin, as well as intermediate levels of protein and water (Rhoades, 1977). This behavior is apparently fixed evolutionarily; the grasshoppers feed exclusively at night (when birds are not active) and their preferences appear quite rigid (Schultz, personal observation, 1978).

Astroma females undergo a dramatic increase in body size as they mature; aviary experiments indicate that they become nearly immune to common bird species when they reach a bodysize threshold (Schultz, 1981). The apparent cause for this immunity is that birds find it necessary to remove the grasshopper's gut and gut contents before feeding on it. As a result of handling times that can exceed 5 min, other prey are comparatively more valuable, and these grasshoppers are avoided when adults (but not as nymphs, whose gut contents compose a smaller fraction of their mass and a smaller resin dose) (Schultz, 1981). Risk of predation for adult females is reduced by virtue of the resinous plant material in their guts.

Grasshoppers consuming the youngest *Larrea* leaf tissues would ingest higher resin concentrations than they do from their preferred leaves. This could provide superior protection from birds, but would slow growth and lengthen the time necessary to achieve the size threshold for "immunity". Individuals consuming older tissues would ingest less resin and grow faster, but their guts may not contain enough resin to deter birds. Unfortunately, we do not know the impact of resin on grasshopper reproduction.

According to Rhoades (1977), consuming the youngest leaves can slow grasshopper growth twofold to fourfold. In terms of the Werner *et al.* (1983b) model, this would be the equivalent of at least doubling mortality rates. Feeding on older tissues could increase growth rates by 50–100%. To complete this calculation, we need data on the risk associated with having too little resin in the gut; these are not available. However, it is clear that this grasshopper selects leaves of a particular age from an array of leaf age classes on which it could develop successfully, not necessarily to maximize growth, but as a compromise between advantages of rapid growth and the need for protection against predators.

The restriction of *Astroma*'s feeding to a small subset of leaves on any plant reduces the availability of food resources to the insect. Less than 20% of the total leaves on a plant are "available" to grasshoppers on this basis (Schultz, unpublished data 1978). This reduced availability depends on an interaction between the plant and the grasshopper's natural enemies. Were there no natural enemy effect, it is possible that five times more leaf material would be available to support grasshopper populations than is really the case.

Abiotic factors probably also play a role. *Larrea cuneifolia* is a desert shrub, and the growth of congeners is influenced by water and nutrient availability (Lightfoot and Whitford, 1987). Rapidly growing plants may have leaf populations with shifted age structures, and may have reduced resin concentrations overall (Schultz, personal observation, 1978). Hence, water availability could alter the array and amount of leaf tissues available to *Astroma* or other herbivores. Nonetheless, the actual availability of leaves would still be considerably less than their number or biomass, and it would not be accurate to claim that either abiotic factors or the host plant were the most significant, or *driving* influences on insect populations.

Predation risk models developed for fish appear promising for helping us understand the interaction between plant variation and the action of natural enemies on herbivores. However, we need quantitative estimates of risk on each potential food resource; all the observations presently available in the literature are qualitative. A promising system in which these models are presently being applied is Stamp's buckmoth–*Spiraea*–wasp interaction (N. Stamp, personal communication, 1991).

B. When Is It the Plant, and When Is It the Enemy?

Depending on one's view (or biases), any of the examples discussed so far could be described as influenced mainly by the plant, enemies, or abiotic environment. This is expected when the interactions are complex and the underlying mechanisms obscure. I believe that some simple graphical descriptions of these interactions reveal the fact that all imaginable possibilities are likely to occur.

Let us depict a simple relationship between a hypothetical plant tissue trait and herbivore growth or fecundity as in Figure 1 (solid line). In this first example, there is no functional relationship between the two; their slope is zero. In this case, the insect is insensitive to variation in some plant allelochemical, or nutritional, or physical trait. Similarly, there is no impact of variation in this trait on survivorship in the face of natural enemy attack (dashed lines). The position of the dashed survivorship line below that of the fecundity line indicates that population growth is constrained more by enemies than by food. In effect, we can think of the area below the lowest line(s), b, as the carrying capacity (K) for our herbivore. Here, it is set by enemies, and this graph depicts the hypothesis of Hairston *et al.* (1960): herbivores are always limited by enemies, not food. The position of the dotted survivorship line in Figure 1 depicts the "anti-Hairston *et al.*" (1960) view in which food is always more limiting than are enemies, producing carrying capacity "a".

In Figure 2, I describe the more common situation in which insect growth–fecundity varies along a gradient of a varying plant trait; Figure 2A depicts an increase. This might be a typical response to a nutrient's increase in concentration. I have also depicted survivorship as unresponsive to diet. Believing that enemies are always either more or less important than diet would position the survivorship line completely below or above (respectively) the fecundity line. However, the examples described thus far (and the literature in general) suggest that this is rarely, if ever, the case. Hence, the survivorship and fecundity lines cross at some point (Fig. 2A). To the left of this point, food is the most important influence on herbivore population

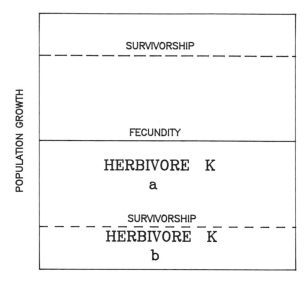

Figure 1   Relationship between a variable plant trait (e.g., nutrient content, allelochemical concentration, tissue toughness, stature, morphology) and the contribution of herbivore fecundity (solid line) and survivorship in the face of attack by natural enemies (predators, parasites, pathogens). In this case, neither fecundity nor the effectiveness of natural enemies is influenced by the particular plant trait in question. In this and subsequent graphs, the lowest line or combination of lines sets the lower limit on herbivore population growth (Herbivore K, a and b). Hence, the space so defined represents the impact of this plant trait (via its two effects) on herbivore carrying capacity.

size (via fecundity). To the right of the intersection, enemies are more important. The herbivore's K is represented by the area below the lowest segments of both lines.

In Figure 2A, the availability of plant resources is set by the plant to the left of the intersection, and by enemies to the right of the intersection. In such a case, whether the plant or the enemy appears to be the dominant force depends upon the state of the variable plant trait. In Figure 2B, I have reversed the impact of a plant trait on fecundity; it declines along a gradient of variation while natural enemy impacts remain constant. In this case, enemies are more important to population growth and K on the left, food more important via fecundity on the right.

Since plant tissue varies on many scales in space and time, these graphs may represent many scenarios. For example, continuous variation in a plant trait from left to right as depicted in Figure 2A could represent increasing leaf quality (e.g., protein content) along a gradient of soil quality. Thus, on poor soils (left) food's impact on herbivore populations may be dominant, but on better soils (right) enemies may become more influential. Figure 2B

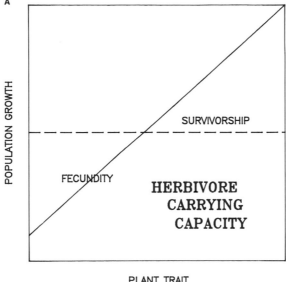

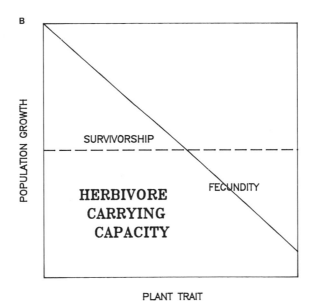

Figure 2 Here there is a positive (A) or negative (B) effect of plant trait variation on herbivore fecundity, and no effect on natural enemy effectiveness. The plant trait variable sets the herbivore's carrying capacity at the left end of its variation in A (e.g., low nutrient concentrations), at the right end in B (e.g., high allelochemical concentrations). Natural enemies are more effective at the opposite ends of plant variation range, by default.

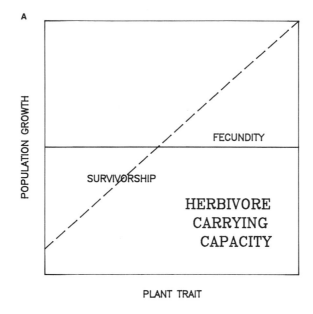

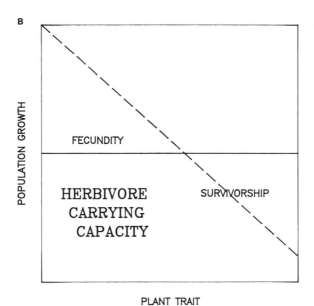

Figure 3   Here there is a positive (A) or negative (B) effect of plant trait variation on survivorship under attack by natural enemies, with no effect of the same trait on fecundity. Herbivore carrying capacity is set by natural enemies to the left of the plant trait gradient in A, to the right in B.

could depict variation in a plant trait that has an increasingly negative impact on fecundity or growth as phenological time goes on; natural enemies are more important early (left). This case resembles maturation of forest tree leaves, during which toughness increases while nutrients decrease. Insect growth would rarely be limited by food quality on the youngest tissues, but could be so on the older leaves. Other factors that could produce continuous variation in leaf traits include induction by damage (in which case the depicted impacts may be insect density dependent), abiotic conditions, plant genotype, or changes in herbivore physiology with age.

Figures 3A and B depict survivorship as more responsive to plant trait variation than is fecundity/growth. Again, zones of plant variation in which the plant or the enemies appear to have dominant influences on the herbivore can be identified. Figures 3A and B may represent cases in which there are differences among host plant species in terms of risk, but not in terms of food quality. For example, alternative host plants confer differential survivorship on *Pieris napi* in the face of parasitoid attack, even though the alternative plant species are suitable food (Ohsaki and Sato, 1990). These host plants could be arrayed in order of the safety they provide from parasitoid mortality from left to right to produce Figure 3A. The case described by Read *et al.* (1970) in which flea beetle susceptibility to parasitoid attack (but evidently not beetle growth) varied with the shadiness of the host plant habitat may be depicted by Figure 3B, in which the plant trait would be shade. Continuously varying interactions with enemies could also arise from enemies whose activity is facilitated by plant variation, or density-dependent mortality (with highest herbivore densities on the best host plant tissues).

The literature suggests that there are many variable plant traits which influence both fecundity/growth and survivorship under natural enemy attack. Hence, these two factors must frequently vary as functions of a single plant trait. Naturally, there could be an infinite array of possible relationships, but they must generally fall into the four classes depicted in Figure 4. I have depicted cases in which both fecundity and survivorship are influenced positively (A,B) or negatively (C,D) by the same trait. The steepness of the slopes indicates the strength of the plant trait's influence on each factor (greater impact on survivorship than fecundity in A,C; greater impact on fecundity in B,D). Figures 4A and B represent cases in which variation in food quality, e.g., a nutrient, has a positive influence on growth and also on the insect's ability to resist enemies, for example, disease (Hare and Andreadis, 1983; see review in Schultz and Keating, 1991). Figures 4C and D represent cases in which variation in food quality, e.g., an allelochemical, has a negative effect on both growth and disease resistance.

The dynamics of gall makers (*Euura lasiolepis*, Tenthridinidae) on desert willows may represent a good example of Figure 4A (or a curvilinear variation), with plant vigor as the plant trait axis, and the fecundity/survivorship intersection well to the left. Galler growth and fecundity increase with plant vigor, as does survivorship in the face of parasitoid attack (Chapter 6). Because this is a desert system, in which plant growth frequently may be limited by abiotic factors (e.g., water), variation in plant traits could be particularly extreme so that the impact of variation in plant traits on fecundity is driven by abiotic factors. But the outcome in terms of gallmaker populations depends on a weak natural enemy impact. There also are many examples in which natural enemy impacts on galling insects are profound (as in Figs. 4C, 6) and/or interact with host plant traits (Chapter 6, Clancy and Price, 1989; Lawton and Strong, 1981). The impression that the plant is the main driving force in the *Euura*–willow system depends on the particular mechanisms by which the plant and enemies play their roles and the

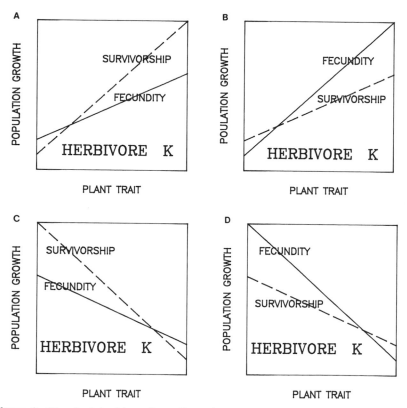

Figure 4   Here both herbivore fecundity and survivorship under natural enemy attack are influenced by the same variable plant trait(s). The impact on both is positive in A and B, negative in C and D. Survivorship is more strongly influenced by the plant in A and C; fecundity is more strongly influenced in B and D. The relative slopes of these lines—the strength of the plant's influence via these two routes—will determine their intersections, and hence over what range of plant variation the plant or natural enemies will have a dominant effect on herbivore carrying capacity.

strength of abiotic factors. This is a good example of a system in which underlying mechanisms are well enough known to permit us to infer causality and describe the net result of complex interactions.

Figures 3 and 4 could also represent leaf quality changes arising in a damage-density-dependent way from defoliations (see Schultz, 1988a). In the gypsy moth/oak system, we know that induced changes in leaves reduce growth and fecundity (Rossiter *et al.*, 1988). Recently, we have shown that induced changes in leaves also increase survivorship in the face of viral attack (Hunter and Schultz, unpublished data, 1991). Figure 4A suggests that the virus should be an important influence on populations when little defoliation has occurred, but food quality's effect on fecundity should become more important as defoliation proceeds. The larvae become increasingly immune to the virus as an outbreak proceeds. However, the confamilial Douglas fir tussock moth (*Orgyia menziesii;* Lymantriidae) becomes more *susceptible* to its nuclear polyhedrosis virus as outbreaks proceed (Fig. 4C). Even in these closely related insects, the nature and direction of the relationships among plant tissue traits, insect performance, and susceptibility to natural enemies can be very different.

Figure 5 presents the most extreme tradeoff situations, in which diet variation works in opposite directions on fecundity/growth and survivorship in the face of natural enemies. The direction of the differences are completely system specific. All of these crossed effects (Figs. 4 and 5) of diet on growth and resistance to enemies are well represented in the literature for diseases (Kushner and Harvey, 1962; Schultz and Keating, 1991), parasitoids (Barbosa and LeTourneau, 1988), and predators (Zalucki *et al.*, 1990). In some cases, high quality food (good growth) increases susceptibility to enemies, in others high quality food decreases susceptibility. The outcome depends on the mechanisms determining susceptibility and the nature and strength of the diet's impact on it and fecundity. For example, increasing phenolic concentrations decrease gypsy moth growth and fecundity (Rossiter *et al.*, 1988), but increase survivorship in the face of viral attack (Schultz *et al.*, 1990) (Fig. 5A). Evidence from the plant disease (Pridham, 1960) and pharmacological (Vanden Berghe *et al.*, 1985) literatures suggests that the mode of action of phenolics on viruses is the same in a wide variety of systems and may frequently reduce host susceptibility. However, dietary phenolics can reduce survivorship under microbial attack in other systems, where the interaction between phenolics and the particular microbes or host is different or absent (e.g., Steinly and Berenbaum, 1985) (Fig. 5B).

In the tobacco hornworm, dietary alkaloids can increase survivorship under attack by generalist parasitoids, but reduce it (or play no role) under attack by specialist species (Barbosa *et al.*, 1986). The difference lies in specific biochemical mechanisms possessed by the specialist and generalist parasitoids. In this system, we are still learning what the effects of dietary alkaloids are on the insect's growth and fecundity, but they appear to be negative (Appel and Martin, 1991).

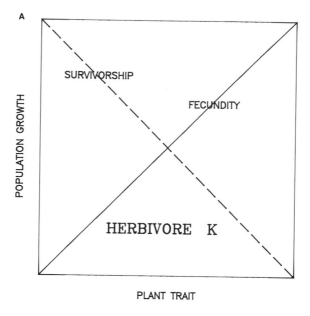

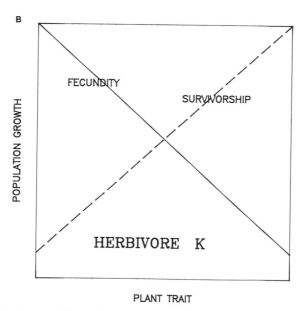

Figure 5   The impact of host plant traits can be opposite on fecundity and survivorship; both possible opposite relationships are depicted here. The maximal direct plant effect and maximal enemy effects occur at opposite ends of the plant trait spectrum.

Figures 1–5 are oversimplified for many systems. It is unlikely that relationships between plant traits and either growth/fecundity or survivorship are linear. Figure 6 is based on what we know about the gypsy moth and oak leaf quality. Fecundity appears to decline only after a threshold change in leaf quality (probably hydrolyzable tannin concentrations), while inhibition of viral infection is increased linearly. In this case, or in a case in which both relationships are curvilinear, the points along a plant trait continuum at which the plant or enemies appear to be the dominant force may be difficult to identify. There could be multiple switches in dominance (as in Fig. 6), depending on the shapes of the relationships involved.

It is increasingly apparent that host-plant impacts on insect growth are nonlinear and frequently involve thresholds. Determining which is more important, the plant or the enemies, can be quite complex. I can identify few, if any, generalizations about these relationships at present. There is no reason to expect one or a few of the kinds of interactions suggested by Figures 1–6 to be more common than others. One of the most likely, in which poor food quality weakens the herbivore and decreases survivorship under disease attack (Fig. 3C,D) (*physiological stress hypothesis*, Steinhaus,

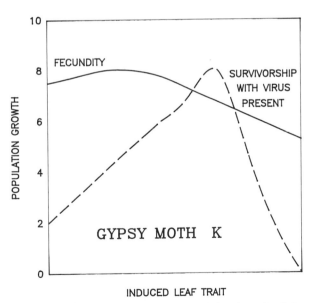

Figure 6 Graphical depiction of the impact of variation in red oak leaves induced by defoliation on gypsy moth fecundity and survivorship when exposed to a baculovirus. These changes cause about a 30% decrease in fecundity at their maximum, beyond a threshold. Over most of this range, larvae become increasingly resistant to the virus. Beyond a leaf quality (larval density) threshold, this resistance fails. The virus has the major impact on population growth at both ends of the leaf trait gradient.

1958) now competes with an equal or greater number of counterexamples (Schultz and Keating, 1991). The number of participants is too great, the possible outcomes too numerous, and the mechanisms involved too sketchy at present.

IV. Obese Generalizations: To Reduce or Not?

The search for generalizations and attempts to develop predictive power are important aspects of a maturing science. They are certainly major goals of ecology. However, as a science matures, the ease with which generalizations arise and their validity increase as a function of the knowledge underlying them. This comes about as reductionist approaches integrate with conceptual approaches. I feel that our understanding of host plant–insect, insect–enemy, and tritrophic interactions can support only the slimmest of generalizations.

We know the identities of only a fraction of a percent of the potential allelochemicals in plant tissues, even in well-studied systems. We do not know the nutritional requirements of the majority of pest insects, much less nonpests. We are just beginning to appreciate interactions between nutrition and allelochemical activity in insects (Slansky, 1991). We do not know the mode of action of the vast majority of potential plant allelochemicals we have identified. The mechanisms governing tritrophic interactions and the resulting dynamics remain mysteries.

It is clear that we cannot understand the impact of host plants on herbivores without including at least a third trophic level (Price *et al.*, 1980; Schultz, 1983b, 1988b). The net impact of any plant trait on an insect herbivore usually seems to be altered by its interaction with a predator, parasitoid, or pathogen. Looking at it another way, impacts of natural enemies on herbivores often seem to depend on or be modified by plant traits. Access by insects to plant resources is governed by an interaction between plant impacts and enemy impacts. In some cases, the plant impact may appear to dominate; in others, the enemy impact seems more important. Although the plant's traits are central to these interactions, I think it an overstatement to describe the plant as the sole "driving" factor in every case; its influence often depends on the presence and activity of other factors, mainly enemies. Like Karr *et al.* (Chapter 9), I believe most systems are probably structured by bottom-up and top-down effects acting together.

So how should we proceed? Even with the sorry state of knowledge about underlying mechanisms, it remains worthwhile to erect testable, generalizable hypotheses. It is only by refuting these that our science can advance. However, hypotheses must be based on as broad and detailed an understanding of the mechanisms actually governing tritrophic interactions as we can obtain. These mechanisms may produce congruent or opposite effects on the participants, even in similar or phylogenetically related systems. In my mind, this mandates an integration of conceptual with mechanistic research. Reductionist studies are needed to prevent obese generalizations that are too broad for support by current knowledge. Creative conceptual development is needed to provide appropriate context for mechanistic studies.

Access to plant resources is sometimes denied to herbivores by plants, sometimes by enemies, sometimes by abiotic factors, but usually by some combination of these. Which of these factors appears to dominate depends on the strengths of the others and (as far as we know now) system-specific underlying mechanisms. The same plant trait that improves growth in insect species A may reduce it for species B. The same plant trait that makes insect species A immune to disease may make species B more susceptible. The same trait that slows growth under one set of conditions may accelerate it under others. The net effect of a plant trait in any case could be nil over the entire range of trait variation, and in interaction with other factors (e.g., competitors, weather, other enemies). We are still in the example-gathering stage of developing understanding about these interactions. It is clear that access to plant resources is a function of plant variation and its interactions with other factors, and that these interactions, more than the variation itself, restrict the availability of plant tissues to insect herbivores and their populations.

Acknowledgments

Thanks to Mark Hunter and Heidi Appel for helpful comments on the manuscript. This work supported by NSF Grant BSR-8918083 (JCS and MDH), and BSR-8813433 (JCS and Mark Abrams).

References

Appel, H. M. and Martin, M. M. (1991). The significance of metabolic load in the evolution of host specificity of *Manduca sexta* (Lepidoptera; Sphingidae). *Ecology*, in press.

Barbosa, P. and Letourneau, D. K. (1988). "Novel Aspects of Insect–Plant Interactions." J. Wiley, New York.

Barbosa, P., Saunders, J. A., Kemper, J., Trumbule, R., Olechno, J., and Martinat, P. (1986). Plant allelchemicals and insect parasitoids. Effects of nicotine on *Cotesia congregata* (Say) (Hymenoptera: Braconidae) and *Hyposoter annulipes* (Cresson) (Hymenoptera: Ichneumonidae). *J. Chem. Ecol.* **12**, 1319–1327.

Berenbaum, M. (1978). Toxicity of a furanocoumarin to armyworms: A case of biosynthetic escape from insect herbivores. *Science* **201**, 532–534.

Bernays, E. and Graham, M. (1988). On the evolution of host specificity in phytophagous arthropods. *Ecology* **69**, 886–892.

Brower, L. P. (1958). Bird predation and food plant specificity in closely related procryptic insects. *Am. Natur.* **92**, 183–187.

Clancy, K. M. and Price, P. W. (1989). Effect of plant resistance, competition, and enemies on a leaf-galling sawfly (Hymenoptera: Tenthredinidae). *Environ. Entomol.* **18**, 284–290.

Denno, R. F. and McClure, M. S. (1983). "Variable Plants and Herbivores in Natural and Managed Ecosystems," Academic Press, New York.

Ehrlich, P. R. and Birch, L. C. (1967). The "balance of nature" and "population control." *Am. Natur.* **101**, 97–107.

Gilliam, J. F. and Fraser, D. F. (1987). Habitat selection under predation hazard: Test of a model with foraging minnows. *Ecology* **68**, 1856–1862.

Hairston, N. G., Smith, F. E., and Slobodkin, L. B. (1960). Community structure, population control, and competition. *Am. Natur.* **94**, 421–425.

Hare, J. D. and Andreadis, T. G. (1983). Variation in the susceptibility of *Leptinotarsa decemlineata* (Coleoptera: Chrysomelidae) when reared on different host plants to the fungal pathogen, *Beauvaria bassiana* in the field and laboratory. *Environ. Entomol.* **12**, 1892–1897.

Holmes, R. T. and Schultz, J. C. (1988). Food availability for forest birds: Effects of prey distribution and abundance on bird foraging. *Can. J. Zool.* **66**, 720–728.

Jeffries, M. J. and Lawton, J. H. (1984). Enemy-free space and the structure of ecological communities. *Biol. Linnean Soc.* **23**, 69–286.

Keating, S. T. and Yendol, W. G. (1987). Influence of selected host plants on gypsy moth (Lepidoptera: Lymantriidae) larval mortality caused by a baculovirus. *Environ. Entomol.* **16**, 459–462.

Kushner, D. J. and Harvey, G. T. (1962). Antibacterial substances in leaves: Their possible role in insect resistance to disease. *J. Insect Path.* **4**, 155–184.

Lawton, J. H. and Strong, D. R. (1981). Community patterns and competition in folivorous insects. *Am. Natur.* **118**, 317–338.

Lightfoot, D. C. and Whitford, W. G. (1987). Variation in insect densities on desert creosotebush: Is nitrogen a factor? *Ecology* **68**, 547–557.

Mittelbach, G. G. (1981). Foraging efficiency and body size: A study of optimal diet and habitat use by bluegills. *Ecology* **62**, 1370–1386.

Ohsaki, N. and Sato, Y. (1990). Avoidance mechanisms of three *Pieris* butterfly species against the parasitoid wasp *Apanteles glomeratus*. *Ecol. Entomol.* **15**, 169–176.

Oyeyele, S. and Zalucki, M. P. (1990). Cardiac glycoside and oviposition by *Danaus plexippus* on *Asclepias fruticosa* in southeast Queensland (Australia), with notes on the effect of plant nitrogen content. *Ecol. Entomol.* **15**, 177–185.

Price, P. W., Bouton, C. E., Gross, P., McPheron, B. A., Thompson, J. N., and Weis, A. E. (1980). Interactions among three trophic levels: Influence of plants on interactions between insect herbivores and natural enemies. *Annu. Rev. Ecol. Syst.* **11**, 41–65.

Pridham, J. B. (1960). "Phenolics in Plants in Health and Disease." Pergamon, New York.

Read, D. P., Feeny, P. P., and Root, R. B. (1970). Habitat selection by the aphid parasite *Diaeretiella rapae* (Hymenoptera: Braconidae) and hyperparasite *Charips brassicae* (Hymenoptera: Cynipidae). *Can. Entomol.* **102**, 1567–1578.

Rhoades, D. F. (1977). Integrated antiherbivore, antidesiccant, and ultraviolet screening properties of creosotebush resin. *Biochem. Syst. Ecol.* **5**, 281–290.

Rossiter, M. (1987). Use of a secondary host by non-outbreak populations of the gypsy moth. *Ecology* **68**, 857–868.

Rossiter, M., Schultz, J. C., and Baldwin, I. T. (1988). Relationships among defoliation, red oak phenolics and gypsy moth growth and reproduction. *Ecology* **69**, 267–277.

Schultz, J. C. (1977). Competition, predation, and the structure of phytophilous insect communities. Ph.D. dissertation, Dept. of Zoology, University of Washington, Seattle, Washington.

Schultz, J. C. (1980). Asymptotic species accumulation models of Larrea insect communities. *In* "Larrea: A Vast Resource of the American Desert" (E. C. Gomez and T. J. Mabry, eds.), Centro de Investigacion en Quimica Aplicada, Saltillo, Mexico.

Schultz, J. C. (1981). Adaptive changes in the antipredator behavior of a grasshopper during development. *Evolution* **35**, 175–179.

Schultz, J. C. (1983a). Impact of variable plant chemical defenses on insect susceptibility to parasites and diseases. *In* "Plant Resistance to Insects" (P. A. Hedin, ed.), pp. 37–55. American Chemical Society, Washington, D.C.

Schultz, J. C. (1983b). Habitat selection and foraging tactics of caterpillars in heterogeneous trees. *In* "Variable Plants and Herbivores in Natural and Managed Systems" (R. F. Denno and M. S. McClure, eds.), pp. 61–90. Academic Press, New York.

Schultz, J. C. (1988a). Plant responses induced by herbivores. *Trends Ecol. Evol.* **3**, 45–49.

Schultz, J. C. (1988b). Many factors influence the evolution of herbivore diets, but plant chemistry is central. *Ecology* **69**, 1364–1369.

Schultz, J. C., Otte, D., and Enders, F. (1977). Larrea as a habitat component for desert invertebrates. *In* "Creosote Bush: Biology and Chemistry of Larrea in New World Deserts" (T. J. Mabry, J. Hunziker, and D. DiFeo, eds.), pp. 176–208. Dowden, Hutchinson and Ross, Stroudsburg, Pennsylvania.

Schultz, J. C., Foster, M. A., and Montgomery, M. E. (1990). Host plant–mediated impact of pathogens on gypsy moth populations. *In* "Population Dynamics of Forest Insects" (A. Watt and M. D. Hunter, eds.) pp. 303–313. NERC Institute of Terrestrial Ecology, Edinburgh, Scotland.

Schultz, J. C. and Keating, S. T. (1991). Host plant–mediated interactions between the gypsy moth and a baculovirus. *In* "Microbial Mediation of Plant–Herbivore Interactions" (P. Barbosa, V. A. Krischik, and C. G. Jones, eds.), pp. 325–337. J. Wiley, New York.

Slansky, Jr., F. (1991). Allelochemical/nutrient interactions in herbivore nutritional ecology. *In* "Herbivores: Their Interaction with Secondary Plant Metabolites," 2nd Ed. (G. A. Rosenthal and M. R. Berenbaum, eds.), Academic Press, San Diego, in press.

Stamp, N. E. (1990). Growth versus molting time of caterpillars as a function of temperature, nutrient concentration, and the phenolic rutin. *Oecologia* **82**, 107–113.

Stamp, N. E. and Bowers, M. D. (1990). Variation in food quality and temperature constrain foraging of gregarious caterpillars. *Ecology* **71**, 1031–1039.

Steinhaus, E. A. (1958). Stress as a factor in insect disease. *Proc. 10th Intern. Congr. Entomol.* **4**, 725–730.

Steinly, B. A. and Berenbaum, M. (1985). Histopathological effects of tannins on the midgut epithelium of *Papilio polyxenes* and *Papilio glaucus*. *Entomol. Exp. Appl.* **39**, 3–9.

Strong, D. R., Lawton, J. H., and Southwood, T. R. E. (1984). "Insects on Plants. Community Patterns and Mechanisms." Blackwells, London.

Vanden Berghe, D. A., Vlietinck, A. J., and Van Hoof, L. (1985). Present status and prospects of plant products as antiviral agents. *In* "Advances in Medicinal Plant Research," 32nd Intern. Congr. Medicinal Plant Research (A. J. Vlietinck and R. A. Dommisse, eds.), pp. 47–99. Wissenschaftliche Verlagsgesellschaft, Stuttgart, Germany.

Werner, E. E. and Hall, D. J. (1974). Optimal foraging and the size selection of prey by the bluegill sunfish (*Lepomis macrochirus*). *Ecology* **55**, 1042–1052.

Werner, E. E., Mittelbach, G. G., Hall, D. J., and Gilliam, J. F. (1983a). Experimental tests of optimal habitat use in fish: The role of relative habitat profitability. *Ecology* **64**, 1525–1539.

Werner, E. E., Gilliam, J. F., Hall, D. F., and Mittelbach, G. G. (1983b). An experimental test of the effects of predation risk on habitat use in fish. *Ecology* **64**, 1540–1548.

Whitham, T. G. (1981). Individual trees as heterogeneous environments: Adaptation to herbivory or epigenetic noise? *In* "Insect Life History Patterns: Habitat and Geographic Variation" (R. F. Denno and H. Dingle, eds.), pp. 9–27. Springer, New York.

Whitham, T. G. (1983). Host manipulation of parasites. *In* "Variable Plants and Herbivores" (R. F. Denno and M. S. McClure, eds.), pp. 15–41. Academic Press, New York.

Wiegert, R. G. and Owen, D. F. (1971). Trophic structure, available resources, and population density in terrestrial vs. aquatic ecosystems. *J. Theor. Biol.* **30**, 69–81.

Zalucki, M. P., Brower, L. P., and Malcolm, S. B. (1990). Oviposition by *Danaus plexippus* in relation to cardenolide content of three *Asclepias* species in the southeastern U.S.A. *Ecol. Entomol.* **15**, 231–240.

8

Resource Limitation on Insect Herbivore Populations

Takayuki Ohgushi

Faculty of Agriculture
Shiga Prefectural Junior College
Kusatsu, Shiga 525, Japan

I. Introduction

A number of long-term studies of natural insect populations, stimulated by the well-known debate surrounding density-dependent population regulation in the late 1950s (Nicholson, 1954; Lack, 1954; Andrewartha and Birch, 1954; Milne, 1957), have made a major contribution to the development of contemporary theories in population ecology (see Southwood, 1968; and den Boer and Gradwell, 1971). Insect population ecologists have mainly focused on natural enemies as principal agents of population regulation (Varley *et al.*, 1973; Dempster, 1975). In contrast to the emphasis on the role of food resources in population dynamics in mammals and birds (Lack, 1966; Ostfeld, 1985; Sinclair, 1989), little attention has been paid to possible effects of plant resources on dynamic patterns in insect herbivore populations, except for resource depletion (Denno and McClure, 1983; Crawley, 1983; Strong *et al.*, 1984). Indeed, few long-term population studies have directly measured the plant resource available to insect populations (Dempster and Pollard, 1981). However, recent reviews of the literature on population dynamics of many herbivorous insects have suggested that intraspecific competition for food resources is important as a limiting factor in insect populations (Dempster, 1983; Stiling, 1988). But food shortage due to heavy herbivory is merely one among many possible effects that are related to plant resources.

The traditional argument in insect population studies has largely overlooked population consequences of host plant characteristics on insect herbivores for several reasons: (1) plants have been considered a superabundant and homogeneous resource; (2) plant attributes have not been incorporated into life table analysis; (3) theoretical studies have mainly concentrated on systems consisting of an insect and its predators or parasitoids; and (4) applications of pest control have emphasized the predominant role of natural enemies. However, recent studies of insect–plant interactions have revealed that a plant is a much more variable and heterogeneous resource for insects than previously thought, and have noted that dynamic features of host plants have significant impacts on insect performance in terms of survivorship and reproduction, and thus population dynamics (Crawley 1983, 1989; Denno and McClure, 1983; Whitham, 1983; Kareiva, 1983; Price, 1983, 1984, 1991; Strong *et al.*, 1984; Wratten *et al.*, 1988; Price *et al.*, 1990). These studies have emphasized the dynamic nature of plant resources on various scales in space and time, and contrast with the previously accepted static view that a plant is an abundant, constant, and homogeneous resource for herbivores (e.g., Hairston *et al.*, 1960).

This chapter addresses the importance of plant variability as a possible agent in limiting insect herbivore populations. Plant resources may not always be predominant regulators in every herbivorous insect system, but direct and indirect effects that plants mediate have long been ignored in empirical studies of herbivorous insect populations. By incorporating plant variability, natural enemies will be placed in their proper perspective in the context of insect population dynamics. I shall emphasize the need for a connection between individual and population ecology (see Chapters 2 and 3) for a more exact understanding of the real effects of insect–plant interactions on the population dynamics of herbivorous insects. Last, I shall briefly illustrate resource limitation on a population of herbivorous lady beetles and its causal mechanism.

II. Resource Limitation

A resource is an environmental factor that is directly used by an organism and that potentially influences individual fitness (Wiens, 1984). In this sense, a plant is a resource for many insect species, providing food, accommodation, and shelter from enemies or environmental stress (Strong *et al.*, 1984). Since individual fitness greatly affects properties at the population level through the processes of survival and reproduction, simple deduction tells us that the plant resource plays an important role in determining population dynamics of insect herbivores. Surprisingly, this idea has received little attention in the long-standing ecological debate on how herbivore insect populations are regulated far below the level of apparent plant defoliation (Strong *et al.*, 1984).

Having considered control agents in populations at different trophic levels, Hairston, Smith, and Slobodkin (1960) hypothesized that herbivores are not resource limited because obvious depletions of green plants by herbivores are exceptions to the general picture, in which the plants are abundant and largely intact. Until recently, the Hairston–Smith–Slobodkin hypothesis has endorsed the conventional wisdom that because the world is green, it is not possible that insect herbivores can be resource limited. The Hairston–Smith–Slobodkin hypothesis, however, hopelessly simplifies the complexities of biological reality (Wiens, 1984), and it is invalid to conclude that the plant resource is not a limiting factor of herbivore populations merely by observing that plants are abundant and remain largely intact (Murdoch, 1966; Ehrlich and Birch, 1967; Rockwood, 1974; Wratten *et al.*, 1988). Crawley (1989) summarized these criticisms as follows: (1) the world is not always green; (2) all that is green is not edible; (3) what is edible is not necessarily of sufficiently high quality to allow increase of the herbivore population. In other words, an apparently abundant plant resource may, in fact, represent a food supply already fully exploited in terms of palatable, nutritious, nontoxic plant material. Although the notion in the Hairston–Smith–Slobodkin hypothesis that plants are generally underutilized food resources has been criticized, the central assertion on the relative importance of predation and parasitism as a limiting agent has not been refuted (Karban, 1986).

An increasing interest in insect–plant interactions has thrown light on the dynamic nature of plant resources. The following evidence obtained from these studies is essential when considering possible roles of host plants in demographic properties of insect populations: (1) plants are highly variable and heterogeneous resources in various scales in space and time (Denno and McClure, 1983; Whitham, 1983; Kareiva, 1983; Price, 1984); (2) plant defense strategies increase resource inaccessibility (Feeny, 1976; Rhoades and Cates, 1976; Schultz, 1983a, 1988; Haukioja and Neuvonen, 1987); and (3) resource availability is not equivalent to resource abundance (Wiens, 1984; Price, 1984). Furthermore, plant resources have not only direct but also indirect effects on survivorship and reproduction of insect herbivores through enhancing or limiting the efficacy of natural enemies (Lawton and McNeill, 1979; Price *et al.*, 1980; Schultz 1983a). This newly emerged dynamic view of plant resources has also noted that heterogeneity specific to plant species greatly reduces the availability of acceptable resources to herbivorous insects, and that various plant characteristics, including nutrients, chemical toxicants, spatial dispersion, and phenological variation, may maintain populations at low densities relative to perceived plant resources (Lawton and McNeill, 1979; Price *et al.*, 1980; Schultz, 1983a; Pimentel, 1988).

In this chapter, I use the term *resource limitation*, following Sinclair (1989), as the process which sets the equilibrium point of an herbivore population. Thus, any factors causing a change in production (inputs of births and immigrants) or loss (outputs of deaths and emigrants) in the population are limiting factors. Note that limiting factors are not necessarily density dependent, which is a necessary condition for regulating factors. Here, I will emphasize particularly the role of plant resources as an agent limiting the equilibrium point of herbivorous insect populations.

III. Possible Causes of Resource Limitation

This section will briefly illustrate possible roles of host plant attributes in limiting insect herbivore populations. I note that a wide variety of plant attributes have significant impacts on insect performance even when a large proportion of the host plants remain intact. You will notice some overlap between the factors that I consider may limit herbivore populations, and those considered by Hunter (Chapter 10) to influence herbivore community structure. The differences reflect Hunter's discussion of resources as influenced by other members of the herbivore community, whereas interactions between species are not a necessary condition of resource limitation as defined here. Since direct and indirect influences by plants will not be fully discussed here; for more details one should refer to some excellent reviews of insect–plant interactions (e.g., Feeny, 1976; McNeill and Southwood, 1978; Mattson, 1980; Price *et al.*, 1980; Scriber and Slansky, 1981; Price, 1983, 1984; Kareiva, 1983; Schultz, 1983a; Faeth, 1987; Haukioja and Neuvonen, 1987; Karban and Myers, 1989).

A. Plant Depletion

Host plant depletion is highly noticeable in some systems and undoubtedly has a substantial impact on insects that utilize those plants. Intraspecific competition from food shortage involves starvation, dispersal, and reduced fecundity (Dempster,1983). Accordingly, the view that intraspecific competition due to food shortage ultimately limits an infinite increase of insect numbers was established in the early stages of insect population studies (Milne, 1957; Harcourt, 1971; Dempster, 1975; Strong, 1984). In particular, several authors have argued that intraspecific competition for food resources at high densities is the only density-dependent agent regulating herbivore populations (Milne,1957; Dempster,1983).

A well-known example of resource depletion and its consequences for population dynamics is the cinnabar moth on tansy ragwort (Dempster, 1971, 1982; Myers and Campbell, 1976; Myers, 1980; Crawley and Gillman,1989). Cinnabar moth populations frequently reach such a high density that host plants are highly exploited, and thereby larvae often face severe intraspecific competition due to food shortage. This resource depletion causes greater larval death from starvation, density dependent larval loss during dispersal to search for food, and produces smaller adults with lower fecundity. Consequently, the abundance of the moth population is principally determined by the biomass of the host plant (Dempster and Pollard, 1981). Increased larval dispersal due to food shortage also leads to overexploitation of the host, resulting in larger population fluctuations (Myers and Campbell, 1976; Myers and Post, 1981). Similarly, strong population limitation, caused by larval starvation and reduced fecundity, results from host depletion by the moth *Coleophora alticolella* in northern England (Randall, 1982). In comparing 17 local populations of the checkerspot butterfly, *Euphydryas editha*, feeding on different food plants, White (1974) found that host defoliation was quite common and often extensive, and concluded that starvation was limiting in at least several local populations. Dempster (1983) summarized the literature on population studies of lepidopteran insects and found that in thirteen out of sixteen cases in which density dependence was demonstrated, some degree of intraspecific competition for resources at high densities was involved.

B. Plant Phenology

Some herbivorous insects that use ephemeral plants or plant parts are often subjected to heavy immature mortality due to a rapid seasonal change in food quantity and quality (Rockwood, 1974; Feeny, 1976; Thompson and Price, 1977; Whitham, 1978, 1980; Futyuma and Wasserman, 1980; Solomon, 1981; Raupp *et al.*, 1988; Aide, 1988; Aide and Londoño, 1989; Hunter, 1990). A slight shift of phenological matching between insects and the host plants suitable for them may have a profound effect on insect survivorship and reproduction. Phenological asynchrony and its consequences for larval survival and growth have been described in moths (Thompson and Price, 1977; Raupp *et al.*, 1988) and aphids (Whitham, 1978, 1980). The impact of phenological asynchrony on survival of newly hatched larvae has been well illustrated in the winter moth *Operophtera brumata* (Feeny,1976). The newly hatched larvae cannot feed on unopened oak buds, nor can they survive on young leaves that have begun to toughen. When larvae hatch more than a few days before bud-burst, they suffer heavy mortality due to food shortage. In contrast, if bud-burst occurs much before egg-hatch, larvae suffer heavy mortality owing to leaf toughness. Similarly, Aide and Londoño (1989) reported the effects of rapid changes in leaf quality on insect performance. Leaf quality of an understory tropical tree *Gustavia superba* quickly declines following rapid leaf expansion (producing a fully expanded leaf from a small bud in 6 to 8 days). A rapidly expanding leaf that matures quickly employs the defenses of leaf toughness and decreased water and nutrient content. As a result, larvae of the lepidopteran herbivore *Entheus priassus* that hatch the day before the leaf is fully expanded are four times more likely to survive than larvae that hatch on the day the leaf reaches full size.

Leaf abscission is a further phenological variable that can be an important source of larval mortality, especially for highly sedentary insects such as leafminers, leaf gallers, and leaf rollers (Faeth *et al.*, 1981; Williams and Whitham, 1986; Clancy and Price, 1986; Simberloff and Stiling, 1987; Auerbach and Simberloff, 1989). Auerbach and Simberloff (1989) noted the relative importance of early leaf abscission as a dominant mortality factor during the larval stage of the leaf-mining moth *Lithocolletis quercus* on the oak *Quercus calliprinos*. Leaf abscission accounted for 43% of the larval mortality of the leaf miner, compared to only 5.8% from natural enemies. A similar high larval mortality (more than 30%) due to induced leaf abscission has been described in the leaf-mining moth *Stilbosis quadricustatella* on the evergreen oak *Quercus geminata* (Simberloff and Stiling, 1987). By reducing larval survival, leaf abscission can strongly limit insect populations. Williams and Whitham (1986) examined the responses of two species of cottonwoods, *Populus*, to gall aphid species in the genus *Pemphigus*, and found that leaf abscission was rapidly induced by gall aphid attack even at low aphid densities, and reduced the aphid population by 25% on narrowleaf cottonwood and by 53% on Fremont cottonwood. However, leaf abscission has little effect on insect performance when larvae are fully developed at the time of abscission (Stiling and Simberloff, 1989: Kahn and Cornell, 1989).

C. Plant Quality

1. Nutrition

As animals consist mainly of protein, whereas plants consist mainly of carbohydrates, nitrogen is the key limiting nutrient for many herbivorous insects (McNeill and Southwood, 1978; Mattson, 1980; Crawley, 1983; Strong *et al.*, 1984; Brodbeck and Strong, 1987). Increased availability of nitrogen often improves larval survival and growth rate (McNeill, 1973; Scriber and Slansky, 1981; Tabashnik, 1982; Cates *et al.*, 1987; Bryant *et al.*, 1987), and female pupal weight and fecundity (McNeill, 1973; Myers, 1981; Minkenberg and Ottenheim, 1990). For example, Bryant *et al.* (1987) experimentally demonstrated that leaf nitrogen of the aspen, *Populus tremuloides*, increased when fertilized which, in turn, resulted in higher larval survival of the large aspen tortrix *Choristoneura conflictana*. Ohmart *et al.* (1985), however, noted that there was a nitrogen-concentration threshold above which increased nitrogen no longer improved pupal weight and developmental time in the leaf beetle *Paropsis atomaria* feeding on *Eucalyptus blakelyi*.

On the other hand, there is evidence that foliar nitrogen is not correlated with species diversity or population density of herbivorous insects (e.g., Faeth *et al.*, 1981). Also, different types of insects tend to respond differentially to total plant nitrogen (Mattson, 1980; Scriber and Slansky, 1981; Crawley, 1983). For sucking insects, nitrogen compounds are likely to increase larval performance, while foliar nitrogen sometimes decreases the performance of chewing insects.

Several authors have suggested that foliar nitrogen or amino acid contents determine population densities of herbivorous insects, by reducing survival and reproductive rates (Feeny, 1970; Dixon, 1970; Onuf *et al.*, 1977; Webb and Moran, 1978; McNeill and Southwood, 1978; Stiling *et al.*, 1982; Brodbeck *et al.*, 1990; but see Faeth *et al.*, 1981). For example, the abundance of herbivorous insects associated with oak trees falls with decreasing nitrogen content of oak leaves (Feeny, 1970). A similar result was obtained in the leaf miner, *Hydrellia valida*, on salt marsh cord grass, *Spartina alterniflora* (Stiling *et al.*, 1982). Webb and Moran (1978) showed a ten-fold increase in population level of the indigenous psyllid *Acizzia russellae* on the thorn tree *Acacia karroo* when the trees were previously pruned, and concluded that the nutrition of the host plant imposes the major limitation on psyllid population growth, which is likely to explain the permanently low endemic population levels of the insect. Leaf nitrogen may also influence population stability. Myers and Post (1981) showed that improved larval survival and moth fecundity in areas with food plants high in nitrogen led the moth populations to periodically overexploit their food supply, which enhanced population fluctuations.

The nutrient contents of host plants are not constant through time. In general, there is a decrease in water and nitrogen content with the aging of plants (McNeill, 1973; Stiling *et al.*, 1982; Johnson *et al.*, 1984). This age-dependent nutrient deficiency often limits population density of herbivores attacking old plants or plant parts (Rockwood, 1974; Potter and Redmond, 1989). In their artificial defoliation experiment on American holly, *Ilex opaca*, Potter and Redmond (1989) illustrated that a second leaf flush induced by an artificial defoliation early in the season offered a high-quality resource of high nitrogen and low leaf toughness available to larvae of the leafminer *Phytomyza ilicicola*, causing a six- to thirteen-fold increase in herbivory. They suggested that a decline in the concentration of nutrients and a rapid increase in leaf toughness with plant age were most likely to limit the population density of the leafminer.

Several studies have shown that foliar water can be limiting for chewing insects, and that low levels reduce larval growth (Scriber, 1977, 1984; Scriber and Slansky, 1981; Tabashnik, 1982). Scriber (1977) examined the effects of leaf water content on the performance of *Hyalophora cecropia* larvae, which fed on leaves of wild cherry, *Prunus serotina*, and found that in spite of no difference in leaf consumption, larvae on low-water leaves had lower growth rates and assimilation rates than those on fully water-supplemented leaves. On the other hand, White (1974, 1978, 1984) emphasized that water-stressed plants become more susceptible to and suitable for their adapted insects because the water stress response of plants can induce certain biochemical changes in plants. He hypothesized that these changes in plant quality allow higher survival of young larvae and higher fecundity, and thus lead to population outbreaks. Evidence that plant water deficits promote insect outbreaks is, however, largely circumstantial, and there are complex differences among taxonomic groups (Larsson, 1989); thereby the hypothesis is still in debate (Mattson and Haack, 1987; Preszler and Price, 1988; Waring and Price, 1990).

2. Constitutive Chemical Defenses

A wide variety of plant chemical defenses can have significant effects on the performance of insect herbivores on their hosts (Feeny, 1976; Rhoades and Cates, 1976; Rhoades, 1983). Plant chemical defenses are classified into two categories: constitutive defenses and induced defenses. Constitutive defenses are the permanent protection of a plant species. Feeny (1976) and Rhoades and Cates (1976) distinguished two categories of defensive chemical substances: qualitative toxins (e.g., alkaloids, terpenoids, and hydrogen cyanide) and quantitative digestion inhibitors (e.g., tannins, lignins and phenols). Qualitative chemicals are effective at low concentrations against nonadapted insects, while quantitative chemicals, operating in a dose-dependent manner, inhibit the digestive process of a wide range of insects. Having emphasized a concept of *plant apparency* (the vulnerability of a plant to discovery by herbivores), Feeny (1976) hypothesized that apparent plants that are easy for herbivores to find should invest heavily in quantitative digestibility reducers that provide generalized protection against all herbivores, while unapparent plants that are difficult to find should rely on escape in space and time, and on small amounts of qualitative toxins. Quantitative defenses are employed by woody plants or permanent woody tissues (e.g., mature leaves), characterizing a late succession. Qualitative defenses are associated with early successional herbaceous plants and ephemeral tissues (e.g., young leaves and buds) (Feeny, 1976; Rhoades and Cates, 1976; McKey, 1979).

Several studies have reported negative correlations between larval performance, in terms of survival and growth rates, and quantitative substances (Feeny, 1976; Rhoades and Cates, 1976; Johnson *et al.*, 1984; Bryant *et al.*, 1987; Lindroth and Peterson, 1988; Karowe, 1989), or qualitative substances (Erickson and Feeny, 1974; Miller and Feeny, 1983; Cates *et al.*, 1987). For example, Feeny (1970) found that the pedunculate oak, *Quercus robur*, received more leaf damage and supported more lepidopteran species at times when tannin levels were relatively low. Also, the presence in the artificial diet of as little as 1% of leaf tannin, extracted from September oak leaves, reduced significantly larval growth rate and pupal weight of the winter moth, *Operophtera brumata*. Similarly, larval weight of the large aspen tortrix, *Choristoneura conflictana*, was significantly lowered when the larvae were reared on artificial diets containing condensed tannin and the phenolic glycosides in aspen leaves (Bryant *et al.*, 1987). However, leaf tannin does not always have negative effects on larval performance (Fox and Macauley, 1977; Taper and Case, 1987; Bernays *et al.*, 1990). Taper and Case (1987) found that leaf tannin levels in oak species were positively correlated with diversity and gall density of leaf-galling cynipid wasps, and suggested that tannin serves a protective function for gall wasps in decreasing larval mortality due to fungal infection.

On the other hand, some specialist herbivores that are adapted to particular plant species are able to detoxify defensive chemicals in host plants and utilize these chemical substances as host finding, feeding, and oviposition stimulants (Crawley, 1983; Howe and Westley, 1988).

Although effects of plant chemical substances on insect performance are relatively diverse and complicated (Howe and Westley, 1988; Bernays *et al.*, 1990), it is likely that defensive chemicals of plant species often affect the population dynamics of insect herbivores, by altering directly larval survival, growth rate, and fecundity, and indirectly the efficacy of natural enemies (Lawton and McNeill, 1979; Rhoades, 1983, 1985; Schultz, 1988). It has been suggested, for example, that temporal changes in the intensity of chemical defenses, which are largely influenced by environmental stress, cause cyclic fluctuations of insect populations (Rhoades,1983, 1985). Consequences of plant chemical defenses for population fluctuations in herbivorous insects will be also discussed in the next section.

Indeed, apparency theory has made a great contribution to the recent development of chemical ecology in insect–plant interactions. Increasing evidence, however, suggests that the distinction between qualitative and quantitative defenses and their effects on insect performance are not always so clear cut as the theory predicts. Also, it has been emphasized recently that plant physiological constraints, which apparency theory has little considered, play a significant role in the evolution of plant defensive systems employing chemical substances (Coley, *et al.*, 1985). Hence, the ecological and evolutionary implications of antiherbivore chemical defense are still under debate (Fox, 1981; Howe and Westley, 1988; Bernays *et al.*, 1990).

3. Induced Chemical Defenses

There is accumulating evidence that physical damage by herbivores to leaves of a number of plant species induces changes in their secondary chemistry and nutritional quality, which adversely affect growth rate, pupal weight, fecundity, and vulnerability to enemies of insect herbivores that subsequently attack the plant (Haukioja *et al.*, 1985; Edwards, *et al.*, 1986; Neuvonen *et al.*, 1987; Silkstone, 1987; Rossiter *et al.*, 1988; Gibberd *et al.*, 1988; but see Myers and Williams, 1987). These changes are detectable within a few hours, days, or weeks, and last a few hours, days, weeks, or years (Haukioja and Neuvonen, 1987; Schultz, 1988).

Red oak trees that received heavy herbivory in the previous year exhibited significantly higher concentrations of tannins and phenols, and these chemicals had negative effects on larval growth of the gypsy moth *Lymantria dispar* (Schultz and Baldwin, 1982). Rossiter *et al.* (1988) also examined the relationship between oak phenolic chemistry and defoliation, and survival and reproduction of the gypsy moth. Pupal mass and fecundity were negatively correlated with an increased level of total phenolics, hydrolyzable tannins, and protein-binding capacity of the host trees, which were associated with greater defoliation. Similarly, Raupp and Denno (1984) found that previously defoliated willows, *Salix babylonica*, resulted in a prolonged larval period and smaller adults with reduced fecundity in the willow leaf beetle, *Plagiodera versicolora*. However, previous herbivore attack does not always have detrimental effects on insect performance (Myers, 1981; Williams and Myers, 1984; Roland and Myers, 1987).

These increased chemical defenses of plants induced by herbivory have been presumed to influence greatly herbivore populations through changes in survivorship and reproduction (Rhoades, 1983, 1985; Wratten *et al.*, 1988; Edwards and Wratten, 1987; Haukioja and Neuvonen, 1987; but see Fowler and Lawton, 1985). An induced defense may limit herbivore insect populations at a low density level relative to plant biomass (Rhoades, 1983;

Karban, 1986; Schultz, 1988; Edelstein-Keshet and Rausher, 1989), or lead to cyclic or eruptive population fluctuations in forest defoliators (Baltensweiler *et al.*, 1977; Haukioja, 1980; Rhoades, 1983; Haukioja and Neuvonen, 1987; but see Karban and Myers, 1989). Since plant quality deteriorates in a density-dependent way as a consequence of an insect attack, induced plant defense may contribute to the stability of insect populations (Haukioja and Neuvonen, 1987). A rapid inducible resistance may cause negative feedback on insect population growth and will tend to stabilize populations, while a long-term inducible defense may decrease an insect population in a delayed density-dependent manner, enhancing population fluctuations in subsequent generations. However, there is still considerable debate about the impact of induced defences on population dynamics of herbivores, perhaps because some authors have emphasized the superficial similarity between postulated plant responses and models of density-dependent population regulation (Schultz, 1988).

D. Plant Dispersion

Since there is a high degree of patchiness in plant communities, herbivorous insects are continually exposed to selection pressures on adaptive responses to the distribution and quality of their resources. Therefore, vegetation texture that includes plant density, diversity, and patch size can play a significant role in determining herbivore population densities (Root, 1973; Kareiva, 1983; Stanton, 1983). Having emphasized the connection between host plant dispersion and herbivore community structure, Root (1973) proposed the resource-concentration hypothesis: specialized herbivores are more likely to find and remain on hosts that are growing in pure stands. A number of studies testing this hypothesis, which have been conducted in artificially manipulated communities of plants, have shown correlations between herbivore abundance and plant diversity (Tahvanainen and Root, 1972; Risch, 1980; Bach, 1980; Lawrence and Bach, 1989; Andow, 1990), patch size (Cromartie, 1975; Solomon, 1981; Kareiva, 1985; Bach, 1984, 1988; Capman *et al.*, 1990), and growth form (Bach, 1981), although the results were not always straightforward (see reviews by Risch *et al.* 1983, Kareiva, 1983, and Stanton, 1983). In general, increased host density and patch size result in higher herbivore density per plant, whereas increased plant diversity is linked with reduced herbivore attack.

For example, Risch (1980, 1981) compared densities of six species of chrysomelid beetles in monocultures and polycultures of maize, beans, and squash in Costa Rica, and found that the numbers of a given beetle species per host plant in the intercrop with nonhost plants were significantly reduced relative to those in the monoculture. When a beetle fed on all plant species in a polyculture, this effect was reversed. This implies that the resource-concentration hypothesis is best applied to specialist herbivores. He also suggested that different movement patterns of adult beetles are likely to cause the observed difference in beetle abundances. Similarly, Bach (1980) found that population densities of the striped cucumber beetle *Acalymma vittata*, a specialist herbivore of cucumber, *Cucumis sativus*, were 10–30 times greater in cucumber monocultures than in polycultures. Differences in residence time and movement patterns were likely to account for these differences in beetle abundances. Host plant growth form also strongly affected beetle abundances (Bach, 1981).

Although there is no doubt that spatial dispersion of plant resources greatly affects herbivore population densities, it is still difficult to clarify underlying mechanisms causing the observed pattern of herbivore densities on different plant dispersions because several attributes of plants (growth rate, growth form, leaf size, fruit and flower production) are largely dependent on plant diversity or plant density in a highly complex way (Bach, 1980; Kareiva, 1983; Stanton, 1983). Only recent studies have concentrated on herbivore movements in response to vegetation texture and quality as the most important factor underlying reduced herbivore populations in diverse vegetation or small patches of the host plant (Bach, 1984, 1988, 1990; Kareiva, 1985; Turchin, 1986; Lawrence, 1988; Lawrence and Bach, 1989).

E. Plant-Mediated Species Interactions

Interspecific interactions have potentially great influences on the performance and abundance of the insects concerned. Recent studies on insect–plant interactions have revealed that these interactions are often indirect, asymmetrical and subtle, and that morphological, phenological, and chemical changes in the host plant alter the success of predation or parasitism by natural enemies (Price *et al.*, 1980; Schultz, 1983a; Price, 1987; Faeth, 1987; Mopper *et al.*, 1990). Even insects feeding at different times or on different parts of a plant may have a substantial effect on the quality or quantity of resources available to one another (Stamp, 1984; Karban, 1986; Faeth, 1987; Crawley and Pattrasudhi, 1988). Such indirect effects mediated by the host plant are more common than previously thought, and are especially common in insect–plant systems (Faeth, 1985; Moran and Whitham, 1990).

1. Three-Trophic-Level Interactions

Three-trophic-level interactions have recently received much attention in insect–plant interactions (Price *et al.*, 1980; Schultz, 1983a, 1988; Weis and Abrahamson, 1985; Faeth, 1986, 1987; Clancy and Price, 1986; Price, 1987; Gross and Price, 1988; Barbosa, 1988; Denno *et al.*, 1990; Hare *et al.*, 1990). This view addresses a significant role of the third trophic level (natural enemies) as part of a plant's battery of defenses against herbivores. It also recognizes that changes in host plant quality directly or indirectly affect the efficacy of parasitoids or predators by altering insect host location or vulnerability (Vinson, 1976; Price *et al.*, 1980; Schultz, 1983a).

The gall-making sawflies on the arroyo willow, *Salix lasiolepis,* provide a good example of the important role of plant phenotype in enhancing the attack of natural enemies (Price and Clancy, 1986; Clancy and Price, 1986; Craig *et al.,* 1990). Mortality of a shoot-galling sawfly, *Euura lasiolepis,* caused by the parasitoid *Pteromalus* sp. was a decreasing function of gall size, because of a short ovipositor of the parasitic wasp relative to gall diameter of the sawfly. As gall size was principally determined by willow phenotype (clone), the host clones regulated gall size of the sawfly, which, in turn, generated differential parasitism in sawfly larvae with different gall size. In addition, the sawfly is vulnerable for a time to attack by the parasitoid *Lathrostizus euurae.* The commencement of the susceptible period depended on sawfly development, and the vulnerable period ended as gall toughness became too limiting. Willow clones differentially determined the rate of gall toughening and parasitism in each clone by changing the susceptible period (Craig *et al.,* 1990).

Plant morphology may be critical in interactions among three trophic levels. Kareiva and Sahakian (1990) experimentally demonstrated a significant interaction between peas and lady beetles in determining aphid population growth, using normal and leafless varieties of the common pea, *Pisum sativum.* Aphids on normal peas showed significantly higher population growth than those on leafless peas. The increased population growth was owing to better escape by aphids from lady beetle predation on normal peas, simply because lady beetles fall off the normal variety twice as frequently as they do the leafless variety. Accordingly, the leafless peas can reduce the rate of aphid population growth by 50%, when assisted by *Coccinella* predators.

However, host plant quality does not always enhance parasitism or predation (Rothschild, 1973; Barbosa, 1988). For example, monarch butterflies and chrysomelid beetles incorporate plant toxins into their own defense against natural enemies such as avian predators (Duffey, 1970; Pasteels *et al.,* 1988).

2. Asymmetric Interactions

Species interactions among guild members sharing the same host plants are often asymmetrical. Although previous studies have focused mainly on interspecific competition between closely related species, asymmetric interactions occur among herbivorous insects, through changes in quality and quantity of host plants, that are very distinct taxonomically and utilize different parts of the shared host plant in very different manners (Lawton and Hassell, 1981; Stamp, 1984; West, 1985; Karban, 1986; Faeth, 1987; Crawley and Pattrasudhi, 1988; Hunter and Willmer, 1989; Moran and Whitham, 1990).

The meadow spittlebug, *Philaenus spumarius,* and the plume moth, *Platyptilia williamsii,* are two abundant herbivorous insects, both of which spend their immature development on new leaves of *Erigeron glaucus.* Karban

(1986) experimentally demonstrated that while the presence of spittlebugs had little impact on the moth, the presence of the plume moth reduced persistence of spittlebugs by 40%. He suggested that the highly asymmetric interaction is due to the different responses of the host plant to leaf herbivory. Although the feeding of the spittlebugs had little effect on subsequent leaf production of the host plant, the moth caterpillars consumed the terminal bud, which caused the rosette to produce fewer new leaves available to the spittlebug as food and refuge. Similarly, the free-living defoliator, *Operophtera brumata*, and the leaf-rolling caterpillar, *Tortrix viridana*, share the pedunculate oak, *Quercus robur*. The leaf-roller caterpillars had lower survival when mixed with *O. brumata* than when alone, but the reverse effect was much smaller (Hunter and Willmer, 1989). This asymmetrical competition is caused by a difference in habitat use by the two species. Since leaf damage by the defoliator disrupts the shelters made by the leaf-rolling tortrix, the latter species can suffer high mortality by desiccation.

Indirect interactions can occur between herbivores that utilize different parts of the same host plant. There were two coexisting aphid species on the lamb's quarter, *Chenopodium album*. *Pemphigus betae* feeds underground on roots, while *Hayhursita atriplicis* feeds above ground where it forms leaf galls, and the species never encounter one another directly. Moran and Whitham (1990) noted that the outcome of the species interaction between the two aphids changes in accord with the level of host resistance to the leaf-galler. On susceptible plants, leaf-galling colonies reduced *P. betae* numbers by a mean of 91% and body length by 22%, often eliminating the root feeders entirely. In contrast, on plants that are resistant to the leaf-galler, the galler colonies were smaller and did not affect the root feeder. In both cases, however, root aphids had little effect on the performance of leaf-galling aphids.

3. Interactions between Temporally Separated Guilds

Interactions between temporally separated guilds may be critical in determining the distribution and survivorship, and thus population dynamics, of insect herbivores under low levels of herbivory (Faeth, 1986, 1987). There is some evidence to support the view that one species attacking the host plant early in the season can change the performance or abundance of another species attacking late in the season, mediated by changes in host quality (West, 1985; Faeth, 1986, 1987; Harrison and Karban, 1986; Hunter, 1987; but see Williams and Myers, 1984).

One example can be seen in the interaction between the ranchman's tiger moth, *Platyprepia virginalis*, and the western tussock moth *Orgyia vetusta*, larvae of which feed on the bush lupine *Lupinus arboreus* . Despite the fact that the larvae of the two moths appear at different times of the year, herbivory by the tiger moth early in the season had a substantial effect on the tussock moth, significantly reducing female pupal weight, fecundity, and daily growth rate (Harrison and Karban, 1986). It was suggested that lower foliage quality resulting from an induced resistance of lupine after early herbivory by the tiger moth was responsible for the reduction in the performance of the tussock moth.

Similar effects of early season herbivores on the populations of leaf miners, leaf chewers and sap suckers have been shown on oak species in Britain and in the United States (West, 1985; Faeth, 1986; Hunter, 1987; see Chapter 10 for details).

IV. Preference, Performance, and Population Dynamics

A. Insect Preference in Host Plant Selection

As shown in the previous section, host plant characteristics can greatly affect the survivorship, reproduction, and population dynamics of herbivorous insects through a wide variety of insect–plant interactions. In particular, host plant quality relevant to insect herbivores is important in both direct and indirect interactions, reducing resource accessibility, or enhancing the efficacy of natural enemies. Variable host plant quality can influence herbivore populations in two ways: performance and preference. Performance is an immediate influence by host plant attributes on survival and reproduction of herbivorous insects. In contrast, preference is defined here as the evolutionary responses of insects to variable and heterogeneous plant resources that improve individual fitness (Whitham, 1980; Singer, 1986; Thompson, 1988). Although resource-utilization tactics have been a central issue in the study of life history evolution of insect herbivores (Denno and Dingle, 1981), little is known about the role of resource utilization in insect population dynamics. However, recent studies of insect–plant interactions have argued for the importance of life history traits in resource use as a principal determinant of the population dynamics in herbivorous insects (Preszler and Price, 1988; Craig *et al.*, 1990; Price, 1992).

I believe that an evaluation of the impact of plant resources on insect performance is the first step toward a comprehensive understanding of the role of insect–plant interactions in population dynamics of herbivorous insects. Then we should concentrate on tactics by which herbivores select suitable resources, which are evolutionary responses of insects to highly variable plant resources. This is because behavioral and physiological constraints in the process of resource utilization in herbivorous insects can greatly alter the consequences of host plant quality for offspring performance and population dynamics (Whitham, 1980; Rausher, 1983). For many herbivorous insects, the searching abilities of larvae are poor relative to those of adults; therefore, oviposition preference of adult females is of paramount importance in the process of selecting suitable host plants or plant parts for their offspring. This is especially true when the newly hatched offspring are not capable of searching for additional hosts until they have fed on the individual chosen by their mother. There is increasing circumstantial evidence to support the view that resource-use tactics of adult insects are critical in population dynamics of herbivorous insects (Price *et al.*, 1990). For example, recent reviews on insect population dynamics demonstrated that density-dependent processes, or key factors, occur in the adult stage in many insect herbivores (Dempster, 1983; Stiling, 1988), suggesting the important role of the reproductive process in the adult stage in determining the temporal pattern of population fluctuations. Furthermore, the importance of the searching behavior of adult insects for favorable resources in determining subsequent population densities in different vegetation textures has been described in a number of studies testing the resource concentration hypothesis mentioned earlier (Kareiva, 1983; Stanton, 1983).

Since natural selection favors individuals that have higher lifetime reproductive success, it can be expected that there is a strong evolutionary correlation between offspring performance and adult preference in resource use. In particular, female herbivorous insects whose offspring develop at the oviposition site are strongly favored by natural selection to optimize oviposition site selection. The behavioral responses of insects that give rise to discrimination are, therefore, in part an evolutionary response to the existence of variation in plant quality, because individuals that avoid plants of low quality and feed preferentially on plants of high quality will leave more offspring than individuals that do not exhibit such behavior (Rausher, 1983). The preference–performance relationship has been recently explored in the evolutionary ecology of herbivorous insects (Singer, 1986; Thompson, 1988). Some studies of host plant selection or habitat selection within a plant have revealed a positive correlation between oviposition preference and relative offspring performance in insect herbivores (Whitham 1980, 1986; Rausher, 1980; Via, 1986; Ng, 1988; Singer *et al.*, 1988; Sitch *et al.*, 1988; Craig *et al.*, 1989; see Thompson, 1988 for a review). It should also be noted that a close relationship between preference and performance enhances low availability of plant resources, because adult females are largely restricted to ovipositing in places where their offspring have higher performance. However, females of some lepidopteran insects lay their full complement of eggs rapidly, independent of plant quality. Oviposition behavior of these herbivorous insects is thus unlikely to correlate with larval survival (Price *et al.*, 1990). It is still useful to examine the relationship between oviposition behavior and offspring performance in those species that do not exhibit apparent adult preference. This is because we can concentrate on resource-use tactics in the immature stage, such as habitat selection or foraging behavior of larvae (Schultz, 1983b; Hunter, 1990), when the importance of adult preference in the process of resource choice is removed.

Therefore, an understanding of the interactions between adult preference and offspring performance needs (1) quantitative estimation of components of offspring performance (fitness components) through a lifetime, and (2) clarification of evolutionary responses of insects to variable plant resources, especially preference in host plant selection by adult insects.

B. Connection between Individual and Population Ecology

Since both host plant selection by adult females and offspring performance (fitness) are properties at the individual level, it is essential to understand how these individual attributes are translated into demographic parameters at the population level through survival and reproductive processes. This requires an approach based on the reduction of population ecology to individual ecology, which focuses on behavioral and physiological mechanisms underlying the dynamic features of populations (Schoener, 1986; Price *et al.*, 1990; Higashi and Ohgushi, 1990; see Chapter 2). Having emphasized a mechanistic approach, Schoener (1986) argued that each parameter of population dynamics exhibited at the population level must be completely translatable into behavioral and physiological parameters at the individual level. Also, a comprehensive understanding of the ways in which resources act to influence population dynamics needs a more reductionist approach, dependent upon knowledge of the dynamics of the components of resources at the individual level (Wiens, 1984; see Chapter 2).

In this context, the traditional approach to population dynamics, using principally correlation statistics, is no longer adequate for two reasons. First, the population ecologists long have focused on the demographic consequences of environmental or biological changes, with little attention to behavioral or physiological mechanisms that may underlie changes in demographic features through changes in birth, death, and migration rates (Smith and Sibly, 1985; Lomnicki, 1988). Recent arguments have emphasized that properties at the individual level, such as behavior, physiology, and morphology are critical in determining a general picture of population dynamics (Lomnicki, 1980, 1988; Hassell and May, 1985; Smith and Sibly, 1985; May, 1986; Barbosa and Baltensweiler, 1987; Hassell, 1986; Price *et al.*, 1990). Second, the traditional analysis of population dynamics, represented by key factor analysis, has largely rested on correlations between population densities and factors that potentially affect survival and reproduction, which tells us little about the causal mechanisms generating the observed pattern of population fluctuations (Royama, 1977; Price, 1987). It is thus not surprising that in spite of many population studies of herbivorous insects, most merely describe patterns of population fluctuation, without revealing convincingly their causal mechanisms.

Since behavioral ecology is the study of the evolutionary background of the relationships between fitness and behavior and other variables, including population density (Krebs and Davies, 1984), it will reveal the underlying mechanisms at the individual level that cause a wide variety of population dynamics (Smith and Sibly, 1985). Thus, a strong connection between behavioral ecology and population ecology is essential to incorporate fully the perspective of the preference–performance interaction into population dynamics of herbivorous insects. In other words, an alternative approach to an understanding of population dynamics via key factor analysis, for example, should be based upon a reduction of population ecology to individual ecology, while highlighting demographic consequences of adaptive life history traits in resource use (see also Harper, 1982 for a plant demography). I agree with the assertion by Preszler and Price (1988) and Price *et al.* (1990) that an evolutionary syndrome and its constraints on the process of host plant use are of crucial importance in determining a general picture of population dynamics in insect herbivores.

Since fitness is the degree of demographic difference among phenotypes, it can be measured by the average lifetime contribution to the breeding population of a particular class of phenotypes, relative to the contributions of other phenotypes (Falconer, 1981; Endler, 1986). Also, fitness is a within-generation measure of the process of natural selection, and thus applies only to phenotypic selection (Endler, 1986; Grant, 1986). Accordingly, fitness estimates require detailed information on survival and reproductive processes in different phenotypic classes, such as cohorts or trait groups, within a population. In studies of natural selection, a serious problem arises in estimating fitness in natural populations (see Endler, 1986 for more detail). One may draw a very misleading conclusion if fitness is obtained by incomplete estimates that do not cover the whole lifetime of organisms. Also, partial fitness fails to reveal important costs and benefits of different phenotypes affecting components of fitness and any ecological constraints operating in different stages of the life history (Clutton-Brock, 1988).

Despite the methodological limitations mentioned above, the widely accepted life table approach with field manipulations, involving mark–recapture and cage experiments for studying population dynamics, will potentially contribute to demonstrating natural selection in natural populations, by estimating lifetime fitness if applied to cohorts or phenotypic classes. First, a life table is designed to estimate demographic parameters of survivorship and reproduction over the whole lifetime and to clarify the intensity and variability of mortality factors affecting populations. It thus allows us to measure lifetime fitnesses for phenotypic classes and compare them among different categories of individuals in a population (Woolfenden and Fitzpatrick, 1984; Koenig and Mumme, 1987; Ohgushi, 1991). Second, a mark–recapture experiment for following the fate of individuals, coupled with a cage experiment for evaluating characteristics of individuals, reveals behavioral and physiological episodes of individuals, which are needed to clarify the biological implications of life history traits. In particular, when applying these field experiments to adult insects, we will obtain detailed information on behavioral and physiological responses of individuals in the process of utilizing variable plant resources, which life table statistics alone hardly reveal.

However, most insect population ecologists fail to recognize that the life table approach, when modified appropriately, can be useful for a better understanding of the evolutionary consequences of the life history traits of individuals. In fact, most studies of insect population dynamics have merely used a mark–recapture technique for individuals as a device for estimating population size at a given time.

Consequently, an incorporation of evolutionary perspectives into life table analysis will offer great benefits to both population dynamics and behavioral ecology, by making a close connection between these separate disciplines in ecology.

V. Resource Limitation on the Herbivorous Lady Beetle

In this section I shall show the result of a 5-year population study of the herbivorous lady beetle, *Epilachna niponica* (Coleoptera: Coccinellidae), at two different localities as an example of resource limitation on an insect herbivore population.

A. Study Area, Materials, and Methods

1. Study Area

This study was performed over a 5-year period (1976–1980) at two sites located in different valleys along the Ado River, in the northwestern part of Shiga Prefecture in central Japan. Site A (60 × 30 m) was situated at a 220 m elevation on an accumulation of sandy deposits caused by dam construction in 1968. The surface of the rather flat and open area consisted mainly of unhardened sandy deposits. Floods caused by heavy rainfall have often submerged and washed away the ground flora along the watercourse, and most of the surviving ground flora were annual and perennial herbs. Site F (90 × 15 m) was situated at a 350 m elevation, about 10 km upstream from site A. The more-hardened soil deposits at this site mean that most grasses and shrubs can successfully escape serious flood damage, except for large-scale floods. Vegetation in and around the site included various deciduous trees, such as *Quercus mongolica* and *Q. salicina*.

2. The Lady Beetle

The lady beetle, *Epilachna niponica*, is a specialist herbivore of thistle plants, and is a univoltine species. Overwintering adult females emerge from hibernacula in soil in early May, and begin to lay eggs in clusters on the undersurface of thistle leaves. Larvae pass through four instars. New adults emerge from early July to early September, feeding on thistle leaves through autumn. Then they enter hibernation by early November. Seasonal changes in numbers of adult and immature beetles are given in Ohgushi and Sawada (1981, 1984).

3. The Host Plant

In this study area, *E. niponica* feeds exclusively on leaves of a thistle, *Cirsium kagamontanum*, which is a perennial herb, distributed in patches along the stream side. It grows rapidly from sprouting in late April to late June, becoming full-sized at 1.5 to 1.8 m in height by late August, and then flowers over 2 months from mid-August. Old leaves begin to wither after summer. Although the number of thistle leaves gradually increases until late August, leaf quality in terms of amino acid and water content consistently declines during the growing season (Ohgushi, 1986).

4. Methods

Each population was censused at 1- to 3-day intervals from early May to early November in each year, 1976–1980. All thistle plants growing in the study sites were carefully examined; the numbers of eggs, fourth instar larvae, pupae, pupal exuviae, and adult beetles were recorded separately for each plant. Adult beetles were individually marked with four small dots of lacquer paint on their elytra. Newly marked adults were released immediately on the thistle plant where they had been captured. Sex, body size, and subsequent capture history (date and place) were recorded for individual beetles. A total of 5969 and 3507 beetles were marked at sites A and F, respectively. Detailed life tables were then constructed for every year at the two study sites. The impact of the lady beetle herbivory on thistle leaves was evaluated by visually estimating the amount of leaf area consumed.

B. Resource Limitation on the Beetle Population

Beetle populations changed synchronously with resource abundance over the course of the study at both of the study sites (Fig. 1). Abundance of host plants changed independently at the nearby sites, but the egg populations of the lady beetle closely followed the biomass of host plants at each site. Indeed, host abundance explains 66% and 98% of variation in overall population fluctuations over 5 years (1976–1980) at site A and site F, respectively. A weaker correlation at site A was evidently due to a large reduction in egg numbers in 1979 by a June flood, which washed away most of the ovipositing females. It should be noted that in spite of the large habitat disturbance, egg populations quickly returned to correspond with resource abundance in the next generation in 1980. This close synchrony between beetle population and host plant abundance in the two localities suggests that the beetle populations are strongly limited by their food resource.

To examine when this resource limitation occurred in the beetle's life cycle, I compared population variability, expressed by the standard deviation of the logarithm of the population densities for these 5 years, among different life stages (Fig. 2). At the two sites, population variability in terms

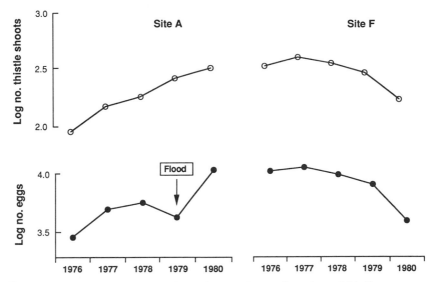

Figure 1 Annual changes in numbers of thistle shoots (○) and eggs laid (●) over a 5-year period at site A and site F. Vertical arrow shows the occurrence of a flood.

of host abundance sharply decreased from reproductive adult to egg stage; thereafter, it tended to increase up to adult emergence. This implies that the population density in relation to host abundance was highly stabilized by the adult beetles during the reproductive process. As a result, egg populations changed only 2.0- and 1.4-fold in density over five generations at sites A and F, respectively, which indicated a much higher level of population stability than that in most insect species so far studied (see Hassell *et al.*, 1976; Wolda, 1978; Connell and Sousa, 1983). In contrast, destabilization of population density increased from the egg to the adult stage.

After rapid population growth from mid-May to mid-June, egg populations reached a plateau at the two sites, suggesting that egg populations were strongly limited in late June (Fig. 3). Note that annual variation in population density sharply declined from mid-May to mid-June. This implies that population stabilization rapidly advanced with increasing population density early in the season. Furthermore, in spite of the different population growth early in the season, the egg density finally reached was almost the same in these two populations.

Dempster and Pollard (1981) illustrated a similar synchronous population fluctuation of the cinnabar moth and its host plant over nine generations. As already mentioned, this temporal resource tracking was brought about by frequent host depletion leading to larval starvation and reduced female size with lower fecundity. However, this scenario based on resource depletion is not applicable to the case of the lady beetle, because leaf damage of thistle

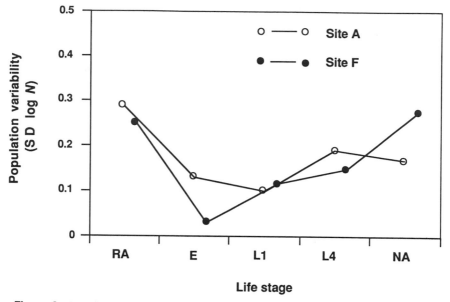

Figure 2 Population variability in each developmental stage. RA, reproductive (overwintering) adults; E, eggs; L_1, first instar larvae; L_4, fourth instar larvae; NA, newly emerged adults. Variability is expressed by the standard deviation of the logarithm of the population densities (individuals per shoot) for 1976 to 1980.

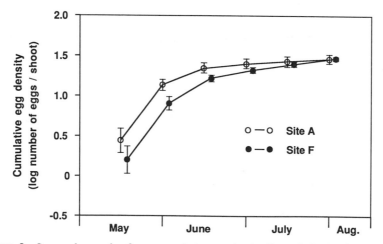

Figure 3 Seasonal growth of egg population at site A (○) and site F (●). Each point represents mean ± SE for 1976 to 1980. Modified from Ohgushi and Sawada (1985a).

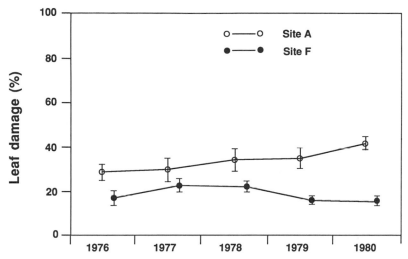

Figure 4 Percentage of leaf damage of host plants due to beetle feeding at site A (○) and site F (●). Each point represents mean ± SE during the study period.

plants still remained at a relatively low level in late June when population limitation apparently occurred (Fig. 4). Furthermore, it is unlikely that larvae could not survive on uneaten leaves, because larvae completed their development when offered the remaining leaves in the laboratory. Consequently, egg populations of the lady beetle are undoubtedly resource limited at far below the level of resource depletion.

Two density-dependent processes were detected during reproduction by the lady beetle. First, there was a significant density-dependent reduction in oviposition rate in June when egg population was strongly limited (Site A: $r = -.66$, $n = 10$, $P < .05$; Site F: $r = -.80$, $n = 10$, $P < .01$). However, this was not evident in May (Site A: $r = -.45$, $n = 5$, not significant (NS); Site F: $r = -.70$, $n = 5$, NS). Second, female survival tended to decline in a density-dependent manner in May, although there was not a statistically significant trend (Site A: $r = -.69$, $n = 5$, NS; Site F: $r = -.81$, $n = 5$, NS). The fact that both oviposition rate and survival of reproductive females were negatively correlated with monthly population density suggests that reproductive processes, such as responses of ovipositing females to egg density, play a key role in limiting the population with respect to the abundance of food.

C. Mechanisms: Responses of Ovipositing Females

What are the underlying mechanisms in maintaining beetle populations at such a low density level relative to resource abundance?

1. Physiological Responses

Since the field observations suggest that density-dependent reduction in reproduction plays an important role in population limitation, I conducted field experiments designed to detect responses of ovipositing females in different host conditions. The first experiment was carried out to examine the egg-laying schedule of each female in the field. Eight thistle plants, approximately the same height and with the same number of leaf nodes, were selected. Individual plants were covered with a nylon cage with a metal frame large enough to allow further foliage growth; one pair of reproductive adults was introduced into each cage for oviposition. Oviposition was followed until death of the female. Five cages were kept for controls; three cages were designed as follows. In the first two cages, each female was transferred to another cage containing an undamaged thistle, 2–3 weeks after the female ended oviposition. In the third cage, two old nymphs of an earwig, *Anechura harmandi*, the predominant egg predator in these study sites, were initially released to remove eggs deposited during the experimental period.

In every cage except for the one with egg predators (cage H), the females refrained from laying eggs around mid-June (Fig. 5). It is notable, however, that they ended oviposition when leaf damage was low enough (less than 50% of the total leaf area) that there was still area available for oviposition. Such females resumed oviposition in a short time when they were transferred into a new cage with an undamaged thistle (cages F and G). In the cage with egg predators (cage H), the leaf damage stayed at a comparatively low level throughout the course of the experiment because of heavy egg predation. Oviposition in this cage thus continued until late July.

Reproductive females in the genus of *Epilachna* readily resorb developing eggs in the ovary when kept under starved conditions (Kurihara, 1975). To evaluate possible impacts of host deterioration on the ovarian status of ovipositing females, I conducted another experiment. On 22 May 1981, I selected six thistle plants and covered them individually with a nylon cage. Different numbers of reproductive adults were then introduced into each cage. On 26 June, 35 days after the experiment started, all females alive in these cages were taken out and their ovaries dissected to determine whether egg resorption had occurred. Egg resorption evidently occurred in cages with high beetle density (cages D, E, and F) at the end of the experiment, when most thistle leaves were highly exploited (Table 1). On the other hand, no females resorbed eggs in cages with only one pair of beetles (cages A, B, and C), where 80% of the total leaf area remained intact.

The cage experiments clearly demonstrated a physiological response of ovipositing females to host deterioration as follows: (1) ovipositing females readily resorb developing eggs in the ovary; (2) deteriorating food resources are the most likely cause leading to the egg resorption; (3) egg resorption occurred even when leaf damage remained at a low level; and (4) the process of resorption is reversible, when the host plant becomes favorable, the resorption immediately ceases, and the ovary again becomes productive. Also, dissection of females from the field sampled at several times during the reproductive season confirmed that the number of adult females having resorbed eggs increased rapidly after mid-June when leaf damage was much lower than that in the experiments (Ohgushi and Sawada, 1985a).

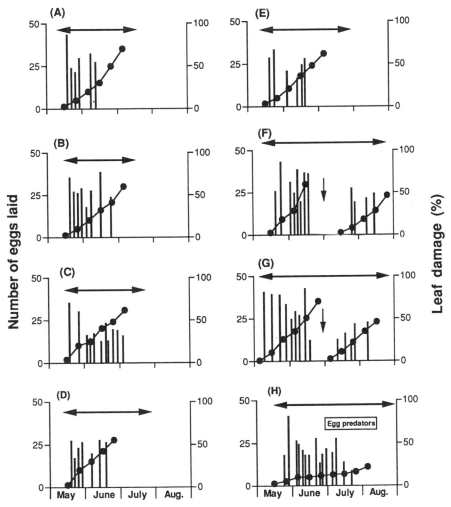

Figure 5   Oviposition schedules for each female in the cage experiment. The number of eggs deposited (vertical bars) and the percentage of leaf damage (●) are shown. The horizontal lines show the female's lifespan. (A)–(E), control; (F)–(G), the females were transferred to another cage with an undamaged thistle at the date indicated by the vertical arrow; (H), a cage with two egg predators. Modified from Ohgushi and Sawada (1985a).

Table 1 Ovarian Status of Adult Females[a]

Cage	No. adults Male	No. adults Female	% Leaf damage	% Females with egg resorption
A	1	1	20	0
B	1	1	20	0
C	1	1	20	0
D	1	2	50	100
E	4	5	80	100
F	4	7	90	100

[a] Modified from Ohgushi and Sawada (1985a).

2. Behavioral Responses

Since the cage experiments did not allow the adult females to choose different plants for oviposition, there remains a possibility that they would move to other plants for oviposition before resorbing eggs if they were left to disperse. To examine this possibility, I assessed movement activity in ovipositing females during the reproductive season. There was a clear seasonal change in movement activity by adult females. Their mobility reached a peak from late May to mid-June, then decreased consistently during the rest of the reproductive season (Fig. 6A). At the same time, cumulative egg densities among different plants were markedly stabilized (Fig. 6B). The density stabilization on a spatial scale is probably brought about by frequent movements of adult females among different plants for oviposition. The increasing spatial stabilization in egg densities also suggests that females tended to avoid oviposition on plants bearing more eggs. Since female movement while searching for a suitable oviposition site was more pronounced when egg density was high early in the season, female losses may be higher in years of high egg density. This is because those females are likely to exhaust energy by frequent movements or to disperse outside the sites in a density-dependent manner.

Consequently, spatial stabilization in egg density by careful oviposition site selection by females in early June, coupled with egg resorption responding to resource deterioration late in the reproductive season, generates a strong limitation on further population growth in terms of host plant quantity. In other words, the behavioral and physiological responses of ovipositing females to resource conditions for oviposition are the causal mechanisms generating limitations on the lady beetle populations.

D. Adaptive Significance of Oviposition Tactics

There is no doubt that the properties of individuals are of crucial importance in determining the fundamental pattern of population dynamics of the lady beetle. Let us turn to the problem of whether these behavioral and physiological responses improve the reproductive success of a female.

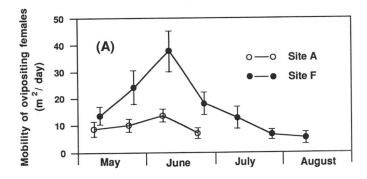

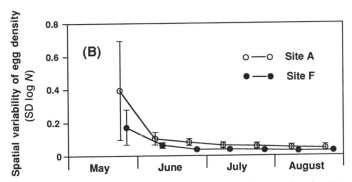

Figure 6 Seasonal changes in movement activity of reproductive females (A) and population variability among different thistle plants (B) at site A (○) and site F (●). Each point represents mean ± SE for 1976 to 1980. Movement activity is expressed by the variance of distances traveled per day of marked adult females. Population variability is expressed by the standard deviation of the logarithm of cumulative egg density (eggs per shoot) among five different thistle groups based on plant size.

1. Selection of an Oviposition Site

The survival of offspring from egg to reproductive age sharply declined with increasing egg density on individual plants (Fig. 7). Survival was particularly high on plants with low egg densities. This implies that the behavior of laying eggs on plants with low egg densities increased the reproductive success of an ovipositing female. Decreasing offspring survival with egg density involves density-dependent larval mortality and adult mortality up to the reproductive age.

Although arthropod predation was an important mortality factor responsible for large larval losses, predation did not act in a density-dependent manner on several spatial scales (Ohgushi, 1988). Since leaf damage by

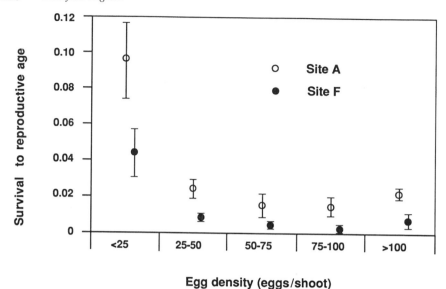

Figure 7 A relationship between egg density and offspring survival from egg to the reproductive age in individual thistle plants at site A (○) and site F (●). Each point represents mean ± SE for 1976 to 1980.

beetle feeding reduced resource quality in terms of amino acid and water content in thistle leaves (Ohgushi, 1986), deterioration in leaf quality, which depends on larval density, is more likely to cause the density-dependent larval mortality. Adults that emerged from plants with high egg densities tended to be small in body size (Ohgushi, 1987). Such smaller adults had a higher mortality during the period from emergence to the reproductive season in the following year, when compared to larger adults. In particular, the size-dependent adult mortality was more apparent during hibernation.

2. Selection During Oviposition Period
Since reproductive females that resorb eggs stop laying eggs, this physiological response will not be adaptive unless the resorbed females are able to oviposit at a future time. There are two possibilities for future oviposition: within-season and between-season oviposition. The former means a resumption of oviposition later in the same reproductive season. This is more likely to occur because egg resorption is a reversible process, as shown in the cage experiment. The latter concerns oviposition in the following reproductive season in the next year.

Ovipositing females may resorb eggs when host plants are seriously damaged by a large flood. This occurred at the end of June in 1979 at site F when a large-scale flood washed away some thistle plants and buried others with

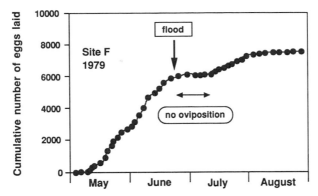

Figure 8 Cumulative number of eggs laid at site F in 1979. Vertical arrow indicates the occurrence of a flood. Note that females stopped laying eggs for half a month following the habitat disturbance.

soil. It should be noted that adult females stopped laying eggs for a half-month following this large habitat disturbance (Fig. 8). However, when most of the damaged thistles recovered and produced new leaves in mid-July, females resumed oviposition. This fact suggests that females refrained from reproduction during the period unfavorable to offspring survivorship, and became reproductive again at a later time when offspring fitness was no longer affected by habitat disturbance.

There is evidence to support second-year oviposition by adult females that stop laying eggs in the first reproductive season. Figure 9 shows seasonal changes in female survival in the reproductive season. At site A, female survival consistently declined with the season, and few individuals survived up until early July. This tendency was also seen at site F until mid-June, but thereafter we can see a clear recovery of subsequent survivals. The apparent improvement of female survival was closely synchronized with egg resorption observed in the field (Ohgushi and Sawada, 1985a), suggesting that there was a trade-off between egg production and subsequent survival.

Then let us examine the possibility of the second-year oviposition of such long-lived reproductive females with increased survival at site F. Over the course of the study, a total of 56 females were alive in mid-August, and nearly 40% of those females survived up to the next spring (Table 2). The fact that only one individual survived to the spring in 1980 was evidently because of heavy autumn floods, which caused heavy adult mortality (Ohgushi, 1986). Hence half of the long-lived females survived up to the second reproductive season, when the 1979 data is omitted. Also, some of them were observed ovipositing in the second reproductive season in 1977, 1978, and 1979. This strongly indicates that long-lived females have a

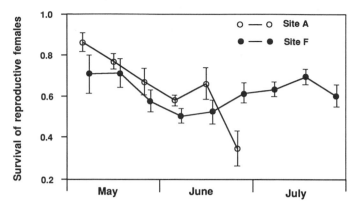

Figure 9 Seasonal changes in survival of reproductive females in every 10-day period at site A (○) and site F (●). Each point represents mean ± SE for 1976 to 1980. Survival rates were estimated by the Jolly-Seber stochastic model using the data from mark–recapture experiments (see Jolly, 1965 and Seber, 1973 for details).

comparatively high possibility of oviposition in the reproductive season of the second year. There was evidence that offspring lifetime fitness at site F was remarkably depressed late in the season (Fig. 10). In particular, lifetime fitness was zero in the late cohorts born after early August, because no adults successfully emerged from those cohorts. Females will have higher reproductive success if they avoid egg-laying late in the reproductive season and postpone oviposition until the second year. It is probable that egg resorption is an adaptive response to improve lifetime reproductive success of long-lived females.

E. Role of Natural Enemies

I have concentrated on the responses of ovipositing females to resource conditions as a causal mechanism underlying the resource limitation on the lady beetle populations. One could argue, however, that natural enemies might also contribute to the population equilibrium of herbivorous lady

Table 2 Number of Reproductive Females Surviving to the Following Reproductive Season in the Next Year at Site F

Females surviving	1976	1977	1978	1979	Total
On 15 August	11	17	14	14	56
In the next spring	7	8	6	1	22
% survivors	63.6	47.1	42.9	7.1	39.3

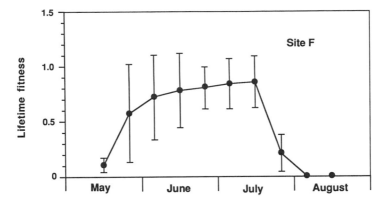

Figure 10 Lifetime fitness of offspring which were born at different times in the reproductive season at site F. Each point represents mean ± SE for each cohort in 1976 to 1980. The lifetime fitness of the i-th cohort (F_i) is defined as follows. $Fi = E_i \times Li \times Ai \times Ri$, where E_i = egg survival to hatching, L_i = larval survival from egg hatch to adult emergence, A_i = survival of adult females from emergence to the reproductive season in the following year, and R_i = lifetime fecundity of females living to average age of i-th cohort.

beetles. Therefore, it is useful to summarize the impact of natural enemies on the population density of the lady beetle in the present study (see Ohgushi and Sawada, 1985b and Ohgushi, 1988 for details). Although eggs and larvae were rarely subjected to parasitism, arthropod predation was a major cause of mortality during the egg and larval stage. The intensity of egg and larval predation greatly differed between the two sites; egg predation was significantly higher at site F [40.0 ± 5.8% of mean ± standard error (SE)] than at site A (11.6 ± 4.9%) for the study period, 1976–1980 (analysis of variance: $F = 36.42$, $P < .01$ for arcsine transformed data). Indeed, the earwig *Anechura harmandi*, which was the most frequent predator of eggs and larvae in the study sites, had significantly higher densities at site F (7.2 ± 0.9/100 shoots of mean ± SE) than at site A (2.0 ± 0.5) for the study period, 1976–1980 (analysis of variance: $F = 71.77$, $P < .01$). As a result, population density of new adults at site A was significantly higher than that at site F (Ohgushi and Sawada, 1985b). In spite of the large difference in adult density between the two sites, the average egg density in terms of resource abundance over the 5 years was almost identical (Fig. 11). Furthermore, natural enemies had little effect on the reproductive processes by which beetle population equilibrium, with respect to resource abundance, was achieved. Therefore, natural enemies did not change the general pattern of population dynamics of the lady beetle.

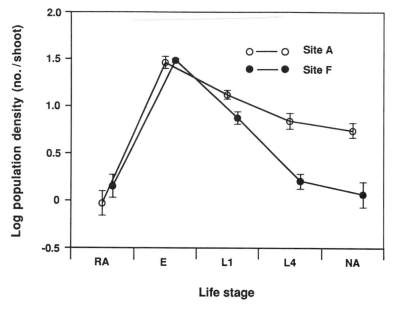

Figure 11 Population density (individuals per shoot) at different life stages at site A (○) and site F (●). Each point represents mean ± SE for 1976 to 1980. RA, reproductive (overwintering) adults; E, eggs; L_1, first instar larvae; L_4, fourth instar larvae; NA, newly emerged adults.

F. Summary

Population studies of *E. niponica* clearly demonstrate that behavioral and physiological responses of ovipositing females, both of which are adaptive life history traits, dominate the population dynamics of the lady beetle. Females search actively for an oviposition site early in the reproductive season, resulting in spatial stabilization of egg density. The increased movement in response to high egg density may reduce female survival and enhance dispersal in a density-dependent manner. Later in the season, females often resorb eggs in response to host deterioration, and then stop laying eggs. The combined effect of these individual responses is a strong density-dependent reduction in the reproductive rate which, in turn, causes temporal resource tracking at the population level and resource limitation on lady beetle populations at a relatively low density.

These behavioral and physiological tactics of ovipositing females also improve their reproductive success. As offspring fitness consistently decreases with increasing egg density on individual plants, selective oviposition, avoiding plants with many eggs, increases the female's reproductive success. Varying oviposition time by egg resorption may be adaptive when the females are engaged in future reproduction. Females stop laying eggs when large-scale disturbance reduces the probability of offspring survival.

Oviposition may resume when the habitat recovers. In addition, long-lived females refrain from laying eggs at the end of the reproductive season when offspring fitness is lower. Egg resorption may enhance subsequent survival of the female and reproduction in the following year. In either case, egg resorption by a reproductive female is an adaptive response, which avoids unfavorable times for offspring and postpones egg-laying to a future time with higher offspring fitness.

VI. Conclusions

Insect population ecologists have long concentrated on natural enemies as a principal factor in population regulation of herbivorous insects. However, increasing evidence from insect–plant interactions has revealed that plant resources are highly variable and heterogeneous in time and space, and has emphasized the possibility of resource limitation on herbivore insect populations at low densities. Life history tactics of resource use by herbivorous insects, and their evolutionary constraints in the selection of host plants or plant parts, should have a significant impact on survivorship and reproduction. This is because herbivorous insects have evolved various resource-use tactics to increase reproductive success using plant resources that are structurally, chemically, and temporally heterogeneous. In particular, tactics in resource choice are of crucial importance in population dynamics, because they may substantially change the effects of plants on survivorship and reproduction of herbivorous insects. Thus, it is essential to recognize that dynamic features of insect herbivore populations are greatly affected not only by plant resources, but also by life history tactics in resource use through survival and reproductive processes. The former can be measured in terms of insect performance, and the latter is expressed as preference.

An alternative approach to population dynamics of herbivorous insects needs an evolutionary perspective based on insect–plant interactions, through the reduction of population ecology to individual ecology. This also focuses on the behavioral and physiological mechanisms, related to the selection of host plants or plant parts, that may underlie population dynamics of herbivorous insects. Future studies of the population dynamics of herbivorous insects will then bridge the gap between population ecology and behavioral ecology, which have rarely interacted.

Acknowledgments

I thank Peter Price and Mark Hunter for their valuable comments on the earlier version of this chapter. Financial support was provided by a Japan Ministry of Education, Science, and Culture Grant-in-Aid for General Scientific Research (#01540555) and for Scientific Research on Priority Areas (#319).

References

Aide, T. M. (1988). Herbivory as a selective agent on the timing of leaf production in a tropical understory community. *Nature* **336**, 574–575.

Aide, T. M., and Londoño, E. C. (1989). The effects of rapid leaf expansion on the growth and survivorship of a lepidopteran herbivore. *Oikos* **55**, 66–70.

Andow, D. A. (1990). Population dynamics of an insect herbivore in simple and diverse habitats. *Ecology* **71**, 1006–1017.

Andrewartha, H. G., and Birch, L. C. (1954). "The Distribution and Abundance of Animals." The University of Chicago Press, Chicago.

Auerbach, M., and Simberloff, D. (1989). Oviposition site preference and larval mortality in a leaf-mining moth. *Ecol. Entomol.* **14**, 131–140.

Bach, C. E. (1980). Effects of plant density and diversity on the population dynamics of a specialist herbivore, the striped cucumber beetle, *Acalymma vittata* (Fab.). *Ecology* **61**, 1515–1530.

Bach, C. E. (1981). Host plant growth form and diversity: Effects on abundance and feeding preference of a specialist herbivore, *Acalymma vittata*. *Oecologia* **50**, 370–375.

Bach, C. E. (1984). Plant spatial pattern and herbivore population dynamics: Plant factors affecting the movement patterns of a tropical cucurbit specialist (*Acalymma innubum*). *Ecology* **65**, 175–190.

Bach, C. E. (1988). Effects of host plant patch size on herbivore density: Patterns. *Ecology* **69**, 1090–1102.

Bach, C. E. (1990). Plant successional stage and insect herbivory: Flea beetles on sand-dune willow. *Ecology* **71**, 598–609.

Baltensweiler, W., Benz, G., Bovey, P., and Delucchi, V. (1977). Dynamics of larch bud moth populations. *Annu. Rev. Entomol.* **22**, 79–100.

Barbosa, P. (1988). Natural enemies and herbivore–plant interactions: Influence of plant allelochemicals and host specificity. *In* "Novel Aspects of Insect–Plant Interactions" (P. Barbosa and D. K. Letourneau, eds.), pp. 201–229. Wiley, New York.

Barbosa, P., and Baltensweiler, W. (1987). Phenotypic plasticity and herbivore outbreaks. *In* "Insect Outbreaks" (P. Barbosa and J. C. Schultz, eds.), pp. 469–503. Academic Press, New York.

Bernays, E. A., Driver, G. C., and Bilgener, M. (1990). Herbivores and plant tannins. *Adv. Ecol. Res.* **19**, 263–302.

Brodbeck, B., and Strong, D. (1987). Amino acid nutrition of herbivorous insects and stress to host plants. *In* "Insect Outbreaks" (P. Barbosa and J. C. Schultz, eds.), 347–364. Academic Press, New York.

Brodbeck, B. V., Mizell, R. F. III, French, W. J., Andersen, P. C., and Aldrich, J. H. (1990). Amino acids as determinants of host preference for the xylem feeding leafhopper, *Homalodisca coagulata* (Homoptera: Cicadellidae). *Oecologia* **83**, 338–345.

Bryant, J. P., Clausen, T. P., Reichardt, P. B., McCarthy, M. C., and Werner, R. A. (1987). Effect of nitrogen fertilization upon the secondary chemistry and nutritional value of quaking aspen (*Populus tremuloides* Michx.) leaves for the large aspen tortrix (*Choristoneura conflictana* (Walker)). *Oecologia* **73**, 513–517.

Capman, W. C., Batzli, G. O., and Simms, L. E. (1990). Responses of the common sooty wing skipper to patches of host plants. *Ecology* **71**, 1430–1440.

Cates, R. G., Henderson, C. B., and Redak, R. A. (1987). Responses of the western spruce budworm to varying levels of nitrogen and terpenes. *Oecologia* **73**, 312–316.

Clancy, K. M., and Price, P. W. (1986). Temporal variation in three-trophic-level interactions among willows, sawflies, and parasites. *Ecology* **67**, 1601–1607.

Clutton-Brock, T. H. (ed.) (1988). "Reproductive Success." The University of Chicago Press, Chicago.

Coley, P. D., Bryant, J. P., and Chapin, F. S., III (1985). Resource availability and plant antiherbivore defense. *Science* **230**, 895–899.

Connell, J. H., and Sousa, W. P. (1983). On the evidence needed to judge ecological stability or persistence. *Am. Nat.* **121**, 789–824.

Craig, T. P., Itami, J. K., and Price, P. W. (1989). A strong relationship between oviposition preference and larval performance in a shoot-galling sawfly. *Ecology* **70**, 1691–1699.

Craig, T. P., Itami, J. K., and Price, P. W. (1990). The window of vulnerability of a shoot-galling sawfly to attack by a parasitoid. *Ecology* **71**, 1471–1482.

Crawley, M. J. (1983). "Herbivory: The Dynamics of Animal–Plant Interactions." Blackwell Scientific Publications, Oxford, England.

Crawley, M. J. (1989). Insect herbivores and plant population dynamics. *Annu. Rev. Entomol.* **34**, 531–564.

Crawley, M. J., and Gillman, M. P. (1989). Population dynamics of cinnabar moth and ragwort in grassland. *J. Anim. Ecol.* **58**, 1035–1050.

Crawley, M. J., and Pattrasudhi, R. (1988). Interspecific competition between insect herbivores: Asymmetric competition between cinnabar moth and the ragwort seed-head fly. *Ecol. Entomol.* **13**, 243–249.

Cromartie, W. J., Jr. (1975). The effect of stand size and vegetational background on the colonization of cruciferous plants by herbivorous insects. *J. Appl. Ecol.* **12**, 517–533.

Dempster, J. P. (1971). The population ecology of the cinnabar moth, *Tyria jacobaeae* L. (Lepidoptera, Arctiidae). *Oecologia* **7**, 26–67.

Dempster, J. P. (1975). "Animal Population Ecology." Academic Press, London.

Dempster, J. P. (1982). The ecology of the cinnabar moth, *Tyria jacobaeae* L. (Lepidoptera: Arctiidae). *Adv. Ecol. Res.* **12**, 1–36.

Dempster, J. P. (1983). The natural control of populations of butterflies and moths. *Biol. Rev.* **58**, 461–481.

Dempster, J. P., and Pollard, E. (1981). Fluctuations in resource availability and insect populations. *Oecologia* **50**, 412–416.

den Boer, P. J., and Gradwell, G. R. (eds.) (1971). "Dynamics of Populations." PUDOC, Wageningen, The Netherlands.

Denno, R. F., and Dingle, H. (1981). "Insect Life History Patterns." Springer-Verlag, New York.

Denno, R. F., and McClure, M. S. (1983). Variability: A key to understanding plant–herbivore interactions. *In* "Variable Plants and Herbivores in Natural and Managed Systems" (R. F. Denno and M. S. McClure, eds.), pp. 1–12. Academic Press, New York.

Denno, R. F., Larsson, S., and Olmstead, K. L. (1990). Role of enemy-free space and plant quality in host-plant selection by willow beetles. *Ecology* **71**, 124–137.

Dixon, A. F. G. (1970). Quality and availability of food for a sycamore aphid population. *In* "Animal Populations in Relation to Their Food Resources" (A. Watson, ed.), pp. 271–287. Blackwell Scientific Publications, Oxford, England.

Duffey, S. S. (1970). Cardiac glycosides and distastefulness: Some observations on the palatability spectrum of butterflies. *Science* **169**, 78–79.

Edelstein-Keshet, L., and Rausher, M. D. (1989). The effects of inducible plant defenses on herbivore populations. 1. Mobile herbivores in continuous time. *Am. Nat.* **133**, 787–810.

Edwards, P. J., and Wratten, S. D. (1987). Ecological significance of wound-induced changes in plant chemistry. *In* "Insects–Plants" (V. Labeyrie, G. Fabres, and D. Lachaise, eds.), pp. 213–218. Dr W. Junk, Dordrecht, The Netherlands.

Edwards, P. J., Wratten, S. D., and Greenwood, S. (1986). Palatability of British trees to insects: Constitutive and induced defences. *Oecologia* **69**, 316–319.

Ehrlich, P. E., and Birch, L. C. (1967). The balance of nature and population control. *Am. Nat.* **101,** 97–107.

Endler, J. A. (1986). "Natural Selection in the Wild." Princeton University Press, Princeton, New Jersey.

Erickson, J. M., and Feeny, P. (1974). Sinigrin: A chemical barrier to the black swallowtail butterfly, *Papilio polyxenes. Ecology* **55,** 103–111.

Faeth, S. H. (1985). Host leaf selection by leaf miners: Interactions among three trophic levels. *Ecology* **66,** 870–875.

Faeth, S. H. (1986). Indirect interactions between temporally separated herbivores mediated by the host plant. *Ecology* **67,** 479–494.

Faeth, S. H. (1987). Community structure and folivorous insect outbreaks: The roles of vertical and horizontal interactions. *In* "Insect Outbreaks" (P. Barbosa and J. C. Schultz, eds.), pp. 135–171. Academic Press, New York.

Faeth, S. H., Connor, E. F., and Simberloff, D. (1981). Early leaf abscission: A neglected source of mortality for folivores. *Am. Nat.* **117,** 409–415.

Faeth, S. H., Mopper, S., and Simberloff, D. (1981). Abundances and diversity of leaf-mining insects on three oak host species: Effects of host-plant phenology and nitrogen content of leaves. *Oikos* **37,** 238–251.

Falconer, D. S. (1981). "Introduction to Quantitative Genetics." 2nd Ed. Longman, London.

Feeny, P. (1970). Seasonal changes in oak leaf tannins and nutrients as a cause of spring feeding by winter moth caterpillars. *Ecology* **51,** 565–581.

Feeny, P. (1976). Plant apparency and chemical defense. *Rec. Adv. Phytochem.* **10,** 1–40.

Fowler, S. V., and Lawton, J. H. (1985). Rapidly induced defenses and talking trees: The devil's advocate position. *Am. Nat.* **126,** 181–195.

Fox, L. R. (1981). Defense and dynamics in plant–herbivore systems. *Am. Zool.* **21,** 853–864.

Fox, L. R., and Macauley, B. J. (1977). Insect grazing on *Eucalyptus* in response to variation in leaf tannins and nitrogen. *Oecologia* **29,** 145–162.

Futuyma, D. J., and Wasserman, S. S. (1980). Resource concentration and herbivory in oak forests. *Science* **210,** 920–922.

Gibberd, R., Edwards, P. J., and Wratten, S. D. (1988). Wound-induced changes in the acceptability of tree-foliage to Lepidoptera: Within-leaf effects. *Oikos* **51,** 43–47.

Grant, P. R. (1986). "Ecology and Evolution of Darwin's Finches." Princeton University Press, Princeton, New Jersey.

Gross, P., and Price, P. W. (1988). Plant influences on parasitism of two leafminers: A test of enemy-free space. *Ecology* **69,** 1506–1516.

Hairston, N. G., Smith, F. E., and Slobodkin, L. B. (1960). Community structure, population control, and competition. *Am. Nat.* **94,** 421–425.

Harcourt, D. G. (1971). Population dynamics of *Leptinotarsa decemlineata* (Say) in eastern Ontario. III. Major population processes. *Can. Entomol.* **103,** 1049–1061.

Hare, J. D., Yu, D. S., and Luck, R. F. (1990). Variation in life history parameters of California red scale on different citrus cultivars. *Ecology* **71,** 1451–1460.

Harper, J. L. (1982). After description. *In* "The Plant Community as a Working Mechanism" (E. I. Newman, ed.), pp. 11–25. Blackwell Scientific Publications, Oxford, England.

Harrison, S., and Karban, R. (1986). Effects of an early-season folivorous moth on the success of a later-season species, mediated by a change in the quality of the shared host, *Lupinus arboreus* Sims. *Oecologia* **69,** 354–359.

Hassell, M. P. (1986). Detecting density dependence. *TREE* **1,** 90–93.

Hassell, M. P., and May, R. M. (1985). From individual behaviour to population dynamics. *In* "Behavioural Ecology" (R. M. Sibly and R. H. Smith, eds.), pp. 3–32. Blackwell Scientific Publications, Oxford, England.

Hassell, M. P., Lawton, J. H., and May, R. M. (1976). Patterns of dynamical behaviour in single-species populations. *J. Anim. Ecol.* **45,** 471–486.

Haukioja, E. (1980). On the role of plant defences in the fluctuation of herbivore populations. *Oikos* **35**, 202–213.

Haukioja, E., and Neuvonen, S. (1987). Insect population dynamics and induction of plant resistance: The testing of hypotheses. *In* "Insect Outbreaks" (P. Barbosa and J. C. Schultz, eds.), pp. 411–432. Academic Press, New York.

Haukioja, E., Niemela, P., and Siren, S. (1985). Foliage phenols and nitrogen in relation to growth, insect damage, and ability to recover after defoliation, in the mountain birch *Betula pubescens* ssp. *tortuosa*. *Oecologia* **65**, 214–222.

Higashi, M., and Ohgushi, T. (1990). Three interconnections of ecology. *In* "Ecology for Tomorrow" (H. Kawanabe, T. Ohgushi, and M. Higashi, eds.), pp. 199–205. Physiology and Ecology Japan, Vol. 27, Special Number, Kyoto, Japan.

Howe, H. F., and Westley, L. C. (1988). "Ecological Relationships of Plants and Animals." Oxford University Press, New York.

Hunter, M. D. (1987). Opposing effects of spring defoliation on late season oak caterpillars. *Ecol. Entomol.* **12**, 373–382.

Hunter, M. D. (1990). Differential susceptibility to variable plant phenology and its role in competition between two insect herbivores on oak. *Ecol. Entomol.* **15**, 401–408.

Hunter, M. D., and Willmer, P. G. (1989). The potential for interspecific competition between two abundant defoliators on oak: Leaf damage and habitat quality. *Ecol. Entomol.* **14**, 267–277.

Johnson, N. D., Chu, C.C., Ehrlich, P. R., and Mooney, H. A. (1984). The seasonal dynamics of leaf resin, nitrogen, and herbivore damage in *Eriodictyon californicum* and their parallels in *Diplacus aurantiacus*. *Oecologia* **61**, 398–402.

Jolly, G. M. (1965). Explicit estimates from capture–recapture data with both death and immigration—stochastic model. *Biometrika* **52**, 225–247.

Kahn, D. M., and Cornell, H. V. (1989). Leafminers, early leaf abscission, and parasitoids: A tritrophic interaction. *Ecology* **70**, 1219–1226.

Karban, R. (1986). Interspecific competition between folivorous insects on *Erigeron glaucus*. *Ecology* **67**, 1063–1072.

Karban, R., and Myers, J. H. (1989). Induced plant responses to herbivory. *Annu. Rev. Ecol. Syst.* **20**, 331–348.

Kareiva, P. (1983). Influence of vegetation texture on herbivore populations: Resource concentration and herbivore movement. *In* "Variable Plants and Herbivores in Natural and Managed Systems" (R. F. Denno and M. S. McClure, eds.), pp. 259–289. Academic Press, New York.

Kareiva, P. (1985). Finding and losing host plants by *Phyllotreta*: Patch size and surrounding habitat. *Ecology* **66**, 1809–1816.

Kareiva, P., and Sahakian, R. (1990). Tritrophic effects of a simple architectural mutation in pea plants. *Nature* **345**, 433–434.

Karowe, D. N. (1989). Differential effect of tannic acid on two tree-feeding Lepidoptera: Implications for theories of plant anti-herbivore chemistry. *Oecologia* **80**, 507–512.

Koenig, W. D., and Mumme, R. L. (1987). "Cooperatively Breeding Acorn Woodpecker." Princeton University Press, Princeton, New Jersey.

Krebs, J. R., and Davies, N. B. (eds.) (1984). "Behavioural Ecology." 2nd Ed. Blackwell Scientific Publications, Oxford, England.

Kurihara, M. (1975). Anatomical and histological studies on the germinal vesicle in degenerating oocyte of starved females of the lady beetle, *Epilachna vigintioctomaculata* Motschulsky (Coleoptera, Coccinellidae). *Kontyû* **43**, 91–105.

Lack, D. (1954). "The Natural Regulation of Animal Numbers." Clarendon Press, Oxford, England.

Lack, D. (1966). "Population Studies of Birds." Clarendon Press, Oxford, England.

Larsson, S. (1989). Stressful times for the plant stress–insect performance hypothesis. *Oikos* **56,** 277–283.

Lawrence, W. S. (1988). Movement ecology of the red milkweed beetle in relation to population size and structure. *J. Anim. Ecol.* **57,** 21–35.

Lawrence, W. S., and Bach, C. E. (1989). Chrysomelid beetle movements in relation to host-plant size and surrounding non-host vegetation. *Ecology* **70,** 1679–1690.

Lawton, J. H., and Hassell, M. P. (1981). Asymmetrical competition in insects. *Nature* **289,** 793–795.

Lawton, J. H., and McNeill, S. (1979). Between the devil and the deep blue sea: On the problem of being a herbivore. *In* "Population Dynamics" (R. M. Anderson, B. D. Turner, and L. R. Taylor, eds.), pp. 223–244. Blackwell Scientific Publications, Oxford, England.

Lindroth, R. L., and Peterson, S. S. (1988). Effects of plant phenols on performance of southern armyworm larvae. *Oecologia* **75,** 185–189.

Łomnicki, A. (1980). Regulation of population density due to individual differences and patchy environment. *Oikos* **35,** 185–193.

Łomnicki, A. (1988). "Population Ecology of Individuals." Princeton University Press, Princeton, New Jersey.

Mattson, W. J., and Haack, R. A. (1987). The role of drought stress in provoking outbreaks of phytophagous insects. *In* "Insect Outbreaks" (P. Barbosa and J. C. Schultz, eds.), pp. 365–407. Academic Press, New York.

Mattson, W. J. Jr. (1980). Herbivory in relation to plant nitrogen content. *Annu. Rev. Ecol. Syst.* **11,** 119–161.

May, R. M. (1986). The search for patterns in the balance of nature: advances and retreats. *Ecology* **67,** 1115–1126.

McKey, D. (1979). The distribution of secondary compounds within plants. *In* "Herbivores: Their Interaction with Secondary Plant Metabolites" (G. A. Rosenthal and D. H. Janzen, eds.), pp. 56–134. Academic Press, New York.

McNeill, S. (1973). The dynamics of a population of *Leptopterna dolabrata* (Heteroptera: Miridae) in relation to its food resources. *J. Anim. Ecol.* **42,** 495–507.

McNeill, S., and Southwood, T. R. E. (1978). The role of nitrogen in the development of insect/plant relationships. *In* "Biochemical Aspects of Plant and Animal Coevolution" (J. B. Harborne, ed.), pp. 77–98. Academic Press, London.

Miller, J. S., and Feeny, P. P. (1989). Interspecific differences among swallowtail larvae (Lepidoptera: Papilionidae) in susceptibility to aristolochic acids and berberine. *Ecol. Entomol.* **14,** 287–296.

Milne, A. (1957). The natural control of insect populations. *Can. Entomol.* **89,** 193–213.

Minkenberg, O. P. J. M., and Ottenheim, J. J. G. W. (1990). Effect of leaf nitrogen content of tomato plants on preference and performance of a leafmining fly. *Oecologia* **83,** 291–298.

Mopper, S., Whitham, T. G., and Price, P. W. (1990). Plant phenotype and interspecific competition between insects determine sawfly performance and density. *Ecology* **71,** 2135–2144.

Moran, N. A., and Whitham, T. G. (1990). Differential colonization of resistant and susceptible host plants: *Pemphigus* and *Populus*. *Ecology* **71,** 1059–1067.

Murdoch, W. W. (1966). Community structure, population control, and competition—a critique. *Am. Nat.* **100,** 219–226.

Myers, J. H. (1980). Is the insect or the plant the driving force in the cinnabar moth–tansy ragwort system? *Oecologia* **47,** 16–21.

Myers, J. H. (1981). Interactions between western tent catapillars and wild rose: A test of some general plant herbivore hypotheses. *J. Anim. Ecol.* **50,** 11–25.

Myers, J. H., and Campbell, B. J. (1976). Distribution and dispersal in populations capable of resource depletion. A field study on cinnabar moth. *Oecologia* **24,** 7–20.

Myers, J. H., and Post, B. J. (1981). Plant nitrogen and fluctuations of insect populations: A test with the cinnabar moth–tansy ragwort system. *Oecologia* **48**, 151–156.

Myers, J. H., and Williams, K. S. (1987). Lack of short- or long-term inducible defenses in the red alder–western tent caterpillar system. *Oikos* **48**, 73–78.

Neuvonen, S., Haukioja, E., and Molarius, A. (1987). Delayed inducible resistance against a leaf-chewing insect in four deciduous tree species. *Oecologia* **74**, 363–369.

Ng, D. (1988). A novel level of interactions in plant-insect systems. *Nature* **334**, 611–613.

Nicholson, A. J. (1954). An outline of the dynamics of animal populations. *Aust. J. Zool.* **2**, 9–65.

Ohgushi, T. (1986). Population dynamics of an herbivorous lady beetle, *Henosepilachna niponica*, in a seasonal environment. *J. Anim. Ecol.* **55**, 861–879.

Ohgushi, T. (1987). Factors affecting body size variation within a population of an herbivorous lady beetle, *Henosepilachna niponica* (Lewis). *Res. Popul. Ecol.* **29**, 147–154.

Ohgushi, T. (1988). Temporal and spatial relationships between an herbivorous lady beetle *Epilachna niponica* and its predator, the earwig *Anechura harmandi*. *Res. Popul. Ecol.* **30**, 57–68.

Ohgushi, T. (1991). Lifetime fitness and evolution of reproductive pattern in the herbivorous lady beetle. *Ecology*, in press.

Ohgushi, T., and Sawada, H. (1981). The dynamics of natural populations of a phytophagous lady beetle, *Henosepilachna pustulosa* under different habitat conditions I. Comparison of adult population parameters among local populations in relation to habitat stability. *Res. Popul. Ecol.* **23**, 94–115.

Ohgushi, T., and Sawada, H. (1984). Inter-population variation of life history characteristics and its significance on survival process of an herbivorous lady beetle, *Henosepilachna niponica* (Lewis) (Coleoptera, Coccinellidae). *Kontyû* **52**, 399–406.

Ohgushi, T., and Sawada, H. (1985a). Population equilibrium with respect to available food resource and its behavioural basis in an herbivorous lady beetle, *Henosepilachna niponica*. *J. Anim. Ecol.* **54**, 781–796.

Ohgushi, T., and Sawada, H. (1985b). Arthropod predation limits the population density of an herbivorous lady beetle, *Henosepilachna niponica* (Lewis). *Res. Popul. Ecol.* **27**, 351–359.

Ohmart, C. P., Stewart, L. G., and Thomas, J. R. (1985). Effects of food quality, particularly nitrogen concentrations of *Eucalyptus blakelyi* foliage on the growth of *Paropsis atomaria* larvae (Coleoptera: Chrysomelidae). *Oecologia* **65**, 543–549.

Onuf, C. P., Teal, J. M., and Valiela, I. (1977). Interactions of nutrients, plant growth, and herbivory in a mangrove ecosystem. *Ecology* **58**, 514–526.

Ostfeld, R. S. (1985). Limiting resources and territoriality in microtine rodents. *Am. Nat.* **126**, 1–15.

Pasteels, J. M., Rowell-Rahier, M., and Raupp, M. J. (1988). Plant-derived defense in chrysomelid beetles. *In* "Novel Aspects of Insect–Plant Interactions" (P. Barbosa and D. K. Letourneau, eds.), pp. 235–272. Wiley, New York.

Pimentel, D. (1988). Herbivore population feeding pressure on plant hosts: Feedback evolution and host conservation. *Oikos* **53**, 289–302.

Potter, D. A., and Redmond, C. T. (1989). Early spring defoliation, secondary leaf flush, and leafminer outbreaks on American holly. *Oecologia* **81**, 192–197.

Preszler, R. W., and Price, P. W. (1988). Host quality and sawfly populations: A new approach to life table analysis. *Ecology* **69**, 2012–2020.

Price, P. W. (1983). Hypotheses on organization and evolution in herbivorous insect communities. *In* "Variable Plants and Herbivores in Natural and Managed Systems" (R. F. Denno and M. S. McClure, eds.), pp. 559–596. Academic Press, New York.

Price, P. W. (1984). Alternative paradigms in community ecology. *In* "A New Ecology" (P. W. Price, C. N. Slobodchikoff, and W. S. Gaud, eds.), pp. 353–383. Wiley, New York.

Price, P. W. (1987). The role of natural enemies in insect populations. *In* "Insect Outbreaks" (P. Barbosa and J. C. Schultz, eds.), pp. 287–312. Academic Press, New York.

Price, P. W. (1992). Plant resources as the mechanistic basis for insect herbivore population dynamics. *In* "Effects of Resource Distribution and Animal–Plant Interactions" (M. D. Hunter, T. Ohgushi, and P. W. Price, eds.), pp. 139–173, Academic Press, San Diego.

Price, P. W., and Clancy, K. M. (1986). Interactions among three trophic levels: Gall size and parasitoid attack. *Ecology* **67**, 1593–1600.

Price, P. W., Bouton, C. E., Gross, P., McPheron, B. A., Thompson, J. N., and Weis, A. E. (1980). Interactions among three trophic levels: Influence of plants on interactions between insect herbivores and natural enemies. *Annu. Rev. Ecol. Syst.* **11**, 41–65.

Price, P. W., Cobb, N., Craig, T. P., Fernandes, G. W., Itami, J. K., Mopper, S., and Preszler, R. W. (1990). Insect herbivore population dynamics on trees and shrubs: New approaches relevant to latent and eruptive species and life table development. *In* "Insect–Plant Interactions" (E. A. Bernays, ed.), Vol. 2, pp. 1–38. CRC Press, Boca Raton, Florida.

Randall, M. G. M. (1982). The dynamics of an insect population throughout its altitudinal distribution: *Coleophora alticolella* (Lepidoptera) in northern England. *J. Anim. Ecol.* **51**, 993–1016.

Raupp, M. J., and Denno, R. F. (1984). The suitability of damaged willow leaves as food for the leaf beetle, *Plagiodera versicolora*. *Ecol. Entomol.* **9**, 443–448.

Raupp, M. J., Werren, J. H., and Sadof, C. S. (1988). Effects of short-term phenological changes in leaf suitability on the survivorship, growth, and development of gypsy moth (Lepidoptera: Lymantriidae) larvae. *Environ. Entomol.* **17**, 316–319.

Rausher, M. D. (1980). Host abundance, juvenile survival, and oviposition preference in *Battus philenor*. *Evolution* **34**, 342–355.

Rausher, M. D. (1983). Ecology for host-selection behavior in phytophagous insects. *In* "Variable Plants and Herbivores in Natural and Managed Systems" (R. F. Denno and M. S. McClure, eds.), pp. 223–257. Academic Press, New York.

Rhoades, D. F. (1983). Herbivore population dynamics and plant chemistry. *In* "Variable Plants and Herbivores in Natural and Managed Systems" (R. F. Denno and M. S. McClure, eds.), pp. 155–220. Academic Press, New York.

Rhoades, D. F. (1985). Offensive–defensive interactions between herbivores and plants: Their relevance in herbivore population dynamics and ecological theory. *Am. Nat.* **125**, 205–238.

Rhoades, D. F., and Cates, R. G. (1976). Toward a general theory of plant antiherbivore chemistry. *Rec. Adv. Phytochem.* **10**, 168–213.

Risch, S. (1980). The population dynamics of several herbivorous beetles in a tropical agroecosystem: The effect of intercropping corn, beans, and squash in Costa Rica. *J. Appl. Ecol.* **17**, 593–612.

Risch, S. J. (1981). Insect herbivore abundance in tropical monocultures and polycultures: An experimental test of two hypotheses. *Ecology* **62**, 1325–1340.

Risch, S. J., Andow, D., and Altieri, M. A. (1983). Agroecosystem diversity and pest control: Data, tentative conclusions, and new research directions. *Environ. Entomol.* **12**, 625–629.

Rockwood, L. L. (1974). Seasonal changes in the susceptibility of *Crescentia alata* leaves to the flea beetle, *Oedionychus* sp. *Ecology* **55**, 142–148.

Roland, J., and Myers, J. H. (1987). Improved insect performance from host-plant defoliation: Winter moth on oak and apple. *Ecol. Entomol.* **12**, 409–414.

Root, R. B. (1973). Organization of a plant–arthropod association in simple and diverse habitats: The fauna of collards (*Brassica oleracea*). *Ecol. Monogr.* **43**, 95–124.

Rossiter, M. C., Schultz, J. C., and Baldwin, I. T. (1988). Relationships among defoliation, red oak phenolics, and gypsy moth growth and reproduction. *Ecology* **69**, 267–277.

Rothschild, M. (1973). Secondary plant substances and warning colouration in insects. *In* "Insect/Plant Relationships" (H. F. van Emden, ed.), pp. 59–83. Blackwell Scientific Publications, Oxford, England.

Royama, T. (1977). Population persistence and density dependence. *Ecol. Monogr.* **47**, 1–35.

Schoener, T. W. (1986). Mechanistic approaches to community ecology: A new reductionism? *Am. Zool.* **26**, 81–106.

Schultz, J. C. (1988). Plant responses induced by herbivores. *TREE* **3**, 45–49.

Schultz, J. C. (1983a). Impact of variable plant defensive chemistry on susceptibility of insects to natural enemies. *In* "Plant Resistance to Insects" (P. Hedin, ed.), pp. 37–54. American Chemical Society, Washington, D.C.

Schultz, J. C. (1983b). Habitat selection and foraging tactics of caterpillars in heterogeneous trees. *In* "Variable Plants and Herbivores in Natural and Managed Systems" (R. F. Denno and M. S. McClure, eds.), pp. 61–90. Academic Press, New York.

Schultz, J. C., and Baldwin, I. T. (1982). Oak leaf quality declines in response to defoliation by gypsy moth larvae. *Science* **217**, 149–151.

Scriber, J. M. (1977). Limiting effects of low leaf-water content on the nitrogen utilization, energy budget, and larval growth of *Hyalophora cecropia* (Lepidoptera: Saturniidae). *Oecologia* **28**, 269–287.

Scriber, J. M. (1984). Host-plant suitability. *In* "Chemical Ecology of Insects" (W. J. Bell and R. T. Cardé, eds.), pp. 159–202. Chapman and Hall, London.

Scriber, J. M., and Slansky, F., Jr. (1981). The nutritional ecology of immature insects. *Annu. Rev. Entomol.* **26**, 183–211.

Seber, G. A. F. (1973). "The Estimation of Animal Abundance and Related Parameters." Griffin, London.

Silkstone, B. E. (1987). The consequences of leaf damage for subsequent insect grazing on birch (*Betula* spp.). A field experiment. *Oecologia* **74**, 149–152.

Simberloff, D., and Stiling, P. (1987). Larval dispersion and survivorship in a leaf-mining moth. *Ecology* **68**, 1647–1657.

Sinclair, A. R. E. (1989). Population regulation in animals. *In* "Ecological Concepts" (J. M. Cherrett, ed.), pp. 197–241. Blackwell Scientific Publications, Oxford, England.

Singer, M. C. (1986). The definition and measurement of oviposition preference in plant-feeding insects. *In* "Insect–Plant Interactions" (J. R. Miller and T. A. Miller, eds.), pp. 65–94. Springer-Verlag, New York.

Singer, M. C., Ng, D., and Thomas, C. D. (1988). Heritability of oviposition preference and its relationship to offspring performance within a single insect population. *Evolution* **42**, 977–985.

Sitch, T. A., Grewcock, D. A., and Gilbert, F. S. (1988). Factors affecting components of fitness in a gall-making wasp (*Cynips divisa* Hartig). *Oecologia* **76**, 371–375.

Smith, R. H., and Sibly, R. (1985). Behavioural ecology and population dynamics: Towards a synthesis. *In* "Behavioural Ecology" (R. M. Sibly and R. H. Smith, eds.), pp. 577–591. Blackwell Scientific Publications, Oxford, England.

Solomon, B. P. (1981). Response of a host-specific herbivore to resource density, relative abundance, and phenology. *Ecology* **62**, 1205–1214.

Southwood, T. R. E. (ed.) (1968). "Insect Abundance." Blackwell Scientific Publications, Oxford, England.

Stamp, N. E. (1984). Effect of defoliation by checkerspot caterpillars (*Euphydryas phaeton*) and sawfly larvae (*Macrophya nigra* and *Tenthredo grandis*) on their host plants (*Chelone* spp.). *Oecologia* **63**, 275–280.

Stanton, M. L. (1983). Spatial patterns in the plant community and their effects upon insect search. *In* "Herbivorous Insects" (S. Ahmad, ed.), pp. 125–157. Academic Press, New York.

Stiling, P. (1988). Density-dependent processes and key factors in insect populations. *J. Anim. Ecol.* **57**, 581–593.

Stiling, P., and Simberloff, D. (1989). Leaf abscission: Induced defense against pests or response to damage? *Oikos* **55**, 43–49.

Stiling, P. D., Brodbeck, B. V., and Strong, D. R. (1982). Foliar nitrogen and larval parasitism as determinants of leafminer distribution patterns on *Spartina alterniflora*. *Ecol. Entomol.* **7**, 447–452.

Strong, D. R. (1984). Density-vague ecology and liberal population regulation in insects. *In* "A New Ecology" (P. W. Price, C. N. Slobodchikoff, and W. S. Gaud, eds.), pp. 313–327. Wiley, New York.

Strong, D. R., Lawton, J. H., and Sir R. Southwood (1984). "Insects on Plants: Community Patterns and Mechanisms." Blackwell Scientific Publications, Oxford, England.

Tabashnik, B. E. (1982). Responses of pest and non-pest *Colias* butterfly larvae to intraspecific variation in leaf nitrogen and water content. *Oecologia* **55**, 389–394.

Tahvanainen, J. O., and Root, R. B. (1972). The influence of vegetational diversity on the population ecology of a specialized herbivore, *Phyllotreta cruciferae* (Coleoptera: Chrysomelidae). *Oecologia* **10**, 321–346.

Taper, M. L., and Case, T. J. (1987). Interactions between oak tannins and parasite community structure: Unexpected benefits of tannins to cynipid gall-wasps. *Oecologia* **71**, 254–261.

Thompson, J. N. (1988). Evolutionary ecology of the relationship between oviposition preference and performance of offspring in phytophagous insects. *Entomol. Exp. Appl.* **47**, 3–14.

Thompson, J. N., and Price, P. W. (1977). Plant plasticity, phenology, and herbivore dispersion: Wild parsnip and the parsnip webworm. *Ecology* **58**, 1112–1119.

Turchin, P. B. (1986). Modelling the effect of host patch size on Mexican bean beetle emigration. *Ecology* **67**, 124–132.

Varley, G. C., Gradwell, G. R., and Hassell, M. P. (1973). "Insect Population Ecology: An Analytical Approach." Blackwell Scientific Publications, Oxford, England.

Via, S. (1986). Genetic covariance between oviposition preference and larval performance in an insect herbivore. *Evolution* **40**, 778–785.

Vinson, S. B. (1976). Host selection by insect parasitoids. *Annu. Rev. Entomol.* **21**, 109–133.

Waring, G. L., and Price, P. W. (1990). Plant water stress and gall formation (Cecidomyiidae: *Asphondylia* spp.) on creosote bush. *Ecol. Entomol.* **15**, 87–95.

Webb, J. W., and Moran, V. C. (1978). The influence of the host plant on the population dynamics of *Acizzia russellae* (Homoptera: Psyllidae). *Ecol. Entomol.* **3**, 313–321.

Weis, A. E., and Abrahamson, W. G. (1985). Potential selective pressures by parasitoids on a plant–herbivore interaction. *Ecology* **66**, 1261–1269.

West, C. (1985). Factors underlying the late seasonal appearance of the lepidopterous leafmining guild on oak. *Ecol. Entomol.* **10**, 111–120.

White, R. R. (1974). Food plant defoliation and larval starvation of *Euphydryas editha*. *Oecologia* **14**, 307–315.

White, T. C. R. (1974). A hypothesis to explain outbreaks of looper caterpillars, with special reference to populations of *Selidosema suavis* in a plantation of *Pinus radiata* in New Zealand. *Oecologia* **16**, 279–301.

White, T. C. R. (1978). The importance of a relative shortage of food in animal ecology. *Oecologia* **33**, 71–86.

White, T. C. R. (1984). The abundance of invertebrate herbivores in relation to the availability of nitrogen in stressed food plants. *Oecologia* **63**, 90–105.

Whitham, T. G. (1978). Habitat selection by *Pemphigus* aphids in response to resource limitation and competition. *Ecology* **59**, 1164–1176.

Whitham, T. G. (1980). The theory of habitat selection: Examined and extended using *Pemphigus* aphids. *Am. Nat.* **115**, 449–466.

Whitham, T. G. (1983). Host manipulation of parasites: Within-plant variation as a defense against rapidly evolving pests. *In* "Variable Plants and Herbivores in Natural and Managed Systems" (R. F. Denno and M. S. McClure, eds.), pp. 15–41. Academic Press, New York.

Whitham, T. G. (1986). Costs and benefits of territoriality: Behavioral and reproductive release by competing aphids. *Ecology* **67,** 139–147.

Wiens, J. A. (1984). Resource systems, populations, and communities. *In* "A New Ecology" (P. W. Price, C. N. Slobodchikoff, and W. S. Gaud, eds.), pp. 397–436. Wiley, New York.

Williams, A. G., and Whitham, T. G. (1986). Premature leaf abscission: An induced plant defense against gall aphids. *Ecology* **67,** 1619–1627.

Williams, K. S., and Myers, J. H. (1984). Previous herbivore attack of red alder may improve food quality for fall webworm larvae. *Oecologia* **63,** 166–170.

Wolda, H. (1978). Fluctuations in abundance of tropical insects. *Am. Nat.* **112,** 1017–1045.

Woolfenden, G. E., and Fitzpatrick, J. W. (1984). "The Florida Scrub Jay." Princeton University Press, Princeton, New Jersey.

Wratten, S. D., Edwards, P. J., and Winder, L. (1988). Insect herbivory in relation to dynamic changes in host plant quality. *Biol. J. Linn. Soc.* **35,** 339–350.

9

Bottom-Up versus Top-Down Regulation of Vertebrate Populations: Lessons from Birds and Fish

James R. Karr
Institute for Environmental Studies,
Engineering Annex FM-12,
University of Washington,
Seattle, Washington

Isaac J. Schlosser[1]
Department of Biology
University of North Dakota
University Station
Grand Forks, North Dakota

Michele Dionne
The Wells Reserve,
R. R. #2,
Box 806,
Wells, Maine

[1] Present address: U.S. Forest Service, Redwood Sciences Laboratory, Arcata, California

I. Introduction

According to the classic paper of Hairston, Smith, and Slobodkin (1960), plant biomass is abundant because predators keep herbivore populations at low densities (top-down influence). Thus, the world is green because herbivore populations consume only a small fraction of primary production. Plants, in this scenario, are not regulated by top-down forces. Rather, they are limited by bottom-up influences, especially the availability of nutrients (Tilman, 1988). Recent work demonstrates that the relative influence of bottom-up and top-down regulation on plants varies along gradients of primary productivity (Oksanen, 1990), with top-down forces dominating in productive and extremely barren areas, while the reverse seems true in areas of intermediate productivity. Interest in bottom-up versus top-down control of populations continues unabated today (Fretwell, 1987; Oksanen, 1988, 1990; Vadas, 1989; this volume).

The study of plant–animal interactions (Fig. 1, left), perhaps more than any other, was stimulated by the logic of the Hairston *et al.* (1960) paper. Most papers in this volume explore the merit of this theory from a bottom-up perspective with herbivores, especially insects, as the ecological group of interest. We take a different approach. First, we selected vertebrates, a group with comparatively longer life spans and larger body sizes than is characteristic of most taxa considered in this volume. Second, rather than select a single habitat or a productivity gradient, we present a comparative analysis of bottom-up and top-down dynamics in selected terrestrial and aquatic habitats: forests, streams, and lakes.

Plants play a key but variable role among the habitats (Fig. 1, right). They provide food for vertebrates (fruits, seeds, nectar, leaves; Fig. 1, line a), or provide food for vertebrate prey (herbivorous insects, molluscs, zooplankton; Fig. 1, line b). In addition, plants may constitute the dominant structure of a habitat (vegetation structure in forest or littoral habitats) and, thus, be a major influence on the biology of vertebrates. In other habitats (e.g., small streams), living plants may be a minor component. Finally, vertebrates and a few invertebrates may be predators on vertebrates (Fig. 1, line c).

Trophic influences on vertebrates may be direct (predation event on prey) or indirect (alteration of feeding behavior in the presence of a predator). Presence of vascular macrophytes in the littoral zone of a lake may have an indirect effect on predation by providing cover to small planktivorous fish.

Trophic influences also affect many aspects of the biology of vertebrates, including physiology, behavior, morphology, demography, and ecology. Seasonal changes in the food supply and food consumed influence physiological and metabolic processes (Levey and Karasov, 1989). Temporal variation in food supply stimulates the evolution of food-storing behaviors (e.g., caching and fat deposition). Changes in distribution and abundance of food

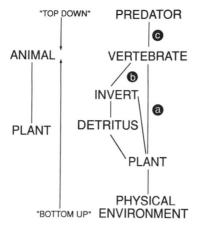

Figure 1 General diagram for studies of plant–animal interactions and for food webs influencing vertebrates.

resources alters aggressive behavior and spacing (territorial systems) in nectarivorous birds (Feinsinger and Colwell, 1978) and foraging tactics and locations in insectivores (Holmes *et al.*, 1978; Morse, 1990). Density, aggression, and food use vary between canopy and undergrowth flocks (Munn, 1985), apparently as a result of bottom-up influences. Flocking itself may be a response to top-down influences. Fish morphology is likewise influenced by bottom-up and top-down processes (Keast and Webb, 1966; Werner, 1984; Crowder, 1986). Survival and growth rates vary with predation intensity (Werner and Hall, 1988). Food-web interactions may temporarily influence vertebrates, or they may mold population and community responses in evolutionary time. In this chapter we use selected examples of other vertebrates' attributes, but we concentrate on population- and community-level processes and vertebrates' responses.

Our analysis of ecological interactions across vertebrate taxa among three major habitats illustrates that the dichotomy between bottom-up and top-down regulation is much like the discussion of whether a glass is half empty or half full. Ecologists may see bottom-up or top-down depending on the experimental design or duration of the study, on the systems or species under investigation, and on specific circumstances at the time of the study. Trophic regulation in vertebrates results from a shifting mosaic of biotic and abiotic influences.

II. Forest Birds

A. Introduction

Associations between vegetation structure and avian diversity (MacArthur and MacArthur, 1961; Karr and Roth, 1971) have long been attributed to bottom-up (food resource) influences, often mediated by competitive interactions. Plant growth form, the dominant element of forest habitat, alters the physical environment and provides a diversity of substrates that supply food, nest sites, and physiological refuges. Many forest birds feed directly on plant material, generally fruits and flowers (See Chapter 12). In tropical forests, for example, 50–80% of tropical forest plants have birds as their primary dispersal agents (Gentry, 1990). Few forest birds feed directly on nonreproductive plant tissues, but many consume herbivorous and carnivorous invertebrates that are dependent on plant tissues for food or shelter. These associations suggest a bottom-up influence on birds. However, forest structure also serves to provide refuges from predation on eggs, nestlings, and adults. We show that assumptions about a dominance of bottom-up influences on the ecology of birds deserve more careful scrutiny.

B. Bottom-Up Effects on Forest Birds

Bottom-up influences on birds may come directly from limited availability of food or indirectly when competitors usurp all or a portion of the food supply. Study of interspecific competition for limited food supplies dominated avian ecology in recent decades, but a broader perspective is emerging (Wiens, 1989a,b). The use of neutral models has expanded and challenges to the dominance of competition are common. Recher (1990, p. 336) recently noted that "selection is at least as likely to be for efficient foraging with the necessary flexibility to adjust to short-term changes in resources as it is to avoid competition." Perhaps the most important lessons from these and other challenges of the competition dogma is that the type, abundance, and pattern of availability of a food resource in space and time have major impacts on the influence of bottom-up factors. As a result, the success of a species may be determined by the absolute quantity of food available to it, the species' efficiency at harvesting that food, or the extent to which its harvesting is challenged by competitors. Environmental attributes [e.g., microclimate (Grubb, 1977; Karr and Freemark, 1983; Karr and Brawn, 1990), and regional landscapes (MacLintock *et al.*, 1977; Turner *et al.*, 1989)] may influence and even prevent a bird from obtaining sufficient food to maintain itself and reproduce successfully. Similarly, the attributes of a community are determined by, among other things, the integration of these factors in a multispecies context. Because of the growing recognition of diversity of bottom-up influences on avian ecology, we organize our discussion according to specific food resources.

1. Frugivory

Frugivore densities and movements are often closely correlated with local fruit abundance (Worthington, 1982; Levey, 1988; Blake and Loiselle, 1991; Loiselle and Blake, 1989, 1991; Fleming, this volume). Seasonal changes in fruit availability stimulate various responses in birds. In time of fruit shortage, the frugivorous red-capped manakin (*Pipra mentalis*) stops breeding, spends more time foraging, and eats more insects (Worthington, 1982). This manakin also moves among forest types to track fruit abundance (Martin and Karr, 1986; Levey, 1988). Finally, altitudinal migrations of manakins and other frugivores result from seasonal changes in fruit availability. In Costa Rica, low fruit abundances at the end of the year in highland areas, a time of fruiting peak in the lowlands, coincide with the time of bird migration from highland to lowland areas (Blake and Loiselle, 1991; Loiselle and Blake, 1991). Year-to-year variation in the magnitude of altitudinal movements also occurs. For example, alteration of seasonal rainfall pattern during the 1983 El Niño year altered flowering and fruiting phenologies, and migrant frugivores were unusually abundant (Stiles and Clark, 1989). New altitudinal migrants also appeared at La Selva in the Atlantic coastal lowlands of Costa Rica in 1983 (Stiles and Clark, 1989).

Perhaps the most comprehensive work on the connections between fruits and the ecology of frugivorous birds is the work of Moermond and his colleagues (Moermond and Denslow, 1985; Levey, 1988). Bird morphology and behavior interact with fruit display, availability, distribution on trees and shrubs, and nutritional quality. In Costa Rica, white-ruffed manakins (*Corapipo alteri*) consumed 43 of 95 species of fruits available locally, but only six species were consumed sufficiently often to affect timing of seasonal movements by the manakin (Stiles and Clark, 1989).

Plants and the fruits they produce affect the distribution and abundances of forest birds throughout the world (Gautier-Hion and Michaloud, 1989); however, the details of the bird–fruit interaction vary considerably among the continents. Frugivory by Old World birds is reduced in comparison to similar habitats in neotropical areas (Gentry, 1990), apparently because the evolutionary responses of plants in those areas involve coevolutionary interactions with mammals. Indeed, most of the species-rich, woody plant families in Africa and Asia are mammal dispersed. At the extreme, forests of southeast Asia have few bird species or individuals that are undergrowth frugivores (Karr, 1980; Wong 1986).

In contrast, 70% of woody plant species in New Zealand are probably dispersed by birds. Recent extinctions of birds threaten the future of plants dependent on bird dispersal of their seeds (Clout and Hay, 1989), illustrating the close association between frugivorous birds and the plants whose fruits they consume.

2. Nectarivory

Nectarivorous birds illustrate numerous adaptations to bottom-up influences; some are general to all nectarivorous birds, while others are not. Nectar feeders throughout the world have a common set of morphological and physiological adaptations (Collins and Patton, 1989): elongated and decurved bills; trough-like tongues; muscles around the oral cavity that permit rapid in and out movement of the tongue; short, simple digestive tracts that permit rapid absorption of sugars. Even capture rate of nectar (5 to 10 μl/sec) remains relatively constant across taxa that differ in size (hummingbirds, 6–18 g; honeyeaters, 8–180 g; sunbirds and honeycreepers, 6–18 g). Clearly, plants and nectarivores produce strong reciprocal selective pressures.

However, other aspects of nectarivore biology vary geographically. Nectar limitation appears common in the Neotropics (hummingbirds), episodic in Hawaii (honeycreepers), but absent in Australia (honeyeaters, Carpenter, 1978). Resource partitioning and community structure vary accordingly. Australian honeyeater communities are diverse, and partitioning is subtle, perhaps along a gradient associated with the quantity of insects in their diet. Partitioning among hummingbirds occurs along gradients of flower size and resource defensibility, often resulting in resource-defined population equilibria (Feinsinger and Colwell, 1978). The spatial scale of these dynamics may be complicated. Regional changes in food abundance may determine the intensity of local competition by affecting recruitment of competitors to localized, rich feeding areas (F. L. Carpenter, personal communication, 1990). High nectar availability for many hummingbirds allows them to feed only 5–30% of the day (Collins and Patton, 1989) and frees males from parental responsibilities, permitting males to sing on courtship areas (leks).

Overall, highly specialized bird-pollination systems have evolved in the hummingbirds, but plants are less tied to birds in Australia and Hawaii where mammals and insects, respectively, are more important to the plants.

Nectar is only one plant exudate. The eucalyptus forests of Australia offer a more diverse array of foods to resident birds, and the population and community ecology of Australian exudate feeders is unique among forest avifaunas (Keast *et al.*, 1985). In addition to nectar, eucalypts produce abundant sap, manna in response to insect attack, and lerps and honeydew are secreted by sapsucking insects. All are consumed by birds, illustrating a clear and unique bottom-up influence on avian ecology at the community level.

3. Insectivory

Nearly all species of temperate forest birds are insectivores during the breeding season, although Australia provides an important counterpoint to the trend, in which only 18% of species of the open forest are insectivorous (Pyke, 1985). Insectivores constitute 45–52% of the forest species at four neotropical forest sites (Karr *et al.*, 1990). Nonetheless, strong associations between arthropod abundance and bird populations are rarely detected, possibly owing to sampling difficulties. A recent exception shows that brown creeper (*Certhia familiaris*) abundances were correlated significantly ($p < 0.01$) with abundances of spiders 6–11 mm long, as evaluated with crawl traps on tree bark in the northwest United States (Mariani and Manuwal, 1990).

Generally, researchers have not been successful in showing clear associations between abundances of insectivorous birds and their prey because abundance and availability (Fig. 2A) may not be highly correlated (Emlen, 1977; Wiens and Rotenberry, 1979; Karr and Brawn, 1990; Pyke, 1985). Indeed, prey should be more abundant where safer and, thus, less available (Hutto, 1990). Karr and Brawn (1990) sampled foliage and litter arthropods at about 60 sampling sites in the undergrowth of forest in central Panama. Activity rates of 20 species of forest birds and five foraging guilds (Fig. 2B) at those sites (based on mist-net capture rates) revealed widely varying consumer–resource associations. Positive correlations between some species and arthropod abundances were balanced by species that were less common when and where their presumed food resources were relatively abundant. Microclimatic conditions (e.g., humidity) also influenced the nature of bird/food interactions. For example, dry sites appeared to be unsuitable habitat for certain bird species despite high arthropod abundances at those sites (Karr and Brawn, 1990). Pyke (1985) found that the correspondence between insectivore density and the biomass of flying insects in Australia was poor except in the September to November period when both were at their highest densities.

Population effects of food abundances vary in space and time. When food abundances are manipulated (naturally or in experiments), the frequency of renesting attempts, second brood attempts, growth rates of young, fledging and hatching success, clutch size, and nestling starvation change in New England forest birds (Holmes, 1990). Overall, Holmes found food to be abundant only once every 10–20 years. Periods between those years were prolonged periods of low food abundance that limited bird populations, a contrast with the stable high food availability in shrub grasslands that experience occasional crunch years with limited availability of food (Wiens, 1989a). Even clutch size often declines seasonally in insectivorous birds, which is apparently related to seasonal changes in food quality and food quantity (Foster, 1974).

Neotropical birds that follow army-ants (Willis and Oniki, 1978) and African birds that follow driver-ants (Chapin, 1932; Brosset, 1969; Willis, 1986) illustrate how unique approaches to obtaining foods open new opportunities to exploit abundant insect resources that are not readily available. Ant raids attract birds (and a number of parasitic insects), which eat insects flushed from the leaf litter and lower trunks, branches, and foliage of forest plants. Without army-ant foraging swarms, most arthropods in forest undergrowth, especially arthropods of leaf litter, are simply not available to birds. Passage of an ant swarm and its avian attendants lowers the abundance of their primary food (crickets, roaches) by 50%. While populations of these taxa recover in as little as a week, other arthropods recover to only half their original level in 100 days (Franks, 1982).

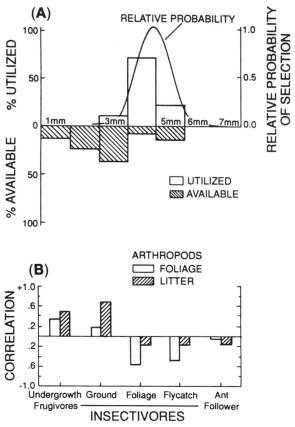

Figure 2 (A) Relative probability of selection of leafhoppers by Wilson's warblers plotted as a function of prey length. The graph is superimposed on the distributions of leafhopper lengths in the samples of available and selected prey. From McDonald *et al.* (1990). (B) Correlations (Spearman's Rho) of capture rates for five avian foraging guilds with abundances of foliage and litter arthropods for 1983 to 1985 in central Panama. From Karr and Brawn (1990). (C) Birds (left) and lerps (right) counted at an experimental site. From Loyn *et al.* (1983). (D) Numbers of undepredated nests relative to numbers of days exposed for off-ground nests in three groups: artificial nests, moss-covered artificial nests, and real (natural) nests. From Martin (1988b). (*Figure continues.*)

Perhaps the most elegant example of the complexity of interactions among birds and their primary food is illustrated by the bell miner (a bird, *Manorina melanophrys*) in Australia (Loyn *et al.*, 1983). Experimental removal of bell miners results in eruptions of psyllid insects followed by colonization and growth of populations of bird species that were behaviorally excluded by the aggressive bell miner (Fig. 2C). Foraging by the colonizing species reduces psyllid populations, and the birds decline in abundance and disap-

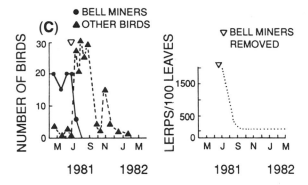

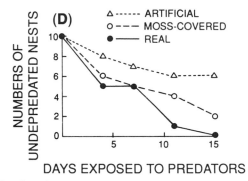

Figure 2   (Continued)

pear with the loss of their abundant food supply. In the undisturbed system, the bell miner behaviorally regulates the population of other birds, and insect abundance governs its population. In the absence of bell miners, other birds are ultimately limited by the loss of their psyllid food.

4. Omnivory

About one fourth of the forest birds at each of four neotropical forest sites (La Selva, Costa Rica; central Panama; Manu, Peru; Manaus, Brazil) consume substantial amounts of both insects and fruits (Karr *et al.*, 1990). These species include birds that forage on insects and fruits throughout the year as well as temporal omnivores, species that switch between insects and fruits during the annual wet and dry cycle. High species richness and high abundance per species in many of these birds suggest that omnivory may serve to buffer changing food availability, and thus limit bottom-up control. The difficulty of tracking the details of food consumption and food availability makes it unlikely that empirical and experimental studies will provide more detail on the ecology of omnivory in the near future.

5. Summary

Bottom-up effects are common among at least three of the four major food resources. Knowledge of the precise nature of those effects is limited by the difficulty involved in design of experimental studies to examine them. Because nectar and fruit resources are more easily manipulated than insect resources, the extent of bottom-up influences is best known for species feeding on those resources. Geographic (intercontinental) variation in food–vertebrate dynamics alters the influence of bottom-up factors, often as a result of historical differences in the balance among major vertebrate taxa.

C. Top-Down Effects on Forest Birds

Behavioral and population ecology of birds is also molded by the top-down effects of predation. Most predation is on eggs and nestlings (Welty and Baptista, 1988), but adults are not immune to mortality from predation.

Predation influences where and when adults feed (Lima, 1987a; Lima and Dill, 1990). Forest birds may flee into woody vegetation when predators are detected (Grubb and Greenwald, 1982), while grassland birds exhibit a reverse pattern (Lima, 1990). Flocking by many forest birds is likely to be a response to the threat of predation. Brosset (1990) even suggests that day-to-day interrelations among species seem more related to antipredator strategies than to competitive interactions.

Some finches are excluded from certain habitats by the threat of predation (Schluter, 1988), while in other areas they balance proximity to cover (too near or too far), energy needs, vigilance behavior, and the risk of predation (Lima, 1988; Lima and Dill, 1990).

Many aspects of nesting ecology are clearly molded by the threat of predation (Martin, 1988c). Selective pressures from predators influence intraspecific and interspecific placement of nests (height and macrohabitat; Martin, 1988a,c) and clutch size (Foster, 1974; Slagsvold, 1982; Lima, 1987b; Brawn *et al.*, submitted). Hermit thrush nests, for example, are concentrated in sites with high densities of small white fir (Martin and Roper, 1988), where reproductive success is highest and predation rates are reduced. Raccoon foraging efficiency on artificial nests declines in areas with high undergrowth foliage density (Bowman and Harris, 1980).

Experimental studies show that predators modify their search behavior when nest densities are altered experimentally (Martin, 1988b). Predators also find natural nests containing eggs more readily than artificial nests containing eggs (Fig. 2D), suggesting that selection has increased their ability to recognize the nests of resident breeders as potential food sources.

Predator removal also results in increases in populations of game birds (see Martin, 1991 for review), although the interpretation of results of those experiments may require careful scrutiny. In a classic study, Edminster (1939) removed predators from ruffed grouse habitat and decreased nest predation rates from about 60 to 30%. However, compensatory changes in the mortality of chicks and adults resulted in similar population levels, complicating interpretation of the population-level effects.

Interactions of food supply, social behavior, and predation rates often complicate the interpretation of top-down influences. In Panama, clay-colored robins (*Turdus grayi*) nesting at high density in an area of low food supply, but low predation, produced more young than members of a low-density population with higher food supply, but higher predation (Dyrcz, 1983). In Europe, fieldfare (*Turdus pilaris*) breeding success varied directly in relation to the size of the breeding colony (Slagsvold, 1980). Artificial nests of fieldfares in colonies were lost at lower rates to predation than were similar nests outside colonies.

For two decades, studies of avian communities concentrated on the role of habitat structure in influencing avian community diversity, generally assuming that the foliage profile served as a measure of the diversity of food resources available to the resident community. Recent work (Wilcove, 1985; Martin, 1988a,b,c; Tomialojc and Wesolowski, 1990; Holmes, 1990; Brosset, 1990) clearly demonstrates predation also plays a major role in influencing avian adaptations. Seasonal changes in foraging have even been associated with changing vulnerability to predation (Ford *et al.,* 1990).

D. Role of Environmental Variability

Virtually all environments, including the interior of tropical forests, experience temporal variation in physical and biological factors. This variability commonly alters the balance between bottom-up and top-down influences. Populations of eastern bluebirds change from year to year, apparently owing to climatic influences; populations are low following severe winters and severe spring storms (Sauer and Droge, 1990). Populations of *Parus* on Corsica are limited by effects of the hot, dry summer on young in the nest, while the key limiting factor for mainland populations seems to be winter survival (Blondel *et al.,* submitted). Winter conditions (both temperature and days of snow cover) are an important determinant of populations in Polish forest birds (Tomialojc and Wesolowski, 1990). Spring weather conditions with a 1-year time lag were more important than current year food in determining population sizes for canopy insectivores in the same Polish forest.

Phenotypic plasticity in reproductive traits may be an important adaptation allowing populations to survive environmental variation. Clutch size in western bluebirds remained constant over a range of environmental conditions, but timing of clutch initiation, percentage of fledging success, and frequency of second nest attempts varied in response to environmental conditions (Brawn, 1991). Correlations between traits varied widely, often changing sign among samples from different habitats and years. Variation in these demographic traits represented responses of nesting adults to changing environmental conditions. Late spring snows delayed reproduction, increasing the importance of a limited time available for breeding. Feeding conditions were more influential following a dry spring. Such short-term ecological shifts of birds in response to variation in climate and food can underly the evolution of life histories (Partridge and Harvey, 1988).

Because systematic studies of the influence of environmental heterogeneity have not been undertaken in a wide diversity of habitats, the development of general principles is difficult. In perhaps the most detailed studies to date, Grant and Grant (1989) found rainfall to be more important than temperature in regulating Galapagos finch populations; climatic extremes have obvious evolutionary effects on population and community levels (Grant and Grant, 1989). Cyclical and random fluctuations in environmental factors interact with biogeographic constraints (faunal access, regional habitat mosaic) to play primary roles in the structuring of wetland bird communities (Kushlan *et al.*, 1985).

Habitat-based (plant structure) models are not likely to be good predictors of bird abundances from year to year (Blake *et al.*, submitted; Brawn, 1991), because of the influence of weather-related factors on bird abundances. While definitive evidence is not yet available, this conclusion reinforces the notion that habitat associations of birds in central Panama are not solely food-resource mediated (Karr and Freemark, 1983; Karr and Brawn, 1990).

Stormy weather also influences the population biology of birds. In the presence of wet foliage, foraging spruce-woods warblers shift to interior areas of trees where they seldom forage at other times (Morse, 1990). Loss of eggs and nestlings in spruce-woods warblers is increased during storms. Storm duration also influences foraging ecology. In north temperate forest, rainfall usually occurs as part of the movement of frontal systems and low pressure cells (average rainfall duration, 6.2 hr), in contrast to tropical rainfall, which occurs as short downpours (1.9 hr), often in the late afternoon (MacArthur, 1972). Tropical birds typically stop foraging during these rainfall periods, while temperate breeding species often continue to forage after their arrival in Panama from their temperate breeding grounds (J. Karr, personal observation in routine field work, 1968–1989).

Finally, food limitation for birds can occur at any season, not just the breeding season, the primary focus of most studies (Martin and Karr, 1990). Early spring and late fall may be especially critical for migrant birds. Although seasonal shifts in percentage use of various foraging maneuvers exist, overall rankings of maneuvers are consistent over time, suggesting that species are plastic only within limits set by their evolutionary history.

Clearly, the magnitude and pattern of variation in the physical environment acts alone and in concert with bottom-up and top-down influences to mold avian biology.

III. Temperate Stream Fishes

A. Introduction

In contrast to the forest environments just described, the structure of stream habitats is not dominated by living plants. The energy flux associated with the downhill movement of water and particulates in the stream channel, along with characteristics of the drainage basin, are the predominant influences on the physical structure of streams (Richards, 1982). Channel obstructions, such as boulders, bends, and woody debris also increase stream complexity, particularly in headwater areas (Keller and Swanson, 1979; Sullivan *et al.*, 1987). These hydrologic interactions produce extreme spatial heterogeneity at small and large spatial scales, including substrate sorting, pool-riffle development, channel meandering, and changes in channel slope, stream width, and floodplain development (Sullivan *et al.*, 1987). Additionally, most streams exhibit substantial temporal variation in the physical environment (e.g., temperature, discharge), and supply of organic matter. In this section, we examine spatial and temporal variation in the influence of bottom-up processes, top-down interactions, and abiotic factors on stream fishes.

B. Bottom-Up Effects on Temperate Stream Fishes

Foraging success is critical to stream fishes, especially during early life stages, because rapid growth enhances the likelihood of escaping size-selective predation (Werner and Gilliam, 1984), and some types of harsh abiotic conditions (Schlosser, 1985; Harvey, 1987). Rapid growth also enhances the potential for gamete production once sexual maturity is reached (Bagenal, 1978). Is the bottom-up supply of organic and invertebrate resources critical in controlling resource utilization and growth rates of fishes in headwater streams and floodplain rivers?

1. Headwater Streams

The trophic structure of fish communities in small, undisturbed temperate streams is relatively simple, with most fish feeding on invertebrates (Pflieger, 1975; Becker, 1983; Angermeier, 1982, 1985). Studies of small insectivorous fishes indicate they normally consume a wide variety of invertebrate taxa (Pasch and Lyford, 1972; Mendelson, 1975) and exhibit extensive interspecific overlap in types of prey taken (Angermeier, 1982).

Considerable temporal variation in abundance of invertebrate resources occurs in headwater streams because of the dynamic nature of the physical environment. For example, nonlinear relationships exist between discharge and invertebrate abundance (Schlosser and Ebel, 1989; Stock and Schlosser, 1990). Invertebrate abundances are low under very low flow conditions, probably because of reduced habitat space, low current velocity, and reduced transport of organic material (Schlosser and Ebel, 1989). Benthic and drifting invertebrate densities increase as discharge increases, particularly among filtering-collectors in riffle habitats. At very high discharges, invertebrate abundances are dramatically reduced in all habitats because of substrate scouring (Elwood and Waters, 1969; Stock and Schlosser, 1990).

Temporal variation in invertebrate abundance in headwater streams also occurs owing to natural seasonal cycles in temperature, the supply of organic matter, and invertebrate growth and development. Typically, the abundance of benthic and drifting insects in these streams is at a maximum in spring, with a minimum in summer, and a secondary increase in autumn (Hynes, 1970; Schlosser, 1982; Angermeier, 1982, 1985; Schlosser and Toth, 1984). This pattern coincides with leaf inputs in autumn, leaf processing and invertebrate growth throughout winter and spring, emergence of invertebrate adults in spring and early summer, and recruitment of smaller larval stages in autumn (Hynes, 1970).

In association with this seasonal variation in invertebrate abundance, many fish exhibit significant variation in the number and nature of invertebrate prey taken (Angermeier, 1982; Schlosser and Toth, 1984; Schlosser and Angermeier, 1990). Usually, fish guts contain the most aquatic invertebrate prey in late spring and early summer (May–June) and the least in early autumn (October). Late summer declines in number of aquatic invertebrate prey are associated with increased use of terrestrial prey and detritus by some taxa, particularly Cyprinidae (Angermeier, 1982, 1985; Schlosser and Angermeier, 1990). Furthermore, diet breadths of many insectivorous fishes increase as food levels decline (Angermeier, 1982). Overall, these patterns suggest the availability of aquatic invertebrate prey declines from early summer through autumn in natural headwater streams, and most fish species experience a resource depression at this time (see also Mason, 1976; Zaret and Rand, 1971).

Is there any indication that this apparent seasonal depression in invertebrate resources limits growth rates of small insectivorous fishes? Analyses of seasonal variation in growth rates suggest it does, at least for some species. Growth rates for a number of fish species in an Illinois stream increased less between spring and summer than would have been predicted based on seasonal temperature differences (Schlosser, 1982), suggesting resource limitation was occurring during periods of low invertebrate abundance in summer.

Stronger evidence that the summer resource depression limits growth rates of younger age classes of headwater fishes was provided by Mason (1976). He supplemented the food available to juvenile coho salmon (*Onchorhynchus kisutch*) in stream segments during summer low flow periods on Vancouver Island, while keeping other segments as controls. Supplemental feeding canceled any density effects on survival and emigration, and accelerated growth rates and deposition of energy reserves (Fig. 3A).

In addition to temporal variation in resource abundance, fishes in head-

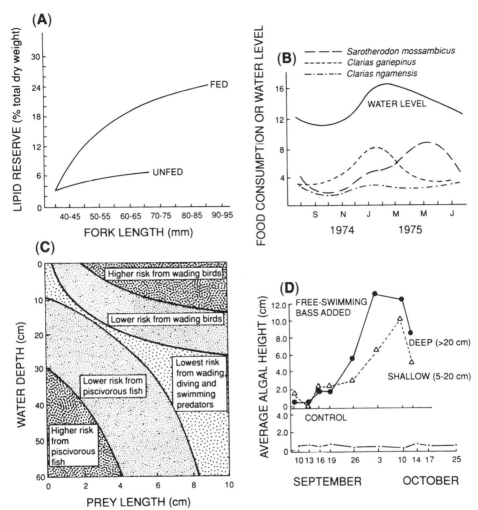

Figure 3 (A) Influence of presence (fed) or absence (unfed) of supplemental feeding on lipid reserves of underyearling coho salmon (*Oncorhynchus kisutch*) in Vancouver. Adapted from Mason (1976). (B) Seasonal variation in water level (m) and daily food consumption (g food · kg fish⁻¹· day⁻¹) by three species of fish from a floodplain river. Adapted from Welcomme (1985). (C) Hypothesized distribution of depth- and size-specific risk of predation in streams from aquatic (e.g., fish) and terrestrial (e.g., heron) piscivores. From Power (1987). (D) Algal standing crop in a single pool inhabited by algal-grazing stoneroller minnows (*Campostoma anomalum*), following addition of bass (*Micropterus* spp.). From Power *et al.* (1985). Control was a separate pool upstream with a naturally occurring school of *Campostoma*, to which bass were not added.

water streams also encounter considerable spatial variation in resource availability. Substrate size and current velocity are especially important factors determining the abundance of aquatic invertebrates (Rabeni and Minshall, 1977). Fine substrates such as silt and sand are unstable (Luedtke and Brusven, 1976), and generally support fewer large invertebrates than do coarse, stable substrates (Wene and Wickliff, 1940; Resh, 1977; Waters, 1982). Substrate stability and large particle size also enhance detrital accumulation (Rabeni and Minshall, 1977) and periphyton production (Hynes, 1970). Studies of spatial variation in resource use by fishes over different substrates (Angermeier, 1985) reveal several patterns, suggesting fish experience a greater resource depression over silt–sand than gravel–cobble substrates (Angermeier, 1985).

2. Floodplain Rivers

Trophic complexity of the fish community is considerably greater in large floodplain rivers than in headwater streams because of the greater habitat diversity and organic richness provided by the floodplain (Welcomme, 1985). Although river fishes exhibit considerable opportunism in prey taken, and the different species frequently overlap extensively in the types of food consumed (Welcomme, 1985), they can generally be placed into one of several trophic groups (Horwitz, 1978; Welcomme, 1985), including mud eaters, herbivores, detritivores, omnivores, planktivores, insectivores, and piscivores.

As in headwater streams, large river fishes appear to experience seasonal depressions in resource availability and resource utilization. Because of the increased complexity of the trophic structure of the fish community and the physical environment, broad generalizations as to which trophic groups experience resource depression, and when the depressions occur, are not easily made. Excluding the strong effects of temperature on resource-utilization patterns in north temperate areas, the strongest link between food availability, food utilization, and growth in fishes from large rivers is associated with seasonal inundation of floodplains (Bayley, 1988; Welcomme, 1985; Junk *et al.*, 1989; Quiros and Cuch, 1989). At this time, increased availability of organic materials and invertebrates, along with expansion of physical habitat for feeding, favors increased food intake and growth rate of most river fishes (Welcomme, 1985). During reduced flow conditions, decreased availability of organic substrates and invertebrate abundance, along with contraction of the physical environment, result in reduced food intake and growth rate (Welcomme, 1985).

As one might expect, because of the high trophic complexity in large rivers, the pattern of abundant feeding during inundation of the floodplain and reduced feeding during low stream discharge is complex. For example, within a river system, some species exhibit maximal food consumption just before peak water levels, others exhibit peak feeding as the floodplain is draining, and still others feed at fairly constant rates throughout the year (Fig. 3B). Nevertheless, the resource depression and reduction in food intake is usually most severe during low-water periods (Welcomme, 1985).

C. Top-Down Effects on Stream Fishes

What evidence is there that predation by higher trophic levels on lower trophic levels regulates resource abundance, habitat use, species interactions, and/or survival of stream fishes? Herbivorous fishes often have strong effects on instream plant abundance, and thus can accentuate depressions in these resources and the invertebrates associated with them (Power and Matthews, 1983; Power, 1984). In contrast, insectivorous stream fishes have relatively weak but variable effects on insect abundance. Neither density of drifting invertebrates (Allan, 1982) nor density of benthic insects in coarse substrates (Allan, 1982; Reice, 1983; Flecker and Allan, 1984) exhibit strong responses to vertebrate predation. However, insect abundance in pools (Wilzbach *et al.*, 1986; Schlosser and Ebel, 1989) or habitats with sand and silt substrates (Angermeier, 1985) appear to be reduced by fish predation. These differences among habitats in the influence of vertebrate predators on the abundance of invertebrate prey are likely to be influenced by differences in prey exchange rates (immigration/emigration) among the habitat patches (Cooper *et al.*, 1990).

Vertebrate piscivores have strong effects on habitat use by stream fishes. Effects of vertebrate piscivores are particularly interesting because two groups of vertebrate piscivores differ in spatial distribution and size-selectivity of prey (Power, 1987). Terrestrial piscivores, such as herons, are most efficient at capturing fish in shallow habitats (Power, 1987), are usually not gape limited, and select the largest prey. Aquatic piscivores, such as bass (*Micropterus*), are usually gape limited (Werner, 1977) and select smaller prey (Schlosser, 1988). Also, because aquatic piscivores tend to be larger fishes, they are restricted to deeper habitats by terrestrial piscivores (Power, 1987).

Differences in the size-selective distribution of predation risk by terrestrial versus aquatic piscivores have important implications for stream fishes. First, mortality rates from predation are likely to be highest for large fishes in shallow habitats and small fishes in deeper habitats (Fig. 3C). Because of spatial variation in predation risk for large and small fishes, a strong complementarity frequently occurs in streams in the distribution of these two size groups. Juveniles of all taxa and adults of taxa with small maximal body size, tend to be found at high densities in shallow upstream areas, stream edges, floodplains, shallow pools, or riffles. Adults of taxa with large maximal size tend to be found in deeper habitats associated with pools, downstream, or midchannel areas (Power, 1984; Welcomme, 1985; Lowe-McConnell, 1987; Schlosser, 1987a,b).

Second, restrictions placed by vertebrate piscivores on habitat use by fishes alters their interaction with lower trophic levels. For example, Power *et al.* (1985) examined the interaction among bass (*Micropterus* spp.), an algal-grazing minnow (*Campostoma anomalum*), and attached algae. In the absence of bass, *Campostoma* were sighted most frequently grazing on algae on cobble and bedrock substrates in the deeper parts of pools. Within 1 hr of bass addition to pools, large and small *Campostoma* were restricted to the shallow areas of the pool. Furthermore, when bass were added to the pools, algal standing crop increased in pools and decreased in shallow areas because of the shift in habitats used by *Campostoma* (Fig. 3D). Over the following weeks, however, algae increased in both shallow and deep habitats because of predation and predator avoidance (emigration) by *Campostoma*.

Third, the restriction of small fishes to shallow habitats by aquatic piscivores results in increased intra- and interspecific overlap in habitat use among small fishes (Schlosser, 1987a). The impact of this increased overlap in habitat use on growth and survival of stream fishes has yet to be rigorously examined. However, theoretical considerations (Werner and Gilliam, 1984) and empirical results from lakes and ponds (Werner *et al.*, 1983a,b; Mittelbach, 1984) suggest such restrictions on habitat use are likely to affect juvenile recruitment and the population dynamics of stream fishes.

Finally, vertebrate piscivores can potentially regulate fish abundance in streams directly, through prey consumption. Unfortunately, because it is logistically difficult to experimentally exclude piscivores over a large spatial scale, little conclusive evidence exists concerning the direct effect of vertebrate piscivores on the population sizes of stream fishes. Descriptive studies suggest, however, that the relative influence of aquatic piscivores on fish abundance is likely to change from headwater streams to large rivers. Comparisons of predation rates by aquatic piscivores with temporal variation in migration and recruitment rates suggest colonization and recruitment dynamics, as mediated by abiotic factors, are likely to be more important than aquatic piscivory in regulating fish abundance in upstream areas (Moyle and Li, 1979; Lowe-McConnell, 1987; Schlosser and Ebel, 1989). In contrast, an analysis of differences between fish production and fish yield in large rivers (Bayley, 1989) suggests aquatic piscivores may regulate fish abundance in these areas (see also Welcomme, 1979,1985; Lowe-McConnell, 1987). Similar assessments of spatial variation in the role of terrestrial piscivores have yet to be made, but they will also probably reveal interesting differences between upstream and downstream areas and between floodplain and main channel areas.

D. Direct Effects of Temporal Abiotic Variability on Stream Fishes

The survival and population dynamics of stream fishes are also directly influenced by temporal variation in the abiotic environment. Three abiotic factors are especially influential: dissolved oxygen, water temperature, and stream discharge.

1. Oxygen and Temperature

Streams usually have sufficient turbulence to maintain adequate oxygen levels even if high levels of respiration are occurring in association with organic decomposition (Hynes, 1960, 1970). However, very low oxygen levels (<1 ppm) may cause fish mortality if excessive decomposition occurs in conjunction with intermittent stream flow or with isolation of the habitat from the main channel (Larimore *et al.*, 1959; Tramer, 1978; Welcomme, 1985; Lowe-McConnell, 1987).

Similarly, the normal magnitude of temporal variation in water temperature rarely exceeds the upper thermal tolerance of most stream fishes (Matthews and Maness, 1979; Beschta *et al.*, 1987; Welcomme, 1985). In fact, these temperature fluctuations are necessary to initiate spawning in many taxa (Becker, 1983). However, if habitats are fully exposed to solar radiation, have a small volume, and are stagnant, then extremely high water temperatures can cause direct mortality of fishes (Larimore *et al.*, 1959; Welcomme, 1979, 1985). Probably a more critical problem for fishes in north temperate streams is exposure to temperatures well below their optimum. The very cold (0–4°C) water temperatures during winter, along with absence of warmwater refugia, results in severe metabolic stress for fishes. Reduced feeding rate at low temperatures, in conjunction with continued expenditure of energy for routine maintenance activities, can directly cause mortality of individuals. Winter periods are particularly stressful for smaller size classes of fishes because of their low metabolic reserves (Oliver *et al.*, 1979; Shuter *et al.*, 1980; Cunjak *et al.*, 1987; Cunjak, 1988).

2. Stream Discharge

The most dramatic aspects of temporal abiotic variability in streams are associated with fluctuations in discharge. Since discharge is strongly influenced by the timing and nature of precipitation in the drainage basin, fluctuating water levels are characteristic of most stream ecosystems.

Both high and low stream discharges directly influence the survival and population dynamics of stream fishes. Furthermore, the magnitude of the effect of fluctuations in discharge on stream fishes exhibits considerable spatial variation in the drainage basin. For example, extended periods of high discharge during spawning decreases fish reproductive success in headwater streams by physically displacing eggs and fry (Harvey, 1987), and by covering eggs with sediment (Starrett, 1951; Winn, 1958; Everest *et al.*, 1987). Severe floods in headwater streams also reduce the abundance of older age classes of fishes (Elwood and Waters, 1969) because debris can fill pools and cover riffles, reducing their preferred habitat (Minckley and Meffe, 1987). In contrast, flooding in large rivers frequently enhances fish reproduction (Bayley, 1988, 1989) by increasing food and habitat availability for juvenile fishes.

Very low stream discharges also have pronounced effects on fishes, particularly in upstream and floodplain habitats. Fishes frequently respond to low discharge by small-scale shifts in habitats or large-scale alterations in their distribution in the drainage basin. As stream discharge decreases, species inhabiting upstream areas frequently move downstream (Ross *et al.*, 1985), and floodplain species move to the main channel (Lowe-McConnell, 1987). During periods of extremely low discharge, small streams and floodplains dry up completely or become isolated stagnant pools with high temperatures and low oxygen levels (Larimore *et al.*, 1959; Lowe-McConnell, 1987). These harsh physical conditions frequently result in the nearly complete decimation of the fish fauna. For fish populations to persist under these conditions, suitable refugia must be available. Isolated pools and larger streams usually serve as refugia during these periods, harboring a recolonization source of fishes (Paloumpis, 1958; Whiteside and McNatt, 1972; Lowe-McConnell, 1987). As a result, the relationship among frequency of faunal decimation by low discharge, presence of refugia from harsh physical conditions, and rate of fish recolonization is likely to be particularly critical in determining the dynamics of fish communities in headwater and floodplain areas where disappearance of aquatic habitat is common (Welcomme, 1979; Kushlan, 1976; Horwitz, 1978; Lowe-McConnell, 1987; Schlosser, 1987b).

IV. Temperate Lake Fishes

A. Introduction

In temperate lakes, living plants form the base of the food web in pelagic (open water) and littoral (shallow, near shore) habitats, but create physical habitat structure only in the littoral zone. In open water, suspended algal cells are available to zooplankton and fish. In littoral habitat, rooted vascular plants (macrophytes) provide substrate for epiphytic algae grazed by invertebrate consumers. Plant tissues may also enter the food web as detritus.

Although a few temperate lake fishes feed on phytoplankton as adults (Gophen *et al.*, 1982; Johansson and Persson, 1986; Lazzaro, 1987), most open-water fishes feed on zooplankton or other fish. Direct feeding on plant material is even less frequent in the littoral zone, where most fish consume invertebrates that graze epiphytes and detritus from macrophyte substrates, or on those grazers' invertebrate predators.

How do changes in plant quality and quantity influence fish ecology? What is the importance of plant resources for fish, relative to the effects of predation and the physical environment? Here, with our focus on control of vertebrate populations, we are interested in ways that plants and predators influence fish, rather than in ways that nutrients and herbivores influence plants (i.e., top-down versus bottom-up control of lake primary productivity; Crowder *et al.*, 1988; McQueen *et al.*, 1989).

B. Bottom-Up Effects

1. Pelagic Habitats

Temporal variation in phytoplankton abundance (Harris and Piccinin, 1990; Reynolds, 1990) can influence the ecology of planktivorous fish, but surprisingly, the extent of influence is not well known. Correlations between indicators of primary productivity and fish yield suggest bottom-up, mechanistic linkages between primary production and fishes (McConnell *et al.*, 1977; Oglesby, 1977; Hanson and Leggett, 1985). In artificial pond experiments, zooplankton and planktivore biomass increased in response to nutrient additions (Hall *et al.*, 1970). Others show positive correlations between phytoplankton and zooplankton abundances (McCauley and Kalff, 1981; Mills and Schiavone, 1982; Hanson and Peters, 1984), bottom-up control of zooplankton by phytoplankton (Ferguson *et al.*, 1982; Vanni, 1987), and quantitatively relate food intake to individual fish growth (Elliott, 1976; Kitchell *et al.*, 1977; Ney, 1990). Finally, bottom-up control of planktivorous fish has been directly documented in a number of systems (Kohler and Ney, 1981; Werner and Hall, 1988; Mittelbach, 1988; Osenberg *et al.*, 1988; Persson and Greenberg, 1990).

In spite of these documented relationships, few have studied the specific mechanisms connecting phytoplankton, zooplankton, and planktivorous fishes. An exception is a persistent 2-year, two-trophic-level cycle in a eutrophic English lake (Cryer *et al.*, 1986, Townsend and Perrow, 1989). Nutrients determined phytoplankton abundance, and competition for zooplankton during years of high planktivore recruitment led in the following year to poor planktivore recruitment, but good individual growth and reproduction, when fish were fewer and zooplankton, abundant. Little is known of the importance to trophic interactions of spatial patchiness in phytoplankton, zooplankton, planktivores, and piscivores in lakes. Some planktivores switch from particulate feeding to filter feeding in response to zooplankton concentrations in the water column (Lazzaro, 1987; Ehlinger, 1989). The phenomenon of zooplankton vertical migration separates zooplankton prey in space and time from actively feeding planktivore predators (Lampert, 1987).

Studies of cultural eutrophication provide additional evidence of bottom-up effects on lake fish communities. For percids, growth and biomass increase up to some threshold level of eutrophication, and then decline, in response to changes in parasitism, feeding and spawning habits, and dissolved oxygen (Leach *et al.*, 1977). In lakes where eutrophication progresses to the extent that percids begin to decline, cyprinids increase (Biro, 1977; Leach *et al.*, 1977). Dense blooms of blue-green algae often have deleterious effects on herbivorous zooplankton (Haney, 1987), and may cause fish kills (Olrik *et al.*, 1984).

2. Littoral Habitats

Temporal changes in macrophyte abundance and quality translate into changes in prey availability for foraging fishes. Fish growth mirrored seasonal patterns of macrophyte abundance in a well-vegetated Wisconsin lake, where fish diets were composed largely of plant-associated invertebrates (either on plants, or in sediments under plants; Engel, 1987). Pumpkinseed sunfish (*Lepomis gibbosus*) and rock bass (*Ambloplites rupestris*) diets changed with plant abundance (French, 1988). Large bluegill sunfish (*Lepomis macrochirus*) forage in vegetated habitats until net energy gain per unit time falls below foraging returns in open habitats (Mittelbach, 1981). Bluegill cause shifts in plant-associated prey through their own foraging activities (Crowder and Cooper, 1982; Morin, 1984; Mittelbach, 1988). Flooding of terrestrial grasses in reservoirs during the fish-spawning period (Strange *et al.*, 1982) leads to increased fish-spawning success, prey availability, and growth of young-of-the-year (YOY) fish.

Spatial variation of littoral macrophytes also has profound effects on fish in lakes. Typically, macrophyte species are distributed heterogeneously (both patch density and patch size) in the littoral zone (Hutchinson, 1975; Sand-Jensen and Sondergard, 1979; Chambers, 1987). Spatial variation of macrophytes also occurs in lakes along the depth gradient from shore to open water. Moreover, patches may contain canopy, understory, and benthic plant species (Engel, 1985).

This spatial variation can have bottom-up effects on foraging fish. Diets of bluegills resembled the species distribution and abundance of plant-associated rather than benthic invertebrates in Florida lakes (Schramm and Jirka, 1989), and bluegills displayed positive selection for those prey that were most abundant in association with plants. Species-specific plant abundance influences abundance of associated invertebrate prey (Cyr and Downing, 1988), and the foraging success of fish predators. Crowder and Cooper (1982) found that prey per unit of vegetated habitat increased monotonically with plant density, but that fish foraging rates peaked at intermediate plant density, demonstrating the interaction between prey abundance and plant interference. Many researchers have studied the influence of artificial plant density on fish foraging success in the laboratory at controlled levels of prey abundance. Typically, high plant density reduces fish foraging success, even exhibiting a density-threshold effect on fish feeding rate (Gotceitas and Colgan, 1989). Finally, when plants are arranged in discrete patches, foraging sunfish moved between patches using a capture-rate–based rule (DeVries *et al.*, 1989).

The presence of littoral zone vegetation provides spawning habitat for many freshwater fishes and habitat for many species after hatching (Scott and Crossman, 1973; Keast *et al.*, 1978; Gregory and Powles, 1985). Larvae of some species migrate from the littoral to the pelagic zone, then return to the littoral vegetation after several weeks (Whiteside *et al.*, 1985; Werner and Hall, 1988). Abundance and growth of YOY largemouth bass (*Micropterus salmoides*) is correlated with the abundance of littoral vegetation (Aggus and Elliot, 1975). Fish larvae in Chemung Lake, Ontario, show species and stage-specific preferences for depth and density of littoral vegetation (Gregory and Powles, 1985).

Macrophytes also exert bottom-up control by mediating competitive interactions among littoral fishes. Sunfish species partition vegetated and unvegetated habitats in response to competition based on habitat-specific differences in foraging efficiencies (Werner and Hall, 1976, 1979; Mittelbach, 1984). Differences in foraging success between pumpkinseeds and rock bass in vegetation (French, 1988), the segregation of roach (*Rutilus rutilus*, open water), and rudd (*Scardinius erythrophthalmus*, vegetated habitats) in many European lakes (Johansson, 1987), and partitioning of vegetated habitats between perch (*Perca fluviatilis*), roach, and bream (*Abramis brama*, Diehl, 1988) also appear to result from competition through differences in habitat-specific foraging abilities.

Spatial variation in vegetated littoral habitats also occurs at the level of plant growth form (size, number, and configuration of leaves and stems). Plant-dwelling invertebrate prey vary with plant growth form (Rooke, 1986; Dvorak, 1987; Dionne, 1990), and growth form influences the availability of these prey to fish predators. Pumpkinseed sunfish fed at higher rates on cladoceran and damselfly prey in laboratory habitats of leafless, thick-stemmed emergent plants than in habitats of leafy, thin-stemmed submergent plants (Fig. 4A). In the littoral zone of a New Hampshire lake, pumpkinseeds occurred at different densities in adjacent sites of emergent and submergent plants. They exhibited differences in diet, prey selectivity, and overall predation effects on plant-associated prey, owing to differences in plant- and sediment-associated prey and plant growth form between the sites.

Studies of littoral fish assemblages suggest relationships between plant community structure and fish community structure. On a regional scale, the distribution of sunfishes in lakes of the New Jersey coastal plain coincides with the availability of open-water and macrophyte-associated prey (Graham and Hastings, 1984). Within lakes, different fish species associate with macrophytes occurring at different depths (Hall and Werner, 1977), with longitudinal macrophyte production gradients (Gascon and Leggett, 1977), or with habitats characterized by submergent, emergent, and mixed littoral macrophytes (Holland and Huston, 1985). A drastic loss of fish species in Lake Mendota, Wisconsin, occurred with replacement of the lake's native macrophytes by the exotic *Myriophyllum spicatum* (Lyons, 1989). Fish diversity in lakes has also been related to measures of horizontal (Tonn and Magnuson, 1982) and vertical vegetation complexity (Eadie and Keast, 1984).

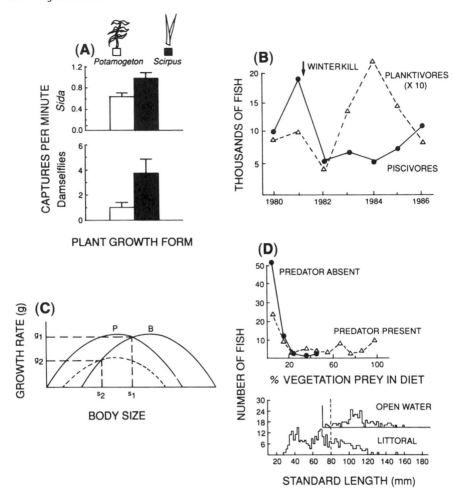

Figure 4. (A) Feeding rates of pumpkinseed sunfish in vegetated laboratory habitats structured by different plant growth forms. From Dionne and Folt (1991). (B) Inverse relationship between piscivores and planktivores in Lake St. George before, during, and after piscivores recover from winterkill reduction in numbers. From McQueen *et al.* (1989). (C) Model of perch growth trajectories depicting observed influence of competition with roach on perch habitat shifts from pelagic (P) to benthic (B) feeding. Solid curves indicate relationship between growth rate and body size for perch in the two habitats when roach are absent. s_1 indicates the body size at which perch should switch foraging habitats. Broken curve indicates the situation when roach, which feed only in open water, are present, and s_2 shows that perch should switch to benthic feeding at a smaller body size. From Persson and Greenberg (1990). (D) Upper panel shows a shift by bluegills to foraging in vegetated habitat when bass (predators) are present. From Werner *et al.* (1983a). Lower panel shows size distribution of planktivorous bluegill in open water, on same scale with size distribution of bluegill feeding in littoral habitat in Deep Lake, Michigan. From Werner and Hall (1988).

C. Top-Down Effects

1. Pelagic Habitats

Top-down control of planktivores has been shown or can be inferred from studies of piscivore-induced changes in phytoplankton or zooplankton, via interactions passing through planktivore trophic levels. These *trophic cascades* have been described in natural systems (Mills and Schiavone, 1982; McQueen *et al.*, 1989; Johannes *et al.*, 1989; McQueen, 1990) and through the introduction or removal of piscivores (Shapiro and Wright, 1984; Scavia *et al.*, 1986; Wagner, 1986; Carpenter *et al.*, 1987; McQueen, 1990). Piscivore manipulation demonstrates predator control only for the duration of the study, while the nonexperimental studies suggest patterns that persist over longer periods. In a survey of 13 New York lakes, Mills and Schiavone (1982) determined that northern pike (*Esox lucius*) and walleye (*Stizostedion vitreum*) controlled yellow perch (*Perca flavescens*) and black crappie (*Pomoxis nigromaculatus*). A winterkill of piscivores in an Ontario lake led to a doubling of planktivore numbers, indicating predator-release; after a number of years both predators and prey returned to prewinterkill levels (Fig. 4B). Planktivores also show adaptive behavioral responses to predators (schooling) that reduces planktivore predation risk (see Pitcher, 1986 for review).

Given the evidence that both bottom-up and top-down control of planktivores occurs in lake systems, we now focus on factors that determine which mode of control prevails at a particular site. One major factor is the phenomenon of the size-dependent niche (Werner and Gilliam, 1984; Gilliam and Fraser, 1988). Fish body size increases over the entire lifespan. As body size changes, so do natural history attributes such as prey size-selection, foraging habitats, risk of predation, and fecundity. Thus, any factor influencing fish growth rates can have a major impact on many aspects of fish ecology.

Body size becomes important from the moment a larva begins exogenous feeding. Through gape limitation, the relative size of zooplankton, plant-associated prey, and YOY fish, has an overriding influence on fish year-class success. The more suitable the prey size-distribution, the faster YOY fish can grow, the larger the size of prey they ingest (allowing faster growth), the lower the risk of being eaten by predators, and the greater the chance of acquiring sufficient energy reserves to survive the winter. Because most planktivores produce enormous numbers of larvae, they are controlled during the first year by the processes (either bottom-up or top-down) that determine early cohort survival and growth (of both planktivores and piscivores). Several recent studies support this view.

In Oneida Lake, New York, the outcome of trophic interactions is established annually by the year-class strength of the dominant planktivore (yellow perch; Mills *et al.*, 1987). Early larval survival is weather dependent (Clady, 1976), and later survival is determined by predation from walleye.

In years when predation is not sufficient to drive YOY perch below 20 kg/ha, intense perch feeding leads to collapse of their larger cladoceran prey, slowing perch growth and increasing mortality from walleye. Thus, bottom-up and top-down effects interact to maintain a food web that has persisted for 30 years.

Predation on shad (*Dorosoma* spp.) by largemouth bass hinges on the timing of shad and bass spawning, which varies with weather effects on water temperature (Adams and DeAngelis, 1987). If bass hatch too late, they are too small to prey effectively upon YOY shad. Consequently, bass grow slowly, and experience high overwinter mortality. In this example, the relative body size of YOY piscivores and planktivores determines whether the planktivore trophic level can be controlled from the top down.

In Lake Svedborg, planktivorous roach escape control from predatory perch by competing with YOY perch for zooplankton (Persson and Greenberg, 1990). In response, perch switch to feeding on macroinvertebrates at an earlier stage, and experience slower growth probably due to both interspecific and intraspecific competition. This will reduce the number of perch that grow large enough to prey on roach (Fig. 4C). When roach were experimentally reduced, roach–perch competition was eliminated, and perch switched to piscivory earlier (Persson, 1986). In south Scandinavian lakes, piscivores control planktivores in less-productive lakes (Persson *et al.*, 1988), but with higher productivity, roach escape control through competitive effects on perch growth.

A final example of size-dependent interactions that influence trophic control of planktivores is from two Ohio reservoirs (DeVries *et al.*, 1991). In Clark Lake, larval threadfin shad (*Dorosoma petenense*) and bluegill did not compete for zooplankton because shad spawned later than bluegill. Thus, shad did not influence the abundance of bluegill as forage for YOY bass (restricted to littoral habitats). In Stonelick Lake, on the other hand, the larval peaks of bluegill and shad overlapped, coincident with zooplankton decimation, and bluegill that hatched with shad simply disappeared. Thus, the major peak of larval bluegill never migrated to the littoral zone, and bass showed much lower growth.

In these studies, the time of spawning and the temporal pattern of zooplankton abundance also influence planktivore–piscivore interactions. Apparently, the nature of lake systems provides ample opportunity for weather (via light, temperature, and wind effects on phytoplankton, zooplankton, and fish reproduction) to set the stage for biological interactions.

2. Littoral Habitats

In littoral habitats, the physical structure of the vegetation influences predator–prey interactions. In artificial pond experiments, predation by largemouth bass was reduced because small planktivorous bluegill foraged on plant-associated invertebrates in the less-profitable littoral habitat (Fig. 4D), where they experienced reduced growth. In natural Michigan lakes, bluegill show similar shifts in habitat (Fig. 4D), and the size which bluegill switched to feeding on plankton was related to bass density. Fish in littoral habitat experienced much slower growth than fish of the same size caged in pelagic habitat. Thus, a top-down, predator-induced movement of fish into the vegetation leads to bottom-up control.

Since the pioneering work of Werner, Mittelbach, Gilliam, and Hall, others have detected predator-mediated use of plant cover by lake fish. In laboratory habitats, bluegill select the density of vegetation sufficient to provide refuge from predation by largemouth bass (Gotceitas and Colgan, 1987), at the cost of reduced foraging rates (Gotceitas, 1990a). In the absence of predators, foraging bluegill selected the vegetated habitat that maximized their energy intake (Gotceitas, 1990b), but in the presence of bass, they selected foraging habitat that minimized the mortality–foraging rate ratio (described by Gilliam and Fraser, 1987). From this work, a picture emerges of predator-controlled planktivore foraging success, through shifts by planktivores to habitats offering better short-term survival, but reduced food availability. Piscivore foraging behavior is also altered in the presence of vegetation (Schramm and Zale, 1985; Savino and Stein, 1989).

D. Environmental Variation

Direct effects of weather on fish are little studied except for wind and temperature effects on fish eggs and larvae (Eipper, 1975; Summerfelt, 1975; Clady, 1976), water temperature effects on fish growth and habitat selection (Crowder *et al.*, 1981; Coutant, 1985), and winter fish kills.

Fish are ectotherms, so their metabolic rates are determined by the ambient temperature. Consequently, fish growth rates are temperature dependent. Mean smallmouth bass (*Micropterus dolomieui*) growth rates throughout North America are correlated with mean surface water temperature (Coble, 1967). Free-ranging largemouth bass select natural habitat that is closest to optimal temperature for growth (Coutant, 1975). Temperature can influence year-class success of smallmouth bass (Serns, 1982), yellow perch (Eshenroder, 1977), and alewives (*Alosa pseudoharengus* - Henderson and Brown, 1985). Striped bass (*Morone saxatilis*) can succumb to physiological stress in summer when their temperature and oxygen requirements restrict them to habitats with insufficient prey (Coutant, 1985). Winter temperature also dictates the body size required for successful overwintering in some species (Oliver *et al.*, 1979), and this size appears to determine the northern limit of percid and smallmouth bass distribution (Shuter and Post, 1990).

A fish winterkill is a dramatic demonstration of an abiotic force affecting fish populations (Schneberger, 1970). Since winterkills can be selective (Casselman and Harvey, 1975; Tonn and Magnuson, 1982), they have been used as natural experiments to examine the nature of trophic control (McQueen *et al.*, 1989; Johannes *et al.*, 1989). In some systems, winter oxygen concentrations are the prime determinant of fish community structure (Tonn and Magnuson, 1982; Rahel, 1984).

V. Synthesis

The complexity of trophic interactions varies widely, as does the role of plants in our study communities. Because of that variation, it is difficult to document general principles that apply throughout. In a real sense, we are caught between the devil of oversimplification and a deep blue sea of endless unrelated facts (Lawton and McNeill, 1979). The situation is especially complex when, on the one hand, experimental studies suggest simple trophic relationships, while detailed studies in natural environments suggest far more complex local dynamics. The addition of intercontinental comparisons, as described here for tropical forest birds (e.g., nectarivores), illustrates the importance of historical influences on the trophic relations of vertebrates. Despite this complexity, several major principles emerge from our analyses, and we suggest several questions that remain unanswered.

1. Neither bottom-up nor top-down factors predominate as influences on vertebrates in forest, stream, or lake habitats. Rather, abiotic and biotic processes interact to affect the nature of trophic regulation. Trophic influences may be direct (a predation event) or indirect (alteration in habitat use due to threat of predation). Short-term dynamics of regulatory processes are influenced in ecological time, and life-history and community attributes in fishes and birds are molded in evolutionary time.

2. The nature of trophic interactions is influenced by differences in the life-history and behavioral characteristics of the organisms. Early development of endotherms, such as in forest birds, involves rapid development to adult size, while ectothermic fish develop more slowly, often with a prolonged juvenile period. Both are susceptible to predation as eggs and recent hatchlings, but the relatively large recent hatchlings of birds develop rapidly, allowing them to escape predation risk associated with nesting. In contrast, hatchling fish are small and attain adult size relatively slowly, prolonging vulnerability to predation and abiotic variability and creating size-structured populations. Dramatic changes in body size throughout the life cycle of fish result in pronounced ontogenetic changes in habitat and/or resource use. As a result, factors influencing growth and survival of early life stages of fishes seem particularly critical in determining their population and community dynamics.

Because fish are ectotherms, energy requirements vary with ambient temperature; thus, energetic needs in fish vary seasonally more than in birds. Birds shift seasonally among food resources more than do fish, or migrate long distances (perhaps coupled with changes in food types) to ensure an adequate supply of food. Seasonal movements of fish seem more tied to a search for locations with reduced rates of predation on eggs and larvae as well as to ensure larval food supplies.

Variation in behavior also influences the ecology of trophic interactions.

Pre- and post-hatching parental care is more common in birds than fish. Clutch sizes of birds are small and vary little with age, while fish produce many eggs and exhibit size-related changes in fecundity. Another behavioral attribute, schooling (in fish) or flocking (in birds), may often be a response to predation. Pair bonds beyond copulation are very rare in fish but dominate in forest birds, with parents devoting considerable time and energy to feeding and protecting a comparatively small number of offspring.

3. Physical habitat structure mitigates the influences of bottom-up and top-down regulation and the abiotic environment. Structure may derive primarily from the growth form of plants as in forest and the littoral zone of lakes, or it may be determined primarily by physical processes (watershed hydrology and geology). These structural attributes often have indirect effects on vertebrates. For example, when vulnerable planktivores select vegetated littoral habitats, or birds select foraging areas or nest sites to avoid predation, the ecological results do not fit neatly into the top-down versus bottom-up dichotomy. Smaller, vulnerable size-classes of planktivores experience higher survival in areas with macrophytes, but growth rates may then be reduced by competition for food.

4. To the extent that trophic controls operate, they result from a shifting array of direct and indirect influences between abiotic and biotic processes. Trophic control of fishes in streams and lakes results from interactions among abiotic forces, supply of organic matter, abundance of invertebrate consumers, relative recruitment success among year-classes, size-dependent competition, and size-dependent predation at invertebrate and vertebrate trophic levels. Recent detailed studies by Power and Schlosser in streams and Mills, Persson, DeVries, and others in lakes describe mechanisms by which these factors interact, and collectively suggest that changes in abiotic factors (temperature, wind, precipitation, water levels) shift the timing of biotic events and alter the relative strength of trophic interactions. As a result, trophic controls of stream and lake fishes should be viewed as a branching hierarchy, with population and community dynamics controlled at certain times and at certain locations by predation, while resource supply might be more important at other times and locations. Moreover, if harsh abiotic conditions, such as winterkill in lakes and severe droughts in streams, cause frequent and extreme fish mortality, then neither bottom-up nor top-down forms of trophic control are likely to prevail in determining population and community attributes. Rather, the occurrence of severe physical–chemical conditions, presence and location of refugia, and dynamics of recolonization processes will be of overriding importance (Tonn and Magnuson, 1982; Schlosser, 1987b; Magnuson, 1991). Although the relative importance of these factors might change between stream and lake environments, the basic nature of their influence on the population and community attributes of fishes seems to be similar.

Research on birds emphasizes the critical role that adult survival, competition for food, and egg production play in controlling population and community attributes (Lack, 1954, 1968). Studies of birds, perhaps more than those of any other group, have concentrated on interspecific competition for limited resources. Two decades of observations and experiments have documented many bottom-up influences on birds. Recently, a more pluralistic approach (Schoener, 1986) has developed in population (Martin, 1986, 1987) and community ecology (Karr, 1980; Karr and Freemark, 1983; Wiens, 1989a,b; Holmes, 1990), clearly demonstrating "the folly of seeking a single factor as the sole determinant of avian community structure" (Karr, 1980; p. 283) or population ecology. Each species is, in a Gleasonian sense, affected by and uniquely responds to numerous factors in its environment (Brawn *et al.*, 1987).

Overall, bird abundances fluctuate because of the influence of many environmental factors such as food availability, floristics, vegetation structure, presence of other bird species, biogeographic history, predation, seasonal mortality pattern, latitude, elevation and microtopography, macro- and micro-climate, parasitism, and disease (Karr, 1976, 1980, 1990b; Holmes, 1990; Wiens, 1989a,b; Cooper, 1990; Keast, 1990). Because abiotic and biotic processes vary geographically, the nature of adaptive responses is not consistent across biogeographic regions for the same major food resource, (e.g., nectar) nor are they consistent among guilds within the same region (e.g., frugivores versus omnivores versus insectivores in neotropical forest). The apparent consistency in dynamics in lake and stream systems may be more a function of the geographically narrow range of studies relative to more frequent intercontinental comparisons in birds.

5. Generalizations that cross these habitats and taxa are likely only when the relevant organisms have similar natural histories and find themselves in similar environments. This important point, first made by MacArthur (1972) in the introduction of *Geographical Ecology*, bears repeating here. While MacArthur's caveat should not divert us from seeking general principles, it reinforces the limitations of seeking principles so general that they lose utility. For example, sources of energy to support food webs differ among habitats. Most birds and lake fish feed either directly on living plant material (nectar and fruits, often with coevolutionary dependencies) or on organisms that feed on living plants. Birds that feed on leaf-litter arthropods (including hanging dead leaves in tropical forest) or detritus feeding fishes are the primary exception; these species typically represent a small proportion of the resident vertebrate fauna. In contrast, most stream fish, especially those of small streams, feed on invertebrates that depend on a detritus food chain nourished by leaves that fall from plants.

It is appropriate, as well, to note that the kinds of ecological questions that can be explored vary among habitats and taxa. Detailed analyses of food consumption are easier for fish because population densities tend to be larger, and there is a relative lack of constraints on collection of specimens. Fish are also easier to use in controlled, manipulative experiments, while behavioral observations on free-living individuals may be easier with birds.

6. Advances in understanding of trophic control of vertebrate populations and communities is critical to both basic and applied ecology. Study of trophic and environmental control of vertebrates is of interest because it satisfies our natural curiosity about the functioning of the biological world. In addition, that knowledge is critical to informed decisions about the use and protection of Earth's biological diversity. As human influences become more pervasive and involve more-complex cumulative impacts, the challenges we face in protecting and restoring that diversity expand (Karr, 1990a, 1991). Classic wildlife biology seeks to restore *habitat* for target species without regard to the complete trophic context, while fishery harvest quotas are determined by examination of stock-recruitment curves. As the studies described in this paper clearly show, these approaches are too simplistic.

Efforts to restore a stream system must first restore watershed hydrology and channel hydraulics to create the appropriate physical habitat, including leaf packs and accumulation of woody debris in the channel. For most streams in North America, the role of woody debris has declined in the past century, especially in large rivers, because of active efforts to rid streams of such debris (Sedell *et al.*, 1988). Destruction of wetland and littoral habitats in lakes is often done without consideration of biological consequences. Removal of snags from forest, the terrestrial equivalent of woody debris in the stream channel, influences the distribution and abundances of many wildlife species as well.

Thus, advances in expanding fields such as environmental protection, conservation biology, fish and wildlife management, restoration ecology, and landscape ecology depend on a more informed and pluralistic approach to understanding and using knowledge of trophic regulation.

7. Although much has been learned about the nature of trophic control of vertebrate populations in the past three decades, we have at best partial answers to many questions. Three such questions follow:

(A) *What food web characteristics (e.g., predator:prey ratios, trophic position of taxa) trigger a shift in relative importance of regulation from bottom-up to top-down?*

(B) *How frequent and intense do harsh physical conditions need to be before abiotic variability, refugia, and extinction and colonization processes become of paramount importance in determining population and community attributes?*

(C) *How frequently do shifts from bottom-up to top-down (or biotic to abiotic) control occur in natural and human influenced systems?*

Fretwell and Oksanen (Oksanen *et al.*, 1981; Fretwell, 1987; Oksanen, 1988, 1990) suggest a promising approach to the question of control (question A) in their model of food chain dynamics; control from above or below depends on position of the trophic level in the food chain. However, the model leaves many things unexplained, especially those processes that determine when a top predator can achieve a large enough population to exert control over its prey. In contrast, questions B and C are largely unexplored.

VI. Summary

Trophic control in forest, lake, and stream vertebrates is a shifting mosaic of abiotic and biotic forces acting on seasonal and year-to-year time scales and at a diversity of spatial scales. Control mechanisms at any instant result from the accumulation of past events and current influences. These represent pressures impinging on the organism from within, from other organisms at the same and higher and lower trophic levels, and from the physical environment. Climatic or other irregularities (e.g., anthropogenic disturbances) tend to shift the balance among controlling processes. In short, proximate mechanisms regulating abundances of species result from variation in absolute and relative strength of many forces acting in space and time. The challenge is in understanding and predicting the ecological effects of these processes.

Acknowledgments

This paper would not have been possible without the support of many individuals and agencies. Financial support was provided by the National Science Foundation (DEB 82-06672 to JRK; BSR 88-04926 to IJS), U. S. Environmental Protection Agency (R806391 to JRK), Tennessee Valley Authority (TV80095T to JRK), and the Environmental Sciences Program of the Smithsonian Tropical Research Institute (to JRK). Isaac J. Schlosser was supported on a sabbatical leave by the Department of Biology at the Virginia Polytechnic Institute and State University in Blacksburg, Virginia. J. Brawn, C. Kellner, B. Kerans, and M. Power provided valuable comments on an earlier version of the manuscript.

References

Adams, S. M., and DeAngelis, D. L. (1987). Indirect effects of early bass–shad interactions on predator population structure and food web dynamics. *In* "Predation: Direct and Indirect Impacts on Aquatic Communities." (W. L. Kerfoot and A. Sih, eds.), pp. 103–117. University Press of New England, Hanover, New Hampshire.

Aggus, L. R., and Elliott, G. V. (1975). Effects of cover and food on year-class strength of largemouth bass. *In* "Black Bass Biology and Management." (H. Clepper, ed.), pp. 317–322. Sport Fishing Institute, Washington, D.C.

Allan, J. D. (1982). The effects of reduction in trout density on the invertebrate community of a mountain stream. *Ecology* **63,** 1444–1455.

Angermeier, P. L. (1982). Resource seasonality and fish diets in an Illinois stream. *Environ. Biol. Fishes* **7,** 251–264.

Angermeier, P. L. (1985). Spatio-temporal patterns of foraging success for fishes in an Illinois stream. *Am. Midl. Nat.* **114,** 342–359.

Bagenal, T. B. (1978). Aspects of fish fecundity. *In* "Ecology of Freshwater Fish Production" (S. D. Gerking, ed.), pp. 75–101. Wiley, New York.

Bayley, P. B. (1988). Factors affecting growth rates of young tropical fishes: Seasonality and density-dependence. *Environ. Biol. Fishes* **21,** 127–142.

Bayley, P. B. (1989). Aquatic environments in the Amazon Basin with an analysis of carbon sources, fish production, and yield. *Canadian Special Publication in Fisher. Aquatic Sci.* **106,** 399–408.

Becker, G. C. (1983). "Fishes of Wisconsin." University of Wisconsin Press, Madison, Wisconsin.

Beschta, R. L., Bilby, R. E., Brown, G. W., Holtby, L. B., and Hofstra, T. D. (1987). Stream temperature and aquatic habitat. *In* "Streamside Management: Forestry Fisheries Interactions." (E. O. Salo and T. W. Cundy, eds.), pp. 191–232. Contribution No. 57, Institute of Forest Resources, University of Washington, Seattle, Washington.

Biro, P. (1977). Effects of exploitation, introductions, and eutrophication on percids in Lake Balaton. *J. Fisher. Res. Bd. Can.* **34,** 1678–1683.

Blake, J. G., and Loiselle, B. A. (1991). Variation in resource abundance affects capture rates of birds in three lowland habitats in Costa Rica. *Auk* **108,** 114–130.

Blake, J. G., Hanowski, J. M., and Niemi, G. J. Correlations between birds and habitat: Annual variation in species-habitat relationships. Manuscript in revision.

Blondel, J., Pradel, R., and Lebreton, J.-D. Survival rates in two populations of blue tit *Parus caerulens*. A test of theory on population biology on an island. Submitted for publication.

Bowman, G. B., and Harris, L. D. (1980). Effect of spatial heterogenicity onground-nest predation. *J. Wild. Manag.* **44,** 806–813.

Brawn, J. D. (1991). Environmental effects on variation and covariation in reproductive traits of western bluebirds. *Oecologia,* in press.

Brawn, J. D., Boecklen, W. J., and Balda, R. P. (1987). Investigations of density interactions among breeding birds in ponderosa pine forests: Correlative and experimental evidence. *Oecologia* **72,** 348–357.

Brawn, J. D., Karr, J. R., and Kaufmann, K. W. Evolution of small clutch size in tropical birds: evidence of a benign or harsh environment? Submitted for publication.

Brosset, A. (1969). La vie sociale des oiseaux dans une foret equatoriale du Gabon. *Biologica Gabonica* **5,** 29–69.

Brosset, A. (1990). A long-term study of the rain forest birds in M'Passa (Gabon). *In* "Biogeography and Ecology of Forest Bird Communities." (A. Keast, ed.), pp. 259–274. SPB Academic Publishing, The Hague, The Netherlands.

Carpenter, F. L. (1978). A spectrum of nectar-eater communities. *Am. Zool.* **18,** 809–819.

Carpenter, S. R., Kitchell, J. F., Hodgson, J. R., Cochran, P. A., Elser, J. J., Elser, M. M., Lodge, D. M., Kretchmer, D., He, X., and von Ende, C. N. (1987). Regulation of lake primary productivity by food web structure. *Ecology* **68,** 1863–1876.

Casselman, J. M., and Harvey, H. H. (1975). Selective fish mortality resulting from low winter oxygen. *Verhandlengen Internationale Vereinigung fuer Theoretischeund Angewandte Limnologie* **19,** 2418–2429.

Chambers, P. A. (1987). Light and nutrients in the control of aquatic plant community structure II. *In situ* observations. *J. Ecol.* 75, 621–628.

Chapin, J. P. (1932). The birds of the Belgian Congo. Part 1, *Bull. Am. Mus. Nat. Hist.* **65,** 1–756.

Clady, M. D. (1976). Influence of temperature and wind on the survival of early stages of yellow perch, *Perca flavescens. J. Fisher. Res. Bd. Can.* **33,** 1887–1893.

Clout, M. N., and Hay, J. R. (1989). The importance of birds as browsers, pollinators and seed disperses in New Zealand forests. *N. Z. J. Ecol.* **12** (Suppl.), 27–32.

Coble, D. W. (1967). Relationship of temperature to total annual growth in adult smallmouth bass. *J. Fisher. Res. Bd. Can.* **24,** 87–99.

Collins, B. G., and Paton, D. C. (1989). Consequences of differences in body mass, wing length, and leg morphology for nectar-feeding birds. *Aust. J. Ecol.* **14,** 269–289.

Cooper, J. E. (1990). Birds and zoonoses. *Ibis* **132,** 181–191.

Cooper, S. D., Walde, S. J., and Peckarsky, B. L. (1990). Prey exchange rates and the impact of predators on prey populations in streams. *Ecology* **71,** 1503–1514.

Coutant, C. C. (1975). Responses of bass to natural and artificial temperature regimes. *In* "Black Bass Biology and Management." (H. Clepper, ed.), pp. 273–285. Sport Fishing Institute, Washington, D.C.

Coutant, C. C. (1985). Striped bass, temperature, and dissolved oxygen: A speculative hypothesis for environmental risk. *Trans. Am. Fisher. Soc.* **114,** 31–61.

Crowder, L. B. (1986). Ecological and morphological shifts in Lake Michigan fishes: Glimpses of the ghost of competition past. *Environ. Biol. Fishes* **16,** 147–157.

Crowder, L. B., and Cooper, W. E. (1982). Habitat structural complexity and the interaction between bluegills and their prey. *Ecology* **63,** 1802–1813.

Crowder, L. B., Magnuson, J. J., and Brandt, S. B. (1981). Complementarity in the use of food and thermal habitat by Lake Michigan fishes. *Can. J. Fisher. Aquat. Sci.* **38,** 662–668.

Crowder, L. B., Drenner, R. W., Kerfoot, W. C., McQueen, D. J., Mills, E. L., Sommer, U., Spencer, C. N., and Vanni, M. J. (1988). Food web interactions in lakes. *In* "Complex Interactions in Lake Communities" (Carpenter, S. R., ed.), pp. 141–160. Springer-Verlag, New York.

Cryer, M., Peirson, G., and Townsend, C. R. (1986). Reciprocal interactions between roach, *Rutilus rutilus,* and zooplankton in a small lake: Prey dynamics and fish growth and recruitment. *Limnol. Oceanogr.* **31,** 1022–1038.

Cunjak, R. A. (1988). Physiological consequences of overwintering in streams: The cost of acclimatization. *Can. J. Fisher. Aquat. Sci.* **45,** 443–452.

Cunjak, R. A., Curry, R. A., and Power, G. (1987). Seasonal energy budget of brook trout in streams: Implications of a possible energy deficit in early winter. *Trans. Am. Fisher. Soc.* **116,** 817–828.

Cyr, H., and Downing, J. A. (1988). Empirical relationships of phytomacrofaunal abundance to plant biomass and macrophyte bed characteristics. *Can. J. Fisher. Aquat. Sci.* **45,** 976–984.

DeVries, D. R., Stein, R. A., Miner, J. G., and Mittelbach, G. G. (1991). Threadfin shad as supplementary forage: Consequences for young-of-year fishes. *Trans. Am. Fisher. Soc.,* in press.

DeVries, D. R., Stein, R. A., and Chesson, P. L. (1989). Sunfish foraging among patches: The patch-departure decision. *Anim. Behav.* **37,** 455–464.

Diehl, S. (1988). Foraging efficiency of three freshwater fishes: Effects of structural complexity and light. *Oikos* **53,** 207–214.

Dionne, M. (1990). How littoral macrophyte growth form influences foraging ecology of pumpkinseed sunfish. Ph.D. dissertation, Dartmouth College, Hanover, New Hampshire.

Dionne, M. and Folt, C. L. (1991). An experimental analysis of macrophyte growth forms as fish foraging habitat. *Can. J.Fisher. Aquat. Sci.* **48,** 123–131.

Dvorak, J. (1987). Production–ecological relationships between aquatic vascular plants and invertebrates in shallow waters and wetlands—a review. *Archiv fur Hydrobiologie, Beiheft: Ergebnisse der Limnologie* **27,** 181–184.

Dyrcz, A. (1983). Breeding ecology of the clay-colored robin *Turdus grayi* in lowland Panama. *Ibis* **125,** 287–304.

Eadie, J. M., and Keast, A. (1984). Resource heterogeneity and fish species diversity in lakes. *Can. J. Zool.* **62,** 1689–1695.

Edminster, F. C. (1939). The effect of predator control on ruffed grouse populations in New York. *J. Wild. Manag.* **3,** 345–352.

Ehlinger, T. J. (1989). Foraging mode switches in the golden shiner (*Notemigonus crysoleucas*). *Can. J. Fisher. Aquat. Sci.* **46,** 1250–1254.

Eipper, A. W. (1975). Environmental influences on the mortality of bass embryos and larvae. *In* "Black Bass Biology and Management" (H. Clepper, ed.), pp. 295–305. Sport Fishing Institute, Washington, D.C.

Elliott, J. M. (1976). The energetics of feeding, metabolism and growth of brown trout (*Salmo trutta* L.) in relation to body weight, water temperature and ration size. *J. Anim. Ecol.* **45**, 923–948.

Elwood, J. W., and Waters, T. F. (1969). Effects of floods on food consumption and production rates of a stream brook trout population. *Trans. Am. Fisher. Soc.* **98**, 253–262.

Emlen, J. T. (1977). Land bird communities of Grand Bahama Island: The structure and dynamics of an avifauna. *Ornithol. Monogr.* **24**, 1–129.

Engel, S. (1985). Aquatic community interactions of submerged macrophytes. Technical Bulletin #156, Wisconsin Department of Natural Resources, Madison, Wisconsin.

Engel, S. (1987). The impact of submerged macrophytes on largemouth bass and bluegills. *Lake Reserv. Manag.* **3**, 227–234.

Eshenroder, R. L. (1977). Effects of intensified fishing, species changes, and spring water temperatures on yellow perch, *Perca flavescens*, in Saginaw Bay. *J. Fisher. Res. Bd. Can.* **34**, 1830–1838.

Everest, F. H., Beschta, R. L., Scrivener, J. C., Kaski, K. V., Sedell, J. R., and Cederholm, C. J. (1987). Fine sediment and salmonid production: A paradox. *In* "Streamside Management: Forestry and Fishery Interactions" (E. O. Salo and T. W. Cundy, eds.), pp. 98–142. Contribution No. 57. University of Washington, Institute of Forest Resources, Seattle, Washington.

Feinsinger, P., and Colwell, R. K. (1978). Community organization among neotropical nectar-feeding birds. *Am. Zool.* **18**, 779–795.

Ferguson, A. J. D., Thompson, J. M., and Reynolds, C. S. (1982). Structure and dynamics of zooplankton communities maintained in closed systems, with special reference to algal food supply. *J. Plankton Res.* **4**, 523–543.

Flecker, A. S., and Allan, J. D. (1984). The importance of predation, substrate and spatial refugia in determining lotic insect distributions. *Oecologia* **64**, 306–313.

Ford, H. A., Bridges, L., and Noske, S. (1990). Interobserver differences in recording foraging behavior of fuscous honeyeaters. *Stud. Avian Biol.* **13**, 199–201.

Foster, M. S. (1974). A model to explain molt–breeding overlap and clutch size in some tropical birds. *Evolution* **28**, 182–190.

Franks, N. (1982). Ecology and population regulation in the army ant *Eciton burchelli*. *In* "The Ecology of a Tropical Forest: Seasonal Rhythms and Long-term Changes" (E. G. Leigh, Jr., A. S. Rand, and D. M. Windsor, eds.), pp. 389–395. Smithsonian Institution, Washington, D.C.

French, J. R. P., III. (1988). Effect of submersed aquatic macrophytes on resource partitioning in yearling rock bass (*Ambloplites rupestris*) and pumpkinseeds (*Lepomis gibbosus*) in Lake St. Clair. *J. Great Lakes Res.* **14**, 291–300.

Fretwell, S. D. (1987). Food chain dynamics: The central theory of population ecology. *Oikos* **50**, 291–301.

Gascon, D. and Leggett, W. C. (1977). Distribution, abundance, and resource utilization of littoral zone fishes in response to a nutrient/production gradient in Lake Memphremagog. *J. Fisher. Res. Bd. Can.* **34**, 1105–1117.

Gautier-Horn, A., and Michaloud, G. (1989). Are figs always keystone resources for tropical frugivorous vertebrates? A test in Gabon. *Ecology* **70**, 1826–1833.

Gentry, A. (1990). Tropical forests. *In* "Biogeography and Ecology of Forest Bird Communities" (A. Keast, ed.), pp. 35–43. SPB Academic Publishing, The Hague, The Netherlands.

Gilliam, J. R., and Fraser, D. F. (1987). Habitat selection under predation hazard: Test of a model with foraging minnows. *Ecology* **68**, 1856–1862.

Gilliam, J. F., and Fraser, D. F. (1988). Resource depletion and habitat segregation by competitors under predation hazard. *In* "Size-Structured Populations" (B. Ebenman and L. Persson, eds.), pp. 173–184. Springer-Verlag, New York.

278 *James R. Karr et al.*

Gophen, M., Drenner, R. W., and Vinyard, G. L. (1982). Cichlid stocking and the decline of the Galilee Saint Peter's fish (*Sarotherodon galilaeus*) in Lake Kinneret, Israel. *Can. J. Fisher. Aquat. Sci.* **40**, 983–986.

Gotceitas, V. (1990a). Variation in plant stem density and its effects on foraging success of juvenile bluegill sunfish. *Environ. Biol. Fishes* **27**, 63–70.

Gotceitas, V. (1990b). Foraging and predator avoidance: A test of a patch choice model with juvenile bluegill sunfish. *Oecologia* **83**, 346–351.

Gotceitas, V., and Colgan, P. (1987). Selection between densities of artificial vegetation by young bluegills avoiding predation. *Trans. Am. Fisher. Soc.* **116**, 40–49.

Gotceitas, V., and Colgan, P. (1989). Predator foraging success and habitat complexity: Quantitative test of the threshold hypothesis. *Oecologia* **80**, 158–166.

Graham, J. H., and Hastings, R. W. (1984). Distributional patterns of sunfishes on the New Jersey coastal plain. *Environ. Biol. Fishes* **10**, 137–148.

Grant, B. R., and Grant, P. R. (1989). Evolutionary dynamics of a natural population: The large cactus finch of the Galapagos. University of Chicago Press, Chicago, Illinois.

Gregory, R. S., and Powles, P. M. (1985). Chronology, distribution, and sizes of larval fish sampled by light traps in macrophytic Chemung Lake. *Can. J. Zool.* **63**, 2569–2577.

Grubb, T. C., Jr. (1977). Weather-dependent foraging behavior of some birds wintering in a deciduous woodland: Horizontal adjustments. *Condor* **79**, 271–274.

Grubb, T. C., Jr., and Greenwald, L. (1982). Sparrows and a brushpile: Foraging responses to different combinations of predation risk and energy cost. *Anim. Behav.* **30**, 637–640.

Hall, D. J., and Werner, E. E. (1977). Seasonal distribution and abundance of fishes in the littoral zone of a Michigan lake. *Trans. Am. Fisher. Soc.* **106**, 545–555.

Hall, D. J., Cooper, W. E., and Werner, E. E. (1970). An experimental approach to the production dynamics and structure of freshwater animal communities. *Limnol. Oceanogr.* **15**, 839–928.

Hairston, N. G., Smith, F. E., and Slobodkin, L. B. (1960). Community structure, population control, and competition. *Am. Nat.* **94**, 421–425.

Haney, J. F. (1987). Field studies on zooplankton–cyanobacteria interactions. *N. Z. J. Mar. Freshw. Res.* **21**, 467–475.

Hanson, J. M., and Leggett, W. C. (1985). Experimental and field evidence for inter- and intraspecific competition in two freshwater fishes. *Can. J. Fisher. Aquat. Sci.* **42**(2), 280–286.

Hanson, J. M., and Peters, R. H. (1984). Empirical prediction of crustacean zooplankton biomass and profundal macrobenthos biomass in lakes. *Can. J. Fisher. Aquat. Sci.* **41**, 439–445.

Harris, G. P., and Piccinin, B. B. (1980). Physical variability and phytoplankton communities. IV. Temporal changes in the phytoplankton community of a physically variable lake. *Archiv für Hydrobiologie* **89**(4), 447–473.

Harvey, B. C. (1987). Susceptibility of young-of-the-year fishes to downstream displacement by flooding. *Trans. Am. Fisher. Soc.* **116**, 851–855.

Henderson, B. A., and Brown, E. H., Jr. (1985). Effects of abundance and water temperature on recruitment and growth of alewife (*Alosa pseudoharengus*) near South Bay, Lake Huron, 1954–1982. *Can. J. Fisher. Aquat. Sci.* **42**, 1608–1613.

Holland, L. E., and Huston, M. L. (1985). Distribution and food habits of young-of-the-year fishes in a backwater lake of the upper Mississippi river. *J. Freshw. Ecol.* **3**, 81–91.

Holmes, R. T. (1990). Ecology and evolutionary impacts of bird predation on forest insects: An overview. *Stud. Avian Biol.* **13**, 6–13.

Holmes, R. T., Sherry, T. W., and Bennett, S. E. (1978). Diurnal and individual variability in the foraging behavior of American redstarts (*Setophaga ruticilla*). *Oecologia* **36**, 141–149.

Horwitz, R. J. (1978). Temporal variability patterns and the distributional patterns of stream fishes. *Ecol. Monogr.* **48**, 307–321.

Hutchinson, G. E. (1975). "A treatise on limnology, Vol. 3. Limnological Botany." John Wiley, New York.

Hutto, R. L. (1990). Measuring the availability of food resources. *Stud. Avian Biol.* **13,** 20–28.

Hynes, H. B. N. (1960). "The Biology of Polluted Waters." University Press, Liverpool, England.

Hynes, H. B. N. (1970). "Ecology of Running Waters." University of Toronto Press, Ontario, Canada.

Johannes, M. R. S., McQueen, D. J., Stewart, T. J., and Post, J. R. (1989). Golden shiner (*Notemigonus crysoleucas*) population abundance: Correlations with food and predators. *Can. J. Fisher. Aquat. Sci.* **46,** 810–817.

Johansson, L. (1987). Experimental evidence for interactive habitat segregation between roach (*Rutilus rutilus*) and rudd (*Scardinius erythrophthalmus*) in a shallow eutrophic lake. *Oecologia (Berlin)* **73,** 21–27.

Johansson, L., and Persson, L. (1986). The fish community of temperate eutrophic lakes. *In* "Carbon Dynamics in Eutrophic, Temperate Lakes" (B. Riemann and M. Sondergaard, eds.), pp. 237–266. Elsevier Science, New York.

Junk, W. J., Bayley, P. B., and Sparks, R. E. (1989). The flood pulse concept in river–floodplain systems. *Can. Spec. Publ. Fisher. Aquat. Sci.* **106,** 110–127.

Karr, J. R. (1976). Seasonality, resource availability, and community diversity of neotropical lowland habitats. *Ecol. Monogr.* **46,** 457–458.

Karr, J. R. (1980). Geographical variation in the avifaunas of tropical forest undergrowth. *Auk* **97,** 283–298.

Karr, J. R. (1990a). Biological integrity and the goal of environmental legislation: Lessons for conservation biology. *Conserv. Biol.* **4,** 244–250.

Karr, J. R. (1990b). Interactions between forest birds and their habitats: A comparative synthesis. *In* "Biogeography and Ecology of Forest Bird Communities" (A. Keast, ed.), pp. 379–386. SPB Academic Publishing: The Hague, The Netherlands.

Karr, J. R. (1991). Biological integrity: A long-neglected aspect of water resource management. *Ecol. Appl.* **1,** 66–84.

Karr, J. R., and Brawn, J. D. (1990). Food resources of understory birds in central Panama: Quantification and effects on avian populations. *Stud. Avian Biol.* **13,** 58–64.

Karr, J. R., and Freemark, K. E. (1983). Habitat selection and environmental gradients: Dynamics in the "stable" tropics. *Ecology* **64,** 1481–1494.

Karr, J. R., and Roth, R. R. (1971). Vegetation structure and avian diversity in several New World areas. *Am. Nat.* **115,** 423–435.

Karr, J. R., Robinson, S. K., Blake, J. G., and Bierregaard, R. O., Jr. (1990). Birds of Four Neotropical Forests, *In* "Four Neotropical Rainforests." (A. Gentry, ed.), pp. 237–269. Yale University Press, New Haven.

Keast, A. (ed.). (1990). "Biogeography and Ecology of Forest Bird Communities." SPB Academic Publishing, The Hague, Netherlands.

Keast, A. and Webb, D. (1966). Mouth and body form relative to feeding ecology in the fish fauna of a small lake, Lake Opinicon, Ontario. *J. Fisher. Res. Bd. Can.* **23,** 1845–1874.

Keast, A., Harker, J., and Turnbull, D. (1978). Nearshore fish habitat utilization and species associations in Lake Opinicon (Ontario, Canada). *Environ. Biol. Fishes* **3,** 173–184.

Keast, A., Recher, H. F., Ford, H., and Saunders, D (eds.) (1985). "Birds of Eucalypt Forests and Woodlands: Ecology, Conservation, Management." Surrey Beatty and Sons, Chipping Norton, New South Wales, Australia.

Keller, E. A., and Swanson, F. J. (1979). Effects of large organic material on channel form and fluvial processes. *Earth Surface Processes* **4,** 361–380.

Kitchell, J. F., Stewart, D. J., and Weininger, D. (1977). Applications of a bioenergetics model to yellow perch (*Perca flavescens*) and walleye (*Stizostedion vitreum vitreum*). *Fisher. Res. Bd. Can.* **34,** 1922–1935.

Kohler, C. C., and Ney, J. J. (1981). Consequences of an alewife die-off to fish and zooplankton in a reservoir. *Trans. Am. Fisher. Soc.* **110,** 360–369.

Kushlan, J. A. (1976). Environmental stability and fish community diversity. *Ecology* **57**, 821–825.

Kushlan, J. A., Morales, G., and Frohring, P. C. (1985). Foraging niche relations of wading birds in tropical wet savannas. *Ornith. Monogr.* **36**, 663–682.

Lack, D. (1954). "The Natural Regulation of Animal Numbers." Clarendon, Oxford, England.

Lack, D. (1968). "Ecological Adaptations for Breeding in Birds." Metheun, London.

Lampert, W. (1987). Vertical migration of freshwater zooplankton: Indirect effects of vertebrate predators on algal communities. *In* "Predation: Direct and Indirect Impacts on Aquatic Communities." (W. C. Kerfoot and A. Sigh, eds.), pp 291–299. University Press of New England, Hanover, New Hampshire.

Larimore, R. L., Childers, W. F., and Heckrotte, C. (1959). Destruction and re-establishment of stream fish and invertebrates affected by drought. *Trans. Am. Fisher. Soc.* **88**, 261–285.

Lawton, J. H., and McNeill, S. (1979). Between the devil and the deep blue sea: On the problem of being a herbivore. *In* "Population Dynamics. The 20th Symposium of the British Ecological Society." (R. M. Anderson, B. D. Turner, and L. R. Taylor, eds.), pp. 223–224. Blackwells, London.

Lazzaro, X. (1987). A review of planktivorous fishes: Their evolution, feeding behaviours, selectivities, and impacts. *Hydrobiologia* **146**, 97–167.

Leach, J. H., Johnson, M. G., and Kelso, J. R. M. (1977). Responses of percid fishes and their habitats to eutrophication. *J. Fisher. Res. Bd. Can.* **34**, 1964–1971.

Levey, D. J. (1988). Spatial and temporal variation in Costa Rican fruit and fruit-eating bird communities. *Ecol. Monogr.* **58**(4), 251–269.

Levey, D. J., and Karasov, W. H. (1989). Digestive responses of temperate birds switched to fruit or insect diets. *Auk* **106**, 675–686.

Lima, S. L. (1987a). Vigilance while feeding and its relation to the risk of predation. *J. Theor. Biol.* **124**, 303–316.

Lima, S. L. (1987b). Clutch size in birds: A predation perspective. *Ecology* **68**, 1062–1070.

Lima, S. L. (1988). Initiation of daily feeding in dark-eyed juncoes: Influences of energy reserves and predation risk. *Oikos* **53**, 3–11.

Lima, S. L. (1990). Protective cover and the use of space: Different strategies in finches. *Oikos* **58**, 151–158.

Lima, S. L. and Dill, L. M. (1990). Behavioral decisions made under risk of predation: A review and prospectus. *Can. J. Zool.* **68**, 619–640.

Loiselle, B. A., and Blake, J. G. (1989). Diets of understory fruit-eating birds in Costa Rica: Seasonality and resource abundance. *Avian Biol.* **13**, 89–102.

Loiselle, B. A., and Blake, J. G. (1991). Resource abundance and temporal variation in fruit-eating birds along a wet forest elevational gradient in Costa Rica. *Ecology* **72**, 180–193.

Lowe-McConnell, R. H. (1987). Ecological studies in tropical fish communities. Cambridge University Press, London.

Loyn, R. H., Runnals, R. G., Forward, G. Y., and Tyers, J. (1983). Territorial bell miners and other birds affecting populations of insect prey. *Science* **221**, 1411–1413.

Luedtke, R. J., and Brusven, M. A. (1976). Effects of sand sedimentation on colonization of stream insects. *J. Fisher. Res. Bd. Can.* **33**, 1881–1886.

Lyons, J. (1989). Changes in the abundance of small littoral-zone fishes in Lake Mendota, Wisconsin. *Can. J. Zool.* **67**, 2910–2917.

MacArthur, R. H. (1972). "Geographical Ecology: Patterns in the Distribution of Species." Harper & Row, New York.

MacArthur, R. H., and MacArthur, J. W. (1961). On bird species diversity. *Ecology* **42**, 357–374.

MacClintock, L., Whitcomb, R. F., and Whitcomb, B. L. (1977). Island biogeography and "habitat islands" of eastern forest. II. Evidence for the value of corridors and the minimization of isolation in preservation of biotic diversity. *Am. Birds* **31**, 6–16.

Magnuson, J. J. (1991). Fish and fisheries ecology. *Ecol. Appl.* **1**, 13–26.

Mariani, J. M., and Manuwal, D. A. (1990). Factors influencing brown creeper (*Certhia americana*) abundance patterns in the southern Washington Cascade range. *Stud. Avian Biol.* **13**, 53–57.

Martin, T. E. (1986). Competition in breeding birds: on the importance of considering processes at the individual level. Current Ornithology **4**, 181–210.

Martin, T. E. (1987). Competition in breeding birds: On the importance of considering processes at the level of the individual. Current Ornithology **4**, 181–210.

Martin, T. E. (1988a). Habitat and area effects on forest bird assemblages: Is nest predation an influence? *Ecology* **69**, 74–84.

Martin, T. E. (1988b). On the advantage of being different: Nest predation and the coexistence of bird species. *Proc. Nat. Acad. Sci. U.S.A.* **85**, 2196–2199.

Martin, T. E. (1988c). Processes organizing open-nesting bird assemblages: Competition or nest predation? *Evol. Ecol.* **2**, 37–50.

Martin, T. E. (1991). Food limitation in terrestrial breeding birds: Is that all there is? *Proc. Int. Ornith. Cong.* **20**, In Press.

Martin, T. E., and Karr, J. R. (1986). Patch utilization by migrating birds: Resource oriented? *Ornis Scandinavica* **17**, 165–174.

Martin, T. E., and Karr, J. R. (1990). Behavioral plasticity of foraging maneuvers of migratory warblers: Multiple selection periods for niches? *Stud. Avian Biol.* **13**, 353–359.

Martin, T. E., and Roper, J. J. (1988). Nest predation and nest-site selection of a western population on the hermit thrush. *Condor* **90**, 51–57.

Mason, J. C. (1976). Response of underyearling coho salmon to supplemental feeding in a natural stream. *J. Wild. Manag.* **40**, 775–788.

Matthews, W. J., and Maness, J. D. (1979). Critical thermal maxima, oxygen tolerances, and success of cyprinid fishes in a southwestern river. *Amer. Midl. Natur.* **102**, 374–377.

McCauley, E., and Kalff, J. (1981). Empirical relationships between phytoplankton and zooplankton biomass among lakes. *Can. J. Fisher. Aquatic Sci.* **38**, 458–463.

McConnell, W. J., Lewis, S., and Olson, J. E. (1977). Gross photosynthesis as an estimator of potential fish production. *Trans. Am. Fisher. Soc.* **106**, 417–423.

McDonald, L. L., Manly, B. F. J., and Raley, C. M. (1990). Analyzing foraging and habitat use through selective functions. *Stud. Avian Biol.* **13**, 325–331.

McQueen, D. J. (1990). Manipulating lake community structure: Where do we go from here? *Freshw. Biol.* **23**, 613–620.

Mendelson, J. (1975). Feeding relationship among species of *Notropis* (Pisces:Cyprinidae) in a Wisconsin stream. *Ecol. Monogr.* **45**, 199–230.

McQueen, D. J., Johannes, M. R. S., Post, J. R., Stewart, T. J., and Lean, D. R. S. (1989). Bottom-up and top-down impacts on freshwater pelagic community structure. *Ecol. Monogr.* **59**, 289–309.

Mills, E. L., and Schiavone, A., Jr. (1982). Evaluation of fish communities through assessment of zooplankton populations and measures of lake productivity. *North Am. J. Fisher. Manag.* **2**, 14–27.

Mills, E. L., Forney, J. L., and Wagner, K. J. (1987). Fish predation and its cascading effect on the Oneida Lake food chain. *In* "Predation: Direct and Indirect Impacts on Aquatic Communities" (W. C. Kerfoot and A. Sih, eds.), pp. 118–131. University Press of New England, Hanover, New Hampshire.

Minckley, W. L., and Meffe, G. K. (1987). Differential selection by flooding in stream fish communities of the arid American Southwest. *In* "Community and Evolution Ecology of North American Stream Fishes" (W. J. Matthews and D. C. Heins, eds.), pp. 93–104. University of Oklahoma Press, Norman, Oklahoma.

Mittelbach, G. G. (1981). Foraging efficiency and body size: A study of optimal diet and habitat use by bluegills. *Ecology* **62**, 1370–1386.

Mittelbach, G. G. (1984). Predation and resource partitioning in two sunfishes (Centrarchidae). *Ecology* **65**, 499–513.

Mittelbach, G. G. (1988). Competition among refuging sunfishes and effects of fish density on littoral zone invertebrates. *Ecology* **69**, 614–623.

Moermond, T. C., and Denslow, J. S. (1985). Neotropical frugivores: The influence of nutrition, behavior, and morphology on fruit choice. *Ornithol. Monogr.* **36**, 865–897.

Morin, P. J. (1984). The impact of fish exclusion on the abundance and species composition of larval odonates: Results of short-term experiments in a North Carolina farm pond. *Ecology* **65**, 53–60.

Morse, D. H. (1990). Food exploitation by birds: Some current problems and future goals. *Stud. Avian Biol.* **13**, 134–143.

Moyle, P. B. and Li, H. W. (1979). Community ecology and predator–prey relations in warmwater streams. *In* "Predator-prey Systems in Fisheries Management" (H. Clepper, ed.), pp. 171–190. Sport Fishing Institute, Washington, D.C.

Munn, C. A. (1985). A permanent canopy and understory flock in Amazonia: Species composition and population density. *Ornithol. Monogr.* **36**, 683–712.

Ney, J. J. (1990). Trophic economics in fisheries: Assessment of demand–supply relationships between predators and prey. *Rev. Aquat. Sci.* **2**, 55–81.

Oglesby, R. T. (1977). Relationships of fish yield to lake phytoplankton standing crop, production, and morphoedaphic factors. *J. Fisher. Res. Bd. Can.* **34**, 2271–2279.

Oksanen, L. (1988). Ecosystem organization: Mutualism and cybernetics or plain Darwinian struggle for existence? *Am. Nat.* **131**, 424–444.

Oksanen, L. (1990). Predation, herbivory, and plant strategies along gradients of primary productivity. *In* "Perspective in Plant Competition" (J. B. Grace and D. Tilman, eds.), pp. 445–474. Academic Press, Orlando, Florida.

Oksanen, L., Fretwell, S. D., Arruda, J., and Niemelä, P. (1981). Exploitation ecosystems in gradients of primary productivity. *Am. Nat.* **118**, 240–261

Oliver, J. D., Holeton, G. F., and Chua, K. E. (1979). Overwinter mortality of fingerling smallmouth bass in relation to size, relative energy stores, and environmental temperature. *Trans. Amer. Fisher. Soc.* **108**, 130–136.

Olrik, D., Lunder, S., and Rasmussen, K. (1984). Interactions between phytoplankton, zooplankton, and fish in the nutrient-rich shallow Lake Hjarbaek Fjord, Denmark. *Int. Rev. Gesamten Hydrobiologie* **69**, 389–405.

Osenberg, C. W., Werner, E. E., Mittelbach, G. G., and Hall, D. J. (1988).Growth patterns in bluegill (*Lepomis macrochirus*) and pumpkinseed (*L. gibbosus*) sunfish: Environmental variation and the importance of ontogenetic niche shifts. *Can. J. Fisher. Aquat. Sci.* **45**, 17–26.

Paloumpis, A. A. (1958). Responses of some minnows to flood and drought conditions in an intermittent stream. *Iowa State J. Sci.* **32**, 547–561.

Partridge, L., and Harvey, P. H. (1988). The ecological context of life-history evolution. *Science* **241**, 1449–1454.

Pasch, R. W., and Lyford, J. H. Jr. (1972). The food habits of two species of *Cottus* occupying the same habitat. *Trans. Am. Fisher. Soc.* **101**, 377–381.

Persson, L. (1986). Effects of reduced interspecific competition on resource utilization in perch (*Perca fluviatilis*). *Ecology* **67**, 355–364.

Persson, L., and Greenberg, L. A. (1990). Juvenile competitive bottlenecks: The perch (*Perca fluviatilis*)–roach (*Rutilus rutilus*) interaction. *Ecology* **71**, 44–56.

Persson, L., Andersson, G., Hamrin, S. F., and Johansson, L. (1988). Predator regulation and primary production along the productivity gradient of temperate lake ecosystems. *In* "Complex interactions in Lake Communities" (S. R. Carpenter ed.), pp. 45–65. Springer-Verlag, New York.

Pflieger, W. L. (1975). "The Fishes of Missouri." Missouri Department of Conservation, Jefferson City, Missouri.

Pitcher, T. J. (1986). Functions of shoaling behavior in teleosts. *In* "The Behavior of Teleost Fishes" (T. J. Pitcher, ed.), pp. 294–337. Johns Hopkins University, Baltimore, Maryland.

Power, M. E. (1984). Depth distributions of armored catfish: Predator-induced resource avoidance? *Ecology* **65**, 523–528.

Power, M. E. (1987). Predator avoidance by grazing fishes in temperate and tropical streams: Importance of stream depth and prey size. *In* "Predation: Direct and Indirect Impacts on Aquatic Communities" (W. C. Kerfoot and A. Sih, eds.), pp. 333–351. University of New England Press, Hanover, New Hampshire.

Power, M. E., and Matthews, W. J. (1983). Algae-grazing minnows (*Campostoma anomalum*), piscivorous bass (*Micropterus* spp.) and the distribution of attached algae in a small prairie-margin stream. *Oecologia (Berlin)* **60**, 328–332.

Power, M. E., Matthews, W. J., and Stewart, A. J. (1985). Grazing minnows, piscivorous bass, and stream algae: Dynamics of a strong interaction. *Ecology* **66**, 1448–1456.

Pyke, G. H. (1985). The relationships between abundances of honeyeaters and their food resources in open forest areas near Sydney. *In* "Birds of Eucalypt Forests and Woodlands: Ecology, Conservation, Management" (A. Keast, H. F. Recher, H. Ford, and D. Saunders eds.) pp. 65–77. Surrey Beatty, Chipping Norton, New South Wales, Australia.

Quiros, R., and Cuch, S. (1989). The fisheries and limnology of the lower Plata Basin. *Can. Spec. Publ. Fisher. Aquat. Sci.* **106**, 429–443.

Rabeni, C. F., and Minshall, G. W. (1977). Factors affecting microdistribution of stream benthic insects. *Oikos* **29**, 33–43.

Rahel, F. J. (1984). Factors structuring fish assemblages along a bog lake successional gradient. *Ecology* **65**, 1276–1289.

Recher, H. F. (1990). Specialist or generalists: Avian response to spatial and temporal changes in resources. *Stud. Avian Biol.* **13**, 333–336.

Reice, S. R. (1983). Predation and substratum factors in lotic community structure. *In* "Dynamics of Lotic Ecosystems" (T. Fontaine and S. Bartell, eds.), pp. 325–345. Ann Arbor Science, Ann Arbor, Michigan.

Resh, V. H. (1977). Habitat and substrate influences on population and production dynamics of a stream caddishly, *Ceraclea ancylus* (Leptoceridae). *Freshw. Biol.* **7**, 261–277.

Reynolds, C. S. (1990). Temporal scales of variability in pelagic environments and the response of phytoplankton. *Freshw. Biol.* **23**, 25–53.

Richards, K. (1982). "Rivers: Form and Process in Alluvial Channels". Methuen, New York.

Rooke, B. (1986). Macroinvertebrates associated with macrophytes and plastic imitations in the Eramosa River, Ontario, Canada. *Archiv für Hydrobiologie* **106**, 307–325.

Ross, S. T., Matthews, W. J., and Echelle, A. A. (1985). Persistence of stream fish assemblages: Effects of environmental change. *Am. Nat.* **126**, 24–40.

Sand-Jensen, K., and Sondergaard, M. (1979). Distribution and quantitative development of aquatic macrophytes in relation to sediment characteristics in oligotrophic Lake Kalgaard, Denmark. *Freshw. Biol.* **9**, 1–11.

Sauer, J. R., and Droege, S. (1990). Recent population trends of the eastern bluebird. *Wilson Bull.* **102**, 239–252.

Savino, J. F., and Stein, R. A. (1989). Behavioral interactions between fish predators and their prey: Effects of plant density. *Anim. Behav.* **37**, 311–321.

Scavia, D., Fahnenstiel, G. L., Evans, M. S., and Jude, D. J., and Lehman, J. T. (1986). Influence of salmonine predation and weather on long-term water quality trends in Lake Michigan. *Can. J. Fisher. Aquat. Sci.* **43**, 435–443.

Schlosser, I. J. (1982). Fish community structure and function along two habitat gradients in a headwater stream. *Ecol. Monogr.* **52**, 395–414.

Schlosser, I. J. (1985). Flow regime, juvenile abundance, and the assemblage structure of stream fishes. *Ecology* **66**, 1484–1490.

Schlosser, I. J. (1987a). The role of predation in age- and size-related habitat use by stream fishes. *Ecology* **68,** 651–659.

Schlosser, I. J. (1987b). A conceptual framework for fish communities in small warmwater streams. *In* " The Ecology and Evolution of North American Stream Fishes." (W. J. Matthew and D. C. Heins ed.), pp. 17–24. University of Oklahoma Press, Norman, Oklahoma.

Schlosser, I. J. (1988). Predation risk and habitat selection by two size classes of a stream cyprinid: Experimental test of a hypothesis. *Oikos* **52,** 36–40.

Schlosser, I. J., and Angermeier, P. L. (1990). The influence of environmental variability, resource abundance, and predation on juvenile cyprinid and centrachid fishes. *Polish Arch. Hydrobiol.* **37,** 267–286.

Schlosser, I. J., and Ebel, K. K. (1989). Effects of flow regime and cyprinal predation on a headwater stream. *Ecol. Monogr.* **59,** 41–57.

Schlosser, I. J., and Toth, L. A. (1984). Niche relationship and population ecology of rainbow (*Etheostoma caeruleum*) and fantail (*E. flabellare*) darters in a temporally variable environment. *Oikos* **42,** 229–238.

Schluter, D. (1988). The evolution of finch communities on island and continents: Kenya vs. Galapagos. *Ecol. Monogr.* **58**(4), 229–249.

Schneberger, E. (ed.) (1970). A symposium on the management of midwestern winterkill lakes. North Central Division American Fisheries Society. FAS Publishing, Madison, Wisconsin.

Schoener, T. W. (1986). Overview: Kinds of ecological communities—ecology becomes pluralistic. *In* "Community Ecology" (J. Diamond and T. J. Case, eds.), pp. 467–479. Harper & Row, New York.

Schramm, H. L., Jr., and Jirka, K. J. (1989). Epiphytic macroinvertebrates as a food resource for bluegills in Florida lakes. *Trans. Am. Fisher. Soc.* **118,** 416–426.

Schramm, H. L., Jr., and Zale, A. V. (1985). Effects of cover and prey size on preferences of juvenile largemouth bass for blue tilapias and bluegills in tanks. *Trans. Am. Fisher. Soc.* **114,** 725–731.

Scott, W. B., and E. T. Crossman. (1973). Freshwater fishes of Canada. *Bull. Fisher. Res. Bd. Can.* **184,** 1–966.

Sedell, J. R., Bisson, P. A., Swanson, F. J., and Gregory, S. V. (1988). What we know about large trees that fall into streams and rivers. *In* "From the Forest to the Sea: A Story of Fallen Trees" (C. Maser, R. F. Tarrant, J. M. Trappeand, and J. F. Franklin, eds.), pp. 47–81. General Technical Report, PNW-GTR-229, USDA, Forest Service, Pacific Northwest Research Station, Portland, Oregon.

Serns, S. L. (1982). Relation of temperature and population density to first-year recruitment and growth of smallmouth bass in a Wisconsin Lake. *Trans. Am. Fisher.Soc.* **111,** 570–574.

Shapiro, J., and Wright, D. I. (1984). Lake restoration by biomanipulation: Round Lake, Minnesota, the first two years. *Freshw. Biol.* **14,** 371–383.

Shuter, B. J., and Post, J. R. (1990). Climate, population viability, and the zoogeography of temperate fishes. *Trans. Am. Fisher. Soc.* **119,** 314–336.

Shuter, B. J., Maclean, J. A., Fry, F. E. J., and Regier, H. A. (1980). Stochastic simulation of temperature effects on first-year survival of smallmouth bass. *Trans. Am. Fisher. Soc.* **109,** 1–34.

Slagsvold, T. (1980). Egg predation in woodland in relation to the presence and density of breeding fieldforer (*Turdus pilariss*). *Ornis Scandinavica* **11,** 92–98.

Slagsvold, T. (1982). Clutch size variation in passerine birds: The nest predation hypothesis. *Oecologia (Berlin)* **54,** 159–169.

Starrett, W. C. (1951). Some factors affecting the abundance of minnows in the Des Moines River, Iowa. *Ecology* **32,** 13–27.

Stiles, F. G., and Clark, D. A. (1989). Conservation of tropical rain forest birds: A case study from Costa Rica. International Council for Bird Preservation. *Am. Birds* **43,** 420–428.

Stock, J. D., and Schlosser, I. J. (1990). Short-term effects of a catastrophic beaver dam collapse on a stream fish community. *Environ. Biol. Fishes.* **31,** 123–129.

Strange, R. J., Kittrell, W. B., and Broadbent, T. D. (1982). Effects of seeding reservoir fluctuation zones on young-of-the-year black bass and associated species. *N. Am. J. Fisher. Manag.* **2,** 307–315.

Sullivan, K., Lisle, T. E., Dolloff, C. A., Grant, G. E., and Ried, L. M. (1987). Stream channels: The link between forest and fishes. *In* "Streamside Management: forestry and fishery interactions" (E. O. Salo and T. W. Cundy, eds.), pp. 39–97. Contribution No. 57, University of Washington, Institute of Forest Resources, Seattle, WA.

Summerfelt, R. C. (1975). Relationship between weather and year-class strength of largemouth bass. *In* "Black Bass Biology and Management" (H. Clepper, ed.), pp. 166–174. Sport Fishing Institute, Washington, D.C.

Tilman, D. (1988). "Plant Strategies and the Structure and Dynamics of Plant Communities." Princeton University Press, Princeton, New Jersey.

Tomialojc, L., and Wesolowski, T. (1990). Bird communities of a primaeval temperate forest of Bialowieza, Poland. *In* "Biogeography and Ecology of Forest Bird Communities" (A. Keast, ed.), pp. 141–165. SPB Academic Publishing, The Hague, The Netherlands.

Tonn, W. M., and Magnuson, J. J. (1982). Patterns in the species composition and richness of fish assemblages in northern Wisconsin lakes. *Ecology* **63,** 1149–1166.

Townsend, C. R. (1988). Fish, fleas, and phytoplankton. *New Scient.* **118,** 67–70.

Townsend, C. R., and Perrow, M. R. (1989). Eutrophication may produce population cycles in roach, *Rutilus rutilus* (L.), by two contrasting mechanisms. *J. Fish Biol.* **34,** 161–164.

Tramer, E. J. (1978). Catastrophic mortality of stream fishes trapped in shrinking pools. *Am. Midl. Nat.* **97,** 469–478.

Turner, M. G., Gardner, R. H., Dale, V. H., and O'Neill, R. V. (1989). Predicting the spread of disturbance across heterogeneous landscapes. *Oikos* **55,** 121–129.

Vadas, R. L. Jr. (1989). Food-web patterns in ecosystems: A reply to Fretwell and Oksanen. *Oikos* **56,** 339–343.

Vanni, M. J. (1987). Effects of food availability and fish predation on a zooplankton community. *Ecol. Monogr.* **57,** 61–88.

Wagner, K. J. (1986). Biological management of a pond ecosystem to meet water use objectives. *In* "Lake and Reservoir Management", Vol. II. Proceedings of the Fifth International Symposium on Applied Lake and Watershed Management (NALMS), November 1985, Lake Geneva, Wisconsin. American Lake Management Society, Washington, D.C.

Waters, T. F. (1982). Annual production by a stream brook char population and by its principal invertebrate food. *Environ. Biol. Fishes* **7,** 165–170.

Welcomme, R. L. (1979). "The Fisheries Ecology of Floodplain Rivers." Longman Press, London, England.

Welcomme, R. L. (1985). River fisheries. FAO Technical Paper 262. Food and Agriculture Organization of the United Nations. Rome, Italy.

Welty, J. C., and Baptista, L. (1988). The Life of Birds, 4th Ed. W.B. Saunders College Publishing, Philadelphia, Pennsylvania.

Wene, G., and Wickliff, E. L. (1940). Modifications of a stream bottom and its effect on the insect fauna. *Can. Entomol.* **72,** 131–135.

Werner, E. E. (1977). Species packing and niche complementarity in three sunfishes. *Am. Nat.* **11,** 553–579.

Werner, E. E. (1984). The mechanisms of species interactions and community organization in fish. *In* "Ecological Communities" (D. R. Strong Jr., D. Simberloff, L. G. Abele, and A. B. Thistle, eds.), pp. 360–382. Princeton University Press, Princeton, New Jersey.

Werner, E. E., and Gilliam, J. F. (1984). The ontogenetic niche and species interactions in size structured populations. *Annu. Rev. Ecol. Syst.* **15,** 395–425.

Werner, E. E., and D. J. Hall. (1976). Niche shifts in sunfishes: Experimental evidence and significance. *Science* **191,** 404–406.

Werner, E. E., and Hall, D. J. (1979). Foraging efficiency and habitat switching in competing sunfishes. *Ecology* **60,** 256–264.

Werner, E. E., and Hall, D. J. (1988). Ontogenetic habitat shifts in bluegill: The foraging rate–predation risk trade-off. *Ecology* **69**(5), 1352–1366.

Werner, E. E., Gilliam, J. F., Hall, D. J., and Mittelbach, G. G. (1983a). An experimental test of the effects of predation risk on habitat use in fish. *Ecology* **64**, 1540–1548.

Werner, E. E., Mittelbach, G. G., Hall, D. J., and Gilliam, J. F. (1983b). Experimental tests of optimal habitat use in fish: The role or relative habitat profitability. *Ecology* **64**, 1525–1539.

Whiteside, B. G., and McNatt, R. M. (1972). Fish species diversity in relation to stream order and physicochemical conditions in the Plum Creek drainage basin. *Am. Midl. Nat.* **88**, 90–101.

Whiteside, M. C., Swindoll, C. M., and Doolittle, W. L. (1985). Factors affecting the early life history of yellow perch, *Perca flavescens. Environ. Biol. Fishes* **12**, 47–56.

Wiens, J. A. (1989a). "The Ecology of Bird Communities. Vol. 1. Foundations and Patterns." Cambridge University Press, Cambridge, England.

Wiens, J. A. (1989b). "The Ecology of Bird Communities. Vol. 2. Processes and variations." Cambridge University Press, Cambridge, England.

Wiens, J. A., and Rottenberry, J. T. (1979). Diet niche relationships among North American grassland and shrubsteppe birds. *Oecologia* **42**, 253–292.

Wilcove, D. S. (1985). Nest predation in forest tracts and the decline of migratory song birds. *Ecology* **66**, 1211–1214.

Willis, E. O. (1986). West African thrushes as safari ant followers. *Le Gerfaut* **76**, 95–108.

Willis, E. O., and Oniki, Y. (1978). Birds and army ants. *Annu. Rev. Ecol. Syst.* **9**, 243–265.

Wilzbach, M. A., Cummins, K. W., and Hall, J. D. (1986). Influence of habitat manipulations on interactions between cutthroat trout and invertebrate drift. *Ecology* **67**, 898–911.

Winn, H. E. (1958). Comparative reproductive behavior and ecology of fourteen species of darters (Pisces: Percidae). *Ecol. Monogr.* **28**, 155–191.

Wong, M. L. (1986). Trophic organization of understory birds in a Malaysian dipterocarp forest. *Auk* **103**, 100–116.

Worthington, A. (1982). Population sizes and breeding rhythms of two species of manakins in relation to food supply. *In* "Ecology of a Tropical Forest: Seasonal Rhythms and Long-term Changes" (E. G. Leigh, A. S. Rand, and D. Windsor, eds.), pp. 213–225. Smithsonian Institution Press, Washington, D.C.

Zaret, T. M., and Rand, S. A. (1971). Competition in stream fishes: Support for the competitive exclusion principle. *Ecology* **52**, 336–342.

10

Interactions within Herbivore Communities Mediated by the Host Plant: The Keystone Herbivore Concept

Mark D. Hunter

Department of Entomology
Pennsylvania State University
University Park, Pennsylvania

I. Introduction: Feedback Loops in Communities

One of the sources of variability in plants as a resource for animals that has not so far been considered in this volume is the influence of the herbivores themselves on plants as resources. As animals live on and consume plants, they may change the quantity, quality, and distribution of plants and plant parts. Several chapters have emphasized the dynamic nature of plant resource distribution, and animals themselves are part of this dynamic process.

In Chapter 6, Price presents the thesis that autecological forces, through their effects on plant growth and distribution, act as primary determinants of animal (particularly insect) population dynamics on plants. He argues that the availability of nutrients and climatic patterns influence plant resources for animals by determining variation in resource carrying capacity. His view is that mechanisms of population change are dominated by cascading effects of plant resources up the trophic system, a *bottom-up* rather than *top-down* picture of ecological systems (Chapters 1 and 6).

However, there are *feedback loops* inherent in trophic systems which can influence the patterns we observe in nature. While I agree with Price that "quantity and quality of primary production is likely to affect all upper trophic levels" and "a fundamental reality will be the cascading effects of the plant up the trophic system through the path of energy flow" p. 141, 158. (Chapter 6), I suggest that some proportion of the heterogeneity in plant quantity and quality is the result of herbivory itself. Specifically, I suggest that there are *keystone herbivores* in natural communities which, as the first trophic level to encounter and modify plants as resources, can have an impact on animal–plant communities as a whole (Fig. 1).

Traditionally, ecologists have concentrated on the role of *top carnivores* (Paine, 1966), or *keystone predators* in communities. There is good evidence that, in some systems, keystone predators shape interactions among species by maintaining prey densities below a threshold which would lead to competition and exclusion—they promote species diversity by regulating the abundance of strong competitors at lower trophic levels (Paine, 1966; Nelson, 1979a,b; Carpenter *et al.*, 1987). There has also been some consideration of *keystone mutualists* (Gilbert, 1980; Howe and Westley, 1988), species which dominate a community by providing or distributing some beneficial resource, such as seeds or fruit. Discussion of keystone mutualism has been restricted, however, and has not greatly influenced a general conviction that the distribution and abundance of many species, particularly those at midtrophic levels, are determined primarily by predation.

Although competition for resources has dominated the literature of community organization in some systems (Lack, 1954; Harper, 1977), the view that predators are major determinants of community structure has prevailed in many studies, perhaps because of the general focus on top-down phenomena in ecological systems (Chapter 1). This is despite studies of food web stability and persistence, which show that keystone species are equally likely at all trophic levels (Pimm, 1982), and an emerging empirical literature in support of this view (Gilbert, 1980; Ehrlich and Daily, 1988; Howe and Westley, 1988; Terborgh, 1988; Brown and Heske, 1990; Kerbes *et al.,* 1990). While not seeking to challenge the existence of keystone predators in many communities, it may be timely to consider the evidence for other forces acting on communities, and to seek the balanced view (Faeth, 1987b).

As Price (Chapter 6) points out, such balance is emerging in some aquatic systems (Carpenter and Kitchell, 1988; Hay, 1991). While I agree with Karr (Chapter 9) that natural enemies may dominate community structure in some systems, while the influence of primary producers may dominate in others, I suspect that the relative contributions of these are most easily distinguished by superimposing their effects on a bottom up view of trophic structure (Fig. 1). This makes sense to me, if only because the removal of higher trophic levels leaves lower levels intact (if modified), whereas the removal of primary producers leaves no system at all. A bottom-up view of

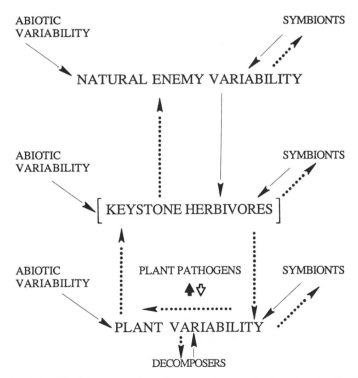

Figure 1   A simplified *bottom-up* view of community organization, showing feedback loops from higher to lower trophic levels.

ecology, however, still allows community structure to be determined at any level in the trophic web, depending on the relative strengths of different feedback loops (Fig. 1).

This book considers the role of natural heterogeneity in animal–plant interactions, and I want to consider the feedback loop between herbivores and their hosts as one generator of that heterogeneity. In the spirit of the current move to clarify the role of host plants in the population dynamics of animals on plants (Price, 1990; Price *et al.*, 1990; Chapter 6), this chapter considers the ways that animals can change plant distribution, abundance, and quality, and how this can feed back up the trophic system as one determinant of animal–plant community structure.

II. The Routes of Feedback: Four Critical Plant Parameters

There is good evidence that four measures of plant resource distribution can affect animal–plant communities. These are plant species diversity, plant abundance, plant structure, and plant chemistry. I have not included plant phenology, plant life history, or plant genotype in this list, because the effects of these are mediated primarily through their influence on plant structure and chemistry. I do not mean to deny their importance as determinants of community structure, and I give some consideration to plant phenology and life history later in this chapter. The effects of variable plant genotype on animal–plant interactions are described in detail in Chapter 4.

There is a wealth of evidence that plant diversity, abundance, structure, and chemistry are prime determinants of herbivore community structure. Plant diversity, for example, can influence community dynamics and plant–herbivore–predator interactions (Root, 1973; Risch *et al.*, 1983); simple communities, both natural and artificially simplified (Redfearn & Pimm, 1987), have dynamics which differ from those of more complex systems (Pimm, 1982). Likewise, the abundance of a particular plant species or plant part in space and time can govern patterns within the community of animals that use that resource (Strong *et al.*, 1984; Crawley, 1983, 1989). The number of arthropod herbivores associated with woody plant species, for example, can be related to the area over which that species is distributed (Southwood, 1960, 1961; Strong, 1979).

Plant growth form or *architecture* can also influence patterns within the communities of animals that use plants (Lawton and Schroder, 1977, 1978). In general, increasing growth form complexity is associated with an increasing number of herbivore species using the plant (Lawton, 1983; Moran, 1980; but see Cornell & Washburn, 1979; Fowler, 1985; and Fig. 3, Section V, below). Finally, plant chemistry can have a range of effects on herbivore communities in evolutionary and ecological time (Denno and McClure, 1983; Strong *et al.*, 1984). A number of theories of plant defensive strategy have been proposed to explain patterns of herbivory, specialization, and changes in animal–plant interactions through succession (Feeny, 1976; Rhoades and Cates, 1976; Reader and Southwood, 1981; Coley *et al.*, 1985).

Both Connor and Simberloff (1979) and Lawton and Strong (1981) correctly identified the null hypothesis for community ecology as one in which animals coexist independently, without effective interaction. This null hypothesis is valuable because it demands evidence rather than supposition of species interactions, but is limited by our understanding of the routes of those interactions. It is critical to my argument that plant diversity, abundance, growth form, and chemistry are seen as interactive variables (Faeth, 1987b), and not as autecological factors (Lawton and Strong, 1981). Modification of these variables by herbivory provides one route by which herbivores can interact within communities, and a suitable test of the role of herbivores in animal–plant community structure.

Diversity, abundance, growth form, and chemistry, therefore, are the potential routes of feedback for keystone herbivores in natural communities. Given that these four parameters can have a major impact on the structure and dynamics of animal–plant communities, it is reasonable that species or guilds which dramatically influence any or all of these parameters have the potential to affect community structure by redirecting the flow of plant resources up the trophic system.

What is the evidence that keystone herbivores can operate in this way? Certainly, herbivory can influence the population dynamics and community structure of plants (Crawley, 1983; Gibson and Hamilton, 1984). Studies have also shown that leaf damage by herbivores can influence the chemistry (Green and Ryan, 1972; Haukioja and Niemela, 1977; Rossiter *et al.*, 1988; Hartley and Lawton, 1990) and structure (Whitham & Mopper, 1985; Hunter, 1987a; Hunter and Willmer, 1989) of plant parts. Changes in the quality and quantity of plants, caused by the animals they support, can have secondary effects on the population quality (Rossiter, 1991; Chapter 2), population dynamics (Karban and Carey, 1984; West, 1985; Faeth, 1985), and community structure (Faeth, 1987b; Hunter and West, 1990) of herbivore species.

Although plant-mediated interactions among herbivore species is a well-established phenomenon, I am less concerned with the impact of one animal species on another than with effects on communities as a whole. With some notable exceptions (e.g., Eltringham, 1974; McNaughton, 1976; Crawley, 1983; Faeth, 1987b; Brown and Heske, 1990), authors have rarely considered the effects of the changes in plants that are caused by animals on herbivore community structure. Toward the balanced view of community organization considered important by many of us (Price *et al.*, 1980; Whitham, 1983; Schultz, 1983; Faeth, 1987b; Strong, 1988; Hawkins, 1990), this chapter attempts to precipitate debate on the existence of keystone herbivores as common features of natural communities.

III. A Definition for Keystone Species

Before considering the possible role of keystone herbivores in communities, I would like to establish a working definition. Keystone species have been defined as "animal or plant species with a pervasive influence on community composition." (Howe and Westley, 1988, p. 218). The best evidence of this *pervasive influence* is when the removal or augmentation of a species has changed the composition of a community, and that is the criterion I will use here where possible.

In considering the evidence for keystone herbivores, this chapter considers only interactions among species mediated by the host plant. Although herbivore species might also influence community structure directly through interspecific territoriality or aggression, or indirectly through changes in some shared natural enemy complex, these are beyond the main focus of this volume—the effects of resource distribution on animal–plant interactions. Discussion is therefore restricted to community interactions mediated by changes in some measure of plant resource distribution.

IV. Plant-Mediated Interactions in Animal–Plant Communities

In 1983, Crawley argued that "By their activity, [herbivores] create or modify much of the spatial heterogeneity which is so important in the dynamics of all the species [in a community]"(p. 290). This section describes some of the heterogeneity introduced into natural communities by animals that use plants. Each of the following subsections considers the evidence (usually strong) that animals can change one of the four plant parameters (diversity, abundance, growth form, and chemistry) described previously. That is followed by a description of the evidence (usually weaker) that such changes influence the shape of animal communities.

Finally, I argue that the evidence for plant-mediated effects on community structure is weak because of a recent emphasis on species by species interactions within systems, at the expense of studies of community-level interactions. This is in spite of early studies that stressed that an ecosystem view of nature was required to capture the complex character of natural communities (Forbes, 1887; Tansley, 1935, 1939).

A. The Influence of Herbivores on Plant Diversity

In a way analogous to the effect of some marine keystone predators on their invertebrate prey (Paine 1966), some generalist herbivores may regulate plant diversity by selective grazing. Darwin (1859) recognized that grazing could maintain otherwise dominant competitors in plant communities at densities which allowed poorer competitors to survive. Although the details of grazing—mediated plant species richness and community dynamics can be complex (Crawley, 1983), it has become almost dogma that grazing by sheep and cattle has a dramatic influence on plant communities in pasture (Jones, 1933; Chadwick, 1960; Harper, 1977; Rawes, 1981), and can, in some cases, prevent the succession of grassland to woodland.

Artificial grazing systems, ancient though some may be, can only illuminate the mechanisms underlying grazing-mediated changes in plant diversity, not their profusion in natural ecosystems. There are, however, some examples of major changes in vegetation structure in natural and semi-natural systems which attest to the influence of some herbivores on plant community dynamics. Rabbits, for example, can convert *Calluna* heath, which is generally very species poor, to grassland (Farrow, 1925), while the occasional invasion of elephants onto the lower slopes (below 4000 m) of Mount Kenya, can convert this tropical alpine environment, usually dominated by giant tree-like forms of *Lobelia* and *Senecio*, into grassland (Mulkey *et al.*, 1984). Natural regeneration in forests can be affected by herbivory, and the exclusion of white-tailed deer, *Odocoileus virginiana*, from climax *Pinus resinosa* forest in Minnesota, dramatically favors re-establishment of trees and shrubs severely inhibited by high deer populations (Ross *et al.*, 1970).

Similar effects are seen in tropical forests as humans encroach on grazing mammal habitat, and the mammals recede. Dirzo and Miranda (1990), for example, have suggested that the greater abundance and lower diversity of seedlings and saplings in the tropical forest at Los Tuxtlas compared with other, less disturbed forests in Mexico, are due to the loss of large mammal herbivores such as tapirs, peccari, and deer.

In marine systems, herbivory on seaweeds is intense, and often the primary factor affecting the distribution and abundance of marine plants (Lubchenco and Gaines, 1981; Hay, 1981a, 1985, 1991; Lewis, 1986; Schiel and Foster, 1986). Generalist feeders such as fishes and sea urchins are the herbivores of greatest importance (Hay, 1985, 1991; Lewis, 1986), and can consume nearly 100% of local algal production (Carpenter, 1986). Other groups, such as the mesograzers (some amphipods, crabs, and polychaetes), appear to have less impact on primary production (Hay *et al.*, 1988; Hay, 1991), except during occasional outbreaks (Tegner and Dayton, 1987). Rather than dominant species, marine systems may support dominant guilds of herbivores (M.E. Hay, personal communication, 1990). Certainly, the diffuse herbivory (Fox, 1981) of fish and urchins can dominate plant community structure on some coral reefs (Carpenter, 1986).

Birds, too, can dominate the vegetation within a community. Lesser snow geese, for example, are considered keystone species on the west coast of Hudson Bay (Kerbes *et al.*, 1990), and they determine the structure and species composition of the coastal plant community. Grubbing for rhizomes by the geese produces patches of bare peat susceptible to erosion, and results in low plant species diversity. In contrast, above-ground grazing by the geese increases plant diversity (Kerbes *et al.*, 1990). Lesser snow geese are the only keystone species of which I am aware that can both increase and decrease the diversity of the trophic level below their own, depending upon where they forage, and what type of foraging they do.

Perhaps one of the most dramatic examples of a herbivore-dominated plant community is the *tortoise turf* of the Aldabra atoll. Aldabra is a coral atoll made up of four discrete islands separated by lagoon channels. The giant tortoise (*Geochelone gigantea* Schwigger) populations on these islands show relatively plastic reproductive strategies determined primarily by density (Gibson and Hamilton, 1984). The unusual lack of feral predators on Aldabra, compared with other islands in the area (Arnold, 1979; Stoddart and Peake, 1979), is probably responsible for the survival of the giant tortoise populations. Rapid growth through the vulnerable hatchling stage, and the large size of adults, means that predation rates are relatively low. Although the populations on different islands differ in size and reproductive strategy, most are apparently regulated by near-scramble competition (Gibson and Hamilton, 1984), with negatively density-dependent reproduction and recruitment.

Tortoise population regulation probably depends on an interaction between weather and primary production (Swingland, 1977; Swingland and Lessels, 1979). Certainly, these herbivores greatly modify their environment, and are responsible for the death of many trees and shrubs, as well as considerable soil erosion (Merton *et al.*, 1976). Their grazing results in the production of the close-cropped tortoise turf (an assemblage of dwarf grass, sedge, and herb species (Grubb, 1971)), which contributes about 61% of the tortoises' food source (Gibson and Hamilton, 1983). Tortoises also depend on shade to prevent death from heat exposure (Swingland and Lessels, 1979), and their foraging may be limited by their own destruction of shade trees (Merton *et al.*, 1976), as well as by food limitation (Swingland, 1977; Gibson and Hamilton, 1984).

The giant tortoises of Aldabra, therefore, have a *pervasive influence* on the vegetation of their habitat, and so fulfill the requirements of a keystone species as described in Section III. Do they, however, influence the distribution and abundance of other animal species as a result of their impact on the plant community? That is, does the effect of *G.gigantea* on vegetation diversity feed back through the trophic system as a major determinant of animal–plant community structure (Sections I,II)? Exclosure experiments have shown that, within 4 years of eliminating tortoise grazing from open mixed scrub, the 1–5 mm high tortoise turf is replaced by a sward 10–20 cm high, with patchy regeneration of shrub seedlings and saplings (Gibson *et al.*, 1983). Yet the effects of this rapid, extensive succession on other inhabitants of open mixed scrub have never been investigated. One might imagine that removal or augmentation of tortoises should have a dramatic effect on invertebrate and other vertebrate herbivores on the atoll, but the impact of exclusion on animal–plant community structure has never been assessed.

The tortoises of Aldabra atoll provide one of the best examples of what I call *first-order* evidence for the impact of keystone herbivores on communities, that is, a dramatic influence on primary producers. There is as yet, however, no *second-order* experimental evidence that shows cascading effects of tortoise feeding up or down the trophic system. Although it may be very likely that such cascading effects exist (Chapters 1 and 6), they have never been quantified (Gibson, personal communication, 1990).

Similarly, although there are considerable data that support the *pervasive influence* of some marine herbivores on algal communities on reefs (Lubchenco and Gaines, 1981; Carpenter, 1986; Hay, 1991), little is known about interactions among marine herbivores mediated by changes in plant diversity. One unusual example is the effect of damselfish on algal diversity (Hixon and Brostoff, 1983). Reef damselfish manage *gardens* of palatable algae from which they selectively remove unpalatable species; algal diversity can be higher in their gardens than in the background community. This increased palatability and diversity has little impact on other herbivores, however, because damselfish aggressively defend their gardens from other grazers.

Overall, one might expect that changes in species diversity would influence specialists more than generalists, since increases or decreases in the relative abundance of a particular alga on a reef should affect any species that relies on that alga more than they would a generalist feeder. It appears, however, that marine generalist herbivores are often food limited (Hawkins and Hartknoll, 1983; Morrison, 1988), and that specialization in mesograzers may have evolved in response to severe predation pressure (examples in Hay, 1991). Hay (1991) argues that the *enemy-free space* provided by feeding on unpalatable species of algae has been selectively advantageous, and that marine mesograzers may be predator limited in a way analogous to terrestrial insect herbivores (Strong *et al.,* 1984; but see Faeth, 1987b; Price, 1990; Section V).

Some studies demonstrate good first- and second-order evidence for herbivore-induced changes in plant species diversity, and subsequent changes in animal communities. Variation in sheep stocking levels, for example, can change both sward diversity and invertebrate diversity in grassland (Hutchinson and King, 1980). High levels of grazing, which might have been expected to maximize plant, and therefore, arthropod diversity, decreased the abundance of all invertebrates except ants. Low grazing levels resulted in low diversity too, presumably because swards were dominated by competitively superior plant species. Only intermediate stocking densities, which maximized both species *and* architectural diversity, caused increases in grassland invertebrate species diversity (Hutchinson and King, 1980).

Rodents and ants, because they often consume seeds and/or small plants, can strongly affect plant recruitment and species diversity (Davidson *et al.,*

1984; Brown and Heske, 1990). In the Chihuahuan desert–grassland transition zone, for example, three species of kangaroo rat are prime determinants of plant species diversity and patterns of soil disturbance. When this keystone rodent guild is excluded from the desert shrub habitat, it is transformed into grassland (Brown and Heske, 1990). In a classic feedback loop, the influence of kangaroo rats on the desert–grassland transition determines the species diversity of both other rodent species (J.H. Brown, unpublished data 1990), and granivorous birds (Thompson *et al.*, 1991).

Exclusion experiments have also highlighted the importance of grazer control of algae in freshwater systems. In an Ozark mountain stream, for example, cyanobacterial felts are overgrown by turfs of benthic diatoms in a matter of days, when grazing fish and invertebrates are excluded (Power *et al.*, 1988). When fish are allowed to recolonize, the diatom turfs are stripped within minutes, and the cyanobacterial felts reemerge in about 11 days. Power *et al.* (1988) have suggested that there may be cascading effects up the trophic system, since the relative abundance of cyanobacteria and diatoms can determine water clarity, the amount of food for suspension feeders, and rates of nitrogen fixation.

As one of the best examples of the effects of a keystone herbivore, Eltringham (1974) reported changes in the large mammal fauna of Rwenzori National Park, Uganda, following eradication of hippopotamus. High population densities caused erosion and loss of vegetation cover to the degree that it appeared that the park would shortly become a desert. Elimination of hippopotamus from one area of the park by shooting caused regeneration of plant diversity (Laws, 1968; Thornton, 1971) and increases in buffalo, elephant, and waterbuck density. Densities of bushbuck and warthog decreased, but, overall, large mammal density increased 167%, and biomass increased 301%.

B. The Influence of Herbivores on Plant Abundance

Very often, more than one species of animal utilizes or is dependent on a single plant species. Herbivores that dramatically change the abundance of a plant species (or some plant part on which other animals specialize), have the potential to act as keystone herbivores by changing the distribution and abundance of community species that depend on that plant resource. It is logical to assume that the populations of specialist herbivores should be more responsive to a decline in the abundance of their host plant resource, but generalists can also be influenced by plant-mediated interactions in communities (Hunter, 1987a; Hunter and Willmer, 1989).

Empirical data on the effects of herbivores on plant population dynamics are rare in terrestrial systems (Crawley, 1983, 1989) and, since little is understood about the influence of herbivory on plant recruitment, it may be too early to reach conclusions about the general prevalence of interactions among herbivores mediated by changes in plant populations. The most obvious cases of herbivore-mediated changes in plant population dynamics are the complete, or almost complete, exclusion of plant species, and this is particularly well documented in marine systems (Hay, 1981a; Lewis, 1986; Carpenter, 1986; Morrison, 1988). Terrestrial plants, too, can be excluded by herbivores. The western pine beetle *Dentroctonus brevicomis,* for example, tends to exclude *Pinus ponderosa* from mixed forests on the western slopes of the Californian Sierras (Stark and Dahlsten, 1970). Likewise, two native species of *Scolytus* beetle almost eradicated the elm from southern England, although the nonnative fungal pathogen *Ceratocystis ulmi* introduced from North America (Gibbs, 1978) facilitated the exclusion (Speight and Wainhouse, 1989).

Plant species replacements, as a result of preferential feeding by herbivores, may be a common phenomenon. Populations of white-tailed deer, for example, have been shown to encourage replacement of *Pinus strobus* by *Betula papyrifera* in ancient *Pinus resinosa* forest in Minnesota (Ross *et al.,* 1970). The replacement is caused by both the preference of the deer for pine, and the greater ability of birch to withstand browsing. White-tailed deer may also inhibit regeneration of hemlock, *Tsuga canadensis,* forest— intense browsing favors growth of sugar maple, *Acer saccharum,* saplings over hemlock (Anderson and Loucks, 1979).

According to Crawley (1989), a recurrent theme in plant–herbivore population dynamics is that plant recruitment is usually not limited by invertebrate herbivores, even when the herbivore populations are themselves food limited (but see Hawkins and Hartnoll, 1983; Lubchenco and Gaines, 1981). Some arthropods, however, can cause the gradual loss of particular plant species, and their replacement with others. There is good evidence that the introduction and spread of the gypsy moth, *Lymantria dispar,* in the northeastern United States is resulting in the replacement of some oak stands with red maple (Collins, 1961; Kegg, 1971, 1973; Campbell and Sloan, 1977), although the spatial scale involved precludes the use of experimental exclosures. Even an extreme polyphage like the gypsy moth, however, has preferred hosts, and larval growth rates and defoliation levels are higher on oak than on red maple (Doane and McManus, 1981). The general resistance of red maple to the gypsy moth, therefore, is considered the best explanation for the apparent replacement of oak in Connecticut.

The nutritional quality of the host tree is not sufficient, however, to explain tree species replacement by the gypsy moth. Early larval growth rates are higher, for example, on bigtooth aspen, *Populus grandidentata,* than on some oak species (Hunter and Schultz, unpublished data, 1989), yet sites susceptible to gypsy moth outbreak (and consequently tree mortality) are not characterized by a high density of aspen (Valentine and Houston, 1981). Allelochemicals abundant in oak foliage inhibit the pathogenicity of a gypsy moth nuclear polyhedrosis virus (Keating and Yendol, 1987; Keating *et al.,* 1990) and this, in combination with other factors, may exacerbate gypsy moth outbreaks (Schultz *et al.*, 1990; Chapter 7), which result in tree mortality, and tree species replacement.

Of course, herbivores need not exclude a plant species to have an impact on its population dynamics. In some cases, herbivores restrict the distribution of plants and, in concert with other ecological forces, reduce them from their *fundamental* to their *realized* niche (*sensu* Hutchinson, 1957). Some marine algae, for example, thought to be *deep-water specialists,* grow up to 400 times better in shallow areas when herbivores are excluded (Hay, 1981a; see Lubchenco and Gaines, 1981 for other examples). Likewise, the distribution of the Saguaro cactus in the Sonoran desert is restricted by an interaction between rodents (especially woodrats) and spiney trees (Turner *et al.*, 1969). The cacti require shade trees for successful germination, but unless those trees are spiney, seedlings are rapidly removed by woodrats. Similarly, the distribution of goldenbush on parts of the California coast is limited by high levels of flower and seed predation that lower plant recruitment (Louda, 1982a,b).

Variation in the effects of herbivores on plants caused by spatial and temporal abiotic and biotic heterogeneity may be common. I have previously described the influence of infrequent elephant invasions on the *Lobelia–Senecio* plant community below 4000 m on Mount Kenya (Section IV,A.). There is an equally dramatic effect caused by the rock hyrax, *Procavia johnstoni,* above 4000 m. Normally, hyrax exert low levels of herbivory on giant *Lobelia* species but, after several years of drought, their other host plants diminish, and they concentrate on *Lobelia.* The result of this herbivore–climate interaction is the occasional devastation of *Lobelia* populations (Young, 1984, 1985; Young and Smith, 1987).

Herbivores may also change the age distribution of their host plants. Defoliators that attack only plants of a particular age are a common feature of insect–plant interactions (Martin, 1966; Niemela *et al.*, 1980; Kearsley and Whitham, 1989; Price *et al.*, 1990), and individuals of an unsuitable age class may be *transparent* to foraging animals (Chapter 1). There are good examples of animal species that can influence the abundance, life history, and diversity of primary producers by, for example, specializing on seeds (Brown and Heske, 1990).

The effect of an herbivore on plant population dynamics is not always direct. Futuyma and Wasserman (1980) have argued that some rare plant species may remain rare because they are defoliated by animal species that exist at high population density on some other food source. They found that the fall cankerworm, *Alsophila pometaria,* maintained high populations on the scarlet oak, *Quercus coccinea,* in New York State. The caterpillars then transferred to the younger foliage of the patchily distributed white oak, *Quercus alba,* trees, causing severe defoliation. They suggest that "it must frequently be the case . . . that uncommon species are prevented from increasing if they are fed on by a generalized predator (herbivore) that is maintained at high density by a common species of prey (host plant)" (p. 921).

These few examples illustrate that some herbivores can influence the abundance of a particular plant species within a community. To generalize, it appears that arthropods are less likely to influence populations of plants in terrestrial systems than are vertebrates, and that generalists may have a greater impact than specialists if only because they can exclude rare plants while maintaining high density on common plants. Specialists may have indirect effects on their plant's dynamics, however, by lowering its competitive ability with other plant species (Crawley, 1989).

Is there any evidence that keystone herbivores can influence animal communities by reducing the population of a plant species that is an important resource for other herbivores? As with effects mediated by changes in plant communities, the evidence is relatively weak. A logical hypothesis might be that the loss of oak species by gypsy moth defoliation, and the replacement of pine with birch and hemlock with sugar maple, as a result of browsing by white-tailed deer (see above), should certainly influence specialist herbivores and possibly influence generalist herbivores that use those species for food or shelter. I know of no studies, however, that have considered, let alone demonstrated, community-level responses to the loss of these tree species. Although a few studies have investigated the effects of extensive defoliation and plant species loss on ecosystem parameters, such as changes in hydrology (Bethlahmy, 1975) and changes in litter composition, nutrient cycling, and above-ground production (Pollard, 1972; Batzer, 1973; McGee, 1975), the ramifications of these changes for animal communities are almost completely unknown.

Redfearn and Pimm (1987) consider it unlikely that the loss of plant species from complex communities, particularly those containing a high proportion of generalist herbivores, will cause significant further losses in the trophic web. As evidence, they cite the near extinction of chestnut, *Castanea dentata*, from eastern North America (Krebs, 1978). Although chestnut used to occupy more than 40% of the canopy in some areas, its loss has had little noticeable impact on the community, other than the probable departure of seven species of insect specialist. Most other chestnut feeders were generalists (Opler, 1978).

However, it seems unlikely that the spread of the gypsy moth through the northeastern United States would have been so rapid if chestnut were still a major component of the deciduous forest canopy (J. C. Schultz, personal communication, 1989). The gypsy moth, in turn, can influence tree species diversity and the community structure of late-season herbivores (author's unpublished data, 1989). Rigorous experiments that mimic herbivore-related losses of plants from natural communities, and measure the effects on other herbivores, are almost unknown. An increasing number of successes in the biological control of weeds, however, may provide some useful answers (Crawley, 1989).

There are some notable exceptions to this regrettable pattern. Blakely and Dingle (1978), for example, report the loss of two species of milkweed bug from a community as a result of defoliation by the monarch butterfly, *Danaus plexippus*. The monarch feeds on *Asclepias curassavica* and *Calotropis procera* on Barbados, and has almost eliminated the former by maintaining high populations on the latter (see Futuyma and Wasserman, 1980). The two species of *Oncopeltus* bug are unable to feed on the seeds of *Calotropis* because the pod walls are too thick. The near loss of *Asclepias*, therefore, has resulted in the demise of two species in the sapsucking guild of herbivores on milkweed in Barbados.

Hails and Crawley (1991) failed to find any effect of the invasion of the acorn gall wasp *Andricus quercuscalicis* into the gall-forming guild on the English oak, *Quercus robur*. Despite its considerable impact on acorn production (up to 60% loss in some years), and strong evidence that *A. quercuscalicis* is food limited in its agamic generation, the authors could demonstrate no interaction between *A. quercuscalicis* and other cynipids on oak. This is at first surprising, given that the community of gall-forming cynipids on *Q. robur* is species rich and highly interconnected (Askew, 1961). However, no other gallers feed directly on acorns, and parasitism on *A. quercuscalicis* by generalists (which might result in *apparent competition* by increasing the level of natural enemies in the community (Holt, 1977)), is almost absent. Crawley has suggested (1983, 1989) that exploitation of acorns by *A. quercuscalicis* or the weevil *Curculio glandium* might affect vertebrates such as jays, rabbits, and wood pigeons, which utilize *Q. robur* acorns, and this has not yet been investigated.

Although the evidence is poor, therefore, that herbivores can influence species up the trophic system by affecting plant population dynamics, that may be simply because "we have so little information on the regulation of plant populations in the wild." (Crawley, 1989, p. 531). It seems premature, then, to make any firm conclusions about community processes dictated by variation in plant abundance. This will not change until experiments compare the death rate and recruitment of plant species in the presence and absence of herbivores, and measure changes in the density of other animal and plant species in the community simultaneously.

C. The Influence of Herbivores on Plant Growth Form

Interactions among animals on plants, mediated by changes in plant growth form, are rarely considered in natural communities (Hunter, 1987a; Hunter and Willmer, 1989). Many authors, however, consider plant structure (e.g., the presence of trichomes, leaf toughness, plant architecture) as important to the colonization and success of animals on plants (Southwood,

1973; Wellso, 1973; Gilbert and Smiley, 1978; Lamb, 1980; and especially Potter and Kimmerer, 1988). Plant growth form can influence the ability of species to reach the plant (Hagley *et al.,* 1980; Kennedy, 1986), consume the plant (Turnipseed, 1977; Berenbaum, 1978; Raupp, 1985; Nicols-Orians and Schultz, 1989; Duffy and Hay, 1990), maintain homeostasis (Tahvanainen, 1972; Willmer, 1982; Hunter and Willmer, 1989), and can dictate their interactions with natural enemies (Heinrich and Collins, 1983; Damman, 1987; Kareiva and Sahakian, 1990).

In one recent study, the aphid *Aphis verbascae* was shown to choose preferentially a relatively novel, nutritionally poorer, but smooth host plant species (*Buddleia davidii*) over its natural trichome-covered host (*Verbascum thapsus*). Although nutritionally poorer, *B.davidii* was more easily colonized by aphids, and experimental removal of the trichomes reversed host plant choice (Keenlyside, 1989).

Herbivores that could dramatically change characteristics of plant growth form as they utilized plants, therefore, would have the potential to influence the distribution and abundance of other community herbivores that used that plant. However, even first-order evidence of this type of interaction (see Section IV,A), seems scarce in terrestrial systems (the production of regrowth foliage is considered in Section IV,D). Changes in plant growth form in response to herbivory are well documented in marine systems (Duffy and Hay, 1990), and it is not clear whether the apparent difference between land and sea reflects a real divergence between systems, or a lack of research effort in terrestrial communities.

A few terrestrial examples of herbivore-mediated changes in plant structure have, however, been documented. Chronic levels of insect herbivory, for example, can modify plant shape in the pinyon pine, *Pinus edulis* (Whitham and Mopper, 1985). Likewise, browsing by elk and deer on *Ipomopsis aggregata* alters the plant architecture, increasing the number of inflorescences from one to four (Paige and Whitham, 1987). Feeding by giant tortoises on Aldabra atoll, as described earlier, produces *tortoise turf,* a plant community comprising 21 species of grasses and herbs (Grubb, 1971; Section IV,A). Tortoise turf contains a high proportion of genotypic and phenotypic dwarf plants (Merton *et al.,* 1976). Both natural selection and phenotypic plasticity have interacted with tortoise grazing to generate herbivore-induced plant growth forms.

Leaf rolling by some Lepidopteran larvae can also change the growth form of plants (Henson, 1958; Damman, 1987; Hunter, 1987a) and, although the creation of leaf rolls may be important in intraspecific interactions (Damman, 1987; Hunter, 1987b), I know of only one system (the phytophagous insects of the pedunculate oak, *Q. robur*) in which a change in leaf architecture caused by leaf rolling has mediated interactions at the community level (Hunter, 1987a; Hunter and Willmer, 1989; Hunter and West, 1990). This system is considered in detail in Section V.

In some marine communities, grazing-mediated changes in plant growth form are considered antiherbivore defenses (Duffy and Hay, 1990). Growth-form plasticity allows some algae to form short, dense, highly branched turfs in grazed areas (Hay, 1981b), and some reef seaweeds have two morphologically distinct forms, the appearance of which depends upon grazing pressure (Lewis *et al.*, 1987). The brown alga *Padina*, for example, changes from a prostrate turf to a rapidly growing blade-like form within 96 hr of excluding herbivorous fish. In its blade-like form, it is a superior space competitor against other benthic algae and sessile invertebrates, and release from grazing can allow it to overgrow and kill coral (Lewis, 1986; Lewis *et al.*, 1987).

Given that there is little evidence of herbivore-mediated changes in plant structure, it is no surprise to find few studies that consider it an important component of community dynamics. The architectural modification caused by insect herbivory on the pinyon pine (Whitham and Mopper, 1985), however, may have cascading effects back up the trophic system. Since the change in pine structure also results in the loss of female function by the tree, insect herbivory can influence the tree's seed dispersal agent, the pinyon jay (Christensen and Whitham, unpublished data, 1991).

Section V considers in detail one case in which changes in plant architecture caused by herbivory play a fundamental role in herbivore community structure. General conclusions on the importance of structural modification by herbivores on animal–plant community processes, however, await further investigation. I would suggest, nonetheless, that they might be more common in marine systems than in terrestrial systems, if only because of the very rapid plastic growth response of some algae.

D. The Influence of Herbivores on Plant Chemistry

Research in animal–plant interactions, particularly that which has focused on terrestrial arthropods, has been dominated by investigations into the role of secondary plant metabolites in determining the distribution, abundance, and evolution of herbivores on plants (Ehrlich and Raven, 1965; Feeny, 1970, 1976; Rhoades and Cates, 1976; Rosenthal and Janzen, 1979; Coley *et al.*, 1985). One of the most important developments in this field has been the discovery that damage to plants by herbivores can induce changes in the quality and quantity of secondary compounds in plant tissue (Green and Ryan, 1972; Haukioja and Niemela, 1977, 1979; Edwards and Wratten, 1982, 1983; Raupp and Denno, 1984; Bergelson *et al.*, 1986; Hartley and Lawton, 1987; Rossiter *et al.*, 1988; Hartley and Firn, 1989; Renaud *et al.*, 1990).

Although the phenomenon of wound-induced change in plant chemistry is now well established, there is still some debate as to its evolutionary and ecological significance (Edwards and Wratten, 1983, 1985, 1987; Baldwin and Schultz, 1983; Fowler and Lawton, 1985; Bergelson and Lawton, 1988;

Hartley and Lawton, 1990). One viewpoint contends that wound-induced change in foliage chemistry is a defensive reaction by plants, which disperses grazing and grazers, thus spreading defoliation load for the plant, and increasing predation risks for herbivores (Edwards and Wratten, 1983, 1985, 1987). The opposing view suggests that the experimental evidence that dispersed grazing and herbivore movement are beneficial for plants and deleterious for animals is weak (Fowler and Lawton, 1985; Bergelson *et al.*, 1986; Bergelson and Lawton, 1988), and that damage-induced change in foliage chemistry is a consequence of wound repair, possibly directed at plant pathogens (Hartley and Lawton, 1990).

Although it may be that some aspects of the debate can be resolved by considering the modular organization and hormonal regulation systems of plants (Haukioja *et al.*, 1990), a resolution is not critical to the current discussion. More important for my purposes here is that wound-induced changes in plants, whatever their evolutionary origins, have been demonstrated to influence the population dynamics of some animals that use plants (Croft and Hoying, 1977; Karban and Carey, 1984; West, 1985; Faeth, 1985, 1986; Harrison and Karban, 1986). While the mechanisms of population change are not always clear, there is evidence both for direct negative effects of chemistry on defoliators (Puritch and Nijholt, 1974; Niemela *et al.*, 1979), and interactions mediated at the third trophic level (Price *et al.*, 1980; Faeth, 1987a; Faeth and Bultman, 1986).

Wound-induced changes in leaf chemistry also can have positive effects on herbivores (Niemela *et al.*, 1984; Williams and Myers, 1984; Kidd *et al.*, 1985; Rhoades, 1985; Haukioja *et al.*, 1990), and Faeth (1987a) suggested that we should not be surprised that species respond differentially to chemical changes in plants. Some studies have shown that the effect of wound-induced change in plant quality on a particular herbivore can be a balance between positive and negative forces (Faeth, 1986; Hunter, 1987a), and that both chemical and physical heterogeneity introduced into the habitat by patterns of defoliation can radically alter interactions among species in a community (Hunter and West, 1990).

One further component of resource heterogeneity introduced into habitats, as a result of grazing by some herbivores, is regrowth foliage. Variable tree phenology and the production of new leaves in response to damage, can be treated as a separate resource dimension influenced by herbivory (Faeth, 1987a,b), but most of the documented effects of regrowth on herbivores relate to leaf chemistry, and it is convenient to consider it here.

By definition, regrowth foliage is younger than primary foliage, and young leaves are usually considered more nutritious than mature leaves (Mattson, 1980). Yet regrowth is often chemically different from primary foliage, at least when sampled at the same point during the season, and the responses of herbivores to regrowth foliage are variable. Both vertebrate and invertebrate herbivores have been shown to perform less favorably on regrowth foliage than on the primary foliage of some plant species (Bryant, 1981; Hunter, 1987a). In other systems, however, regrowth is a far superior food source, and can lead to dramatic increases in animal populations (Carne, 1965; Rockwood, 1974; Webb and Moran, 1978). Some herbivores may depend on grazing to continually generate young, nutritious foliage (Vesey-FitzGerald, 1960; Bell, 1970; McNaughton, 1976).

Whether through wound-induced modification of primary foliage, or through the generation of regrowth foliage, many herbivores appear able to create a chemical mosaic that represents a heterogeneous resource for other animals on plants. Can we isolate keystone herbivores that influence animal communities through their impact on plant chemistry? Most studies of wound-induced change in plants have considered the impact of such change on either the herbivore species that makes the damage (intraspecific effects), or on other species that utilize the same host plant (interspecific effects). Few have considered the possible influence of previous defoliation on herbivore communities as a whole.

Stan Faeth at Arizona State University has led the field in describing chemically mediated interactions among herbivore guilds (*sensu* Root, 1967) on their host plants. The combined defoliation of spring Lepidoptera, Coleoptera, Orthoptera and Hymenoptera on the evergreen oak, *Quercus emoryi*, has an overall negative impact on species in the leaf-mining guild (eight species recorded), which concentrate feeding after most leaf-chewing has finished (Faeth, 1985, 1987a). Although ovipositing females prefer to select leaves within trees which are intact, successful emergence of leaf miners is higher from undamaged than from damaged leaves, primarily because of differential rates of parasitism among leaf types (Faeth, 1987a). Spring defoliation appears to increase the concentration of condensed tannin in leaves (Faeth, 1987a), and artificial application of tannin on leaf surfaces increases parasitism rates (Faeth and Bultman, 1986). Condensed tannins do appear to reduce miner mortality from other causes such as pathogens, but the overall effect of spring defoliation on late-season leaf miners is negative (Faeth, 1987a). Recognition of subtle interactions among temporally separated herbivores in different guilds has prompted Faeth to call for a reassessment of the role of horizontal interactions in structuring communities of animals on plants (Faeth, 1987a,b).

In Faeth's system, the defoliation which influences the distribution and abundance of species of leaf miner on *Q.emoryi*, is done by a number of different spring leaf chewers. It may be inaccurate, therefore, to speak of a single keystone species in the community. This is also a recurring theme in tropical marine systems where several herbivore species, particularly herbivorous fishes, have a pervasive influence on algal growth, defense, distribution, and abundance (M. E. Hay, personal communication, 1990, Section IV,A). These combined effects may lead us to consider leaf-chewing insects and herbivorous fish or urchins as keystone guilds rather than keystone species in some communities.

V. A Case Study: The Search for Keystone Herbivores on the English Oak

The previous section considered the evidence for interactions within communities of animals on plants mediated by changes in four characteristics of plant resource distribution (plant diversity, abundance, growth form, and chemistry). In this section, I describe research carried out on the pedunculate oak, *Quercus robur*, in Wytham Woods, England, in which two of these four characteristics are modified by defoliation. In this system, there are two species of insect herbivore that regularly occur at high densities, causing significant leaf damage to their host plant. Here, I distinguish between their impact on (1) other members of the guild of early season leaf chewers on *Q. robur*, and (2) the community of late-season insect herbivores. I argue that their impact on the former is negligible, and their impact on the latter, overwhelming.

A. The System

Tortrix viridana (Lepidoptera: Tortricidae) and *Operophtera brumata* (Lepidoptera: Geometridae) are abundant spring defoliators on *Q. robur* in Europe. Larvae of both species hatch in late April/May, in approximate synchrony with oak budburst, and feed concurrently until pupation in June. Larvae of *O. brumata* are extremely polyphagous (Tenow, 1972), but their performance is highest on oak species (Wint, 1983). Larvae of *T. viridana* appear restricted to the two species of native oak in Britain (Hunter, 1990). The adults of *T. viridana* emerge in July, and females oviposit on twigs in oak canopies where the eggs remain until the following spring; the adult females are primarily responsible for host location (Du Merle, 1981; Du Merle and Pinguet, 1982). Adult *O. brumata* emerge from November through January, and the wingless females oviposit toward the top of the first vertical object they encounter, usually a tree. Neonates then disperse on silk threads and are almost entirely responsible for host location (Briggs, 1957; Holliday, 1977).

The population dynamics of *T. viridana* and *O. brumata* in Britain are excellent examples of population change dictated by the (temporal) resource distribution of primary producers (Chapters 1 and 6). Fluctuations in numbers are determined primarily by the degree of synchrony between larval eclosion and oak budburst (Schutte, 1957; Satchell, 1962; Varley and Gradwell, 1968). Briefly, in years when caterpillars hatch before budburst, dispersal rates and starvation levels are thought to be high, and populations fall. In years when budburst coincides with, or precedes larval hatch, colonization levels are high and populations rise. Although some natural enemies exert a degree of density dependence in the system (Varley and Gradwell, 1968), the major source of population change for both species is the asynchrony between oak budburst and larval hatch.

Despite variation in population levels of these species, defoliation levels in

Wytham Woods are consistently high. The average leaf area removed from *Q. robur* annually is around 40%, generally higher than the same species in parkland (Crawley, 1985, 1987), and higher than in most other plant–herbivore systems (Howe and Westley, 1988). Leaf damage levels on individual trees can vary from insignificant to complete defoliation within a matter of meters. The phytophagous insect fauna of *Q. robur* in Britain is very species rich (Southwood, 1961; Strong, 1974), and the heterogeneous, high levels of defoliation by *T. viridana* and *O. brumata* make them excellent candidates for keystone species in the oak–insect herbivore community.

B. Interactions within the Guild of Early-Season Leaf Chewers on Oak

Given the close relationship between the population dynamics of both *T. viridana* and *O. brumata,* and *Q. robur* budburst (above), it should be no surprise to find that, within a year, there is a significant relationship between the calendar date of budburst of individual trees and the number of caterpillars of these species they support (Hunter, 1992; Fig. 2). Trees that burst bud early have high densities of *O. brumata* and *T. viridana* and, although the data shown are for mature trees, the relationship is similar for saplings (author's unpublished data, 1987).

If the presence of *T. viridana* or *O. brumata* on trees were determining the distribution or abundance of other spring leaf chewers, one might predict two things:

1. that the density of other chewer species would be negatively correlated with the abundance of *T. viridana* and *O. brumata,* and
2. trees with low densities of the two major defoliators should have a more diverse caterpillar fauna.

Neither of these predictions holds true. Table I summarizes relationships

Table 1 Relationships among the Densities of Early-Season Leaf Chewers on Ten Individual Oak Trees[a]

Leaf chewers	Correlation statistics
T. viridana vs pooled species	r = +0.480, n.s.[b]
O. brumata vs pooled species	r = −0.396, n.s.
T. viridana vs *Gypsonoma dealbana*	r = +0.597, n.s.
T. viridana vs *Coleophora sp.*	r = +0.709, p < 0.02
T. viridana vs *Carcina quercana*	r = +0.025, n.s.
O. brumata vs *Gypsonoma dealbana*	r = −0.100, n.s.
O. brumata vs *Coleophora sp.*	r = −0.029, n.s.
O. brumata vs *Carcina quercana*	r = +0.021, n.s.

[a] Mean densities from 3 samples per tree were regressed using simple correlation procedures after transformation of the original data.
[b] n.s., not significant.

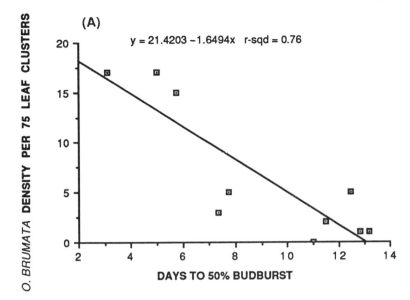

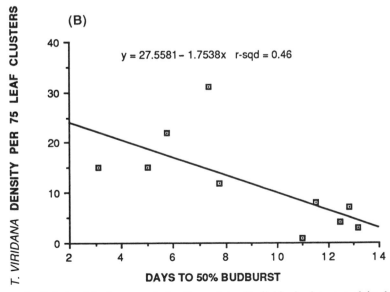

Figure 2 Relationships between the budburst date of individual oak trees and the densities of (A) *O. brumata* and (B) *T. viridana* on those trees. Budburst date is measured as the number of days from the start of budfurst in the oak population to the day when 50% of buds on an individual tree have green tissue exposed.

among early-season leaf chewers on *Q. robur* with respect to density on individual trees. Pooled data demonstrate that, as a guild, early-season chewer density on mature trees is unrelated to densities of either *T. viridana* or *O. brumata*. When the most common of the early-season herbivores are considered individually, no clear patterns emerge. From 15 correlations among species, 11 were positive, four were negative, and only one, statistically significant at the 2% level (given the high number of regressions performed, there is a greater risk that relationships significant at the 5% level will appear by chance alone). The significant relationship was a positive correlation between the density of *T. viridana* and an unidentified species of *Coloephora* (Lepidoptera: Coleophoridae). *Coleophora* is a mobile, case-bearing genus of Lepidoptera, generally considered to be a leaf miner rather than a chewer. The basis of this positive relationship is unknown.

Changes in the density of *T. viridana* and *O. brumata* on trees do not appear to influence chewer species diversity, either. As a natural test of this, I compared log abundance–rank curves (Whittaker, 1972; May, 1975) for mature trees and saplings (Fig. 3). Since log abundance–rank curves describe how a given number of individuals are distributed between different

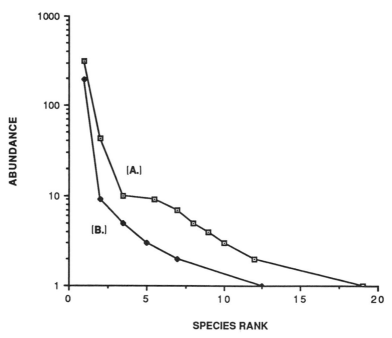

Figure 3   Abundance–rank curves depicting the species diversity of early-season leaf chewers on saplings and mature oak trees. (A) Represents mature trees (B) Represents saplings. Abundance is presented on a log(10) scale.

species in a community, a comparison of curves is a good test for differences in species diversity, and is broadly independent of sample size (Southwood, 1978). Mature trees have higher densities of all leaf chewers than do saplings and, although not so satisfactory as an experimental manipulation, differences in the shapes of the log abundance–rank curves for mature trees and saplings would be consistent with density-related changes in species diversity.

The log abundance–rank curves in Figure 3 confirm the lower density of leaf chewers on saplings. The shapes of the curves are typical of samples of insects on plants (Southwood, 1978)—the community contains a very few species at high density, and a much greater number of rare species. After transformation to straighten the lines, statistical comparison shows that the slopes of the lines are almost identical (F = 0.016, d.f. = 1,12, $p > 0.05$). This suggests that the species diversity of leaf chewers on saplings and mature trees is very closely matched, and therefore independent of overall defoliator density.

Does this mean that early season chewers on *Q. robur* do not interact at all? I tested this for the two major defoliators, *T. viridana* and *O. brumata* (Hunter and Willmer, 1989; Hunter, 1990), and found an unusual result. There was, as might be expected, *occasional competition* (*sensu* Strong *et al.,* 1984) between these species when defoliator density was very high, but it was based on plant-mediated habitat modification (Section IV,C), not food limitation. Larvae of *T. viridana* are leaf rollers and, at high densities, leaf damage by *O. brumata* both destroys the integrity of *T. viridana* leaf rolls and makes it harder for them to build new refuges (Hunter and Willmer, 1989). Most animals that maintain water balance by behavioral mechanisms are poor physiological regulators (Willmer, 1982; Knapp and Casey, 1986; Chauvin *et al.,* 1979), and *T. viridana* is no exception. After a few hours out of leaf rolls, *T. viridana* larvae desiccate and die (Hunter and Willmer, 1989), so the negative impact of *O. brumata* is mediated by wound-induced change in leaf structure, not resource limitation.

Overall, however, interactions within the early-season leaf-chewing guild on *Q. robur* appear to be infrequent, and *T. viridana* and *O. brumata* cannot be said to have a pervasive influence on the spring community of insect phytophages of which they are members. This is surprising, perhaps, given the unusually high levels of defoliation by these species on oak, and suggests that the existence of keystone herbivores in communities cannot be predicted purely on the basis of plant biomass consumed.

C. Interactions between Early-Season Leaf Chewers and Late-Season Herbivores on Oak

Given the effect defoliation can have on leaf chemistry and structure (Sections IV,C,D), it is reasonable to suppose that the extensive leaf damage caused by *T. viridana* and *O. brumata* on *Q. robur* in spring might influence the distribution and abundance of late-season herbivores that feed afterwards. The evidence for this has been reviewed recently (Hunter and West, 1990), and strongly suggests that the two spring defoliators are keystone herbivores in the late-season community. There is good evidence that the guilds of leaf miners (West, 1985), leaf chewers (Hunter, 1987a), and sapsuckers (Silva-Bohorquez, 1987) are influenced in variable and complex ways by spring leaf damage, and that interactions between early- and late-season phytophagous insects can vary from competitive, through neutral, to commensal (Hunter and West, 1990).

The leaf-mining guild, particularly species in the genus *Phyllonorycter*, are negatively influenced by spring leaf chewers (West, 1985). West demonstrated that the late-season phenology of leaf miners is probably selectively advantageous because, although they are forced to feed on nutritionally inferior foliage, it allows miners to avoid direct competition with *T. viridana* and *O. brumata*. Nonetheless, late-season miners still suffer as a result of spring defoliation. Although their distribution among leaves suggests that they avoid damaged leaves where possible, the extensive defoliation by *T. viridana* and *O. brumata* requires that many develop on damaged leaves. Survival to pupation is significantly lower on damaged than undamaged leaves and, unlike other studies on oak leaf miners (Faeth, 1987a; Faeth and Bultman, 1986), seems to be mediated by direct chemical change rather than by increased rates of parasitism (West, 1985).

In contrast, the effect of spring defoliation on the guild of late-season chewers is usually positive (Hunter, 1987a). Although there is a detectable negative effect when late-season chewers are enclosed on foliage that has been damaged by *T. viridana* and *O. brumata* in spring, natural populations of chewers are

1. higher on trees with increasing levels of leaf damage;
2. higher in areas of the canopy with the highest defoliation; and
3. higher on the most damaged leaves within a canopy.

Since the most common late season chewers on *Q. robur* are leaf rollers, and construct their refuges faster on damaged leaves than on undamaged leaves, it would appear that the negative effect of a wound-induced decrease in the nutritional quality of leaves is more than compensated for by a wound-induced increase in habitat quality (Hunter, 1987a).

At the very highest levels of spring defoliation, most of the leaves encountered by late-season chewers are regrowth leaves and, at least for this guild on *Q. robur*, these are nutritionally much poorer (Hunter, 1987a). Populations of late-season chewers are lowest on these trees, demonstrating that the effects of *T. viridana* and *O. brumata* on chewers are nonlinear—they switch from positive to negative at some threshold defoliation level (Hunter and West, 1990).

Lastly, the plant-mediated interaction between spring defoliators and late season sapsuckers on *Q. robur* is the opposite of the leaf-chewer interaction (Silva-Bohorquez, 1987). The aphid *Tuberculoides annulatus,* for example, performs poorly on damaged primary leaves, well on undamaged primary leaves, and best of all on regrowth leaves. Increasing spring defoliation, therefore, decreases the population density of aphids on trees up to some threshold. Above that threshold, the dominance of regrowth leaves favors aphids, and populations increase (Silva-Bohorquez, 1987). The relationship between spring leaf chewers and late-season sapsuckers, therefore, is another example that includes nonlinear interaction terms in community dynamics (Hunter and West, 1990).

The general conclusions of this case study are presented in Figure 4, which summarizes interactions among herbivore guilds on *Q. robur*. In overview, *T. viridana* and *O. brumata* certainly have a pervasive influence on the distribution and abundance of three major insect herbivore guilds on *Q. robur*, and fulfill the criteria of keystone herbivores established in Section III. Their influence on early-season leaf chewers in negligible, and their effects on late-season herbivores are probably not mediated by changes in the *quantity* of plant resource—defoliation levels on oak are compensated for, to some degree, by regrowth in proportion to their loss (Crawley, 1983, although regrowth leaves may be smaller, and their production may be dependent on when the original defoliation occurs). However, the interactions among species that predominate in this community are mediated by changes in resource *quality,* either the nutritional quality of leaves, or the quality of the host plant as a refuge from adverse weather and natural enemies.

VI. Discussion and Conclusions

In a search for keystone herbivores that influence the distribution and abundance of other animal species in a community through plant-mediated interactions, we are required to demonstrate two things. First, we must show that herbivores can influence plants as resources for animals—that by consuming and/or living on plants, herbivores can modify the heterogeneity that is such a feature of plants and plant communities. Second, we have to show that this additional heterogeneity influences the distribution and abundance of other animals, and is a major determinant of community patterns.

I think the case for the former, the first-order evidence for the existence of keystone herbivores, is overwhelming. The examples described above illustrate that herbivores can influence four important parameters: plant species diversity, plant abundance, plant growth form, and plant chemistry, which historically are considered important in animal–plant interactions and community structure (Southwood, 1960; MacArthur and Wilson, 1967;

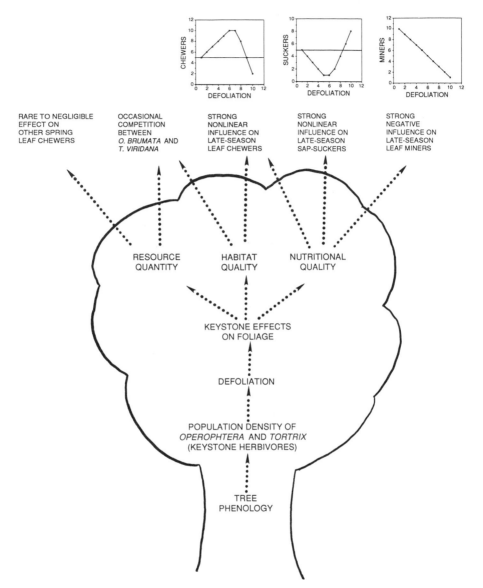

Figure 4 Consequences of defoliation within the community of insect herbivores on *Q. robur* in wytham woods, england. *O. brumata* and *T. viridanar* act as keystone herbivores in the late-season community but not in the early-season community.

McNaughton, 1976; Lubchenco and Gaines, 1981). In some cases, the influence of herbivores is such that their exclusion or introduction to an area alters more than one of these plant parameters (Gibson *et al.*, 1983; Lewis *et al.* 1987; Hunter, 1987a) and these, perhaps, are the most likely places to look for the effects of keystone herbivores.

Second-order evidence implicating herbivores in structuring communities is much rarer. With few exceptions, the responses of herbivore species to plant-mediated changes in resource distribution and abundance are, at best, examined on a species-by-species basis and, at worst, not considered at all. There is no compelling evidence to suggest that keystone herbivores are a common feature of animal–plant communities, and no compelling evidence to the contrary. What is most striking is the general lack of studies considering the issue at all. I suspect this may be owing, in part, to a view that communities are intractable units for experimentation, a view that led to a preponderance of single species-by-species studies in the 1980s. Although this has been valuable in that it fostered experimental rigor and increased statistical accuracy in studies of animal–plant interactions, it may be timely to apply the same degree of rigor to community issues, so that we are not left with a collection of *special case* studies with no means of relating them together. The case study presented here (interactions among the arthropod herbivores of *Q. robur*) shows that detailed mechanistic studies of animal–plant interactions can address issues of community structure.

In what kinds of communities do herbivores have a pervasive influence on other animal species? In this review, I have given examples of fish, reptiles, birds, large and small mammals, and arthropods, which are major determinants of plant heterogeneity and have been shown to influence community dynamics. Although taxonomic boundaries appear to have little relevance, there may be two classes of species to look for. The first may be herbivores that are food limited, and consume a significant percentage of the annual biomass of primary producers in their community. These are *resource quantity* herbivores, most likely to exert an influence by changing the abundance or diversity of host plants consumed by other animals. Examples from this review include giant tortoises on the Aldabra atoll (Merton *et al.*, 1976; Gibson *et al.*, 1983), nesting birds (Kerbes *et al.*, 1990), large mammal grazers such as hippopotamus, tapirs, and peccari (Eltringham, 1974; Dirzo and Miranda, 1990), and the combined effect of some tropical marine fish (Hay, 1991). These species, either by size, isolation, defense, or combined force of numbers, appear to have escaped regulation by natural enemies, and exploit their food source to near its carrying capacity. These are features considered rare among some groups of herbivores (Hairston, *et al.*, 1960; Orians and Paine, 1983; Strong *et al.*, 1984), and phytophagous insects, for example, are unlikely to regularly influence communities in this way.

However, even resource depletion and food limitation are not sufficient criteria for plant-mediated interactions among herbivores. Hails and Crawley (1991) have, as yet, found no impact of female flower limitation in *A. quercuscalicis* on other gall species on *Q. robur.* This may be because oak recruitment is not acorn limited (Crawley, 1989). The influence of herbivory on plant population dynamics, in general, is poorly understood, and the potential effects of keystone herbivores on animal–plant communities mediated by changes in plant abundance are not yet clear.

The second class of herbivores that are promising candidates for community level interactions are those that, even at low density, exert an influence on plant heterogeneity. These are *resource quality* herbivores; they change either the chemistry or structure of plants in such a way as to affect the community of animals that use those plants. It is remarkable, for example, that the distribution and abundance of the leaf-mining guild on *Quercus emoryi* could be determined by defoliation levels that averaged 5% leaf area removed each year (Faeth, 1985). This level of defoliation is well within the range considered normal (3 to 10%) for most plant communities (Bray, 1964; Leigh and Smythe, 1978; Nielsen, 1978; Schowalter *et al.,* 1986; Howe and Westley, 1988), and the occurrence of interactions among herbivores mediated by changes in plant chemistry and structure may be more common than previously realized (Faeth, 1987a,b).

There is a lack of field studies that consider (1) the natural occurrence of physical and chemical changes in plant species and communities caused by herbivores, and (2) the ramifications of these changes for other animals that use plants. If we accept that the distribution of plant resources is a fundamental factor shaping animal–plant interactions and, in turn, that herbivory can influence plant resource distribution patterns, we should endeavor to (1) measure appropriate indices of resources for animals (quality as well as quantity, see Chapter 8), and (2) expand empirical studies of animals on plants to consider changes in distribution and abundance at the community level.

We should not expect the effects of keystone herbivores to be invariant. As with almost all species interactions (Dunson and Travis, 1991), abiotic factors can interact with the effects of keystone species to influence community structure. In one recent example, Hobbs and Mooney (1991) demonstrated that the impact of gopher disturbance on plant recruitment and diversity in serpentine grassland was dependent on levels of rainfall. Future studies should consider the range of environmental variability over which keystone species dominate community structure.

There is still considerable debate among ecologists on the relative importance of natural enemies and plant parameters in determining patterns of plant use by herbivores (Barbosa, 1988; Bernays and Graham, 1988; Courtney, 1988; Fox, 1988; Janzen, 1988; Jermy, 1988; Rausher, 1988; Schultz, 1988; Thompson, 1988), although a balanced view has emerged in some cases (Rossiter, 1987; Carpenter *et al.,* 1985; Carpenter and Kitchell,

1988; Hay, 1991). Given that autecological factors, such as nutrients and weather, are primary determinants of plant distribution, abundance, and carrying capacity, it seems reasonable to overlay other ecological forces on top of these patterns to develop a model of population (Chapter 6) and community (this chapter) processes.

I have argued here that there are feedback loops that generate heterogeneity as energy flows through the trophic web. The first *feedback loop* in the system is presented by herbivores, which can modify the heterogeneity in plant communities that determines the distribution of other herbivore species. This may be the most important loop in some systems (Gibson *et al.*, 1983; Faeth 1985, 1987a; Hunter, 1987a; Hunter and West, 1990), and of little importance in others (Vince *et al.*, 1976; Van Dolah, 1978; Nelson 1979a,b; Stoner, 1979; Edgar, 1983a,b). A second important feedback loop is the action of natural enemies that, by influencing population dynamics and community structure of herbivores (Paine, 1966; Sinclair, 1985), can affect plants, herbivores, and natural enemies alike. The relative strengths of these (and other) feedback loops will determine the structure of communities. It should be no surprise that these ecological forces are complex and interactive (Faeth, 1987b), but a realistic view of animal–plant ecology requires that we integrate lessons learned from different levels in the trophic system into a balanced view of animal–plant communities.

Acknowledgments

I should like to thank Peter Price, Shoichi Kawano, and the organizing committee of the International Congress of Ecology 1990 (Yokohama, Japan), for bringing together many of the authors in this book and for encouraging me to formalize my thoughts on keystone herbivores. I am indebted to Stan Faeth, John Thompson, Mark Hay, Peter Price, and Allan Watt for their comments on an earlier version of this manuscript. I am grateful to the Natural Environment Research Council (UK), NATO (B/RFO/8482) and NSF (BSR-8918083) for their support.

References

Anderson, R. C., and Loucks, O. L. (1979). White-tail deer (*Odocoileus virginineus*) influence on structure and composition of *Tsuga canadensis* forests. *J. Appl. Ecol.* **16,** 855–861.

Arnold, E. N. (1979). Indian Ocean giant tortoises: Their systematics and island adaptations. *Phil. Trans. R. Soc. Lon. B* **286,** 127–145.

Askew, R. R. (1961). On the biology of the inhabitants of oak galls of *Cynipidae* (Hymenoptera) in Britain. *Trans. Soc. Brit. Entomol.* **14,** 237–268.

Baldwin, I. T., and Schultz, J. C. (1983). Rapid changes in tree leaf chemistry induced by damage: Evidence for communication between plants. *Science* **221,** 277–279.

Barbosa, P. (1988). Some thoughts on "the evolution of host range." *Ecology* **69,** 912–915.

Batzer, H. O. (1973). Net effect of spruce budworm defoliation on mortality and growth of balsam fir. *J. For.* **71,** 34–37.

Bell, R. H. V. (1970). The use of the herb layer by grazing ungulates in the Serengeti. *In* "Animal Populations in Relation to their Food Resources" (A. Watson, ed.), pp. 111–124. Blackwell Scientific Publications, Oxford, England.

Berenbaum, M. (1978). Toxicity of a furanocourmarin to armyworms: A case of biosynthetic escape from insect herbivores. *Science* **201,** 532–534.

Bergelson, J. M., and Lawton, J. H. (1988). Does foliage damage influence predation on the insect herbivores of birch? *Ecology* **69,** 434–445.

Bergelson, J., Fowler, S., and Hartley, S. (1986). The effects of foliage damage on casebearing moth larvae, *Coleophora serratella*, feeding on birch. *Ecol. Entomol.* **11,** 241–250.

Bernays, E., and Graham, M. (1988). On the evolution of host specificity in phytophagous arthropods. *Ecology* **69,** 886–892.

Bethlahmy, N. (1975). A Colorado episode: Beetle epidemic, ghost forests, and streamflow. *Northwest Sci.***49,** 95–105.

Blakely, N. R., and Dingle, H. (1978). Competition: Butterflies eliminate milkweed bugs from a Caribbean island. *Oecologia* **37,** 133–136.

Bray, J. R. (1964). Primary consumption in three forest canopies. *Ecology* **45,** 165–167.

Briggs, J. B. (1957). Some features of the biology of the winter moth (*Operophtera brumata*). *J. Hort. Sci.***32,** 108.

Brown, J. H., and Heske, E. J. (1990). Control of a desert–grassland transition by a keystone rodent guild. *Science* **250,** 1705–1707.

Bryant, J. P. (1981). Phytochemical deterrence of snowshoe hare browsing by adventitious shoots of four Alaskan trees. *Science* **213,** 889–890.

Campbell, R. W., and Sloan, R. J. (1977). Forest stand responses to defoliation by the gypsy moth. *For. Sci. Monogr.* **19,** 1–34.

Carne, P. B. (1965). Distribution of the Eucalypt-defoliating sawfly, *Pergs affinis affinis* (Hymenoptera). *Aust. J. Zool.* **13,** 593–612.

Carpenter, R. C. (1986). Partitioning herbivory and its effects on coral reef algal communities. *Ecol. Monogr.* **56,** 345–363.

Carpenter, S. R., and Kitchell, J. F. (1988). Consumer control of lake productivity. *BioScience* **38,** 764–769.

Carpenter, S. R., Kitchell J. F., and Hodgson, J. R. (1985). Cascading trophic interactions and lake productivity. *Bioscience* **35,** 634–639.

Carpenter, S. R., Kitchell, J. F., Hodgson, J. R., Cochran, P. A., Elser, J. J., Elser, M. M., Lodge, D. M., Kretchmer, D., He, X., and von Ende, C. N. (1987). Regulation of lake primary productivity by food web structure. *Ecology* **68,** 1863–1876.

Chadwick, M. J. (1960). *Nardus stricta* L. *J. Ecol.* **48,** 255–267.

Chauvin, G., Vannier, G., and Gueguen, A. (1979). Larval case and water balance in *Tinea pellionella*. *J. Insect Physiol.* **25,** 615–619.

Coley, P. D., Bryant, J. P., and Chapin, F. S. III. (1985). Resource availability and plant antiherbivore defense. *Science* **230,** 895–899.

Collins, S. (1961). Benefits to understory from canopy defoliation by gypsy moth larvae. *Ecology* **42,** 836–838.

Connor, E. F., and Simberloff, D. (1979). The assembly of species communities: Chance or competition? *Ecology* **60,** 1132–1140.

Cornell, H. V., and Washburn, J. O. (1979). Evolution of the richness-area correlation for cynipid gall wasps on oak trees: A comparison of two geographic areas. *Evolution* **33,** 257–274.

Courtney, S. (1988). If it's not coevolution, it must be predation? *Ecology* **69,** 910–911.

Crawley, M. J. (1983). "Herbivory, the Dynamics of Animal–Plant Interactions." Blackwell, Oxford, England.

Crawley, M. J. (1985). Reduction in oak fecundity by low-density herbivore populations. *Nature* **314,** 163–164.

Crawley, M. J. (1987). The effects of insect herbivores on the growth and reproductive performance of English oak. *Proc. 6th Intern. Symp. Insect/Plant Relationships,* Pau, France, 1986, Dr. W. Junk, Dordrecht.

Crawley, M. J. (1989). Insect herbivores and plant population dynamics. *Annu. Rev. Entomol.* **34,** 531–564.

Croft, B. A., and Hoying, S. A. (1977). Competitive displacement of *Panonychus ulmi* (Acarina: Tetranychidae) by *Aculus schlechtendali* (Acarina: Eriophyidae) in apple orchards. *Can. Entomol.* **109,** 1025–1034.

Damman, H. (1987). Leaf quality and enemy avoidance by the larvae of a pyralid moth. *Ecology* **68,** 88–97.

Darwin, C. (1859). "The Origin of Species." John Murray, London.

Davidson, D. W., Inouye, R. S., and Brown, J. H. (1984). Granivory in a desert ecosystem: Experimental evidence for indirect facilitation of ants by rodents. *Ecology* **65;** 1780–1786.

Denno, R. F., and McClure, M. S. (1983). "Variable Plants and Herbivores in Natural and Managed Systems." Academic Press, New York.

Dirzo, R., and Miranda, A. (1990). Altered patterns of herbivory and diversity in the forest understory: A case study of the possible consequences of contemporary defaunation. *In* "Plant–Animal Interactions: Evolutionary Ecology in Tropical and Temperate Regions" (P. W. Price, T. M. Lewinson, G. W. Fernandes, and W. W. Benson, eds.), pp. 273–287. John Wiley and Sons, New York.

Doane, C. C., and McManus, M. L. (1981). "The Gypsy Moth: Research toward Integrated Pest Management." USDA Technical Bulletin 1584, Washington, D.C.

Du Merle, P. (1981). Utilisation de pieges sexuels dans use etude de la tordeuse verte du chene, *Tortrix viridana* L. en montagne Mediterraneene. *Colloques de l'IRNA,* Paris **3,** 125–129.

Du Merle, P. and Pinguet, A. (1982). Mise en evidence par piegeage lumineux de migrations d'adultes chez *Tortrix viridana* L. (Lepidoptera: Tortricidae). *Agronomic* **2,** 81–90.

Duffy, J. E., and Hay, M. E. (1990). Seaweed adaptations to herbivory. *Bioscience* **40,** 368–375.

Dunson, W. A., and Travis, J. (1991) The role of abiotic factors in community organization. *Am. Nat.* in press.

Edgar, G. J. (1983a). The ecology of southeast Tasmanian phytal animal communities. II. Seasonal changes in plant and animal populations. *J. Exp. Mar. Biol. Ecol.* **70,** 159–179.

Edgar, G. J. (1983b). The ecology of southeast Tasmanian phytal animal communities. IV. Factors affecting the distribution of ampithoid emphipods among algae. *J. Exp. Mar. Biol. Ecol.* **70,** 205–225.

Edwards, P. J., and Wratten, S. D. (1982). Wound-induced changes in palatibility in birch (*Betula pubescens* Ehrh. spp. *pubescens*). *Am. Nat.* **120,** 816–818.

Edwards, P. J., and Wratten, S. D. (1983). Wound-induced defenses in plants and their consequences for patterns in insect grazing. *Oecologia* **59,** 88–93.

Edwards, P. J., and Wratten, S. D. (1985). Induced plant defenses against insect grazing: Fact or artefact? *Oikos* **44,** 70–74.

Edwards, P. J., and Wratten, S. D. (1987). Ecological significance of wound-induced changes in plant chemistry. *In* "Insects-Plants" (V. Labeyrie, G. Fabres, and D. Lachaise, eds.), pp. 213–215. W. Junk, Dordrecht, The Netherlands.

Ehrlich, P. R., and Raven, P. H. (1965). Butterflies and plants: A study in coevolution. *Evolution* **18,** 586–608.

Ehrlich, P. R. and Daily, G. C. (1988) Red-naped sapsuckers feeding at willows: Possible keystone herbivores. *Am. Birds* **42,** 357–365.

Eltringham, S. K. (1974). Changes in the large mammal community of Mweya Peninsula, Rwenzori National Park, Uganda, following removal of hippopotamus. *J. Appl. Ecol.* **11,** 855–865.

Faeth, S. H. (1985). Host leaf selection by leaf miners: Interactions among three trophic levels. *Ecology* **66,** 870–875.

Faeth, S. H. (1987a). Indirect interactions between seasonal herbivores via leaf chemistry and structure. *In* "Chemical Mediation of Coevolution " (K. Spencer, ed.), AIBM Symposium, Academic Press, San Diego.

Faeth, S. H. (1987b). Community structure and folivorous insect outbreaks: The role of vertical and horizontal interactions. In "Insect Outbreaks: Ecological and Evolutionary Perspectives." (P. Barbosa and J. C. Schultz, eds.), Academic Press, NY.

Faeth, S. H., and Bultman, T. L. (1986). Interacting effects of increased tannin levels on leaf mining insects. *Entomol. Expt. Appl.* **40,** 297–301.

Farrow, E. P. (1925). "Plant Life on East Anglian Heaths." Cambridge University Press, Cambridge, England.

Feeny, P. (1976). Plant apparency and chemical defense. *Rec. Adv. Phytochem.* **10,** 1–40.

Feeny, P. 1970. Seasonal changes in oak leaf tannins and nutrients as a cause of spring feeding by winter moth caterpillars. *Ecology* **51,** 565–581.

Forbes, S. A. (1887). The lake as a microcosm. *Bull. Illinois Nat. Hist. Survey* **15,** 537–550.

Fowler, S. V. (1985). Differences in insect species-richness and faunal composition of birch seedlings, saplings and trees: The importance of plant architecture. *Ecol. Entomol.* **10,** 159–169.

Fowler, S. V., and Lawton, J. H. (1985). Rapidly induced defenses and talking trees: The devil's advocate position. *Am. Nat.* **126,** 181–195.

Fox, L. R. (1981). Defense and dynamics in plant–herbivore systems. *Am. Zool.* **21,** 853–864.

Fox, L. R. (1988). Diffuse coevolution within complex communities. *Ecology* **69,** 906–907.

Futuyma, D. J., and Wasserman, S. S. (1980). Resource concentration and herbivory in oak forests. *Science* **210,** 920–922.

Gibbs, J. N. (1978). Intercontinental epidemiology of Dutch elm disease. *Annu. Rev. Phytopathol.* **16,** 287–307.

Gibson, C. W. D., and Hamilton, J. (1983). Feeding ecology and seasonal movements of giant tortoises on Aldabra Atoll. *Oecologia* **61,** 230–240.

Gibson, C. W. D., and Hamilton, J. (1984). Population processes in a large herbivorous reptile: The giant tortoise of Aldabra Atoll. *Oecologia* **56,** 84–92.

Gibson, C. W. D., Guilford, T. C., Hamber, C., and Sterling, P. H. (1983). Transition matrix models and succession after release from grazing on Aldabra Atoll. *Vegetation* **52,** 151–159.

Gilbert, L. E. (1980). Food web organization and the conservation of neotropical diversity. *In* "Conservation Biology" (M. E. Soule, and B. A. Wilcox, eds.), pp. 11–33. Sinauer Associates, Sunderland, Massachusetts.

Gilbert, L. E., and Smiley, J. Y. (1978). Determinants of local diversity in phytophagous insects: Host specialists in tropical environments. *Symp. R. Entomol. Soc. Lond.* **9,** 89–104.

Green, T. R., and Ryan, C. A. (1972). Wound-induced proteinase inhibitor in plant leaves: A possible defence mechanism against insects. *Science* **175,** 776–777.

Grubb, P. (1971). The growth, ecology and population structure of giant tortoises on Aldabra. *Phil. Trans. R. Soc. Lond.* **B260,** 327–372.

Hagley, E. A. C., Bronskill, J. F., and Ford, E. J. (1980). Effect of the physical nature of leaf and fruit surfaces on oviposition by the codling moth, *Cydia pomonella* (Lepidoptera: Tortricidae). *Can. Entomol.* **112,** 503–510.

Hails, R. S., and Crawley, J. J. (1991). The population dynamics of an alien insect: *Andricus quercuscalicis* Burgsdorf (Hymenoptera: Cynipidae). *J. Anim. Ecol.,* in press.

Hairston, N. G., Smith, F. E., and Slobodkin, L. B. (1960). Community structure, population control and competition. *Am. Nat.* **44,** 421–425.

Harper, J. L. (1977). "Population Biology of Plants." Academic Press, London.

Harrison, S., and Karban, R. (1986). Effects of an early-season folivorous moth on the success of a later-season species, mediated by a change in the quality of the shared host, *Lupinus arboreus. Oecologia* **69,** 354–359.

Hartley, S. E., and Firn, R. D. (1989). Phenolic biosynthesis, leaf damage, and insect herbivory in birch (*Betula pendula*). *J. Chem. Ecol.* **15**, 275–283.

Hartley, S. E., and Lawton, J. H. (1987). The effects of different types of damage on the chemistry of birch foliage and the responses of birch-feeding insects. *Oecologia* **74**, 432–437.

Hartley, S. E., and Lawton, J. H. (1990). Biochemical aspects and significance of the rapidly induced accumulation of phenolics in birch foliage. *In* "Phytochemical Induction by Herbivores" (D. W. Tallamy, and M. J. Raupp, eds.), John Wiley & Sons, New York.

Haukioja, E., and Niemela, P. (1977). Retarded growth of a geometrid larva after mechanical damage to leaves of its host tree. *Annales zoologici Fennici* **14**, 48–52.

Haukioja, E., and Niemela, P. (1979). Birch leaves as a resource for herbivores: Seasonal occurrence of increased resistence in foliage after mechanical damage of adjacent leaves. *Oecologia* **39**, 151–159.

Haukioja, E., Ruohomaki, K., Senn, J., Suomela, J., and Walls, M. (1990). Consequences of herbivory in the mountain birch (*Betula pubescens* ssp. *tortuosa*): Importance of the functional organization of the tree. *Oecologia* **82**, 238–247.

Hawkins, B. A. (1990). Global patterns of parasitoid assemblage size. *J. Anim. Ecol.*, **59**, 57–72.

Hawkins, S. J., and Hartnoll, R. G. (1983). Grazing of intertidal algae by marine invertebrates. *Oceanogr. Mar. Biol. Annu. Rev.* **21**, 195–282.

Hay, M. E. (1981a). Herbivory, algal distribution, and the maintenance of between-habitat diversity on a tropical fringing reef. *Am. Nat.* **118**, 520–540.

Hay, M. E. (1981b). The functional morphology of turf-forming seaweeds: Persistence in stressful marine habitats *Ecology* **62**, 739–750.

Hay, M. E. (1985). Spatial patterns of herbivore impact and their importance in maintaining algal species richness. *In* "Proc. 5th International Coral Reef Congress," pp. 29–34. Antenne Museum-Ephe, Moorea, French Polynesia.

Hay, M. E. (1991). Seaweed chemical defenses: Their role in the evolution of feeding specialization and in mediating complex interactions. *In* "Ecological Roles for Marine Secondary Metabolites" (V. J. Paul, ed.). Comstock Publishing, Ithaca, New York.

Hay, M. E., Duffy, J. E., Fenical, W., and Gustafson, K. (1988). Chemical defense in the seaweed *Dictyopteris delicatula*: Differential effects against reef fishes and amphipods. *Mar. Ecol. Prog. Ser.* **48**, 185–192.

Heinrich, B., and Collins, S. L. (1983). Caterpillar leaf damage, and the game of hide and seek with birds. *Ecology* **64**, 592–602.

Henson, W. R. (1958). Some ecological implications of the leaf-rolling habit in *Capsolechia niveopulvella* Chamb. *Can. J. Zool.* **36**, 809–818.

Hixon, M. A., and Brostoff, W. N. (1983). Damselfish as keystone species in reverse: Intermediate disturbance and diversity of reef algae. *Science* **220**, 511–513.

Hobbs, R. J., and Mooney, H. A. (1991). Effects of rainfall variability and gopher disturbance on serpentine annual grassland dynamics. *Ecology* **72**, 59–68.

Holliday, N. J. (1977). Population ecology of the winter moth (*Operophtera brumata*) on apple in relation to larval dispersal and time of budburst. *J. Appl. Ecol.* **14**, 803–814.

Holt, R. D. (1977). Predation, apparent competition, and the structure of prey communities. *Theor. Pop. Biol.* **12**, 197–229.

Howe, H. F., and Westley, L. C. (1988). "Ecological Relationships of Plants and Animals." Oxford University Press, Oxford, England.

Hunter, M. D. (1987a). Opposing effects of spring defoliation on late-season oak caterpillars. *Ecol. Entomol.* **12**, 373–382.

Hunter, M. D. (1987b). Sound production in the larvae of *Diurnea fagella* (Lepidoptera: Oecophoridae). *Ecol. Entomol.* **12**, 355–357.

Hunter, M. D. (1990). Differential susceptibility to variable plant phenology and its role in competition between two insect herbivores on oak. *Ecol. Entomol.* **15**, 401–408.

Hunter, M. D. (1992). A variable insect–plant interaction: the relationship between tree budburst phenology and population levels of insect herbivores among trees. *Ecol. Entomol.*, in press.

Hunter, M. D., and West, C. (1990). Variation in the effects of spring defoliation on the late season phytophagous insects of *Quercus robur. In* "Population Dynamics of Forest Insects" (A. D. Watt, S. R. Leather, M. D.Hunter, and N. A. C. Kidd, eds.), *Proc. NERC ITE Conference.*pp. 123–135 Intercept, Edinburgh, Scotland.

Hunter, M. D., and Willmer, P. G. (1989). The potential for interspecific competition between two abundant defoliators on oak: Leaf damage and habitat quality. *Ecol. Entomol.* **14,** 267–277.

Hutchinson, G. E. (1957). Concluding remarks. *Cold Spring Harbor Symp. Quant. Biol.* **22,** 415–427.

Hutchinson, K. J., and King, K. L. (1980). The effects of sheep stocking level on invertebrate abundance, biomass, and energy utilization in a temperate, sown grassland. *J. Appl. Ecol.* **17,** 369–387.

Janzen, D. H. (1988). On the broadening of insect-plant research. *Ecology* **69,** 905.

Jermy, T. (1988). Can predation lead to narrow food specialization in phytophagous insects? *Ecology* **69,** 902–904.

Jones, M. G. (1933). Grassland management and its influence on the sward. *J. R. Agric. Soc.* **94,** 21–41.

Karban, R., and Carey, J. R. (1984). Induced resistance of cotton seedlings to mites. *Science* **225,** 53–54.

Kareiva, P., and Sahakian, R. (1990). Tritrophic effects of a single architectural mutation in pea plants. *Nature* **345,** 433–434.

Kearsley, M. J. C., and Whitham, T. G. (1989) Developmental changes in resistance to herbivory: Implications for individuals and populations. *Ecology* **70,** 422–434.

Keating, S. T., and Yendol, W. G. (1987). Influence of selected host plants on gypsy moth (Lepidoptera: Lymantriidae) larval mortality caused by a baculovirus. *Environ. Entomol.* **16,** 459–462.

Keating, S. T., Hunter, M. D., and Schultz, J. C. (1990). Leaf phenolic inhibition of the gypsy moth nuclear polyhedrosis virus: The role of polyhedral inclusion body aggregation. *J. Chem. Ecol.* **16,** 1445–1457.

Keenlyside, J. (1989). Host choice in aphids. D. Phil. Thesis, University of Oxford, Oxford.

Kegg, J. D. (1971). The impact of gypsy moth: Repeated defoliation in oak in New Jersey. *J. For.* **69,** 852–854.

Kegg, J. D. (1973). Oak mortality caused by repeated gypsy moth defoliations in New Jersey. *J. Econ. Entomol.* **66,** 639–641.

Kennedy, C. E. J. (1986). Attachment may be a basis for specialization in oak aphids. *Ecol. Entomol.* **11,** 291–300.

Kerbes, R. H., Kotanen, P. M., and Jeffries, R. L. (1990). Destruction of wetland habitats by lesser snow geese: A keystone species on the west coast of Hudson Bay. *J. Appl. Ecol.* **27,** 242–258.

Kidd, N. A. C., Lewis, G. B., and Howell, C. A. (1985). An association between two species of pine aphid, *Schizolachnus pineti* and *Eulachnus agilis. Ecol. Entomol.* **10,** 427–432.

Knapp, R., and Casey, T. M. (1986). Thermal ecology, behaviour and growth of gypsy moth and eastern tent caterpillars. *Ecology* **67,** 598–608.

Krebs, C. J. (1978). "Ecology: The Experimental Analysis of Distribution and Abundance." 2nd Ed., Harper & Row, New York.

Lack, D. (1954). The Natural Regulation of Animal Numbers. Oxford University Press, London.

Lamb, R. J. (1980). Hairs protect pods of mustard (*Brassica hista* Gisilba) from flea beetle feeding damage. *Can. J. Plant Sci.* **60,** 1439–1440.

Laws, R. M. (1968). Interactions between elephant and hippopotamus populations and their environments. *E. Afr. Agric. For. J.* **33,** 140–147.

Lawton, J. H. (1983). Plant architecture and the diversity of phytophagous insects. *Annu. Rev. Entomol.* **28,** 23–29.

Lawton, J. H., and Schroeder, D. (1977). Effects of plant type, size of geographical range, and taxonomic isolation on number of insect species associated with British plants. *Nature* **265,** 137–140.

Lawton, J. H., and Schroeder, D. (1978). Some observations on the structure of phytophagous insect communities: The implications for biological control. *Proc. 4th Intern. Symp. Biol. Control of Weeds,* Gainesville, Florida.

Lawton, J. H., and Strong, D. R. (1981). Community patterns and competition in folivorous insects. *Am. Nat.* **118,** 317–338.

Leigh, E. G., Jr., and Smythe, N. (1978). Leaf production, leaf consumption and the regulation of folivory on Barro, Colorado Island. *In* "The Ecology of Arboreal Folivores" (G. G. Montgomery, ed.), Smithsonian Institute Press, Washington, D.C.

Lewis, S. M. (1986). The role of herbivorous fishes in the organization of a Caribbean reef community. *Ecol. Monogr.* **56,** 183–200.

Lewis, S. M., Norris, J. N., and Searles, R. B. (1987). The regulation of morphological plasticity in tropical reef algae by herbivory. *Ecology* **68,** 636–641.

Louda, S. M. (1982a). Distribution ecology: Variation in plant recruitment over a gradient in relation to insect seed predation. *Ecol.Monogr.* **52,** 25–41.

Louda, S. M. (1982b). Limitation of the recruitment of the shrub *Haplopappus squarosus* (Asteraceae) by flower- and seed-feeding insects. *J. Ecol.* **70,** 43–53.

Lubchenco, J., and Gaines, S. D. (1981). A unified approach to marine plant–herbivore interactions. I. Populations and communities. *Annu. Rev. Ecol. Syst.* **12,** 405–437.

MacArthur, R. H., and Wilson, E. O. (1967). "The Theory of Island Biogeography." Princeton University Press, Princeton, New Jersey.

Martin J. L. (1966). The insect ecology of red pine plantations in central Ontario IV. The crown fauna. *Can. Entomol.* **98,** 10–27.

Mattson, W. J. (1980). Herbivory in relation to plant nitrogen content. *Annu. Rev. Ecol. Syst.* **11,** 119–161.

May, R. M. (1975). "Stability and Complexity in Model Ecosystems." 2nd Ed., Princeton University Press, Princeton, New Jersey.

McGee, C. E. (1975). Change in forest canopy affects phenology and development of northern red and scarlet oak seedlings. *For. Sci.* **21,** 175–179.

McNaughton, S. J. (1976). Serengeti migratory wildebeest: Facilitation of energy flow by grazing. *Science* **191,** 92–94.

Merton, L. F. H., Bourn, D. M., and Hnatiuk, R. J. (1976). Giant tortoise and vegetation interactions on Aldabra Atoll. I. Inland. *Biol. Conserv.* **9,** 293–304.

Moran, V. C. (1980). Interactions between phytophagous insects and their *Opuntia* hosts. *Ecol. Entomol.* **5,** 153–164.

Morrison, D. (1988). Comparing fish and urchin grazing in shallow and deeper coral reef algal communities. *Ecology* **69,** 1367–1382.

Mulkey, S. S., Smith, A. P., and Young, T. P. (1984). Predation by elephants on *Senecio keniodendron* (Compositae) in the alpine zone of Mount Kenya. *Biotropica* **16,** 246–248.

Nelson, W. G. (1979a). An analysis of structural pattern in an eelgrass (*Zostera marina* L.) amphipod community. *J. Exp. Mar. Biol. Ecol.* **39,** 231–264.

Nelson, W. G. (1979b). Experimental studies of selective predation on amphipods: Consequences for amphipod distribution and abundance. *J. Exp. Mar. Biol. Ecol.* **38,** 225–245.

Nichols-Orians, C. M., and Schultz, J. C. (1989). Leaf toughness affects leaf harvesting by the leafcutter ant, *Atta cephalotes* (L.) (Hymenoptera: Formicidae). *Biotropica* **21,** 80–83.

Nielsen, B. O. (1978). Above-ground food resources and herbivory in a beech forest ecosystem. *Oikos* **31**, 273–279.

Niemela P., Tuomi J., and Haukioja, E. (1980) Age-specific resistance in trees: Defoliation of tamaracks (*Larix laricina*) by larch budmoth (*Zeiraphera improbana*) (Lepidoptera: Tortricidae). *Rep. Kevo. Subarctic Res. Stat.* **16**, 49–57.

Niemela, P., Aro, E. M., and Haukioja, E. (1979). Birch leaves as resources for herbivores: Damage induced increase in leaf phenols with trypsin-inhibiting effects. *Rep. Kevo. Subarctic Res. Sta.* **15**, 37–40.

Niemela, P., Toumi, J., Mannila, J., and Ojala, P. (1984). The effect of previous damage on the quality of Scots pine foliage as food for Diprionid sawflies. *Z. ange. ent.* **98**, 33–43.

Opler, P. A. (1978). Insects of American chestnut: Possible importance and conservation concern. *Proc. Amer. Chestnut Symp.* **1**, 83–84.

Orians, G. H., and Paine, R. T. (1983). Convergent evolution at the community level. *In* "Coevolution" (D. J. Futuyma and M. Slatkin, eds.), pp. 431–458. Sinauer Associates, Sunderland, Massachusetts.

Paige, K. N., and Whitham, T. G. (1987). Overcompensation in response to mammalian herbivory: The advantage of being eaten. *Am. Nat.* **129**, 407–416.

Paine, R. T. (1966). Food web complexity and species diversity. *Am. Nat.* **100**, 65–75.

Pimm, S. L. (1982). "Food Webs." Chapman & Hall, London.

Pollard, D. F. W. (1972). Estimating woody dry matter loss resulting from defoliation. *For. Sci.* **18**, 135–138.

Potter, D. A., and Kimmerer, T. W. (1988). Do holly leaf spines really deter herbivory? *Oecologia* **75**, 216–221.

Power, M. E., Stewart, A. J., and Matthews, W. J. (1988). Grazer control of algae in an Ozark Mountain stream: Effects of short-term exclusion. *Ecology* **69**, 1894–1898.

Price, P. W. (1990). Evaluating the role of natural enemies in latent and eruptive species: New approaches in life table construction. *In* "Population Dynamics of Forest Insects" (A. D. Watt, S. R. Leather, M. D. Hunter, and N. A. C. Kidd, eds.), pp. 221–232. NERC ITE Conference Proceedings, 1989. Edinburgh, Scotland.

Price, P. W., Bouton, C. E., Gross, P., McPheron, B. A., Thompson, J. N., and Weis, A. E. (1980). Interactions among three tropic levels: Influence of plants on interactions between insect herbivores and natural enemies. *Annu. Rev. Ecol. Syst.* **11**, 41–65.

Price, P. W., Cobb N., Craig T. P., Fernandes, G. W., Itami, J. K., Mopper, S., and Preszler, R. W. (1990). Insect herbivore population dynamics on trees and shrubs: New approaches relevant to latent and eruptive species and life table development. *In:* "Insect–Plant Interactions" (E. A. Bernays, ed.), vol.2. CRC Press, Boca Raton, Florida.

Puritch, G. S., and Nijholt, W. W. (1974). Occurrence of jubabione-related compounds in grand fir and Pacific silver fir infested by balsam wooly aphid. *Can. J. Bot.* **52**, 585–587.

Raupp, M. J. (1985). Effects of leaf toughness on mandibular wear of the leaf beetle *Plagiodera versicolora*. *Ecol. Entomol.* **10**, 73–79.

Raupp, M. J., and Denno, R. F. (1984). The suitability of damaged willow leaves as food for the leaf beetle, *Plagiodera versicolora*. *Ecol. Entomol.* **9**, 443–448.

Rausher, M. D. (1988). Is coevolution dead? *Ecology* **69**, 898–901.

Rawes, M. (1981). Further results of excluding sheep from high-level grasslands in the north Pennines. *J. Ecol.* **69**, 651–669.

Reader, P. M., and Southwood, T. R. E. (1981). The relationship between palatability to invertebrates and the successional status of a plant. *Oecologia* **51**, 271–275.

Redfearn, A., and Pimm, S. L. (1987). Insect outbreaks and community structure. *In* "Insect Outbreaks: Ecological and Evolutionary Perspectives" (P. Barbosa and J. C. Schultz, eds.), Academic Press, San Diego.

Renaud, P. E., Hay, M. E., and Schmitt, T. M. (1990). Interactions of plant stress and herbivory: Intraspecific variation in the susceptibility of a palatable versus an unpalatable seaweed to sea urchin grazing. *Oecologia* **82**, 217–226.

Rhoades, D. F. (1985). Offensive–defensive interactions between herbivores and plants: Their relevance to herbivore population dynamics and community theory. *Am. Nat.* **125,** 205–238.

Rhoades, D. F., and Cates, R. G. (1976). Toward a general theory of plant antiherbivore chemistry. *Rec. Adv. Phytochem.* **10,** 168–213.

Risch, S. J., Andow, D., and Altieri, M. (1983). Agroecosystems diversity and pest control: Data, tentative conclusions, and new research directions. *Environ. Entomol.* **12,** 625–629.

Rockwood, L. L. (1974). Seasonal changes in the susceptibility of *Crescentia alata* leaves to the flea beetle, *Oedionychus* sp. *Ecology* **55,** 142–148.

Root, R. B. (1967). The niche exploitation pattern of the blue-gray gnatcatcher. *Ecol. Monogr.* **37,** 317–350.

Root, R. B. (1973). Organization of plant–arthropod association in simple and diverse habitats: The fauna of collards (*Brassica oleracea*). *Ecol. Monogr.* **43,** 95–124.

Rosenthal, G. A., and Janzen, D. H. (1979). (eds.) "Herbivores: Their Interaction with Secondary Plant Metabolites." Academic Press, New York.

Ross, B. A., Bray, J. R., and Marshall, W. H. (1970). Effects of long-term deer exclusion on a *Pinus resinosa* forest in north-central Minnesota. *Ecology* **51,** 1088–1093.

Rossiter, M. C. (1987). Use of secondary hosts by nonoutbreak populations of the gypsy moth. *Ecology* **68,** 857–868.

Rossiter, M. C. (1991). Environmentally based maternal effects: A hidden force in insect population dynamics? *Oecologia,* in press.

Rossiter, M. C., Schultz, J. C., and Baldwin, I. T. (1988). Relationships among defoliation, red oak phenolics, and gypsy moth growth and reproduction. *Ecology* **69,** 267–277.

Satchell, J. E. (1962). Resistance in oak (*Quercus* spp.) to defoliation by *Tortrix viridana* L. in Roudsea Wood National Nature Reserve. *Ann. Appl. Biol.* **50,** 431–442.

Schiel, D. R., and Foster, M. S. (1986). The structure of subtidal algal stands in temperate waters. *Oceanogr. Mar. Biol. Annu. Rev.* **24,** 265–307.

Schowalter, T. D., Hargrove, W. W., and Crossley, D. A., Jr. (1986). Herbivory in forested ecosystems. *Annu. Rev. Entomol.* **31,** 177–196.

Schultz, J. C. (1983). Habitat selection and foraging tactics of caterpillars in heterogeneous trees. *In* "Variable Plants and Herbivores in Natural and Managed Systems." Academic Press, New York.

Schultz, J. C. (1988). Many factors influence the evolution of herbivore diets, but plant chemistry is central. *Ecology* **69,** 896–897.

Schultz, J. C., Foster, M. A., and Montgomery, M. E. (1990). Hostplant-mediated impacts of a baculovirus on gypsy moth populations. *In* "Population Dynamics of Forest Insects" (A. D. Watt, S. R. Leather, M. D. Hunter, and N. A. C. Kidd, eds.), NERC ITE Conference Proceedings, Edinburgh 1989.

Schutte, F. (1957). Untersuchungen uber die population dynamik des Eichenwicklers *Tortrix viridana* L. *Ztschr. f. ang. Ent.* **40,** 1–36.

Silva-Bohorquez, I. (1987). Interspecific interactions between insects on oak trees, with special reference to defoliators and the oak aphid. D.Phil. Thesis, University of Oxford, Oxford, England.

Sinclair, A. R. E. (1985). Does interspecific competition or predation shape the African ungulate community? *J. Anim. Ecol.* **54,** 899–918.

Southwood, T. R. E. (1960). The abundance of the Hawaiian trees and the number of their associated insect species. *Proc. Hawaiian Ent. Soc.* **17,** 299–303.

Southwood, T. R. E. (1961). The number of species of insect associated with various trees. *J. Animal Ecol.* **30,** 1–8.

Southwood, T. R. E. (1973). The insect plant relationship—an evolutionary perspective. *Symp. R. Entomol. Soc. Lond.* **6,** 143–155.

Southwood, T. R. E. (1978). "Ecological Methods." Chapman & Hall, London.

Speight, M. R., and Wainhouse, D. (1989). "Ecology and Management of Forest Insects." Clarendon Press, Oxford.

Stark, R. W., and Dahlsten, D. L. (eds.) (1970). "Studies on the Population Dynamics of the Western Pine Beetle, *Dendroctonus brevicomis* Le Conte (Coleoptera: Scolytidae)." Univ. of California Div. of Agric. Sci., Berkeley, California.

Stoddart, D. R., and Peake, J. F. (1979). Historical records of Indian Ocean giant tortoise populations. *Phil. Trans. R. Soc. Lond.* **B286**, 147–162.

Stoner, A. W. (1979). Species-specific predation on amphipod Crustacea by the pinfish *Lagodon rhomboides:* Mediation by macrophyte standing crop. *Mar. Biol.* **55**, 201–207.

Strong, D. R. (1974). Asymptotic species accumulation in phytophagous insect communities. *Science* **185**, 1064–1066.

Strong, D. R. (1979). Biogeographic dynamics of insect–host plant communities. *Annu. Rev. Entomol.* **24**, 89–119.

Strong, D. R. (1988). Special Feature: Insect Host Range. *Ecology* **59**(4).

Strong, D. R., Lawton, J. H., and Southwood, T. R. E. (1984). "Insects on Plants. Community Patterns and Mechanisms." Blackwells, London.

Swingland, I. R. (1977). Reproductive effort and life history strategy of the Aldabran giant tortoise. *Nat. Lond.* **269**, 402–404.

Swingland, I. R., and Lessells, C. M. (1979). The natural regulation of giant tortoise populations on Aldabra Atoll: Movement polymorphism, reproductive success, and mortality. *J. Anim. Ecol.* **48**, 639–654.

Tahvanainen, J. O. (1972). Phenology and microhabitat selection of some flea beetles (Coleoptera: Chrysomelidae) on wild and cultivated crucifers in central New York. *Entomol. Scand.* **3**, 120–138.

Tansley, A. G. (1935). The use and absence of vegetational concepts and terms. *Ecology* **16**, 284–307.

Tansley, A. G. (1939). "The British Islands and Their Vegetation." Cambridge University Press, London.

Tegner, M. J., and Dayton, P. K. (1987). El Nino effects on southern California kelp forest communities. *Adv. Ecol. Res.* **17**, 243–279.

Tenow, O. (1972). The outbreaks of *Oporinia autumnata* Bleh. and *Operophtera* Spp. (Lepidoptera: Geometridae) in the Scandanavian mountain chain and northern Finland, 1862–1968. *Zool. Bid. f. Uppsala (suppl.)* **2**, 1–107.

Terborgh, J. W. (1988). The big things that run the world—a sequal to E. O. Wilson. *Conserv. Biol.* **2**, 402–403.

Thompson, D. B., Brown, J. H., and Spencer, W. D. (1991). Indirect facilitation of granivorous birds by desert rodents and ants: Experimental evidence from foraging patterns. *Ecology*, in press.

Thompson, J. N. (1988). Coevolution and alternative hypotheses on insect/plant interactions. *Ecology* **69**, 893–895.

Thornton, D. D. (1971). The effect of complete removal of hippopotamus on grassland in the Queen Elizabeth National Park, Uganda. *E. Afr. Wildl. J.* **9**, 47–55.

Turner, R. M., Alcorn, S. M., and Olin, G. (1969). Mortality of transplanted saguaro seedlings. *Ecology* **50**, 835–844.

Turnipseed, S. G. (1977). Influence of trichome variations on populations of small phytophagous insects on soybean. *Env. Entomol.* **6**, 815–817.

Valentine, H. T., and Houston, D. R. (1981). Stand susceptibility to gypsy moth defoliation. *In* "Hazard-Rating Systems in Forest Insect Pest Management: Symposium Proceedings" (R. L. Hedden, S. J. Barras, and J. E. Coster, eds.), pp. 137–144. USDA For. Serv. Gen. Tech. Rept. WO–27, Washington, D.C.

Van Dolah, R. F. (1978). Factors regulating the distribution and population dynamics of the amphipod *Gammarus palustria* in an intertidal salt marsh community. *Ecol. Monogr.* **48**, 191–217.

Varley, G. C., and Gradwell, G. R. (1968). Population models for the winter moth. *In* "Insect Abundance" (T. R. E. Southwood, ed.), *Symp. R. Ent. Soc. Lond.* **4**, 132–142.

Vesey-FitzGerald, D. F. (1960). Grazing succession among East African game. *J. Mammal.* **41,** 161–171.

Vince, S., Valiela, I., Backus, N., and Teal, J. M. (1976). Predation by the salt marsh killifish *Fundulus heteroclitus* (L.) in relation to prey size and habitat structure: Consequences for prey distribution and abundance. *J. Exp. Mar. Biol. Ecol.* **23,** 255–266.

Webb, J. W., and Moran, V. C. (1978). The influence of the host plant on the population dynamics of *Aciia russellae* (Homoptera: Psyllidae). *Ecol. Entomol.* **3,** 313–321.

Wellso, S. G. (1973). Cereal leaf beetle: Feeding, orientation, development and survival on four small-grain cultivars in the laboratory. *Ann. Entomol. Soc. Am.* **66,** 1201–1208.

West, C. (1985). Factors underlying the late-seasonal appearance of the lepidopterous leaf mining guild on oak. *Ecol. Entomol.* **10,** 111–120.

Whitham, T. G. (1983). Host manipulation of parasites. In "Variable Plants and Herbivores" (R. F. Denno and M. S. McClure, eds.), Academic Press, New York.

Whitham, T. G., and Mopper, S. (1985). Chronic herbivory: Impacts on architecture and sex expression of pinyon pine. *Science* **228,** 1089–1091.

Whittaker, R. H. (1972). Evolution and measurement of species diversity. *Taxonomy* **21,** 213–251.

Williams, K. S., and Myers, J. H. (1984). Previous herbivore attack of red alder may improve food quality for fall webworm larvae. *Oecologia* **63,** 166–170.

Willmer, P. G. (1982). Microclimate and the environmental physiology of insects. *Adv. Insect. Physiol.* **16,** 1–17.

Wint, G. R. W. (1983). The role of alternative host plant species in the life of a polyphagous moth, *Operophtera brumata* (Lepidoptera: Geometridae). *J. Anim. Ecol.* **52,** 439–450.

Young, T. P. (1984). The comparative demography of semelparous *Lobelia telekii* and iteroparous *Lobelia keniensis* on Mount Kenya. *J. Ecol.* **72,** 637–650.

Young, T. P. (1985). *Lobelia telekii* herbivory, mortality, and size at reproduction: Variation with growth rate. *Ecology* **66,** 1879–1883.

Young, T. P., and Smith, A. P. (1987). Alpine herbivory on Mount Kenya. In "Tropical Alpine Environments: Plant Form and Function" (P. Rundel, ed.), pp. 110–135. Springer-Verlag, Berlin.

11

Loose Niches in Tropical Communities: Why Are There So Few Bees and So Many Trees?

David W. Roubik

Smithsonian Tropical Research Institute
Balboa, Panamá

I. Introduction

When competition involves species using similar mixtures, selection favors the specialization of species feeding on the same general resources but living in slightly different
habitats [*spatial or temporal divergence, density specialization*]. Such species are fine-grained.
When there is a considerable initial difference, selection favors the accentuation of the
differences, so that the species tend to use only their own preferred resource and are
coarse-grained. Comparison of the animals of higher taxa such as orders or families
often indicates fine-grained and coarse-grained component groups [*as seen among pollinators and angiosperms*].
—G. E. Hutchinson (1978, p. 211); bracketed italics added here.

In the real world, niche relationships are a kind of controlled chaos. Our
predicament is that field studies can show a cross section of an evolutionary
process, at best within a small biogeographic region, but nothing more. How

valid can this be? Furthermore, apparent generalists may, on occasion, be extreme specialists, and specialists may be fine-grained in space or time. The relationships that seem well defined (i.e., stable, reciprocal, or fairly exclusive) may be either transient or consistent over a period of several years or seasons, and both could change over biogeographic and evolutionary time. Thus in plant/pollinator and seed/disperser systems—mutualisms in which diffuse coevolution must be the rule (Boucher *et al.*,1982; Janzen,1983; Jordano,1987; Roubik, 1989; Horvitz and Schemske,1990; Toft and Karter, 1990; Fleming, Chapter 12 in this volume)—the mutualist part of a species' niche may be loose and somewhat flexible. Facultative mutualisms might one day be recognized as common in the temperate zone (May,1982). They now seem widespread in the tropics. This chapter deals with loose niches and considers why the ratio of tree species to pollinating bees is so great in the tropics. If the behavioral and spatio–temporal flexibility in niche use that stand out in these systems is characteristic of many others, then community ecology is in deep trouble (hints of the situation are found in Wiens,1989; Price *et al.*,1984). More likely, perhaps, is the prospect that recognizing the importance of spatio–temporal changes in niche use will improve our understanding of *core* and *peripheral* species in diffuse interactions. A correlate is that field data must be examined and applied with great caution. My thesis, with support from plant–pollinator studies, is that core species change over space and time, creating loose niches, and interchangable mutualists. With greater spatio–temporal variability in plant resources come looser niches in tropical plant/pollinator communities, and broader niches for certain generalists. In guild assemblages of bees, it so happens that the generalists often are social and maintain perennial colonies. I shall also suggest that relentless competition for food can produce community structure, even when resources are plentiful, the impact of competition relatively slight, and the numbers of competing species relatively low, which is especially applicable to tropical bee communities where social generalists predominate.

II. Problems of Niches and Diversity

In communities of generalists and specialists, what plants and animals do, and an appreciation for their realized niches, is probably far more important than knowing how many taxonomic units they compose. A community of many flexible members of few species is just as diverse as one made up of many more species that have rigid roles. As ecological literature becomes increasingly specialized, a common ground is often sought by paying little attention to the component species beyond their size, weight, name, or numbers. Genera are often the limit to which a *reductionist* approach is applied, rather than species or their variability. Behavior and life histories are largely ignored, especially their variability in space and time. Yet these are among the details that prove decisive in understanding population and community characteristics. Just as standing biomass is, to the study of trophic networks, relatively meaningless without data on recruitment, taxonomic data are of little value without information on behavior, ecological relationships, and phenology. In this sense, an understanding of biodiversity is equivalent to an understanding of realized niches in their fullness, rather than in caricature. During an age justly preoccupied with taking stock of natural resources and the importance of conserving them, there should be renewed interest both in niche theory and in natural history.

The general qualities of bees, most of them related to feeding and nesting niches, can be sorted into categories resembling those applied to other groups, like birds or ants (Wiens, 1989; Hölldobler and Wilson, 1990; Roubik, 1989 and Table 1). Natural guilds are found through comparative study. Processes that link them together, and the nature of the guilds, however, seem to matter a great deal in determining community structure and the results of ecology or evolution (Abrams, 1990). For example, high spatio–temporal diversity and variation among flowering plants is proposed to explain unusually high bumblebee species richness in a northern temperate habitat (Ranta and Vepsäläinen, 1981), the converse of the hypothesis discussed here (low bee richness caused by flowering variability in the tropics). Another study of a bee community in short-grass prairie (Tepedino and Stanton, 1981) suggests spatio–temporal unpredictability in flowering erodes the competition between foragers. In direct contrast, my understanding of tropical communities is that unpredictability in flowering strongly favors certain bees to win in preemptive competition, and prevents many bees and flowers from evolving close mutualisms.

Considering the world biota of angiosperms and the bee pollinators that roughly half of them possess, there can be no doubt that the concept of biodiversity as revealed in species richness has failed. Tropical communities are not restricted in diversity, as their moderate or low bee richness seems to indicate; they are in fact more biologically diverse than bee communities that have many more species (Tables 1, 2). One key element is the perennial abundance of many social bee species (Roubik, 1979, 1989). An individual tropical species probably plays more roles and has a greater share in maintaining angiosperm communities than does the average temperate bee (Michener, 1954, 1979). Not only are the social bees better represented in terms of species, they are numerically the dominant bees (Heithaus, 1979; Roubik, 1989; Inoue *et al.*, 1990; Bawa, 1990).

Many or more species, genera, and higher groups of bees live in the middle latitudes than at lower latitudes (Michener,1979 and Table 2). In contrast, there is frequently an order-of-magnitude difference in floral communities of these regions. A 50-ha plot of old forest in the Neotropics contains some 300 tree species, whereas a similar plot in Malaysia may have

Table 1 Life History Guilds of Tropical and Temperate Bees[a]

Bee guild	Taxa and distinguishing life history trait	Region
Oil bees	Anthophorinae, Ctenoplectridae, Milittidae; collect floral oils from Malpighiceae and other families, used for larval provisions	Worldwide, largely neotropical
Orchid bees	Euglossine apids, males gather chemicals from orchid and other inflorescences, unknown function	Neotropical
Buzz collectors	Anthophoridae, some Apidae, some Halictidae, Colletidae, Oxaeidae females collect pollen by vibrating poricidal anthers	Worldwide, largely tropical
Necrophages	Some Apidae (facultative), three *Trigona* (obligate); obtain protein by consuming dead animals	Largely neotropical
Trapliners	Some Apidae, Anthophoridae, Megachilidae, other? Fly widely between persistent, small resources	Largely tropical
High-density specialists	Many Apidae (honey bees, stingless bees); recruit to resources in large numbers, usually inflorescences of many small flowers	Largely tropical
High-sugar specialists	Some Apidae (honey bees, *Melipona*); strongly prefer concentrated sugar ($>60\%$) in floral nectar	Worldwide
Destructive visitors	Some Apidae, Anthophoridae, Andrenidae, Oxaeidae, Megachilidae; perforate flowers to harvest pollen or nectar	Worldwide, largely tropical
Resin bees	Some Apidae, some Megachilidae; make all or most of nest with resin mined from wood, or gathered at flowers (*Clusia, Dalechampia*)	Tropical
Mason bees	A few Apidae, some Megachilidae; make "exposed" nests from resin, sometimes mixed with mud or diverse material	Worldwide, largely tropical
Digger bees	Many Anthophoridae, Halictidae, Colletidae, Andrenidae, Oxaeidae excavate nests in hard substrate, usually the ground or in wood	Worldwide
Lodger bees	Some Apidae, some Anthophoridae, some Megachilidae; nest in preexisting cavities, often left by other animals	Worldwide, largely tropical
Termite bees	Some Apidae, some Anthophoridae, some Megachilidae; excavate nests in active termitaria	Tropical
Ant bees	Some Apidae; excavate nests in active ant nests	Tropical
Robber bees	Some Apidae; forage exclusively in nests of other social bees	Tropical
Parasitic bees	Roughly 20% of all bees, all major families	Worldwide
Honeydew bees	Some Apidae (honey bees and stingless bees); collect sap and sugary secretions from Homopteran bugs and fungi	Worldwide, largely tropical?
Fruit bees	Some Apidae; collect sap from ripe fruit	Worldwide?
Long-tongued bees	Apidae, Megachilidae, Anthophoridae, Fideliidae, some tropical Colletidae, Halictidae, some temperate Andrenidae, Stenotritidae	Worldwide
Short-tongued bees	Colletidae, Melittidae, Halictidae, Andrenidae, Ctenoplectridae, Oxaeidae, Stenotritidae	Worldwide

Table 1 (*Continued*)

Bee guild	Taxa and distinguishing life history trait	Region
Perennial colonies	Many Apidae (stingless bees, honey bees); obligately eusocial[b]	Tropical
Temporary colonies	Occasional or facultative eusociality in some species, others obligate[b]	Worldwide, largely tropical

[a] From Roubik (1989).

[b] The term eusocial refers to a mother and daughters coexisting as adults in the same nest; if obligate, then the production of reproductive females can only be achieved by colonies, not by lone females.

more than 800 (Foster and Hubbell, 1990). Furthermore, even in the subtropics of south Africa, where there is a rich and biologically diverse bee fauna, there is still an enormous flora with no counterpart seen in the numbers of bee species (Michener, 1979; Raven and Axelrod, 1974). The number of understory plant species does not, however, display a latitudinal gradient (Smith, 1987). Thus, the tree guild is presumably subject to a different set of selection pressures and routes to speciation.

High representation of bats, birds, beetles, and other pollinators exists in the tropics, but their numbers seem in no way to compromise the feeding and pollinating niches available to bees (Roubik, 1989, 1990; Bawa, 1990). Nor do the types of seed dispersers seem to determine the extent to which speciation occurs among angiosperms (Herrera, 1989). Thus, the generality seems to hold, that there are far too many species of flowering trees in the tropics, relative to the numbers of bee species that pollinate them. Temporal and spatial characteristics of plants and pollinators are of the essence if we are to understand this type of community structure. Rather than emphasize mechanics of species interactions and speciation, I shall use biogeographic comparisons and limited case studies to outline features of loose niches in the richest terrestrial communities—the lowland tropical forests.

III. Communities Structured around Variable Components

Certain concepts have general applicability to plant/pollinator and seed/ disperser assemblages. Consider first that resources are permanently out of phase in bee/flower assemblages. The standing crop of adult bees, like the

Table 2 Bee Genera and Subgenera in Major Geographic Regions[a]

Neotropical	315	Nearctic	260	African	175
Oriental	89	Palearctic	243	Australian	127

[a] Michener (1979). Note that many African and Australian groups exist in dry, temperate habitats, and many of the neotropical groups exist in temperate Andean and S. Brazilian floral regions.

standing crop of flowers, usually has nothing to do with the current abundance of the complementary guild. Shifts in adult bee abundance are usually due to past resource conditions related to brood production (Tepedino and Stanton, 1981; Roubik, 1989). The extent and timing of flowering, similarly, is tied to diverse variables that can even preclude pollinator tracking (Zimmerman *et al.*, 1989). Flowers are likely to have either too few or too many *potential* pollinators, and bees are likely to have either too few or too many accessible flowers *of any given species*. Being too specialized is obviously a drawback for either guild. The eventual outcome of competition in relatively stable environments is purportedly a finer division of resources among consumers (Leigh, 1990), which suggests that the more stable mutualist guild will support a larger number of coevolved species. Specialist pollinators like male orchid bees have the most stable known insect populations (Roubik and Ackerman, 1987; Roubik, 1989). A plausible explanation is that these specialists have alternative resources, also noted among oil bees of southern Africa, another morphologically highly specialized group (Manning and Brothers, 1986). A second major feature of the bee–flower interaction relies on behavioral responses of bees that have access to more than one flower species at a time. Bees can *major* on one, while they *minor* at others, and adjust their foraging choice according to changing resource availability (Heinrich, 1975; Oster and Heinrich, 1976). Thus, the abundance of particular bees and of particular flowers is probably mismatched in any period, year-to-year temporal correlations in abundance should often be low (Wolda and Roubik, 1986; Horvitz and Schemske, 1990), and adult bees will use their mobility to survey changing floral landscapes and switch among flowers.

Such flexible foraging behavior has never been matched in formal models of resource utilization. Adding more spatial or temporal features to niche dimensions seems useful, although it is viewed with some skepticism owing to the increased mathematical difficulties inherent in the approach (Hutchinson, 1978; Roughgarden and Diamond, 1986; Thompson, 1988). Some of the *noise* in ecological systems may indeed be intractable or relatively unimportant, but we are currently dealing with much simpler phenomena that throw models into jeopardy. Even the simple observation of where a symbiotic specialist lives within its host has shown that fundamental or *preinteractive* niches may be common in biological communities, rather than the product of competitive displacement (Price, 1984), or even of coevolution with a host (Toft, 1986). Whereas communities have been pictured as either composed of species that are dynamic in their interactions, or of relatively static composition with little or no interaction (Fig. 1), many biological communities may fit neither model.

Within tropical plant and pollinator communities there is ample, if not overwhelming, evidence that spatial and temporal variation have key effects (Gentry, 1974; Appanah, 1985; O'Malley and Bawa, 1987; Ashton *et al.*,

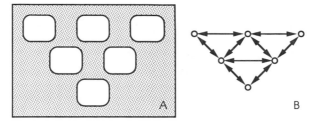

Figure 1   Schematic depictions of (A) *noninteractive* and (B) *interactive* communities.

1988; Roubik, 1989; Horvitz and Schemske, 1990). A staggered succession of flowering can provide temporal separation, allowing one pollinator to service many similar plant species. Physical or chemical changes in nectar, along with changes in the abundance and distribution of pollen and nectar in flowers, can promote a succession of consumers on a single floral resource. Flight ranges of bees often exceed a few kilometers (Roubik, 1989), so that their resources can usually be located. For the flowering plants, a fixed locality is probably not a significant constraint, unless there is relatively great separation of conspecifics (Janzen, 1986). Tropical tree species dispersion is often aggregated (Hubbell, 1979).

Floral resources are, to widely foraging bees, fine-grained in space, while the pollinators are, to the trees, fine-grained in time. A tree has many years in which to reproduce, and a foraging bee has many accessible flower species. Either group may therefore appear coarse-grained in contemporary time, reflecting temporal specialization due to learning, pollinator behavior, or general abundance. Apparent specialization may not be a good indication of dependency. In practice, methodology affects sampling and bias. The unstated assumption in most surveys is that the *average* association is the one that was encountered. And in theory, it is the number of visits flowers receive (i.e., quantity rather than quality of pollination) that determines evolution of some floral traits and the mechanisms of pollinator attraction and manipulation (Harder and Thomson, 1989). Tropical trees might try out all concievable types of pollinator visits during their lifetimes. Plasticity may allow new adaptive zones to be entered (West-Eberhard, 1989), but whether this results in speciation, morphological changes, or other modifications, either for bees or for angiosperms, depends on the regularity with which adaptive shifts occur.

The flexibility of bees visiting flowers is, without doubt, far greater than that of angiosperm flowering phenology and floral traits. Mismatched speciation potential among two such partners is evident in the large ratios of flowering plants relative to the bees that pollinate them. As mentioned above, this might mean that bee populations are more stable than the flower populations, but the flexible foraging behavior of bees could accomplish the same evolutionary end. Schemske (1983) mentions that there are about 4.2 times as many flowering plants as bee species in North America (although many wind-pollinated trees were included). The real ratio of pollinating bees to flowering plants may lie closer to 3 : 1 in this part of the world, whereas in the neotropics the ratio is about 5 : 1, and in southeast Asia it is nearly 10 : 1. (My estimate stipulates one half the primary forest and native second-growth species are pollinated by bees). This situation seems the logical outcome of the interplay between spatial and temporal variability in plant life histories and the behavioral ecology of bees, discussed in more detail later in this chapter.

The extent of flowering within a species and among individuals can vary widely from year to year, as shown by all of the above authors, and in many works cited by them. A special case is found within the rainforests of Malaysia, where community-wide mass flowering years occur at 2- to 10-year intervals. The Dipterocarpaceae, an extremely species-rich group of Paleotropical trees, have many species on this schedule, as well as many that flower every year (Appanah, 1987). Throughout the seasonal tropics, flowering seasons are well defined (Baker *et al.*, 1983; Terborgh, 1986), and even in wet lowland forests with no definite dry season, periodicities in rainfall have been detected (Inoue and Nakamura, 1990). But the number and location of flowers, both at the individual and community levels, shift substantially between years (Gentry, 1974; Appanah, 1985). This range of variation should set limits for phenological schedules and produce a degree of opportunism that increases with spatio-temporal variability.

IV. Loose and Tight Niches among Specialist Guilds: Orchids, Oil Flowers, and Long-Corolla Flowers

Variation in relatively specialized animal/flower relationships is demonstrated by neotropical oil bees and orchid bees, as well as in certain hummingbird/bee/flower systems (Tables 3–6). These examples show the degrees of *looseness* or *tightness* that exist in flower/visitor relationships. Consider these premises: (1) When several to many species visit a flower, only a few are important visitors; (2) if only one or two species visit flowers, they are central in the pollination ecology and evolution of that species, and (3) self-incompatible flowering plants must have tighter relationships with their pollinators than do self-compatible flowers (Simpson and Neff, 1985; Bullock *et al.*, 1989; Bawa, 1990). The first point seems intuitive for any diffuse system, but it would be extremely difficult to demonstrate as *the* explanation for a flower/visitor relationship from a single locality or a single season's data, as would points two and three. A quantitative basis for discussion of loose niches can be built upon some systems, about which we now have a broad information base. These studies are the product of several independent surveys over large geographic areas, or they are from repeated

Table 3 The Oil-Bee Guild Associated with Neotropical
Byrsonima crassifolia (Malpighiaceae)[a]

Region, source	Bees associated with Byrsonima
São Luis, Brazil (2° S. lat.) Rêgo and Albuquerque 1989	*Centris (Centris) caxiensis, C. (C.) spilopoda, C. (Hemisiella) flavifrons, C. (Heterocentris) analis, C. (Paremisia) byrsonimae, C. (P.) fuscata, C. (Trachina) longimana, C. (Hemisiella) trigonoides, Epicharis (Epicharis) flava, E. (Xanthepicharis) bicolor, Paratetrapedia (Lophopedia) tarsalis, P. (Paratetrapedia) testacea, P. (P.) nasuta, Paratetrapedia spp. (3), Tetrapedia aff. diversipes, Augochloropsis aff. crassigena, Trigona pallens, T. fuscipennis, T. fulviventris, Tetragona dorsalis beebei, Dicranthidium arenarium*
Belém, Brazil (0° lat.) Ducke 1902 in Roubik 1979	*Centris, Epicharis, Melipona*
Kourou, French Guiana (5° N. lat.) Roubik 1979	*Centris, Epicharis, Paratetrapedia, Augochloropsis*
Panama & Costa Rica (9–11° N. lat.) Heithaus 1979, Snelling 1984	*Centris (Centris) adanae, C. (C.) aethyetera, C. (C) obscurior, C. (Heterocentris) difformis, C. (Melanocentris) sp., C. (Trachina) fuscata, C. (T.) heithausi, C. (Xanthemisia) lutea, C. (Xanthemisia) rubella, Epicharis (Epicharoides) maculata, E. (Hoploepicharis) lunulata, E. (Parepicharis) metatarsalis, Exomalopsis sp., Tetragona dorsalis, Nannotrigona testaceicornis, Eulaema polychroma*
Veracruz, Mexico (18° N. lat.) Delgadillo, pers. comm.	*Centris, Epicharis, Tetrapedia, Trigona, Apis*

[a] Floral oil, produced in lieu of nectar, provides an essential part of larval nutrition collected by females of all Anthophorine species listed—*Centris, Epicharis, Tetrapedia, Exomalopsis,* and *Paratetrapedia.*

observations during several years. While they leave no doubt concerning the breadth of potential interactions that exist in nature, the studies are still very much limited to qualitative observations. The core and peripheral flower-visiting species are nonetheless evident.

For instance, anthophorine bee genera are clearly pollinators of the self-compatible Malpigh, *Byrsonima crassifolia* (Table 3). The tree is associated with savanna and open woodland vegetation maintained by occasional fires. Female bees constituting most species visiting the flower *must* collect floral oils to feed their larvae, but the oils come from a variety of Malpighiaceae and some other plant families (Roubik, 1989). Use of *B.crassifolia* does not appear essential to any one bee species, and some of its visitors only collect the pollen, such as *Augochloropsis* in Brazil (Rêgo and Albuquerque, 1989), or Africanized *Apis mellifera* in Mexico (R. Delgadillo, personal communication, 1990). *Centris* and *Epicharis* are the principal pollinators of *Byrsonima crassifolia* from eastern Amazonia to Mexico (Table 3). Other bees such as *Paratetrapedia* and *Tetrapedia* are widespread and abundant at flowers, but they are much smaller and slow-moving, thus probably not responsible for much pollination. In addition, the halictid and apid bees that visit flowers, while very abundant (Rêgo and Albuquerque, 1989), also have

Table 4   Flower–Visitor Relationships
Observed during 10 Years in Central Panama
at *Lopimia dasypetala (Malvaceae)*[a]

Flower–visitor	Relationship	Linkage
Phaethornis superciliosus	Pollinator	Tight
Phaethornis longuemareus	Robber	Tight
Trigona ferricauda	Robber	Tight
Phaethornis guy	Pollinator	Loose?

[a] From Roubik (1982).

a minor or a negative role in pollination. What we know about this guild of bees as flower visitors applies well to many other systems. About all that can be said is that certain species are abundant, and the main pollinators; others are abundant, but relatively unimportant as pollinators; and some are parasites or scavengers in this system. Others are opportunists that have little contact with the system, either in space or, apparently, in time (oil flowers have existed at least since Miocene times in tropical America; Simpson and Neff, 1985). Throughout the range where *Byrsonima* and its core visitors occur, their relationship seems assured. But the bees do not particularly need *B. crassifolia*, and the tree does not depend on any single species of bee. The mutualist guild is generic, tribal, subfamilial, or familial, but its species seem free to come and go. Their niches are loose.

Somewhat tighter niches are apparent for flowers with extremely long corollas that can be pollinated or exploited effectively by only a few long-billed hermit hummingbirds and nectar robbers; two examples are given in Tables 4 and 5. The flowers of *Lopimia* and *Quassia* are self-compatible, and the former is autogamous; they are found at forest edges. If a pollinator or

Table 5   Shifts in Flower Visitation at Mainland and Island Communities of *Quassia amara* (Simaroubaceae)[a]

Habitat and years separated	Pollinators	Robbers
Mainland, source	*Phaethornis superciliosus*	*Trigona fulviventris*
Large Island, 75 years	*Phaethronis superciliosus*	*Trigona fulviventris*
	Thalurania colombica	*Thalurania colombica*
		Damophila julie
Small Island, 75 years	*Lepidopyga coeruleogularis*	*Coereba flaveola*
		Lepidopyga coeruleogularis
Large Island, 10,000 years	*Amazilia edward*	
	Ceratina laeta	

[a] From Roubik *et al.* (1985). *Trigona* and *Ceratina* are social bees, the rest are hummingbirds and a bananaquit.

destructive nectarivore is absent, there are numerous ancillary visitors waiting in the wings, but none of the pollinators is then strictly legitimate or as effective as *Phaethornis superciliosus* (Roubik, 1982; Roubik *et al.*, 1985). Of the more abundant visitors, it would be easy to imagine niche shifts occurring during relatively short timespans owing to vicariance events or local changes in abundance (Table 5). In Table 5, the changes following fragmentation of mainland communities are traced from 75 to 10,000 years. Pollinators and destructive flower visitors associated with *Quassia amara* have been replaced or completely lost, and niche shifts have occurred. New pollinators, or robbers, arrived when the original mainland species were either no longer present or less abundant at the flowers. Preemptive competition seems very likely to operate normally in larger, mainland assemblages. On the mainland, no qualitative changes have been noted in the main flower visitors, *Phaethornis superciliosus* and *Trigona fulviventris*, during the past 10 years, and this is also the case for pollinating *P. superciliosus*, the robbing stingless bee *Trigona ferricauda*, and robbing hummingbird *P. longuemareus* associated with *Lopimia*. However, on year 10, I saw *Phaethornis guy* pollinate *Lopimia* in a wetter forest at slightly higher elevation. Flowers there were still robbed for nectar by *T. ferricauda;* thus, a parasite may have a tighter relationship with a host than does a mutualist.

Neotropical orchids have a large assortment of specialized flowers, floral attractants, and self-incompatible breeding systems; they can be thought of as tightly coevolved with their male euglossine bee pollinators (Simpson and Neff, 1985). Half of the 50 strictly euglossine-pollinated orchids studied in central Panama by Ackerman and Roubik have 2–11 species of euglossine pollinators (Roubik and Ackerman, 1987). The remaining orchids have been associated with only one bee species, though at all three study sites, roughly 25% of the bees have never been associated with *any* orchid species. It is interesting that the number of euglossine bee species and the number of orchid species they pollinate are about equal in central Panama. This could be an example of the turns that 1 : 1 coevolution might take. The bee species that have never been seen carrying orchid pollinaria at one site are found in other not-too-distant habitats carrying pollinaria (Roubik, 1989). Ackerman's original suggestion regarding use of alternative resources (usually aroids) by euglossine bees is still the most likely explanation for their patchy association with orchids. About half of all the euglossine bee species in central Panama are associated with 2–12 orchid species, and at any site, 50% are associated with one or none. Persistent flexibility may characterize all such apparently remarkable examples of coevolution, if studied in sufficient depth. For euglossine bees, the association with orchids does not appear to be obligate.

The loose and tight mutualist niches in Table 6 suggest that spatio–temporal variation exists among specialists. The surprise is that tight mutualisms are often multiple for a species. For example, tight relationships exist between several pollinators and one plant species, as well as between one

Table 6   Relationships of Euglossine Bees and Orchids in Central Panama.[a]

Flower–pollinator	Linkage
Aspasia principissa–El. meriana, Ex. frontalis	Tight
Catasetum bicolor–Eg. cybelia, dissimula, dodsoni, mixta, tridentata	Tight
Eg. allosticta, cognata, gorgonensis, hemichlora, variabilis	Loose
Catasetum viridiflavum–El. cingulata	Tight
El. nigrita	Loose
Clowesia warscewitzii–El. bombiformis, meriana	Tight
Cochleanthes lipscombiae–El. bombiformis, meriana	Tight
El. cingulata	Loose
Coryanthes maculata–Eg. azureoviridis, despecta, tridentata	Tight
Coryanthes speciosa–Eg. dressleri	Tight
Eg. cognata	Loose
Cycnoches guttulatum–Eg. cognata, dissimula	Tight
Eg. cybelia, deceptrix, dodsoni	Loose
Dichaea panamensis–Eg. despecta, dissimula, heterosticta, mixta, tridentata, variaiblis	Tight
Eg. allosticta, cyanaspsis, dressleri	Loose
Gongora quinquenervis–Eg. allosticta, bursigera, deceptrix, heterosticta, mixta, tridentata, variabilis	Tight
Eg. cognata, dissimula, dressleri	Loose
Gongora sp.–Eg. gorgonensis	Tight
Houlettia tigrina–El meriana, nigrita	Tight
El. cingulata	Loose
Kefersteinia costaricensis–Eg. deceptrix	Tight
Eg. mixta	Loose
Kefersteinia lactea–Eg. bursigera, cybelia	Tight
Eg. dodsoni, maculilabris, tridentata	Loose
Mormodes igneaum–Eg. dissimula	Tight
Mormodes powellii–Eg. tridentata	Tight
Notylia albida–Eg. hemichlora	Tight
Notylia barkeri–Eg. dissimula	Tight
Eg. variabilis	Loose
Notylia linearis–Eg. cybelia, dodsoni	Tight
Eg. deceptrix, dressleri	Loose
Notylia pentachne–Eg. cognata, El. cingulata, meriana	Tight
El. bombifornis	Loose
Notylia sp.–Eg. championi	Tight
Eg. mixta, tridentata	Loose
Sobralia sp.–El. nigrita	Tight
Stanhopea costaricensis–El. nigrita, Ef. schmidtiana	Tight
Trichocentrum capistratum–Eg. bursgera, deceptrix, tridentata	Tight
Eg. allosticta, crassipunctata, gorgonensis	Loose
Trichopilia maculata–Eg. tridentata	Tight
Eg. bursigera	Loose
Unidentified–*Eg. imperialis*	Tight
Vanilla planifolia–Eg. tridentata	Tight
Vanilla pompona–El. cingulata, meriana	Tight
El. speciosa	Loose

[a] Tight or loose niches reflect widespread or restricted visitation, respectively, of orchids by bees (see text). Data on orchids and male euglossine bees from Roubik and Ackerman (1987). Three sites were studied at middle-elevation Atlantic and Pacific slopes and the central isthmian lowland. *Eg. = Euglossa, Ex. = Exaerete, El. = Eulaema, Ef. = Eufriesea*

pollinator and one plant species. Further, the same flower or bee can maintain both tight and loose relationships. I should like to stress that the observations are essentially qualitative, but even if the absolute abundance of guild members were known, we could deduce nothing about past abundance, distribution, or relative importance in pollination. When a bee species was associated with a particular orchid at two or more of three study sites, which encompassed Atlantic and Pacific forests from 100 to 800 m elevation, the relationship was considered tight. *For the same orchid species,* other bees were seen carrying its pollinia at only one site. They therefore were loose mutualists. I have intentionally left out the more dubious cases of few observations or rare and geographically restricted species—their relations are likely to seem more exclusive than they really are. The niche parameters are relative, and the bee species in Table 6 occur throughout the three study sites. In the 100 total species under examination, tight niches are about twice as common as loose ones—58 of 92 in Table 6. Moreover, as already stated, 25% of the bee species and 50% of orchid species have been matched by field studies to only one mutualist in central Panama. Both guilds are flexible, but the orchids are less so than the bees. Spatial and temporal variablity in the floral community is likely the basis for the asymmetry, despite an equal ratio of guild members.

V. Component Species, Life Histories, and Behavior

Throughout the tropics, the extreme richness in tree species seems to demand a special explanation. In general terms, many have been offered, among them the following (Hubbell and Foster, 1986, pp. 327–329):

> The spatial structure and dynamics of species-rich tropical forests suggest that chance and biological uncertainty may play a major role in shaping the population biology and community ecology of tropical tree communities . . . It is suggested that a common outcome of spatially and temporally uncertain competitors is likely to be a diffuse coevolution of generalist tree species within a few major life history guilds, rather than the pairwise coevolution of specialists in competitive equipoise.

Leigh (1990) suggests this approach may gloss over important life history traits that distinguish different tree species and are potential causes of community diversity. At present, we simply do not have the knowledge to attempt comprehensive examination of individual species rather than guilds. The next section shows how the tropical bee guild can affect tree diversity, while providing some ideas about the function of different bee groups within communities.

Of all the characteristics that might be chosen to discuss bee assemblages, it is the flight ability, seasonality, nest dispersion, and behavioral plasticity (particularly floral visitation) of the component species that seem to matter most to the angiosperms, which use them as flying genitalia. The number of tree species that can coexist is likely to be determined by how many can be effectively pollinated by the local guild of bees. In this regard, Gentry (1974, p. 68) made an observation that has greatly influenced pollination ecologists:

> Different phenological strategies clearly play an important role in making possible effective competition by many related species of tropical Bignoniaceae for the services of the same pollinators. The role of differences in flowering phenology in other groups should be further investigated to evaluate the overall importance of such mechanisms in maintaining high diversity in tropical plants.

It is noteworthy that tropical bee guilds are no more species-rich than those of temperate forests, and they are considerably less so than those of warm, xeric Mediterranean scrub forests (Table 7). At the outset, it is important to recognize that all the bee/flower surveys include true forest habitats of closed canopy, and also more disturbed and open areas. Sampling effort in these studies is extensive enough to accurately depict species richness in the mixed habitats (Michener, 1979; Roubik, 1989). As shown in Table 1, some life-history traits of bees differ fundamentally between tropical and temperate habitats. Required use of trees and resin for nesting seems much more common among tropical forest bees (Roubik, 1990), and several novel life trait guilds are restricted to the tropics.

It has long been suspected that tropical bees are less often floral specialists than are temperate bees, although there was no quantitative information to back up this assertion (Michener, 1954, 1979; Simpson and Neff, 1985). An important aspect of the most species-rich bee communities of the Americas

Table 7   Bee Community Richness in Lowland Tropical and Temperate Areas[a]

Locality	Number of species	Number of genera	Percent social[b]	References
Central Sumatra	110	20	50	Inoue *et al.* (1990)
Belém, Brazil	250	50	50	Ducke (Michener 1979)
French Guiana	245	50	50	Roubik (1990)
W. Costa Rica	200[c]	65[c]	25[c]	Heithaus (1979)
Central Brazil	230	80	35	Camargo and Mazucato (1986) and pers. comm.
Central Mexico	230	90	20	Ayala (1990)
S. California	500	>100	~15	Timberlake (Michener 1979)
S.W. France	500	>100	~15	Perez (Michener 1979)
Central Japan	170	25	~15	Kakutani *et al.* (1990)
Illinois	300	55	10	Pearson (1933)
N. Dakota	245	45	5	Stevens (Michener 1979)

[a] In rounded numbers.
[b] Most Xylocopinae, most halictines, most apids.
[c] Introduced honey bees excluded, *Trigona* s. 1. split into several genera.

is their extraordinarily high number of panurgine Andrenidae, especially *Perdita* (Michener, 1979). These are among the smallest of all bees and thus possess a very limited flight range. I suggest they are closely tied to host plants both in space (by nesting nearby), and in time (adult emergence triggered by the spring rains that prompt flowering in the deserts of southwestern United States). Michener (1979) comments that if such bees are eliminated from consideration, the degree of specialization is still high in this habitat, relative both to the eastern United States and to the tropics. Species richness then appears more similar to that of the Chicago area or North Dakota (Table 7).

The keystone bee species in tropical plant/pollinator assemblages are big bees that fly considerable distances, and/or social bees that maintain perennial colonies and are present through the year. Despite similar range in bee size in tropical and temperate habitats, the largest bees are tropical. The size range of the tropical species encompasses that of all bees, ranging from some extremely small meliponines to extremely large Xylocopini and Euglossinae (Roubik, 1989). The Old World honey bees vary greatly in size, having species both half as large and nearly twice as large as the garden variety of temperate *Apis mellifera;* the largest other tropical bees are *Centris, Epicharis, Ptilotopus, Creightonella,* and *Chalicodoma.* Flight ranges of the larger bees are fairly well documented and show 10–20 km *from the nest* as the upper limits of typical foraging range. Moreover, the honey bees and stingless bees have species characterized by *daily* forager activity ranging over distances of 3 to 10 km from the nest. The effective reach of these colonies or individual foragers, of central importance to understanding their pollinating potential, is therefore from 30 to over 1200 square kilometers (Roubik, 1989). Proportions of social bees in the local bee assemblages are given in Table 7. They range from 20 to 50% in the lowland tropics, while only from 5 to 15% in almost all temperate latitudes. In the tropics, the most familiar social bees are the stingless bees (Meliponinae) and the honey bees (Apinae), both of which are the only bees that maintain perennial colonies, usually of 1000 to 10,000 workers. Other social bees are the Bombinae (bumble bees—all primitively eusocial, and Euglossinae, about half temporarily eusocial and half solitary), Halictinae and Nomiinae (both Halictidae with some primitively eusocial colonies), the Xylocopinae [temporarily eusocial Allodapini in the Old World and social and solitary Xylocopini throughout the tropics, although for the latter none is obligately social (Michener, 1989)], and also a few unusual social megachilids such as *Chalicodoma* in the Old World tropics (Roubik, 1989). With comparatively few exceptions, the nesting habits of these bees are intimately associated with tropical trees (Roubik, 1990).

The spatial sampling of the environment by bees is thorough (see Seeley, 1987), and their memory of the exact location of forage is acute (summaries in Roubik, 1989). For canopy trees, the major component of tropical floristic richness of interest here, the significance of these facts is immediately apparent in the consistent patrolling and visitation of flowers by the *large bee* guild (Frankie *et al.*, 1983; Bawa *et al.*, 1985; Appanah, 1985; Bullock *et al.*, 1989). The large xylocopines and euglossines may live from a few months to, in exceptional cases, several months, to more than a year (Gerling *et al.*, 1989; Michener, 1989; Ackerman and Montalvo, 1985). Large bees are the most constant visitors to the guild of trees and lianas having large flowers. Although bee specialization on one type of flower during foraging trips is essential for them to achieve the status of pollinators rather than parasites of outcrossing trees, there is a lack of direct evidence. However, two types of indirect evidence suggest they are, in fact, the primary or even sole pollinators of the flowers they visit. First, the known obligately outcrossing trees visited by large bees not only produce fruit, but also present a high degree of genetic heterozygosity, supporting the notion that the widely ranging pollinators are, in fact, ranging widely between genetically compatible flowering individuals (Appanah, 1985; O'Malley and Bawa, 1987). Second, when the pollen used for larval provisions in the bees' nests is examined and identified microscopically, it often consists of primarily one flower species, as does that coming in on bee scopae after one foraging bout (Roubik and Michener, 1984; Vinson *et al.*, 1987; Snow and Roubik, 1987; Lobreau-Callen and Coutin, 1987).

Temporal specialization by honey bees and stingless bees has been noted in extensive studies of pollen utilization in Panama (Roubik, 1988, 1989; Roubik and Moreno, 1990; Roubik *et al.*, 1986). An example of the temporal specialization by Africanized honey bee colonies on pollen sources is given in Table 8. During a 10–14 day period, 60–90% of the pollen harvested by three colonies came from a single plant species, and there were always a few dozen to a few hundred species available to the bees. While these bees have large diets in terms of flower species, encompassing an estimated one fourth of the local angiosperms, they are very selective both in temporal and spatial use of the available flowers. A large proportion of the total pollen they harvest during a year comes from a dozen or so species of flowers, and the same species of flowering plants, particularly palm trees or others having small, apparently unspecialized flowers, are used by many coexisting species of bees (Roubik *et al.*, 1986; see also Lobreau-Callen *et al.*, 1986). The value of these relatively small social bees as pollinators is sometimes compromised by relatively short flight ranges and perhaps a tendency to restrict foraging to a single tree canopy or flower patch. Nevertheless, they are often the only available pollinators for self-incompatible or dioecious trees that have small to tiny flowers (Baker *et al.*, 1983; Appanah, 1985; Ashton *et al.*, 1988). Interestingly, such trees also often have high genetic heterozygosity (Hamrick and Loveless, 1986).

If convergence between independent communities can be taken as evidence for similar competitive processes and consumer–resource relation-

Table 8 Temporal Specialization by Africanized Honey
Bees in Central Panama[a]

Max. colonies' intake	Pollen source	Site[b]
90%	*Paspalum*	La Polvareda
77%	*Zea*	La Polvareda
71%	*Croton*	La Polvareda
60%	*Pseudobombax*	La Polvareda
60%	*Spondias*	La Polvareda
90%	*Guazuma*	Pipeline Road
90%	*Oenocarpus*	Pipeline Road
90%	*Elaeis*	Pipeline Road
80%	*Spondias*	Pipeline Road
80%	*Zea*	Pipeline Road
90%	*Zea*	Buena Vista
77%	*Tetracera*	Buena Vista
72%	*Spondias mombin*	Buena Vista
70%	*Spondias radlkoferi*	Buena Vista
63%	Compositae	Buena Vista

[a] Roubik *et al.* (1984). Pollen-collecting traps on hives continuously sampled pollen harvest by three colonies at each site. Palynological analysis determined pollen types each 10–14 days. Maximal harvests of single species over a 12-month period given here.

[b] La Polvareda was deciduous forest and second growth near the Pacific Coast, 1.4 m annual rainfall; Pipeline Road was at primary and secondary moist forest and some clearings in central Panama, 3 m annual rainfall; Buena Vista was agricultural land and patches of old second growth, Atlantic watershed, 3 m annual rainfall.

ships, then the guild assemblages in the Neotropics and Paleotropics must share a great deal of environmental similarity. There is clearly convergence among unrelated species in South and Central America and southeast Asia (Roubik, 1990). Bees with very long tongues have evolved from the anthophorines and even a halictid in Asia, matching the Neotropical Euglossinae. Stingless bees of the *Lophotrigona* group are aggressive group foragers, matching Neotropical *Trigona* s. str. Some Asian megachilids are extremely large and probably provide long-distance pollination service similar to that of Neotropical Centridini. The general representation of social and solitary bees displays the same latitudinal variation in the Old World and the New World (Fig. 2). This information at least suggests that the bee guild consistently creates its own assemblages. By whatever processes this may entail, the flowering plants, the competitors, the resources, and natural enemies of bees have not caused any prominent differences in the makeup of their tropical assemblages. It is all the more striking, then, that the numbers of species of bees in southeast Asia is only half that found in similar Neotropical habitats (Table 7). Yet in both communities one half of all bees are

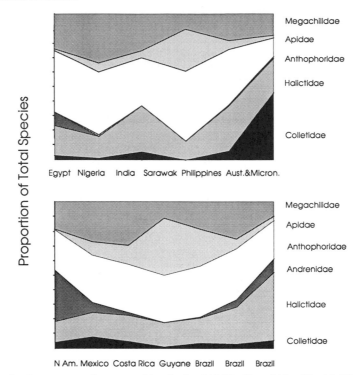

Figure 2 Bee-community composition in the Old World and New World. The five or six major bee families are labeled, showing the relative proportion of total bee species belonging to each at the indicated sites or regions. From Roubik (1989).

social and one fourth of all bees are stingless bees. The extremely variable, community-wide flowering episodes of Malaysia (Appanah, 1985) are likely a major cause for a depauperate bee fauna (Roubik, 1990). But another major difference in Malaysia is the presence of four native honey bees (*Apis koschevnikovi, A. cerana, A. andreniformis,* and *A. dorsata*), which have considerably larger flight ranges and/or colonies than many stingless bees. They are likely to be the ecological equivalent of several stingless bee species, which are in turn equivalent to many solitary or less social bee species (Roubik, 1979, 1989). Greater spatio–temporal variation in floral resources seems to have compressed the entire bee guild, but its composition remained similar to that of other tropical regions.

The platform for testing the hierarchical hypothesis for bee community structure will clearly be that of continued studies of bee biology, pollination ecology, and pollen identification for bee-provisioned brood cells, particularly in the southeast Asian tropics. The role of parasitic bees in community organization is still unclear (but see discussion of robber bees, *Lestrimelitta,* in

Roubik, 1989), and the differential extinction rates of different bee guilds has yet to be examined in detail. The large xylocopines, stingless bees, and honey bees of the Asian forests are presumed to be the chief pollinators of the flowers they visit. Unless the Dipterocarpaceae that Appanah has shown are pollinated solely by tiny thrips (many Dipterocarpaceae are, however, bee-pollinated Appanah, 1987), prove to be the predominant tree species in the southeast Asian forests, then these aseasonal and generalist pollinators thoroughly dominate their pollinator guilds (Inoue *et al.*, 1990).

Hierarchies built on species interactions seem to typify the bee communities, but in contrast to other social insects such as ants, where the bulk of species interact aggressively (Hölldobler and Wilson, 1990), almost all bees forage and nest unaggressively (Roubik, 1989). Parallels are nonetheless evident and point to the still largely unexplored roles of component species' behavior. Hölldobler and Wilson (1990, p. 423) surmise that: "the fewer the ant species in a local community, the more likely the community is to be dominated behaviorally by one or a few species with large, aggressive colonies that maintain absolute territories". Tropical bee hierarchies may depend on interference competitors, well documented in the few stingless bees showing nest overdispersion, group dominance of flower patches, and the use of pheromone odor trails, and aggressive competition at potential nest sites (studies by Hubbell and Johnson, summary in Roubik, 1989). Territory marking and *density specialization* occur, chiefly with stingless bees having huge colonies and aggressive foraging styles—for example *Trigona canifrons* in southeast Asia and *Trigona hyalinata, T. branneri, T. amazonensis, T. fuscipennis, T. spinipes,* and *T. corvina* in the neotropics. Large foragers, relatively immune to the attacks of *Trigona,* displace their competitors at flowers by *incidental interference* due to a larger size. Large xylocopines, anthophorines, and aggressive stingless bees, and the massively recruiting honey bees, can be seen as dominant organizing agents in tropical forest communities. Each has the potential for dominating flowers, and some may aggressively defend their nesting territory, although they do not eliminate competitors in the manner of dominant tropical ants such as *Oecophylla, Crematogaster* or *Eciton.*

Perennially active tropical bee guilds not only encompass the entire size range of bees and their major foraging styles, they also include bees with unusually long tongues that correspond to highly specialized floral morphologies. One strictly neotropical guild, the euglossine bees, provides a means of assessing nectarivore specialization within a pivotal tropical bee guild. Many, if not most, species are active year-round and therefore show relatively little temporal specialization, although clear population peaks occur through the year (Roubik and Ackerman, 1987). More important, none of the bee species obtains its nectar by perforating flowers or using the perforations made by other animals. Their only nectar is obtained legitimately. The types of flowers visited for nectar are long-corolla Marantaceae, Rubiaceae,

Zingiberaceae, Apocynaceae, Bignoniaceae, and Gesneriaceae (Ackerman, 1985). A rather different set of families compose the pollen species visited by females, consisting often of nectarless, *buzz-pollinated* Solanaceae and Leguminosae. Virtually all of the bees use nectar of the same quality, usually near 40% sugar content (D. W. Roubik, unpublished data, 1989). Panama contains the richest species assemblages, including approximately one-third of the species. In central Panama, within cloud forest, Atlantic wet forest, and lowland moist forest, some 53 species of euglossines maintain broad sympatry. I have measured their functional tongue length, using fresh-caught bees only (dried museum specimens, even when rehydrated, cannot be used to measure full tongue extension). The range of tongue lengths is presented in Figure 3. Taking variance of tongue length into consideration (see also Inoue, this volume), gives the impression that, at most, several nectar-feeding niches are used by these bees, certainly nothing resembling their species number. The male and female bees are specialists but probably fine-grained. They may facultatively partition nectar sources in space and time.

Ackerman (1985) demonstrated that euglossines with relatively short tongues were present primarily during the dry and early wet seasons, when most of their host flowers are *big-bang* or *cornucopia* strategists (Gentry, 1974). During mid to late wet seasons, however, many of the bees visit *steady-state* flowering individuals, which present relatively few flowers with relatively large nectar rewards. Euglossine communities studied by Ackerman reflected this difference by shifting in species composition toward larger, longer-tongued bees during the mid to late wet seasons. These species apparently are better able to fly the large distances required to

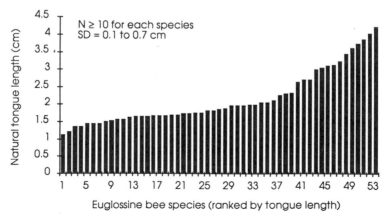

Figure 3 Extended tongue lengths of 53 euglossine bee species sympatric in central Panama. This assemblage exclusively visits flowers legitimately to obtain nectar.

exploit steady-state flowers, and their proboscides match the relatively longer nectar flowers available at those times. However, as shown in Figure 3, there are few euglossines that could be narrowly specialized and isolated from competitors at any time of the year.

VI. Loose Niches and Competition

Competition among bees and flowers, many being fine-grained generalists that have *loose niches*, forces bees to reshuffle their distributions over the available flowers on a daily, seasonal, and yearly basis. For the plants, this process should limit extreme specialization in the attraction of pollinators. Critical to the above argument are two means by which spatial partitioning of resources can be achieved by foraging bees, and both have relatively extensive supporting data: (1) behavioral specialization in space and time, permitted by the learning ability and orientation acuity of large bees in particular, and (2) the various recruitment systems (chemical and behavioral communication) employed by social bee colonies, allowing them to locate and monopolize patches of flowers as they become available (Roubik, 1989). Spatial and temporal partitioning would allow each of the common tropical bee guilds to perform many different pollinating roles through their active seasons, which for the majority of them constitute every daylight hour of the year (Wolda and Roubik, 1986). In behavioral terms, the perennially active tropical bees also include all the major forager subguilds or *foraging syndromes:* Large and forceful bees (Inoue *et al.,* 1990), which can probe within cleistogamic flowers and obtain pollen and nectar unreachable to all but other, destructive flower visitors; bees that regularly search the forest canopy for flowers; small bees that recruit heavily to patches of chasmogamous, unspecialized flowers, and sometimes defend them, or which visit small, highly dispersed flowers; and bees that move long distances between widely spaced, but relatively constant, rich floral resources (Table 1).

When facultative instead of obligate partnerships are the rule, the result will be loose niches (Figs. 4, 5). Diffuse coevolution, as explained by many authors weary of naïve declarations that coevolution has produced each perceived matching of traits (Janzen, 1980), could create such a pattern. Evolution at the community level would be driven by some of the same major forces that shape the degree of adaptation between interacting species—uncertainty in space and time, fluctuations in abundance and density, and immigration and extinction. Although the processes may well be influenced by competitive interactions, the communities are susceptible to invasion by generalists and may be considered to exist in quasi equilibrium only until the next major round of habitat selection, or dispersal from poor to richer resource patches. Ecological and evolutionary effects of competition clearly differ depending on whether the consumers, both fine-grained

Figure 4   Schematic representation of specie's interactive niches fitting together to form a biological community.

and coarse-grained, are flexible in their foraging choice and coevolutionary associations (Abrams, 1990). Diffuse coevolution and loose niches introduce a possibility that guild interactions continue to operate, despite changes in species number, composition, or degree of competition for resources. These communities may never evolve to saturation or stability, but their very nature favors generalists that can specialize, and specialists that can shift between alternative resources. Resources are partitioned to some extent by behavioral mechanisms, and not by rigid niche parameters that purportedly evolve in response to persistent competition or intermittent *crunches.*

   A phenomenon that is unique in demonstrating the invasibility of tropical communities, also reveals the problem of classifying species as diet generalists without considering their spatio–temporal characteristics. The same phenomenon raises some fundamental questions about the validity of models that depict component species as *interlocking pieces of a jigsaw puzzle* (Fig. 4) or the *soap bubbles* used as an analogy by earlier students of biological communities (Hutchinson, 1978). The African honey bee *Apis mellifera scutellata* has, while interbreeding slightly with other strains of temperate *A. mellifera,* colonized approximately 20 million square kilometers of neotropical habitat never before occupied by honey bees (Roubik, 1989; Roubik and Boreham, 1990). This region contains the richest tropical bee communities (Tables 2 and 7), and also tens of thousands of angiosperm species, perhaps one fifth of which the *Africanized honey bee* utilizes as a mutualist, scavenger, and parasite (Roubik, 1988, 1989; Roubik and Moreno, 1990).

Figure 5 Schematic representation of *loose* niches in a biological community. The realized niches are shown as shaded areas within the full interactive niche space of the component species, some of which may undergo modification over time.

The bees are indeed generalists in pollen utilization, and over 200 pollen species have been identified from those harvested by bee colonies in single neotropical sites during the course of a year (Roubik, 1989). This honey bee does not forage aggressively, yet it apparently dominates bee guilds and may drastically alter the pollination ecology of many plant species. Its wide foraging range (10 km in forest), recruitment ability, and relatively large colony size allow it to select among unspecialized flowers and occasionally to become an extreme temporal specialist (Table 8).

For a species as flexible as the honey bee in its native tropical forest habitat, the realized niche in diet items needs to be redefined every few weeks; its function as a pollinator could hardly be static from year to year, and its ability to select among resources available over a wide area will make any ecological relationship with an individual flowering species a loose one. Price (1984) has revealed that ecological literature contains references to guilds of butterflies, marine fishes, bark and ambrosia beetles, ichneumonids, and aphids that are no more specific in their relationships in the tropics compared to temperate areas, even when there is a greater tropical species richness (see also Cody, 1986; Colwell, 1986). One is forced to wonder, however, how much spatio–temporal specialization exists, or whether, as for the bees and tropical plants, we are only beginning to grasp some of the salient features of their life histories and mutualisms.

The potential interchangability of species within guilds of tropical trees (Hubbell and Foster, 1986) and tropical pollinating bees (Simpson and Neff, 1985) has already been suggested. Far from indicating an abundance of empty niches, these hypotheses strengthen the impression that spatio–temporal variation in communities leads to loose niches and diffuse co-evolution. Transient foraging specialization by bees evidently drives relatively high rates of speciation in tropical trees. A general *runaway* process in speciation might occur, whereby maximal use is made of the pollination potential of the local bee fauna, primarily the species that are perennially active and wide ranging. The portion of their niches used by members of an interactive guild should lead to striking variation in the performance of populations from year to year. If the species' niche allows it to persist and reproduce indefinitely, and some of its resources are truly guilds rather than single species, then its niche can, indeed, change in time. The rotation of mutualists is the essence of a loose niche. Relationships across the range of life-history guilds may include similar interactions.

The possible role of large-scale environmental disturbance in the tropics (Colinvaux *et al.*, 1985) cannot be discounted as a factor that would help to maintain pseudoequilibria and relatively loose species packing in some tropical communities. But the depauperate bee fauna and high tree species richness of Malaysia, compared to the Neotropics, suggest that the spatio–temporal variation of flowering has had a major effect on interactive assemblages of generalist bees. The bees, in turn, promote the speciation of plants by their flexible and broadly opportunistic visitation of flowering plants, coupled with temporal specialization.

References

Abrams, P. A. (1990). Ecological vs evolutionary consequences of competition. *Oikos* **57,** 147–151.

Ackerman, J. D. (1985). Euglossine bees and their nectar hosts. *In* "The Botany and Natural History of Panama" (W. G. D'Arcy and M. D. Correa A., eds.), pp. 225–234. *Monogr. Syst. Bot.* 10, Missouri Botanical Garden, St. Louis, Missouri.

Ackerman, J. D., and Montalvo, A. M. (1985). Longevity of euglossine bees. *Biotropica* **17,** 79–81.

Appanah, S. (1985). General flowering in the climax rain forests of Southeast Asia. *J. Trop. Ecol.* **1,** 225–240.

Appanah, S. (1987). Insect pollinators and the diversity of dipterocarps. *In* "Proceedings of the Third Round Table Conference on Dipterocarps" (A. J. G. H. Kostermans, ed.), pp. 277–291. Sannarinda, Indonesia (UNESCO).

Ashton, P. S., Givnish, T. J., and Appanah, S. (1988). Staggered flowering in the Dipterocarpaceae: New insights into floral induction and the evolution of mast fruiting in the aseasonal tropics. *Am. Nat.* **132,** 44–66.

Ayala, R. (1990). Abejas silvestres (Hymenoptera: Apoidea) de chamela, Jalisco, Mexico. *Folia Entomológica Mexicana* No. 77, 395–493 (1988).

Baker, H. G., Bawa, K. S., Frankie, G. W., and Opler, P. A., (1983). Reproductive biology of plants in tropical forests. *In* "Tropical Rain Forest Ecosystems", pp. 183–215, Elsevier Scientific, Amsterdam.

Bawa, K. S. (1974). Breeding systems of tree species of a lowland tropical community. *Evolution* **28**, 85–92.

Bawa, K. S. (1990). Plant–pollinator interactions in tropical rain forests. *Annu. Rev. Ecol. Syst.* **21**, 399–422.

Bawa, K. S., Bullock, S. H., Perry, D. R., Coville, R. E., and Grayum, M. H. (1985). Reproductive biology of tropical lowland rain forest trees. II. Pollination systems. *Am. J. Bot.* **72**, 346–356.

Boucher, D. H., James, S., and Keeler, K. H. (1982). The ecology of mutualism. *Annu. Rev. Ecol. Syst.* **13**, 315–347.

Bullock, S. H., Martinez del Rio, C., and Ayala, R. (1989). Bee visitation rates to trees of *Prockia crucis* differing in flower number. *Oecologia* **78**, 389–393.

Camargo, J. M. F., and Mazucato, M. (1986). Inventario de Apifauna e Flora Apícola de Ribeirão Preto, SP, Brasil. *Dusenia* **14**, 55–87.

Colinvaux, P. A., Miller, M. C., Kim-biu, Liu, Steinitz-Kannan, M., and Frost, I. (1985). Discovery of permanent amazon lakes and hydraulic disturbance in the upper amazon basin. *Nature* **313**, 42–45.

Cody, M. L. (1986). Structural niches in plant communities. *In* "Community Ecology" (J. Diamond, T. J. Case, eds.), pp. 381–405. Harper & Row, New York.

Colwell, R. K. (1986) Community biology and sexual selection: Lessons from hummingbird flower mites. *In* "Community Ecology" (J. Diamond, T. J. Case, eds.), pp. 406–424. Harper & Row, New York.

Foster, R. B., and S. P. Hubbell. (1990). Floristic composition of the Barro Colorado forest. *In* "Four Neotropical Forests" (A. Gentry, ed.), pp. 85–98. Yale University Press, New Haven, Connecticut.

Frankie, G. W., Haber, W. A., Opler, P. A., and Bawa, K. S. (1983). Characteristics and organization of the large bee pollination system in the Costa Rican dry forest. *In* "Handbook of Experimental Pollination Biology" (C. E. Jones and R. J. Little, eds.), pp. 411–448. Van Nostrand Reinhold, New York.

Gentry, A. H. (1974). Flowering phenology and diversity in tropical Bignoniaceae. *Biotropica* **6**, 64–68.

Gerling, D., Velthuis, H. H. W., and Hefetz, A. (1989). Bionomics of the large carpenter bees of the genus *Xylocopa*. *Annu. Rev. Entomol.* **34**, 163–190.

Hamrick, J. L., and Loveless, M. D. (1986). Isozyme variation in tropical trees: Procedures and preliminary results. *Biotropica* **18**, 201–207.

Harder, L. D., and Thomson, J. D. (1989). Evolutionary options for maximizing pollen dispersal of animal-pollinated plants. *Am. Nat.* **133**, 323–344.

Heinrich, B. (1975). Energetics of pollination. *Annu. Rev. Ecol. Syst.* **6**, 139–170.

Heithaus, E. R. (1979). Flower visitation records and resource overlap of bees and wasps in northwest Costa Rica. *Brenesia* **16**, 9–52.

Herrera, C. M. (1989). Seed dispersal by animals: A role in angiosperm diversification? *Am. Nat.* **133**, 309–322.

Hölldobler, B., and Wilson, E. O. (1990). "The Ants." Harvard University Press, Cambridge, Massachusetts.

Horvitz, C. C., and Schemske, D. W. (1990). Spatio–temporal variation in insect mutualists of a neotropical herb. *Ecology* **71**, 1085–1097.

Hubbell, S. P. (1979). Tree dispersion, abundance and diversity in a tropical dry forest. *Science* **203**, 1299–1309.

Hubbell, S. P., and Foster, R. B. (1986). Biology, chance, and history and the structure of tropical rain forest tree communities. *In* "Community Ecology" (J. Diamond, T. J. Case, eds.), pp. 314–329. Harper & Row, New York.

Hutchinson, G. E. (1978). "An Introduction to Population Ecology." Yale University Press, New Haven, Connecticut.

Inoue, T, and Nakamura, K. (1990). Physical and biological background for insect studies in Sumatra. *In* "Natural History of Social Wasps and Bees in Equatorial Sumatra" (S. F. Sakakami, R. Ohgushi, and D. W. Roubik, eds.), pp. 1–11. Hokkaido University Press, Sapporo, Japan.

Inoue, T., Salmah, S., Sakagami, S. F., Yamane, S., and Kato, M. (1990). An analysis of anthophilous insects in central Sumatra. *In* "Natural History of Social Wasps and Bees in Equatorial Sumatra" (S. F. Sakakami, R. Ohgushi, and D. W. Roubik, eds.), pp. 175–200. Hokkaido University Press, Sapporo, Japan.

Janzen, D. H. (1980). When is it coevolution? *Evolution* **34**, 611–612.

Janzen, D. H. (1983). Seed and pollen dispersal by animals: Convergence in the ecology of contamination and sloppy harvest. *Biol. J. Linn. Soc.* **20**, 103–113.

Janzen, D. H. (1986). The future of tropical ecology. *Annu. Rev. Ecol. Syst.* **16**, 305–324.

Jordano, P. (1987). Patterns of mutualistic interactions in pollination and seed dispersal: Connectance, dependence asymmetries, and coevolution. *Am. Nat.* **129**, 657–677.

Kakutani, T., Inoue, T., Kato, M., and Ichihashi, H. (1990). Insect–flower relationship in the campus of Kyoto University, Kyoto: An overview of the flowering phenology and the seasonal pattern of insect visits. *Contrib. Biol. Lab. Kyoto Univ.* **27**, 465–521.

Leigh, E. G. Jr. (1990). Community diversity and environmental stability: A re-examination. *Trends Ecol. Evol.* **5**, 340–344.

Lobreau-Callen, D., and Coutin, R. (1987). Ressources Florales Exploitées par Quelques Apoides des Zones Cultiveés en Savane Arborée Sénégalaise Durant La Saison Des Pluies. *Agronomie* **7**, 231–246.

Lobreau-Callen, D., Darchen, R., and Le Thomas, A. (1986). Apport de la Palynologie a la Connaissance des Relations Abeilles/Plantes en Savanes Arborées du Togo et du Bénin. *Apidologie* **17**, 279–306.

Manning, J. C., and Brothers, D. J. (1986). Floral relations of four species of *Rediviva* in Natal. *J. Entomol. Soc. S. Afr.* **49**, 107–114.

May, R. M. (1982). Mutualistic interactions among species. *Nature* **296**, 803–804.

Michener, C. D. (1954). Bees of Panamá. *Bull. Am. Mus. Nat. Hist.* **104**, 1–175.

Michener, C. D. (1979). Biogeography of the bees. *Ann. Missouri Bot. Gard.* **66**, 277–347.

Michener, C. D. (1989). Caste in xylocopine bees. *In* "Social Insects: An Evolutionary Approach to Castes and Reproduction" (W. Engles, ed.), pp. 120–144. Springer-Verlag, Berlin, Germany.

O'Malley, D. M, and Bawa, K. S. (1987). Mating systems of a tropical rain forest tree species. *Am. J. Bot.* **74**, 1143–1149.

Oster, G., and Heinrich, B. (1976). Why do bumblebees major? A mathematical model. *Ecol. Monogr.* **46**, 129–133.

Pearson, J. F. W. (1933). Studies on the ecological relations of bees in the Chicago region. *Ecol. Monogr.* **3**, 373–441.

Price, P. W. (1984). Communities of specialists: Vacant niches in ecological and evolutionary time. *In* "Ecological Communities" (D. R. Strong, D. Simberloff, L. G. Abele, and A. B. Thistle, eds.), pp. 510–523. Princeton University Press, Princeton, New Jersey.

Price, P. W., Slobodchikoff, C. N., and Gaus, W. S. (eds.) (1984). "A New Ecology: Novel Approaches to Interactive Systems". Wiley-Interscience, New York.

Ranta, E., and Vepsäläinen, K. (1981). Why are there so many species? Spatio-temporal heterogeneity and northern bumblebee communities. *Oikos* **36**, 28–34.

Raven, P. H., and Axelrod, D. I. (1974) Angiosperm biogeography and past continental movements. *Ann. Missouri Bot. Garden* **61**, 539–673.

Rêgo, M. M. C., and Albuquerque, P. M. C. de. (1989). Comportamento das Abelhas Visitantes de Murici, *Byrsonima crassifolia* (L.) Kunth, Malpighiaceae. *Bol. Mus. Para. Emilio Goeldi, sér. Zool.* **5**, 179–193.

Roubik, D. W. (1979). Africanized honeybees, stingless bees, and the structure of tropical plant–pollinator communities. *In* "Proc. IVth Int. Symp. on Pollination" (D. Caron, ed.), pp. 403–417. Maryland Agric. Exp. Sta. Spec. Mis. Publ. 1. College Park.

Roubik, D. W. (1982). The ecological impact of nectar-robbing bees and pollinating hummingbirds on a tropical shrub. *Ecology* **63,** 354–360.

Roubik, D. W. (1989). "Ecology and Natural History of Tropical Bees." Cambridge University Press, New York.

Roubik, D. W. (1988). An overview of Africanized honey-bee populations: reproduction, diet, and competition. *In* "Africanized honeybees and bee mites" (G. R. Needham, R. E. Page, Jr., M. Delfinado-Baker, and C. E. Bowman, eds.), pp. 45–54. Ellis Horwood, Chichester, England.

Roubik, D. W. (1990). Niche preemption in tropical bee communities: A comparison of neotropical and malesian faunas. *In* "Natural History of Social Wasps and Bees in Equatorial Sumatra" (S. F. Sakakami, R. Ohgushi, and D. W. Roubik, eds.), pp. 245–257. Hokkaido University Press, Sapporo, Japan.

Roubik, D. W., and Ackerman, J. D. (1987). Long-term ecology of euglossine orchid-bees in Panamá. *Oecologia* **73,** 321–333.

Roubik, D. W., and Boreham, M. M. (1990). Learning to live with africanized honeybees. *Interciencia* **15,** 146–153.

Roubik, D. W., Holbrook, N. M., and Parra, G. V. (1985). Roles of nectar robbers in the reproduction of the tropical treelet *Quassia amara* (simaroubaceae). *Oecologia* **66,** 161–167.

Roubik, D. W., and Moreno, J. E. (1990). Social bees and palm trees: What do pollen diets tell us? *In* "Social Insects in the Environment" (G. K. Veeresh, B. Mallik, and C. A. Viraktamath, eds.), pp. 427–428. Oxford & IBH Publishing, New Delhi, India.

Roubik, D. W., Moreno, J. E., Vergara, C., and Wittmann, D. (1986). Sporadic food competition with the African honey bee: Projected impact on neotropical social bees. *J. Trop. Ecol.* **2,** 97–111.

Roubik, D. W., and Michener, C. D. (1984). Nesting biology of *Crawfordapis luctuosa* in Panamá. *J. Kansas Entomol. Soc.* **57,** 662–671.

Roubik, D. W., Schmalzel, R. J., and Moreno, J. E. (1984). "Estudio Apibotanico de Panamá: Cosecha y Fuentes de Polen y Nectar Usados por *Apis mellifera*, y sus Patrones Estacionales y Anuales". *Organismo Internacional Regional de Sanidad Agropecuaria (OIRSA). Tech, Bull.* **24,** 1–74.

Roughgarden, J., and Diamond, J. (1986). Overview: The roles of species interactions in community ecology. *In* "Community Ecology" (J. Diamond, T. J. Case, eds.), pp. 333–343. Harper & Row, New York.

Schemske, D. W. (1983). Limits to specialization and coevolution in plant-animal mutualisms. *In* "Coevolution" (M. H. Nitecki, ed.), pp. 67–109. University of Chicago Press, Chicago.

Seeley, T. D. (1987). The effectiveness of information collection about food sources by honey bee colonies. *Anim. Behav.* **35,** 1572–1575.

Simpson, B. B, and Neff, H. L. (1985) Plants, the pollinating bees, and the great American interchange. *In* "The Great American Biotic Interchange" (F. G. Stelhi and S. D. Webb, eds.), pp. 427–452. Plenum, New York.

Smith, A. P. (1987). Respuestas de Hierbas del Sotobosque Tropical a Claros Ocasionados por la Caida de Arboles. *Rev. Biol. Trop.* **35,** 111–118.

Snelling, R. R. (1984). Studies on the taxonomy and distribution of American centridine bees. *Contrib. Science Nat. Hist. Mus. Los Angeles Co.* No. **347,** 1–69.

Snow, A. A., and Roubik, D. W. (1987). Pollen deposition and removal by bees visiting two tree species in Panamá. *Biotropica* **19,** 57–63.

Tepedino, V. J., and Stanton, N. L. (1981). Diversity and competition in bee–plant communities on short-grass prairie. *Oikos* **36,** 35–44.

Terborgh, J. (1986). Community aspects of frugivory in tropical forests. *In* "Frugivores and Seed Dispersal" (A. Estrada and T. H. Fleming, eds.), pp. 371–384. Junk, Dordrecht, The Netherlands.

Thompson, J. N. (1988). Variation in interspecific interactions. *Annu. Rev. Ecol. Syst.* **19,** 65–88.

Toft, C. A. (1986). Communities of species with parasitic life-styles. *In* "Community Ecology" (J. Diamond, T. J. Case, eds.), pp. 445–463. Harper & Row, New York.

Toft, C. A., and Karter, A. J. (1990). Parasite–host coevolution. *Trends Ecol. Evol.* **5,** 326–329.

Vinson, S. B., Frankie, G. W., and Coville, R. E. (1987). Nesting habits of *Centris flavofasciata* in Costa Rica. *J. Kansas Entomol. Soc.* **60,** 249–263.

West-Eberhard, M. J. (1989). Phenotypic plasticity and the origins of diversity. *Annu. Rev. Ecol. Syst.* **20,** 249–278.

Wiens, J. A. (1989). "The Ecology of Bird Communities: Foundations and Patterns." Cambridge University Press, Cambridge, England.

Wolda, H., and Roubik, D. W. (1986). Nocturnal bee abundance and seasonal bee activity in a Panamanian forest. *Ecology* **67,** 426–433.

Zimmerman, J. K., Roubik, D. W., and Ackerman, J. D. (1989). Asynchronous phenologies of a neotropical orchid and its euglossine bee pollinator. *Ecology* **70,** 1192–1195.

12

How Do Fruit- and Nectar-Feeding Birds and Mammals Track Their Food Resources?

Theodore H. Fleming
Department of Biology
University of Miami
Coral Gables, Florida

I. Introduction
II. Resource Variability in Theory and Practice
    A. Resource Variability in Theory
    B. Resource Variability in Practice
III. Responses by Frugivores and Nectarivores to Resource Variability
    A. Demographic Responses
    B. Movements
    C. Social Responses
IV. Conclusions
    Appendix
    References

I. Introduction

Compared with other diet classes, frugivory and nectarivory are rather uncommon feeding specializations in birds and mammals. In a recent review of the evolutionary history of fruits and frugivores, I identified only 12 avian families, containing about 600 species, and eight mammalian families, containing about 460 species, as being principally frugivorous in diet (Fleming, 1991). Similarly, only three major families of birds (Trochilidae, Nectariniidae, and Melaphagidae; about 630 species), plus perhaps 200 additional species in several other families (Collins and Paton, 1989), and only two families of bats (Pteropodidae and Phyllostomidae; less than half of their 310 species are strongly nectarivorous), a handful of primates, and the marsupial *Tarsipes rostratus* regularly include nectar and pollen in their diets. Together, these two diet groups represent about 17% of the current diversity of birds and mammals. Despite this relatively low taxonomic diversity, avian and mammalian fruit- and nectar-eaters play extremely important roles as seed dispersers and pollinators in terrestrial ecosystems, especially in the tropics (Bawa, 1990; Fleming, 1988; Stiles, 1985).

Like all animals, frugivores and nectarivores spend most of their lives finding and eating food. The problems that frugivores and nectarivores face in feeding, however, are not necessarily the same as those faced by herbivores and carnivores (including insectivores), the other major diet classes of mammals and birds. Plants often defend their tissues morphologically, chemically, and phenologically against consumption and assimilation by herbivores (Chapters 7 and 10). Similarly, animal prey species use a wide variety of methods to reduce their detection and consumption by carnivores (Chapter 8). As a result of plant defenses, herbivores probably spend more time processing than locating food, whereas the opposite is usually true of carnivores.

In contrast to herbivores and carnivores, which generally have antagonistic relationships with their food species, frugivores and nectarivores generally have positive, mutualistic relationships with their food species. Plants offer a nutritional reward to animals in the form of seed-filled fruits and nectar- and pollen-containing flowers in exchange for increased mobility for their seeds and pollen grains. Though basically positive, these plant–animal interactions are *uneasy partnerships* (Howe and Westley, 1988) that involve conflicts of interest between plants and their dispersers and pollinators. This conflict results from plants requiring more mobility for their seeds and pollen than animals, which are selected to maximize net energy gain per unit of foraging time, are willing to give. Resolution of these conflicts probably seldom occurs for a variety of reasons, including the effects of climatic seasonality on dietary generalization, interspecific differences in foraging behavior, and the lack of congruence in plant and animal distributions (Howe, 1984; Herrera, 1986). Plants can, however, manipulate frugivore and, to an even greater extent, nectarivore foraging behavior with a variety of morphological, nutritional, and phenological methods (Feinsinger, 1983, 1987; Wheelwright and Orians, 1982).

Although fruits and flowers are often conspicuous and easily located, these foods are generally described as being *patchy* and *ephemeral*. Compared with other kinds of foods, especially insects, fruit and flower densities are usually considered to be quite variable in space and time. Assuming for the moment that this is true, my goal in this chapter is to explore the ways in which this resource variability influences the ecology and behavior of avian and mammalian frugivores and nectarivores. I will pay particular attention to the effect resource variability has on the demography and abundance, daily and seasonal movements, social organization, and mating behavior of these birds and mammals. One of the major points that will emerge from this chapter is that the lives of many species of fruit- and nectar-eaters operate on a large spatial scale. Ephemeral, patchy resources select for high mobility, which, as we will see, can profoundly affect many aspects of the lives of frugivores and nectarivores. An excellent overview of animal reponses to patchy environments can be found in Wiens (1976).

II. Resource Variability in Theory and Practice

A. Resource Variability in Theory

To maximize lifetime fitness, animals should closely *track* resources. This tracking can involve at least four different aspects of their life histories:

1. demography/physiology, including the timing of breeding, molt, and torpor, or hibernation;
2. daily and seasonal foraging movements;
3. social organization, including intra and interspecific social interactions, such as territorial and/or gregarious behavior; and
4. mating systems, i.e., monogamy versus polygamy.

Points 3 and 4 have been discussed in Chapter 3, but on a much finer spatial scale than considered here. The major working hypothesis in this chapter is that these four life-history components of frugivores and nectarivores are affected, at least in part, by patterns of resource variation. A second hypothesis is that the life histories of frugivores and nectarivores differ significantly from those of insectivores, the diet class from which frugivory and nectarivory have evolved.

Resource variation and resource tracking have both spatial and temporal aspects. Although space and time are continuous variables, I find it convenient to recognize at least two dimensions for each of these variables. Space has three important dimensions—latitude, longitude, and elevation—in the lives of many frugivores and nectarivores. As discussed in detail later, the annual cycles of many species involve significant spatial shifts along each of these dimensions. Similarly, the time variable can be subdivided into several dimensions, including daily, weekly, and monthly, or longer, time blocks. In general, because nectar availability can change substantially on an hourly basis (e.g., Hixon *et al.,* 1983; McFarland, 1986; Pyke, 1988), temporal variation probably occurs on a finer scale for nectarivores than it does for frugivores. Frugivores and nectarivores thus live in multidimensional worlds both spatially and temporally, and we expect their life histories to reflect the variability underlying this dimensionality.

What are the expected relationships between resource variability and the life histories of frugivorous and nectarivorous birds and mammals? To begin to answer this question, consider the two environments depicted in Figure 1. Environment A completely lacks spatio–temporal variation in the number of food species (40) available each month over a wide range of latitudes. In contrast, the number of food species available in Environment B varies widely in time and space. Spatio–temporal resource variability (patchiness) is absent in Environment A and is high in B. Resource tracking problems clearly are more acute in Environment B than in A. Compared with an organism living in Environment A, an organism in Environment B should have a much more strongly seasonal and less-sedentary life history at all latitudes. In Environment B, breeding should be more strongly seasonal; socially mediated spacing patterns are likely to vary seasonally; social systems are more likely to be gregarious than intolerant; and individuals are more likely to migrate or switch to another food type (Wiens, 1976). As discussed below, the worlds of real-life frugivores and nectarivores are more like Environment B than A.

The two environments in Figure 1 provide a visual picture of spatio–temporal patchiness, but it would be useful to have a more quantitative definition of the term patchy. Intuitively, such a definition might be obtained by calculating resource state changes in an m by n matrix consisting of m temporal columns and n spatial rows. Patchy environments will have higher rates of change and more extreme values from one cell to another than will non-patchy environments. Evaluating the magnitude of resource levels in each cell of the matrix in relation to an animal's energetic needs for reproductive, social, and migratory purposes would allow us to answer such questions as: When and where should a species breed, defend resources, form pair bonds or mate polygamously, join foraging groups, or migrate?

Spatio–temporal resource variability must be viewed in relation to the body size and locomotory adaptations of frugivores and nectarivores. Sizes of avian frugivores range from about 0.010 kg (small tanagers) to 2 to 3 kg (hornbills), and from 0.005 to 1 kg in bats. Sizes of arboreal or terrestrial avian frugivores range from 0.25 (tinamous) to 58 kg (cassowaries), and in arboreal mammals, from 0.40 (tamarins) to 150 kg (orangutans). Terrestrial mammalian frugivores range from 0.020 (rodents) to 7,500 kg (African elephant). Nectarivores are generally smaller than frugivores. Sizes of avian nectarivores range from 0.002 (hummingbirds) to 0.15 kg (wattlebirds) and from 0.008 to 1 kg in bats. Nectarivorous arboreal mammals range from 0.013 (*Tarsipes*) to about 3 kg (in *Cebus* monkeys, which are only partially nectarivorous).

Size is important because we expect the effects of resource and climatic seasonality to decrease as body size increases in both birds and mammals; small species generally live in more coarse-grained environments than do large species [see discussions in Calder (1984) and Peters (1983)]. Likewise, aerial species can more easily track spatially variable resources and can evade seasonally lean or physiologically harsh times of the year by migrating

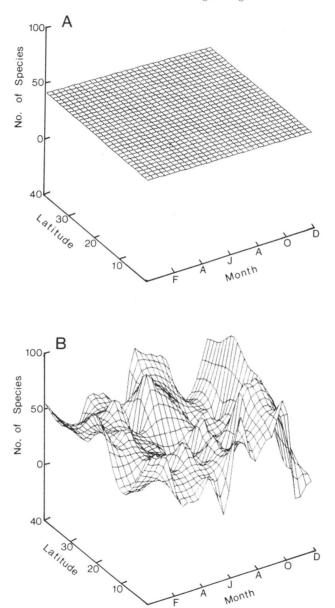

Figure 1 Resource variation, measured as the number of food species available per month, in two idealized environments. Values in 1B were generated by selecting pairs of random numbers for each month at latitudes 0, 10, 20, 30, and 40.

than can terrestrial or arboreal species. Because of their greater mobility, we expect a greater number of aerial birds and mammals to be committed frugivores and nectarivores year-round than are arboreal or terrestrial species.

B. Resource Variability in Practice

How variable in time and space are real-world fruit and nectar resource levels? To more fully understand the evolution of mating systems in New Guinea birds of paradise, Beehler (1983) indicated that more information was needed on spatial dispersion of different food plants, size of fruit crops, intraspecific synchrony in fruit ripening, length of fruiting seasons, annual predictability of fruiting cycles, and the nutritional composition of fruit. Obtaining all of the information in this "wish list," and determining annual variation in fruit/flower crop sizes, would seem to be essential for understanding the behavioral ecology of frugivores and, by substituting flowers for fruits in the list, of nectarivores. Not surprisingly, however, such complete resource information is rarely available for any study system [for an exception, see Herrera (1984)], so we currently have an incomplete picture of the resource environments of most species of frugivores and nectarivores.

Current information about fruit and flower resource environments generally supports the hypothesis that these resources are indeed patchy in time and space. For example, in the handful of studies that have measured temporal and/or spatial variation in animal and plant resources in the same study area, fruit and flower abundance appears to be more variable than insect abundance (Karr, 1976; Martin and Karr, 1986a; Pyke, 1983).

Rather than exhaustively review fruit and flower resource patterns, I shall illustrate general phenological trends with selected sets of data. More extensive reviews of these topics can be found in Baker *et al.* (1983), Feinsinger (1987), Fleming *et al.* (1987), Primack (1987), and Rathcke and Lacey (1985). Two broad latitudinal patterns exist regarding seasonal trends in the number of fruiting and flowering plants:

1. seasonality increases with increasing latitude, and
2. species diversity decreases with increasing latitude (Fig. 2).

At midlatitudes in the temperate zone, the availability of fleshy fruits is highest in the autumn and winter, and lowest in the spring and summer; it is highest during rainy seasons and lowest at wet-to-dry-season transitions in most tropical habitats. Flowering peaks occur in the summer in the eastern United States, but in the winter and spring in southeastern Australia (Paton, 1985a). The dry season (December through March) contains peak flower numbers for nectarivorous bats in Central America (Heithaus *et al.*, 1975), but hummingbird flower peaks occur in both the dry and wet seasons (Fig. 2).

Figs are perhaps the archetypical patchy fruits. Consumed by a wide

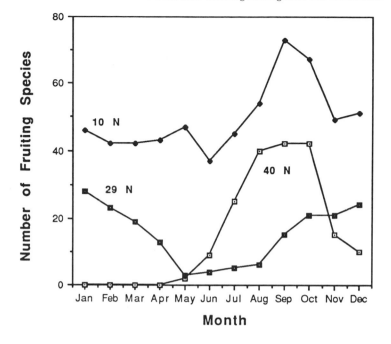

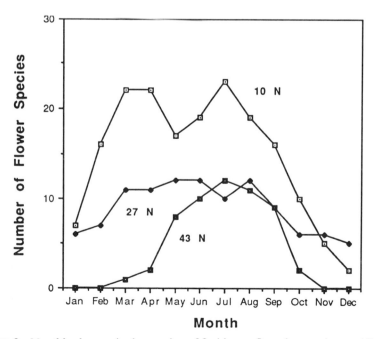

Figure 2 Monthly changes in the number of fruiting or flowering species providing food for birds at three latitudes in the New World. From Austin (1975), Frankie *et al.* (1974), Skeate (1987), Stiles (1980), Thompson and Willson (1979).

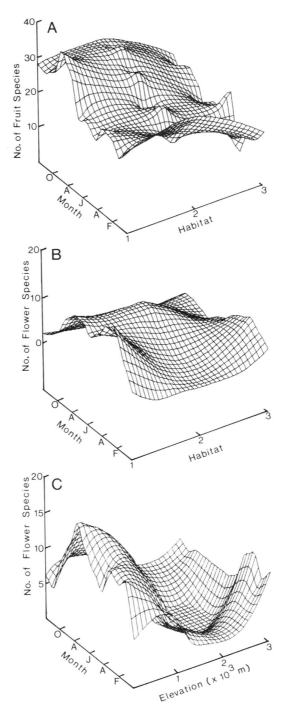

variety of opportunistic and specialized tropical frugivores (Terborgh, 1986), most species of *Ficus* produce enormous but short-lived crops of nutritionally poor fruit. Owing to intraspecific asynchrony in syconia (the technical name for the fruit/flower unit) production and low population densities, fig crops can be widely separated in space and time. Despite being extremely patchy resources, figs serve as keystone fruit resources in some, but not all, tropical forests (Gautier-Hion and Michaloud, 1989; Leighton and Leighton, 1983; Terborgh, 1986). I am unaware of an analogous example of a pantropical, spatio–temporally patchy flower source that serves as a keystone resource for nectarivores.

Using published data, I have attempted to illustrate a *bird's eye view* of spatio–temporal variation in bird-visited fruit and flower species richness at two different spatial scales in Figure 3. In lowland Costa Rican wet forest, fruit diversity appears to be somewhat *bumpier* (i.e., patchier?) than flower diversity in three habitats at La Selva. Seasonal habitat shifts might be expected there, however, in both frugivorous and nectarivorous birds. Seasonal flower richness is higher in both lowland and highland regions of Costa Rica than at middle elevations. This creates a highly patchy landscape for hummingbirds and leads to the prediction that mid-elevation species should migrate either upslope or downslope at certain times of the year. Grant and Grant (1967: Fig. 2) illustrate a similar seasonal shift in the location of hummingbird flowers in California; spring-blooming flowers are concentrated in the lowlands, whereas summer-blooming flowers are concentrated in the Sierra Nevada mountains. Loiselle and Blake (1991a) also document seasonal changes in fruit availability along an altitudinal transect in Costa Rica.

Year-to-year variation in fruit and flower levels can also be substantial. In mediterranean scrublands, annual changes in fruit levels are lower in the lowlands than in the uplands (Herrera, 1984). Skeate (1987) noted that annual variation in fruit crop size was lower in most species of bird-dispersed shrubs, herbs, and vines than in canopy trees in a north Florida hammock community. Flower and nectar levels for hummingbirds, Australian honeyeaters, and Hawaiian honeycreepers sometimes show marked annual variations; Australian *Eucalyptus* is an especially variable flower source (Carpenter, 1978; Keast, 1968; Paton, 1985a; Pyke, 1983, 1988; Stiles, 1978). Mast fruiting, which is particularly common in West Malaysia (Appanah, 1985), is an extreme example of temporal patchiness.

Figure 3 Monthly changes in the number of fruiting or flowering species providing food for birds in three habitats at La Selva, Costa Rica (3A and 3B, respectively) and at four elevations (100, 1000, 1300, and 3000 m) in Costa Rica (3C). Habitats in 3A: 1, edges and second growth; 2, forest understory; 3, forest canopy. Habitats in 3B: 1, second growth; 2, gaps in primary forest; 3, primary forest understory. From Feinsinger (1976), Levey (1988), Stiles (1980, 1985b), Wolf *et al.* (1976).

Because frugivores and nectarivores interact mutualistically with their food plants, we can ask: To what extent are patterns of seasonal and spatial resource variation the products of plant–animal coevolution? Herrera (1984), Skeate (1987), Snow (1971), and Thompson and Willson (1979), among others, have suggested that the fall–winter fruiting peak in temperate latitudes is a response to bird migration. As Rathcke and Lacey (1985) point out, however, disentangling cause and effect relationships in these systems can be difficult, and other explanations are possible (e.g., Herrera, 1984). Similarly, staggered fruiting and blooming seasons that provide birds and mammals with spatio–temporally predictable food supplies have been interpreted as products of interspecific competition for limited dispersers or pollinators, but this hypothesis has seldom been critically examined (Rathcke and Lacey, 1985; see Chapters 5, 11, and 13 for insect pollinator examples). Stiles (1977, 1979) maintained that the staggered blooming times of hummingbird-pollinated plants in Costa Rican wet forest are the product of interspecific competition. In contrast, Murray *et al.* (1987) failed to find the expected character displacement in the blooming times of two guilds of hummingbird flowers at Monteverde, Costa Rica. They attributed this lack of pattern to the spatio–temporal variation these plants experience in both plant–plant and plant–animal interactions [but see Pleasants' (1990) caveat on the statistically conservative nature of their data analysis and Chapter 13 for a discussion of identifying true guilds].

Despite the existence of controversy in this area, it seems reasonable to expect plants to evolve fruiting and/or flowering seasons that coincide with influxes of frugivorous and nectarivorous birds and mammals migrating north or south to avoid physiologically (or biotically?) unfavorable conditions. Such evolution should produce fruit or nectar pathways. One possible example of such a pathway is shown in Figure 4. Large numbers of the nectar-feeding bat *Leptonycteris curasoae* migrate from tropical southern Mexico to spend the spring and summer in the Sonoran desert of northern Mexico and the southwestern United States (Cockrum, 1989). During their northward migration, they feed on nectar and pollen produced by several species of night-blooming cacti. During fall migration, they feed on various species of night-blooming *Agave*. Stable carbon isotope analysis of *Leptonycteris* muscle tissue confirms that this bat specializes on crassulacean acid metabolism (CAM) plants (i.e., plants of the Cactaceae and Agavaceae) during the spring, summer, and early fall, but not during the winter when it feeds on a mixture of C3 (e.g., tropical trees and shrubs) and CAM plants (T. Fleming and L. Sternberg, unpublished data, 1991). Such specialization increases the likelihood of coevolution between these bats and plants.

In summary, frugivorous and nectarivorous birds and mammals feed on food supplies that display considerable variation on a variety of temporal and spatial scales. The availability of these foods appears to be patchier than certain other kinds of foods. It is important to realize, however, that much

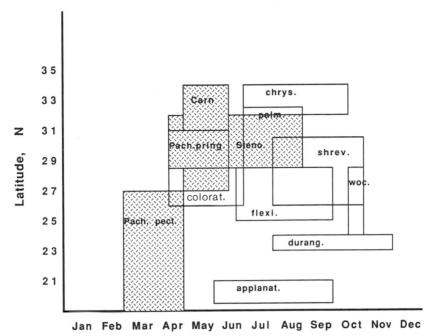

Figure 4   Blooming times and latitudinal distributions of bat- pollinated cacti (filled rectangles) and *Agave* (open rectangles) in western Mexico. Genera of cacti include *Carnegia* (Carn.), *Pachycereus* (Pach.) and *Stenocereus* (Steno.). Species of *Agave* include *applanata, chrysantha, colorata, durangensis, flexispina, palmeri, shrevei,* and *wocomahi.* From Gentry (1982), Shreve and Wiggins (1964).

of this patchiness is temporally and spatially predictable. These animals do not live in chaotic worlds such as depicted in Figure 1B. But neither are their worlds totally constant (i.e., Fig. 1A). To cope with spatio–temporally inconstant, but contingent (i.e., seasonally predictable; Colwell, 1974) worlds, these animals should have spatio–temporally inconstant, but nonetheless predictable life histories.

III. Responses by Frugivores and Nectarivores to Resource Variability

A. Demographic Responses

We expect energetically expensive activities such as reproduction and molt to coincide with energetically favorable times of the year. This is generally the case in most species of frugivores and nectarivores; birth and nesting periods tend to coincide with fruit and flower peaks, respectively. Two well-studied neotropical species can be used to illustrate this point.

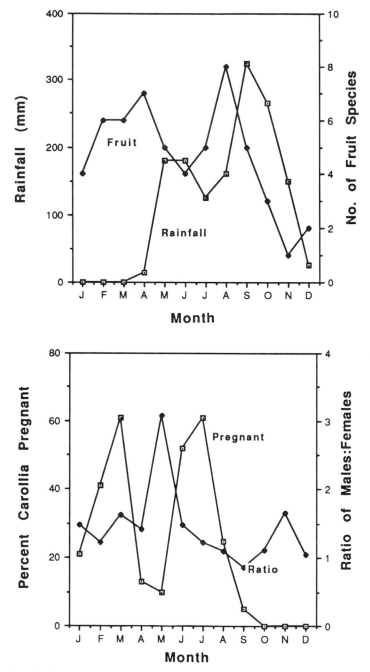

Figure 5   The female reproductive cycle and population sex ratios of the frugivorous bat, *Carollia perspicillata*, in relation to monthly fruit diversity and rainfall at Santa Rosa National Park, Costa Rica. From Fleming (1988).

Like most species of plant-visiting phyllostomid bats (Wilson, 1979), the short-tailed fruit bat (*Carollia perspicillata*) breeds twice a year. Females give birth to a single young in the late dry season (April), and in the middle of the wet season (August). Birth peaks coincide with two annual fruit peaks (Fig. 5), but the young are not weaned at equally favorable times of the year. Babies born in April become independent (at about 5 weeks of age) when fruit levels are increasing, whereas those born in August become independent after the wet-season fruit peak. Surprisingly, survivorship differences between these two cohorts of young do not differ significantly (Fleming, 1988).

An unusual aspect of the demography of *Carollia*, at least in Costa Rican lowland dry tropical forest, is that most females, but not males, migrate from their wet-season cave roosts during the dry season when fruit levels generally are low. As a result, sex ratios in these caves become strongly male-skewed late in the dry season (Fig. 5), and most males in these caves participate in only one mating period per year. When they return to the lowland caves in the wet season, females are pregnant with babies fathered in their (unknown but probably montane) dry season roosts. When they arrive in their dry-season roosts, they are pregnant with babies fathered in their wet-season roosts. In this species, females apparently migrate for energetic reasons, and males are sedentary for social reasons (Fleming, 1988).

The long-tailed hermit (*Phaethornis superciliosus*) is one of the most common hummingbirds in the understory of lowland wet forests in Costa Rica. *Phaethornis* is a lek-breeding species whose annual reproductive cycle is summarized in Figure 6. The main calling and nesting season occurs in January through July when nectar levels, especially those of its main food plant *Heliconia pogonatha,* are highest. Stiles and Wolf (1979) estimate that up to 75% of the annual mortality in this species occurs during October and November when flowers are scarcest. They suggest that this species may be food limited at their La Selva study site.

Many other studies indicate that breeding activity and resource levels are positively correlated in frugivores and nectarivores. In frugivores, birth peaks in New World primates (e.g., *Ateles, Saimiri,* and *Saguinus*) coincide with fruit peaks (Terborgh, 1983; van Roosmalen, 1980), as do the breeding seasons of manakins (*Manacus, Pipra*) [(on Barro Colorado, but possibly not at La Selva; cf. Worthington (1982) and Levey (1988)]], and of the resplendent quetzal (for the scientific names of birds mentioned in the text, see the Appendix) (Wheelwright, 1983). The reproductive cycles of three species of frugivorous bats (*Artibeus jamaicensis, Eidolon helvum,* and *Haplonycteris fischeri*) are unusual among bats because they involve delayed implantation or delayed embryonic development. Heideman (1988) and Sandell (1990) hypothesize that these reproductive cycles have evolved to allow both mating and births to coincide with energetically favorable times of the year. In

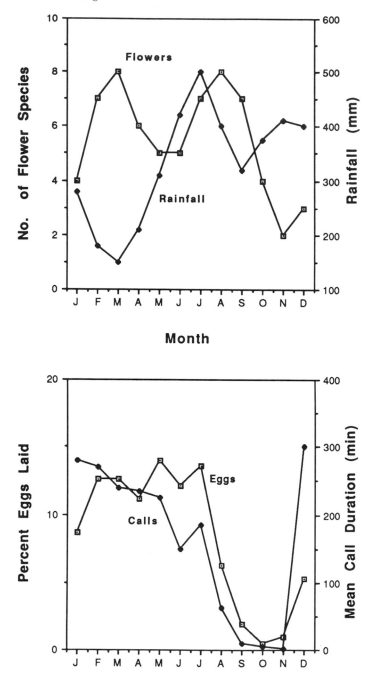

Figure 6 The male calling and female breeding season of the long-tailed hermit hummingbird (*Phaethornis superciliosus*) in relation to monthly flower diversity and rainfall at La Selva, Costa Rica. From Stiles and Wolf (1979).

nectarivores, breeding–flower peak correlations have been reported for Australian honeyeaters (Paton, 1985b; Pyke, 1983; Pyke and Recher, 1986) and New World hummingbirds (Stiles, 1973, 1980, 1985b).

In addition to influencing the timing of breeding, fruit and nectar levels determine patterns of frugivore and nectarivore abundance. Although they are seldom measured simultaneously in the same study area, it is likely that the annual (and often instantaneous) biomass of fruit is far greater than that of nectar (and insects?). For example, at peak fruiting times the wet-weight biomass of 14 species of bat-dispersed fruits in Costa Rican dry tropical forest ranges from 0.1 to 42 kg/ha/day with a median value of 1.5 kg/ha/day. This can be compared with values of 0.01 to 0.11 kg/ha/day of nectar in three species of bat-pollinated columnar cacti in the Sonoran desert, whose bat–flower density is much higher than that of tropical dry forest (Fleming, 1988; T. Fleming and P. Horner, unpublished data, 1990). Similar measurements for insect biomass would be extremely valuable.

These kinds of biomass differences mean that frugivores should have higher population densities, on average, than nectarivores and insectivores. In support of this expectation, Terborgh *et al.* (1990) reported that arboreal frugivore biomass at Manu National Park, Peru, was two orders of magnitude higher than that of nectarivores (26 kg/100 ha, cf. 0.2 kg/100 ha). Mist-net captures of bats at several neotropical locations indicate that frugivorous species of the phyllostomid subfamilies Carolliinae and Stenodermatinae are 4–5 times more abundant than nectarivorous Glossophaginae and insectivorous Phyllostominae (Fleming, 1988). A similar abundance disparity occurs among frugivorous and insectivorous birds in mediterrean scrubland (Herrera, 1984) and tropical habitats (Karr, 1990; Loiselle, 1988; Loiselle and Blake, 1991a).

A second resource-related pattern is temporal variations in abundance, which are often greater in frugivores and nectarivores than in insectivores. Such differences have been reported for temperate fruit-eating birds (Martin and Karr, 1986b) and for Australian honeyeaters (Newland and Wooller, 1985; Ramsey, 1989) (Fig. 7). As I shall discuss in detail, these differences often result from the greater seasonal mobility of frugivores and nectarivores compared with that of insectivores.

A third pattern is seasonal changes in frugivore and nectarivore abundance, which often correlate with seasonal changes in fruit and flower abundances. Positive relationships between abundance and resource levels have been reported in temperate and tropical frugivorous birds (Blake and Hoppes, 1986; Levey, 1988; Loiselle, 1988; Loiselle and Blake, 1991a; Martin and Karr, 1986b), in tropical hummingbirds (Feinsinger, 1976, 1980; Lyon, 1976; Stiles, 1980, 1985b), and in Australian honeyeaters (Collins, 1980; Collins and Briffa, 1982; McFarland, 1986; Paton, 1985b; Pyke, 1985; Ramsey, 1989). Again, much of this variation in abundance can be attributed to immigration rather than to reproduction and mortality.

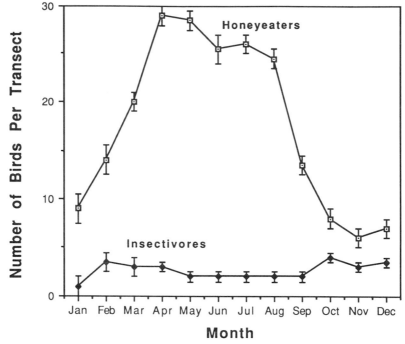

Figure 7   Mean number (± 1 S.E.) of honeyeaters and insectivorous birds seen in monthly transects in *Banksia* woodlands near Perth, Australia. Redrawn with permission from Ramsey (1989).

B. Movements

As Karr (1990) noted for birds, the tracking of food resource abundance drives much of the daily and seasonal movement, including habitat shifts and altitudinal and latitudinal migrations, of frugivores and nectarivores. The importance of plant resources in determining movements in nectarivorous birds is highlighted in the following quotes. "nectarivorous birds as a group show more pronounced migratory or nomadic behavior than do birds of virtually any other trophic category" (Stiles, 1980, p. 340). "The overwhelming majority of seasonal movements in the Meliphagidae [honeyeaters] are directly associated with nectar feeding" (Keast, 1968, p. 198).

1. Daily Movements

How far do frugivores and nectarivores range on a daily basis to harvest their food? In primates, body size and food habits interact to determine daily range lengths. For example, because they are relatively large and specialize on ripe fruits, spider monkeys (*Ateles*) have larger daily range lengths (up to 5 km) than do smaller *Saimiri* and *Cebus* monkeys (up to 3 to

4 km), or larger but more folivorous *Lagothrix* and *Brachyteles* monkeys (up to 1.6 km) (Robinson and Janson, 1987). Similarly, two sympatric hylobatids, the larger, more folivorous siamang (*Hylobates syndactylus*), and the more frugivorous lar gibbon (*H. lar*), differ in daily range length and feeding time; the lar travels twice as far but feeds for only 69% as long as the siamang (Oates, 1987). Waser (1987) noted that a *large patch specialist* primate has evolved to harvest ephemeral but superabundant fruits (e.g., figs) on each tropical continent. Each of these primates (*Hylobates, Pongo, Pan, Cercocebus,* and *Ateles*) has enormous ranges for harvesting these fruits.

Daily range lengths of frugivorous birds and bats can also be substantial. Foraging movements of tropical understory frugivores, such as manakins and other small passerines and *Carollia* bats, vary from a few hundred meters to 3 km (Fleming, 1988; Murray, 1988; Snow 1962a,b). In contrast, canopy feeders such as oilbirds, fruit pigeons, birds of paradise, and pteropodid fruit bats range much more widely (i.e., 10–50 km) between roosts and feeding areas (Beehler and Pruett-Jones, 1983; Bradbury, 1977; Fleming, 1982; Leighton and Leighton, 1983; Snow, 1962c). Terborgh *et al.* (1990) reported that flocks of canopy-feeding frugivorous birds have much larger territories (>20 ha) and use the forest in a patchier fashion than do understory flocks whose territories average about 5 ha. Some chiropteran fig specialists (e.g., *Artibeus, Nyctimene*) forage only a few hundred meters from their day roosts when fruiting plant densities are high (Morrison, 1978; Spencer and Fleming, 1989), whereas others (e.g., *Hypsignathus, Pteropus*) fly 5–50 km from their roosts to feed (Bradbury, 1977; Marshall, 1985). The low density of fig trees in Gabon forests precludes primates and large birds from specializing on them. Instead, figs are eaten by wide-ranging pteropodid bats (Gautier-Hion and Michaloud, 1989).

The daily foraging ranges of nectar-feeding birds and mammals are less well studied than are those of fruit-eaters. Territorial hummingbirds and honeyeaters apparently move relatively short distances (up to 1 km from their territories) to feed, whereas trap-liners, such as the long-tailed hermit, probably are more mobile (Linhart, 1973; Newland and Wooller, 1985; Paton, 1985b; Stiles, 1973; Stiles and Wolf, 1979). The nightly commute distances between roosts and feeding areas of nectarivorous bats range from short (probably < 5 km in *Glossophaga*) to relatively long (10–30 km in *Anoura, Eonycteris,* and *Leptonycteris*) (Helversen and Reyer, 1984; P. Horner and T. Fleming, unpublished data, 1990; Lemke, 1984; Start and Marshall, 1976). Radio-tracking studies of *Leptonycteris curasoae* in the Sonoran desert indicate that these 27 g bats routinely fly 80–100 km during a 7-hr foraging period (P. Horner and T. Fleming, unpublished data, 1990).

2. Seasonal Movements

Seasonal movements of varying spatial magnitudes are common in frugivores and nectarivores. These movements can be broken down into three classes based on spatial scale: habitat shifts, altitudinal migrations, and latitudinal migrations. Habitat shifts appear to be widespread in plant-visiting birds and mammals. In the tropics, such shifts generally involve movements among habitats along successional gradients (Karr, 1989, 1990). For example, Levey (1988) reported that resident frugivorous birds at La Selva, Costa Rica, moved from primary forest into second growth when fruit levels in the former habitat declined relative to those in the latter. Resident hummingbirds at that site also exhibit food-related habitat shifts (Stiles, 1980). In central Panama, frugivorous birds appear to be more likely to change habitats than insectivores or nectarivores (Martin and Karr, 1986a). Many species of Australian honeyeaters undergo habitat changes as they track changing flower distributions (Ford, 1985; Keast, 1968; Paton, 1985a; Pyke, 1983, 1985). Although resource tracking appears to be a major motivating factor behind these habitat shifts, other factors, including seasonal microclimatic constraints and nesting requirements, can also influence these movements (Karr and Brawn, 1990; Skutch, 1950).

Altitudinal migrations have been best studied in Costa Rican birds in which frugivores and nectarivores are more likely to move upslope or downslope than are insectivores (Stiles, 1983). The timing of these movements, which generally occur outside the breeding season, differs among frugivores and nectarivores. Frugivorous altitudinal migrants arrive in the lowlands in October and November and return upslope in January; hummingbirds arrive in the lowlands between April and August and begin to return upslope in August. Arrivals in the lowlands coincide with periods of high fruit or flower abundance (Blake *et al.*, 1990). Migration tendencies in frugivorous birds are complex, and Loiselle and Blake (1991b) identify three classes of altitudinal migrants:

1. complete long distance migrants—species whose entire populations move > 1000 m in elevation between breeding and nonbreeding sites;
2. complete short-distance migrants moving < 1000 m in elevation between breeding and nonbreeding sites; and
3. partial short-distance migrants—species in which only a portion of the population leaves the breeding site.

They also noted that the proportion of migrant species and migrant individuals increases with increasing elevation, and that the intensity and timing of migrations vary between years.

Latitudinal migrants abound among North Temperate zone birds and bats, including all avian nectarivores and frugivores. Hummingbirds breeding in western North America and migrating south along the Sierra Nevadas move in waves and partition limited flower resources inter- and intraspecifically by means of different migration times (Carpenter, 1978; Kodric-Brown and Brown, 1978; Phillips, 1975). Upon arrival at their wintering grounds in Mexico and Central America, these species are socially subordinate to resident tropical species (DesGranges and Grant, 1980; Wolf, 1970). Partially or wholly frugivorous latitudinal migrants become important components of winter bird communities in the Central American tropics, as well as in mediterranean scrublands (Herrera, 1984; Levey, 1988).

Compared with birds, less is known about latitudinal migrations in plant-visiting bats. In the New World, long-distance migrations appear to occur only in nectarivorous bats associated with the Sonoran desert (*Leptonycteris* and *Choeronycteris*) (Barbour and Davis, 1969). It is likely, however, that other neotropical phytophagous bats (e.g., *Phyllostomus discolor*) undergo at least short-distance seasonal movements (Heithaus *et al.*, 1975). Three West African frugivorous bats (*Eidolon helvum, Myonycteris torquata,* and *Nanonycteris veldkampi*) migrate from the forest zone north to the savanna zone early in the wet season. Thomas (1983) postulated that these species migrate against a food-resource gradient, away from an environment rich in fruit and competitors, to one in which food and competition levels are lower because of stronger seasonal food fluctuations. *Pteropus* bats in eastern Australia also migrate hundreds of kilometers, but not necessarily in a latitudinal fashion, to different feeding areas and roost sites in response to changes in the locations of good flower sources (Nelson, 1965; Ratcliffe, 1932).

In addition to habitat, altitudinal, and latitudinal shifts, apparently *nomadic* wanderings are known to occur in some species of nectarivorous and frugivorous birds and mammals. Keast (1968), for example, reported that about 23% of Australian bird species undergo random, nomadic, or otherwise spatially nonrepetitive movements, whereas only about 8% of the species undergo annual north–south movements. He noted that in honeyeaters there is a broad correlation between the proportion of species making seasonal movements in an area and the variability of that area's rainfall; 39% of 65 species make moderately to strongly developed nomadic movements. In Australian eucalypt forests, birds such as honeyeaters, lorikeets, and pardalotes feeding on nectar, manna, honeydew, or lerps are often nomadic, whereas insectivores are sedentary (Ford, 1985). The smallest of Australia's four species of flying foxes (*Pteropus scapulatus*) is strongly dependent on *Eucalyptus* flowers and is the most nomadic species (Richards, 1983a). A number of Old World frugivorous birds, including fruit pigeons in Australia and Borneo, and flowerpeckers and birds of paradise in New Guinea, are also thought to be nomadic (Crome, 1975; Leighton and Leighton, 1983; Pratt and Stiles, 1985).

As Stiles (1973, 1980) has noted, temperate and tropical hummingbirds can be highly nomadic. Tropical hummingbird communities contain three major groups of species: (1) residents that tend to defend the richest nectar sources, (2) altitudinal or latitudinal migrants, and (3) wanderers or nomads that follow the blooming seasons of particular species from one habitat to another (DesGranges and Grant, 1980; Feinsinger, 1976, 1980; Stiles, 1980, 1985b). Three of the 14 species of hummingbirds feeding in successional habitats at Monteverde, Costa Rica (ca. 1500 m) are nomads; an additional five species are altitudinal migrants; and the remaining species are either residents (four species), or migrants from adjacent habitats (Feinsinger, 1976). The ebb and flow of hummingbird species into and out of particular habitats thus make their communities extremely dynamic.

C. Social Responses

1. Spacing Patterns

Many authors (e.g., Wrangham, 1987) have pointed out that the economic defendability and spatio–temporal distribution of food, along with predation pressure, are major factors influencing the evolution of socially mediated spacing patterns in animals. As illustrated in Figure 8, different kinds of spacing patterns or social organizations will be favored, depending on the distribution and defendability of food in space and time. At one end of this defense–cost continuum are aggressive intra- and interspecific territorial systems associated with resources that are relatively uniformly distributed in space. At the other end are nomadic flocks associated with widely

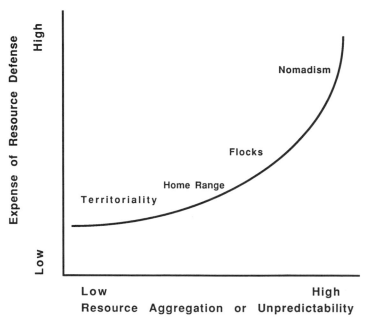

Figure 8   The general relationship between the cost of resource defense, resource aggregation or unpredictability, and spatially defined social organization. Redrawn with permission from Wiens (1976).

spaced resource patches. Spacing patterns in frugivores and nectarivores appear to be sensitive to food distributions. In general, nectarivores tend to occur at the aggressive end, and frugivores tend to occur at the gregarious end of this continuum. According to my second hypothesis (page 357), insectivores should exhibit territorial behavior more frequently than do either frugivores or nectarivores.

a. Patterns in Frugivores   Stiles' (1983) review of the social systems of Costa Rican birds provides an excellent overview of the influence of diet on tropical avian spacing patterns. During the nonbreeding season, most tropical frugivores are nonterritorial, whereas insectivores tend to maintain territories year-round. At all elevations, the modal social system of avian frugivores is single-species flocks, but mixed-species flocks of frugivores are also common (Buskirk, 1976; Morton, 1979; Munn, 1985; Powell, 1985; Remsen, 1985). Flocking is more common in canopy frugivores (e.g., aracaris, tanagers) than in understory frugivores (e.g., manakins, certain tanagers and flycatchers), in part because of the the larger patch sizes and greater interpatch distances of canopy fruits. Compared with mixed-species flocks of insectivores, the foraging locations of tropical frugivores in mixed-species flocks overlap extensively, dominance hierarchies are absent, and flocks are less likely to be territorial (Munn, 1985; Powell, 1985). In contrast, temperate zone avian frugivores and temperate migrants in the tropics tend to be solitary foragers (Herrera, 1984; Stiles, 1983). Among overwintering frugivores in south Florida, gray catbirds and white-eyed vireos are solitary foragers, whereas American robins and tree swallows forage in intraspecific flocks (personal observation, 1990).

Spacing patterns in New Guinea birds of paradise also differ among diet classes. Foraging territories occur in the insectivorous buff-tailed sicklebill, whereas overlapping and nondefended foraging ranges occur in frugivorous species (e.g., magnificent and raggiana birds of paradise, trumpet manucodes) (Beehler, 1987). Flocking behavior does not occur in this family.

Aggressive interactions among avian frugivores feeding in the same tree tend to be infrequent and, when present, tend to be directed toward conspecifics rather than heterospecifics (e.g., Cruz, 1974; Fleming and Williams, 1990; Leck, 1969). Territorial defense of individual fruiting trees, however, has been observed in mistle thrushes in Britain (Snow and Snow, 1984), and in four species of New Guinea forest birds (Pratt, 1984)

Most frugivorous bats roost gregariously by day but feed solitarily and nonterritorially at night. In contrast, some, but not all, tropical insectivorous bats defend feeding territories at night (Bradbury and Emmons, 1974; Bradbury and Vehrencamp, 1976; Fenton and Rautenbach, 1986; Vaughan, 1976; Vaughan and Vaughan, 1986). The feeding behavior of most frugivorous phyllostomid bats generally precludes defense of fruiting trees or areas containing fruiting trees. As described in detail in Bonaccorso and Gush (1987), Fleming (1988), and Morrison (1978), this behavior includes harvesting one fruit or part of a fruit at a time and taking it to a secluded night roost to consume. These bats usually sleep in their night roost when not eating. Large numbers of frugivorous bats often congregate in large fruiting trees (e.g., figs) but enter and leave these trees singly rather than in flocks (e.g., August, 1981; Morrison, 1978). In contrast, *Pteropus* bats often travel from their day roosts to their feeding areas in flocks. Once they arrive at fruiting or flowering trees, however, they sometimes set up individual territories of a few square meters in the tree crown (G. Richards, personal communication, 1987; personal observation, 1987).

Frugivorous primates show a wide range of spacing patterns, which are influenced by food distributions and mating systems. Territoriality is widespread in primates and sometimes reflects the economic defendability of food (Oates, 1987). For example, at Manu National Park and elsewhere in Peru, two species of tamarins (either *Saguinus imperator* or *mystax* and *S. fuscicollis*) travel in mixed-species groups, feed on synchronously fruiting plants bearing small numbers of ripe fruit daily, and defend relatively small

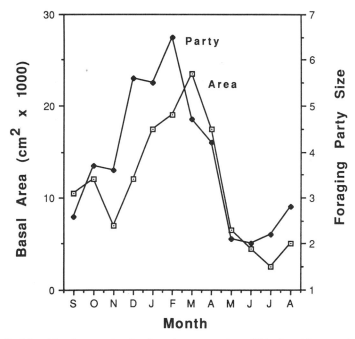

Figure 9 Monthly changes in the foraging-party size of black spider monkeys (*Ateles paniscus*) and resource-patch size, measured as the total basal area of trees and lianas bearing mammal-dispersed fruits, at Manu National Park, Peru. Redrawn with permission from Symington (1988).

territories against other tamarin groups (Garber, 1988; Terborgh, 1983). In contrast, squirrel monkeys (*Saimiri sciureus*) at Manu feed on figs, which are super abundant but ephemeral resources, and travel in large groups whose ranges overlap and are not defended. In general, frugivorous primates that are monogamous, or that possess small home ranges which can be patrolled daily, are usually territorial, whereas those with large overlapping ranges are not (Cheney, 1987).

Group sizes in gregarious primates tend to be correlated with resource patch size. At Santa Rosa National Park in Costa Rica, for example, the social systems of howler (*Allouatta palliata*) and spider (*Ateles geoffroyi*) monkeys can be described as *fusion–fission* because groups coalesce, or split up, depending on the sizes of fruit crops at different times of the year (Chapman, 1988). In spider monkeys (*A. paniscus*) at Manu, foraging-party size tracks food-patch size (Fig. 9). Aggressive behavior and competition for food act to adjust group size to food-patch size (Symington, 1988). Foraging-party sizes tend to be larger in the bonobo (*Pan paniscus*), which feeds in larger trees, than in the chimpanzee (*P. troglodytes*) (White and Wrangham, 1988). Group cohesiveness and rate of movement differs greatly between primates that feed on low-density insects or fruit [*foragers* in Oates' (1987) terminology], compared with those feeding on highly clumped fruits (*banqueters*).

b. Patterns in Nectarivores   Aggressive defense of nectar resources is common in all families of nectarivorous birds (Stiles, 1973; Gill and Wolf, 1975; Pyke, 1980). It is especially common in hummingbirds, both during and outside of the breeding season. Many studies (reviewed in Feinsinger, 1987) have shown that hummingbirds are remarkably flexible in (1) their ability to turn territorial behavior on and off in response to resource levels, and in (2) adjusting territory size to current nectar densities. Carpenter (1987) points out that hummingbirds tend to defend food territories only when *regional* food levels are low, even when *local* food levels are high. She argues that high mobility has been selected for in nectarivores to assess regional as well as local food levels before deciding whether or not to defend a territory. Hawaiian honeycreepers (Drepanididae) are also less likely to defend territories in years when flower densities are low (Carpenter, 1978).

Territorial behavior is not universal in tropical hummingbirds. Stiles (1983) indicates that lowland hummingbirds have two modal foraging (and social) systems:

1. in socially dominant species, which are usually straight-billed, have high wing-disc loading, but are not necessarily large, males defend clumps of flowers against conspecific males and heterospecifics; and
2. in socially subordinate species, which include hermits and short-billed species with low wing-disc loading, birds either forage along *trap lines* of widely spaced plants (hermits), steal nectar from defended patches, or visit flowers not used by dominant species.

In hummingbird communities in Colima, Mexico, migrant species are more likely to set up territories than are resident and *wanderer* species (Des-Granges and Grant, 1980). Only four of six species are territorial during the summer months in montane meadows in Mexico (Lyon, 1976), whereas all four co-occuring species defend certain flowers at particular times of the year in a montane community in Costa Rica (Wolf *et al.*, 1976). Finally, flocking behavior is unreported in hummingbirds except during periods of migration, and then only in certain species (e.g., *Calypte anna*) (Stiles, 1973).

Territorial behavior appears to be less common, and flocking behavior, more common in Australian honeyeaters than in hummingbirds. Pyke (1980) suggested that this is because flower densities are perhaps higher, on average, and predation risks, lower for hummingbirds than for honey-eaters. The presence of territoriality in honeyeaters has been reported by Ford and Paton (1977), Newland and Wooller (1985), and Paton (1985a,b), and its absence by Carpenter (1978), and Collins and Briffa (1982). Intra-specific flocking, which allows birds to find and preempt clumped flower resources, occurs in yellow-faced, white-naped, and scarlet honeyeaters (McFarland, 1986).

Size-based interspecific dominance relationships sometimes determine access to flowers in honeyeater communities (McFarland, 1986; Newland and Wooller, 1985). Because of their large size, wattlebirds (*Anthchaera*) are generally able to control access to rich flower patches except against flocks of small silvereyes (*Zosterops*). Wattlebirds thus often occur at the rich end of nectar resource gradients, whereas small subordinate species (e.g., spinebills (*Acanthorhynchus*)) are often relegated to the low end of such gradients. Medium-sized honeyeaters (e.g., New Holland honeyeaters) are often more *constant* in their presence in honeyeater communities than are larger or smaller species, because they can profitably forage at clumped as well as at dispersed nectar sources (Paton, 1985b). Wattlebirds need rich clumps to forage profitably, whereas small spinebills often migrate to richer resource areas despite being able to persist in areas when resource levels are low.

Territorial behavior appears to be less common in nectarivorous bats than it is in birds. Lemke (1984) reported that in Colombia *Glossophaga soricina* defends flowers of *Agave desmettiana* against conspecifics in the early hours of the evening, before visiting other flowers in trap-line fashion. It is likely that the small pteropodid *Syconycteris australis* is a persistent defender of *Banksia* and other flowers in eastern Australia (G. Pyke and D. Woodside, personal communication, 1987; Richards, 1983b). Gould (1978) noted that individuals of *Pteropus vampyrus* defend parts of the canopy of flowering *Durio zibenthus* trees against conspecifics, and suggested that *Eonycteris spelea* also defends *Parkia speciosa* flowers in Malaya. Defense of flowers has not been observed in other species of flower-visiting bats.

Other than Lemke's (1984) study, trap-line foraging behavior —the usual foraging method in hermit hummingbirds—has not been directly observed in nectarivorous bats, though its suspected occurrence has been mentioned by Baker (1973), Heithaus *et al.* (1974), and Gould (1978). Flock foraging has been better-documented in nectarivorous bats. It occurs in *Leptonycteris curasoae* and *Phyllostomus discolor* in the New World and in *Epomophorous gambianus* in Africa, and *Eonycteris spelea* in Malaya (Fleming, 1982). Group size in *Leptonycteris* and *Phyllostomus* appears to be related to flower density. For example, when they visit individual flowers of the cactus *Pachycereus pringlei*, groups of *Leptonycteris* contain 2–5 bats, whereas they contain 20 or more individuals at panicles of *Agave palmeri* flowers (P. Horner and T. Fleming, unpublished data, 1990; Howell, 1979).

Nectar is an uncommon dietary item in primates and other arboreal mammals. Certain territorial species of nocturnal prosimians as well as diurnal species of cebids and callitrichids are known to consume nectar, especially during tropical dry seasons (Garber, 1988; Hladik *et al.*, 1980; Janson *et al.*, 1981; Terborgh and Stern, 1987). Groups of saddle-backed tamarins (*Saguinus fusicollis*) visit *Combretum, Quararibea,* and *Symphonia* flowers in trap-line fashion. Trap-lining has also been observed in nocturnal species, such as the marsupial *Didelphis marsupialis,* and the procyonids *Potos flavus* and *Bassaricyon alleni* visiting *Quararibea* flowers in Peru, but not in the marsupial *Caluromysiops irrupta* and the night monkey (*Aotus trivirgatus*) (Janson *et al.*, 1981).

2. Mating Patterns

Because resource distributions determine, in part, the *polygyny potential* of any environment, they can play an important role in the evolution of avian and mammalian mating systems (Emlen and Oring, 1977). Environments will have high polygyny potential whenever resources are distributed in defendable clumps that attract several potential mates. Although most fruit- or nectar-eating birds and mammals conform to the modal mating systems of their classes (monogamy and polygamy, respectively), some species differ in spectacular fashion from *expected* mating systems. While not exclusively restricted to frugivorous and nectarivorous species, lek mating systems are more common in these diet classes than in others, and it seems reasonable to postulate that this association is not fortuitous. In some groups of birds and mammals, feeding on fruits or nectar has favored the evolution of lek mating.

a. Patterns in Frugivores   Snow (1971, p. 198) was the first ornithologist to point out the association between frugivory and lek mating in tropical forest birds:

> Hence it may be expected on theoretical grounds that fruit will constitute, at the seasons when it is available, an abundant food supply and one which is easily obtained, whereas insects will constitute a less abundant food supply and one which is less easily located. It is surely for this reason that lek behaviour, which entails the presence of the displaying birds on their display perches for the greater part of the day, has evolved in some groups of frugivorous tropical forest birds, but not in insectivorous birds.

Among frugivorous birds, lek mating is found in neotropical manakins, cotingids, and certain flycatchers and, in the paleotropics, in certain birds of paradise.

Male emancipation from nesting duties is a key step in the evolution of lek mating in birds. Male emancipation will be *permitted* when females can easily find food (usually fruit) for themselves, and can feed their nestlings (usually insects in most polygynous species) without male assistance. In the lek-mating tyrannid flycatcher *Pipramorpha oleaginea,* for example, male emancipation results from a frugivorous diet, a pendant nest that is safe from predators, and female regurgitation of food to the chicks, which is an unusual behavior in flycatchers (Snow and Snow, 1979).

Frugivory is necessary but not sufficient to promote the evolution of lek mating in birds of paradise (BOPs). Beehler and Pruett-Jones (1983) indicate that polygyny and nonterritorial spacing systems only occur in BOPs whose diets include over 50% fruit; insectivorous species are monogamous and territorial. Furthermore, among the frugivores, only species feeding on the nutritious arils of capsular fruit of the Meliaceae or Myristicaceae are polygynous. Fig-eating manucodes are monogamous, and both parents feed these nutritionally poor fruit to their nestlings.

Beehler (1989) identifies three factors that promote the evolution of polygyny and lek mating in birds of paradise. These include:

1. a clumped distribution of fruits which allows displaying males to be exposed to many females with overlapping home ranges;
2. high spatio–temporal predictability and high nutrient content of capsular fruits and insects, which allows females to feed their nestlings without male assistance; and
3. large female home ranges, which are needed to harvest capsular fruits produced in low daily numbers over extended periods of time.

Large overlapping female ranges allow certain males to set up strategically placed display sites, which attract other males as well as females. Dispersion of males within a lek will depend, in part, on the skew in male mating success. A strong skew with one male accruing most of the copulations at a lek will favor large, tightly clumped leks (as in Raggiana BOPs) whereas more equal mating success among males will promote *exploded* leks (as in magnificent BOPs) (Beehler and Foster, 1988). Predation may also be involved because tight leks occur in species living in second growth or along forest edges where raptor densities are high (Beehler and Pruett-Jones, 1983).

Like BOPs, most frugivorous bowerbirds are thought to be polygynous, although this has been confirmed in only a few species (Diamond, 1986). In contrast to manucodes, fig-eating catbirds (*Ailuroedus*) are territorial and monogamous. In most other species, males are nonterritorial except in the immediate vicinity of their bowers, which tend to be evenly spaced to minimize the stealing of bower materials and disruption of mating displays. Only females build nests and care for nestlings.

It should be noted that not all frugivorous birds eating nutritious capsular fruits are polygynous, and not all lek-mating frugivorous birds eat capsular fruits. For example, none of the major consumers of neotropical *Virola* fruits (Myristicaceae), including toucans, a guan, a motmot, and the masked tityra, are polygynous, nor is the resplendent quetzal, a specialist on nutritious fruits of the Lauraceae (Howe, 1987; Wheelwright, 1983). Likewise, the diets of lek-mating manakins, small birds of neotropical forest understories, include mostly berries or small drupes (Worthington, 1982), and polygynous neotropical cotingids (e.g., cock of the rock) do not specialize on capsular fruits (Snow, 1976). Thus, while frugivory *permits* lek-mating in certain groups of tropical birds, diets of lek-maters have not converged on a narrow subset of highly nutritious fruits with a particular set of phenological characteristics. As a result, a single resource scenario cannot be devised to explain the evolution of lek mating in frugivorous birds.

With the possible exception of the insectivore–frugivore *Mysticina tuberculata* (Mystacinidae) of New Zealand (M. Daniel and E. Pierson, personal communication, 1991), lek mating occurs only in frugivorous bats of the epomophorine (or *epauletted*) group of African pteropodids. In species of *Epomops, Epomophorous, Hypsignathus,* and *Micropteropus,* males spend several hours each night during the mating season calling from traditional sites before ranging widely in search of fruit (Bradbury, 1977; Thomas and Marshall, 1984; Wickler and Seibt, 1976). Leks are not located near concentrations of fruit, and lekking species eat the fruit of both primary and secondary forest plants (e.g., of the genera *Ficus, Solanum, Musanga,* and *Anthocleista*). Lek mating is unknown in frugivorous phyllostomid bats. Instead, polygynous mating based on roost site or female defense is the rule in these bats, as well as in many species of frugivorous pteropodid bats (Fleming, 1988).

The mating systems of frugivorous primates include monogamy and polygamy (including polyandry and polygyny). Resource-patch size and predation risk are important components of models of optimal group size and mating systems in primates (Terborgh, 1983; Wrangham, 1987). For example, both Terborgh and Wrangham postulate that resource-patch size determines optimal group size for females, and that male distributions reflect female distributions. According to Terborgh, large food patches permit the formation of groups of females and accompanying males, and optimal group size is determined by a tradeoff between competition for food and protection from predators.

Support for the importance of food-patch size in determining group size comes from a comparison of the social systems of sympatric congeneric primates (e.g., leaf monkeys, macaques, colobus, and lemurs), in which one species lives in small groups with a single adult male and the other lives in larger groups with several adult males (Terborgh and Janson, 1986). In each case, the species living in the small group is more folivorous and has a smaller daily range than the other more frugivorous species. These authors suggest that single male groups occur in 43% of all folivorous primate species compared with only 15% of frugivorous primates, because folivores have more time to defend and monopolize females than do frugivores. Support for the importance of predation as a critical factor comes from the fact that monogamous species tend to be either relatively small and nocturnal (e.g., *Aotus*), or large (i.e., >10 kg) and diurnal (e.g., *Hylobates*, *Symphalangus*) (Terborgh and Janson, 1986). Small groups are favored in nocturnal species to reduce their detection by auditorily orienting predators, whereas large species are too big for most predators to kill.

b. Patterns in Nectarivores

Polygynous mating systems are the rule in hummingbirds, whereas monogamy rules in passerine nectarivorous families (Collins and Paton, 1989). "Clearly the [food] exploitation systems of hummingbirds are intimately related to their social systems; the two are tightly integrated in the ecology of any hummingbird species" (Stiles and Wolf, 1979, p. 71). Two social systems predominate in hummingbirds—food-centered territories and lek mating systems. In the former system, males of socially dominant species defend highly clumped patches of flowers against conspecific males and other species, and females are allowed into territories to mate. In the latter system, males congregate in groups away from concentrated resources and attract females with calls. In both systems, females build nests and care for their two-egg clutches alone. Food-based territories are found in more species than are leks, which occur only in socially subordinate hermit hummingbirds. The absence of polygynous mating systems in other kinds of nectarivorous birds indicates that nectarivory may be necessary but not sufficient for the evolution of polygyny.

The breeding systems of nectarivorous bats are poorly known. Most species probably are polygynous, but none is known to be lek mating. Nectarivorous members of the Phyllostomidae, whose roosting behavior is known, are gregarious roosters, although it is likely that sexual segregation occurs in some species during parturition times (Fleming, 1988). Roost sizes in these species range from dozens to a few hundred individuals in *Glossophaga soricina*, to thousands of individuals in *Leptonycteris curasoae*. Most flower-visiting members of the Pteropodidae are also gregarious roosters, sometimes living in large colonies (e.g., *Eonycteris spelea*). Australian blossom bats (*Syconycteris*), in contrast, are solitary roosters (Richards, 1983b). Except in solitary roosters, mating systems in nectarivorous bats probably involve either female or roost site defense, as is the case in frugivorous bats.

IV. Conclusions

In this chapter I have addressed the hypothesis that various components of the life histories of fruit- and nectar-eating birds and mammals have been significantly influenced by the spatio–temporal variability of their food resources. I have demonstrated that densities of fruits and flowers can be highly variable in space and time and that these resources probably warrant being called patchy. I recognize, however, that this designation needs to be made more operational and that fruit and flower densities need to be more thoroughly quantified, especially with respect to the daily and seasonal energetic needs of frugivores and nectarivores. Careful resource monitoring should be a top priority in any ecobehavioral study of these animals.

My review of the demography, movements, and social organization of frugivores and nectarivores provides strong circumstantial evidence that resource variability has indeed had a strong influence on the evolution of their life histories. Timing of breeding, overall abundance, population fluctuations, daily and seasonal movement patterns, intra- and interspecific social interactions, and mating patterns appear to be sensitive to resource conditions in at least some groups of fruit- or nectar-eating birds and mammals. But statistical correlations do not necessarily allow us to reach strong conclusions regarding cause-and-effect relationships behind these correlations. Strong inferences about underlying causes and effects in biology usually require an experimental approach—an approach that is virtually impossible to pursue when dealing with highly mobile species of birds and mammals.

A comparative approach in which phylogenetic effects are carefully controlled (e.g., Harvey and Read, 1988) would seem to be the next best method for more critically understanding the relationships between resource variability and the evolution of life histories. Throughout this chapter, I have informally used a comparative approach by mentioning how frugivore and nectarivore life histories differ from those of other diet classes. From an evolutionary perspective, the best comparison in birds and phyllostomid bats (and probably also in primates), is between insectivores and frugivores or nectarivores, because the latter two groups appear to be derived from insectivorous ancestors, and/or still have relatives that are insectivorous (Feduccia, 1980; Hill and Smith, 1984). Indeed, many species of nectarivorous or frugivorous birds, bats, and primates still occasionally feed heavily on insects.

Results of this informal comparison indicate that, relative to insectivores, frugivores and nectarivores are more likely to

1. have seasonally variable population sizes and community compositions,
2. undergo seasonal habitat and elevation shifts,
3. be nomadic,

4. travel in intraspecific flocks (frugivores and Australian honeyeaters), and
5. have lek mating systems or be polygynous (birds).
6. be territorial (frugivores only).

While a statistical analysis based on multivariable quantitative data would certainly be desirable as a more rigorous comparative test of my major hypothesis, I believe that data in hand provide more than simply circumstantial support for this hypothesis. I therefore conclude that the life histories of frugivores and nectarivores have been strongly affected by the spatio–temporal variability of their food resources.

As mentioned in the introduction, fruit- and nectar-eating birds and mammals play important roles as seed dispersers and pollinators in many communities. Because of their daily and seasonal movements, these species clearly serve as *mobile links* between plant populations and communities over large geographic areas (e.g., Fig. 4). The destruction of habitats and their resources along altitudinal or latitudinal fruit and nectar pathways clearly will affect the lives not only of the migrants, but also of their food populations elsewhere in their annual cycle. Conservation of habitats within these pathways is vitally important. Otherwise, species as mobile as frugivores and nectarivores are doomed to extinction.

Acknowledgments

My research on frugivorous and nectarivorous bats has been generously supported by the U. S. National Science Foundation, National Geographic Society, National Fish and Wildlife Foundation, and by a Fulbright Fellowship. I thank J. Karr and B. Loiselle for providing me with preprint copies of some of their papers. I also thank J. Karr for critically reading a draft of this paper. This is Contribution No. 368 from the Program in Ecology, Behavior, and Tropical Biology, Department of Biology, University of Miami.

Appendix

Scientific names of the bird species mentioned in the text.

American robin	*Turdus migratorius*
Cock of the rock	*Rupicola rupicola*
Buff-tailed sicklebill	*Epimachus albertisi*
Gray catbird	*Dumatella carolinensis*
Magnificent bird of paradise	*Diphyllodes magnificus*
Masked tityra	*Tityra semifasciata*
Mistle thrush	*Turdus viscivorous*

New Holland honeyeater	*Phylidonyris novaehollandiae*
Raggiana bird of paradise	*Paradisaea raggiana*
Resplendent quetzal	*Pharomachrus mocinno*
Scarlet honeyeater	*Myzomela sanguinolenta*
Tree swallow	*Iridoprocne bicolor*
Trumpet manucode	*Manucodia keraudrenii*
White-eyed vireo	*Vireo griseus*
White-naped honeyeater	*Melithreptus lunatus*
Yellow-faced honeyeater	*Lichenostomus chrysops*

References

Appanah, S. (1985). General flowering in the climax rain forests of Southeast Asia. *J. Trop. Ecol.* **1**, 225–240.

August, P. V. (1981). Fig consumption and seed dispersal by *Artibeus jamaicensis* in the llanos of Venezuela. *Biotropica* (Suppl.) **13**, 70–76.

Austin, D. F. (1975). Bird flowers in the eastern United States. *Florida Sci.* **38**, 1–12.

Baker, H. G. (1973). Evolutionary relationships between flowering plants and animals in American and African tropical forests. *In* "Tropical Forest Ecosystems in Africa and South America: A Comparative Review" (B. J. Meggers, E. S. Ayensu, and W. D. Duckworth, eds.), pp. 145–159. Smithsonian Institution Press, Washington, D.C.

Baker, H. G., Bawa, K. S., Frankie, G. W., and Opler, P. A. (1983). Reproductive biology of plants in tropical forests. *In* "Tropical Rain Forest Ecosystems" (F. B. Golley, ed.), pp. 183–215. Elsevier Scientific, Amsterdam.

Barbour, R. W., and Davis, W. H. (1969). "Bats of America." University of Kentucky Press, Lexington, Kentucky.

Bawa, K. S. (1990). Plant–pollinator interactions in tropical rain forests. *Annu. Rev. Ecol. Syst.* **21**, 399–422.

Beehler, B. (1983). Frugivory and polygamy in birds of paradise. *Auk* **100**, 1–12.

Beehler, B. (1987). Birds of paradise and the evolution of mating systems. *Emu* **87**, 78–89.

Beehler, B. (1989). The birds of paradise. *Sci. Am.* **261** (12), 117–123.

Beehler, B., and Foster, M. S. (1988). Hotshots, hotspots, and female preference in the organization of lek mating systems. *Am. Nat.* **131**, 203–219.

Beehler, B., and Pruett-Jones, S. G. (1983). Display dispersion and diet of birds of paradise: A comparison of nine species. *Behav. Ecol. Sociobiol.* **13**, 229–238.

Blake, J. G., and Hoppes, W. G. (1986). Influence of resource abundance on use of tree-fall gaps by birds in an isolated woodlot. *Auk* **103**, 328–340.

Blake, J. G., Stiles, F. G., and Loiselle, B. A. (1990). Birds of La Selva biological station: Habitat use, trophic composition, and migrants. *In* "Four Neotropical Rain Forests" (A. Gentry, ed.), pp. 161–182. Yale University Press, New Haven, Connecticut.

Bonaccorso, F. J., and Gush, T. J. (1987). An experimental study of the feeding behaviour and foraging strategies of phyllostomid fruit bats. *J. Anim. Ecol.* **56**, 907–920.

Bradbury, J. W. (1977). Lek mating behavior in the hammer-headed bat. *Z. Tierpsychol.* **45**, 225–255.

Bradbury, J. W., and Emmons, L. (1974). Social organization of some Trinidad bats. I. Emballonuridae. *Z. Tierpsychol.* **36**, 137–183.

Bradbury, J. W., and Vehrencamp, S. (1976). Social organization and foraging in emballonurid bats. 1. Field studies. *Behav. Ecol. Sociobiol.* **1**, 337–381.

Buckley, P. A., Foster, M. S., Morton, E. S., Ridgely, R. S., and Buckley, F. G. (eds.). (1985).

"Neotropical Ornithology." *Ornithol. Monogr. No. 36.* American Ornithologist's Union, Washington, D. C.

Buskirk, W. H. (1976). Social systems in a tropical forest avifauna. *Am. Nat.* **110,** 293–310.

Calder, W. A., III. (1984). "Size, Function, and Life History." Harvard University Press, Cambridge, Massachusetts.

Carpenter, F. L. (1978). A spectrum of nectar-eater communities. *Am. Zool.* **18,** 809–819.

Carpenter, F. L. (1987). Food abundance and territoriality: To defend or not to defend? *Am. Zool.* **27,** 387–399.

Chapman, C. (1988). Patch use and patch depletion by the spider and howling monkeys of Santa Rosa National Park, Costa Rica. *Behaviour* **105,** 99–116.

Cheney, D. L. (1987). Interactions and relationships between groups. *In* Smuts, B. B., *et al.* (eds.) pp. 267–281 1987.

Cockrum, E. L. (1989). Seasonal distribution of northwestern populations of the long-nosed bats, genus *Leptonycteris,* family Phyllostomidae, unpublished Manuscript.

Collins, B. G. (1980). Seasonal variations in the abundance and food preferences of honeyeaters (Meliphagidae) at Wongamine, Western Australia. *West. Aust. Nat.* **14,** 207–212.

Collins, B. G., and Briffa, P. (1982). Seasonal variation of abundance and foraging of three species of Australian honeyeaters. *Aust. Wildl. Res.* **9,** 557–569.

Collins, B. G., and Paton, D. C. (1989). Consequences of differences in body mass, wing length, and leg morphology for nectar-feeding birds. *Aust. J. Ecol.* **14,** 269–289.

Colwell, R. S. (1974). Predictability, constancy, and contingency of periodic phenomena. *Ecology* **55,** 1148–1153.

Crome, F. H. J. (1975). The ecology of fruit pigeons in tropical northern Queensland. *Aust. Wildl. Res.* **2,** 155–185.

Cruz, A. (1974). Feeding assemblages of Jamaican birds. *Condor* **76,** 103–107.

Des Granges, J.-L., and Grant, P. R. (1980). Migrant hummingbird's accommodation into tropical communities. *In* "Migrant Birds in the Neotropics" (A. Keast and E. S. Morton, eds.), pp. 395–409. Smithsonian Institution Press, Washington, D.C.

Diamond, J. (1986). Biology of birds of paradise and bowerbirds. *Annu. Rev. Ecol. Syst.* **17,** 17–37.

Emlen, S. T., and Oring, L. W. (1977). Ecology, sexual selection, and the evolution of mating systems. *Science* **197,** 215–223.

Feduccia, A. (1980). "The Age of Birds." Harvard University Press, Cambridge, Massachusetts.

Feinsinger, P. (1976). Organization of a tropical guild of nectarivorous birds. *Ecol. Monogr.* **46,** 257–291.

Feinsinger, P. (1980). Asynchronous migration patterns and the coexistence of tropical hummingbirds. *In* "Migrant Birds in the Neotropics" (A. Keast and E. S. Morton, eds.), pp. 411–419. Smithsonian Institution Press, Washington, D.C.

Feinsinger, P. (1983). Coevolution and pollination. *In* "Coevolution" (D. J. Futuyma and M. Slatkin, eds.), pp. 282–310. Sinauer Assoc., Sunderland, Massachusetts.

Feinsinger, P. (1987). Approaches to nectarivore–plant interactions in the New World. *Rev. Chilena Nat. Hist.* **60,** 285–319.

Fenton, M. B., and Rautenbach, I. L. (1986). A comparison of the roosting and foraging behaviour of three species of African insectivorous bats (Rhinolophidae, Vespertilionidae, and Molossidae). *Can. J. Zool.* **64,** 2860–2867.

Fleming, T. H. (1982). Foraging strategies of plant-visiting bats. *In* "Ecology of Bats" (T. H. Kunz, ed.), pp. 287–325. Plenum Press, New York.

Fleming, T. H. (1988). "The Short-Tailed Fruit Bat." University of Chicago Press, Chicago.

Fleming, T. H. (1991). Fruiting plant–frugivore mutualism: The evolutionary theater and the ecological play. *In* "Plant–Animal Interactions: Evolutionary Ecology in Tropical and Temperate Regions" (P. W. Price, T. M. Lewinsohn, G. W. Fernandes, and W. W. Benson, eds.), pp. 119–144. John Wiley, New York.

Fleming, T. H., Breitwisch, R., and Whitesides, G. H. (1987). Patterns of tropical vertebrate frugivore diversity. *Annu. Rev. Ecol. Syst.* **18,** 91–110.

Fleming, T. H., and Williams, C. F. (1990). Phenology, seed dispersal, and recruitment in *Cecropia peltata* (Moraceae) in Costa Rican tropical dry forest. *J. Trop. Ecol.* **6,** 163–178.

Ford, H. A. (1985). A synthesis of the foraging ecology and behaviour of birds in eucalypt forests and woodlands. *In* Keast, A., *et al.* (eds.) pp. 249–254 1985.

Ford, H. A., and Paton, D. C. (1977). The comparative ecology of ten species of honeyeaters in South Australia. *Aust. J. Ecol.* **2,** 399–407.

Frankie, G. W., Baker, H. G., and Opler, P. A. (1974). Comparative phenological studies of trees in tropical wet and dry forests in the lowlands of Costa Rica. *J. Ecol.* **62,** 881–919.

Garber, P. A. (1988). Diet, foraging patterns, and resource defense in a mixed species troop of *Saguinus mystax* and *Saguinus fusicollis* in Amazonian Peru. *Behaviour* **105,** 18–34.

Gautier-Hion, A., and Michaloud, G. (1989). Are figs always keystone resources for tropical vertebrates? A test in Gabon. *Ecology* **70,** 1826–1833.

Gentry, H. S. (1982). "Agaves of Continental North America." University of Arizona Press, Tucson, Arizona.

Gill, F. B., and Wolf, L. L. (1975). Economics of feeding territoriality in the golden-winged sunbird. *Ecology* **56,** 333–345.

Gould, E. H. (1978). Foraging behavior of Malaysian nectar-feeding bats. *Biotropica* **10,** 184–193.

Grant, K. A., and Grant, V. (1967). Effects of hummingbird migration on plant speciation in the California flora. *Evolution* **21,** 457–465.

Harvey, P. H., and Read, A. F. (1988). How and why do mammalian life histories vary? *In* "Evolution of Life Histories of Mammals" (M. S. Boyce, ed.), pp. 213–232. Yale University Press, New Haven, Connecticut.

Heideman, P. D. (1988). The timing of reproduction in the fruit bat *Haplonycteris fischeri* (Pteropodidae): Geographic variation and delayed development. *J. Zool. Lond.* **215,** 577–595.

Heithaus, E. R., Fleming, T. H., and Opler, P. A. (1975). Patterns of foraging and resource utilization in seven species of bats in a seasonal tropical forest. *Ecology* **56,** 841–854.

Heithaus, E. R., Opler, P. A., and Baker, H. G. (1974). Bat activity and pollination of *Bauhinia pauletia*: Plant–pollinator coevolution. *Ecology* **55,** 412–419.

Helversen, O. V., and Reyer, H.-U. (1984). Nectar intake and energy expenditure in a flower-visiting bat. *Oecologia* **63,** 178–184.

Herrera, C. M. (1984). A study of avian frugivores, bird-dispersed plants, and their interaction in mediterranean scrublands. *Ecol. Monogr.* **54,** 1–23.

Herrera, C. M. (1986). Vertebrate-dispersed plants: Why they don't behave the way they should. *In* "Frugivores and Seed Dispersal" (A. Estrada and T. H. Fleming, eds.), pp. 5–18. Dr. W. Junk, Dordrecht, The Netherlands.

Hill, J. E., and Smith, J. D. (1984). "Bats. A Natural History." University of Texas Press, Austin, Texas.

Hixon, M. A., Carpenter, F. L., and Paton, D. C. (1983). Territory area, flower density, and time budgeting in hummingbirds: An experimental and theoretical analysis. *Am. Nat.* **122,** 366–391.

Hladik, C. M., Charles-Dominique, P., and Petter, J. J. (1980). Feeding strategies of five nocturnal prosimians in the dry forest of the west coast of Madagascar. *In* "Nocturnal Malagasy Prosimians" (P. Charles-Dominique, H. M. Cooper, A. Hladik, C. M. Hladik, E. Pages, G. F. Pariente, A. Petter-Rousseaux, J. J. Petter, and A. Schilling, eds.), pp. 41–73. Academic Press, New York.

Howe, H. F. (1984). Constraints on the evolution of mutualisms. *Am. Nat.* **123,** 764–777.

Howe, H. F. (1987). Consequences of seed dispersal by birds: A case study from Central America. *J. Bombay Nat. Hist. Soc.* **83,** 19–42.

Howe, H. F., and Westley, L. C. (1988). "Ecological Relationships of Plants and Animals." Oxford University Press, New York.

Howell, D. J. (1979). Flock foraging in nectar-feeding bats: Advantages to the bats and to the host plants. *Am. Nat.* **114**, 23–49.

Janson, C. H., Terborgh, J., and Emmons, L. H. (1981). Non-flying mammals as pollinating agents in the Amazonian forest. *Biotropica* (Suppl.) **13**, 1–6.

Karr, J. R. (1976). Seasonality, resource availability, and community diversity in tropical bird communities. *Am. Nat.* **110**, 973–994.

Karr, J. R. (1989). Birds. *In* "Tropical Rain Forest Ecosystems" (H. Lieth and M. J. A. Werger, eds.), pp. 401–416. Elsevier Science, Amsterdam.

Karr, J. R. (1990). Interactions between forest birds and their habitats: A comparative synthesis. *In* "Biogeography and Ecology of Forest Bird Communities" (A. Keast, ed.), pp. 379–386. SPB Academic Publishing, The Hague, The Netherlands.

Karr, J. R., and Brawn, J. D. (1990). Food resources of understory birds in central Panama: Quantification and effects on avian populations. *Stud. Avian Biol.* **13**, 58–64.

Keast, A. (1968). Seasonal movements in the Australian honeyeaters (Meliphagidae) and their ecological significance. *Emu* **67**, 159–209.

Keast, A., Recher, H. F., Ford, H., and Saunders, D. (eds.). (1985). "Birds of Eucalypt Forests and Woodlands, Ecology, Conservation, and Management." Royal Australian Ornithologists Union and Surrey Beatty and Sons, Sydney, Australia.

Kodric-Brown, A., and Brown, J. H. (1978). Influence of economics, interspecific competition, and sexual dimorphism on territoriality of migrant rufous hummingbirds. *Ecology* **59**, 285–296.

Leck, C. F. (1969). Observations of birds exploiting a Central American fruit tree. *Wilson Bull.* **81**, 264–269.

Leighton, M., and Leighton, D. R. (1983). Vertebrate responses to fruiting seasonality within a Bornean rain forest. *In* "Tropical Rain Forest: Ecology and Management" (S. L. Sutton, T. C. Whitmore, and A. C. Chadwick, eds.), pp. 181–196. Blackwell Scientific, London.

Lemke, T. O. (1984). Foraging ecology of the long-nosed bat, *Glossophaga soricina*, with respect to resource availability. *Ecology* **65**, 538–548.

Levey, D. J. (1988). Spatial and temporal variation in Costa Rican fruit and fruit-eating bird abundance. *Ecol. Monogr.* **58**, 251–269.

Linhart, Y. B. (1973). Ecological and behavioral determinants of pollen dispersal in hummingbird-pollinated *Heliconia*. *Am. Nat.* **107**, 511–523.

Loiselle, B. A. (1988). Bird abundance and seasonality in a Costa Rican lowland forest canopy. *Condor* **90**, 761–772.

Loiselle, B. A., and Blake, J. G. (1991a). Temporal variation in birds and fruit along an elevational gradient in Costa Rica. *Ecology* **72**, 180–193.

Loiselle, B. A., and Blake, J. G. (1991b). Patterns of altitudinal migration by frugivorous birds along a wet forest elevational gradient in Costa Rica. Unpublished manuscript.

Lyon, D. L. (1976). A montane hummingbird territorial system in Oaxaca, Mexico. *Wilson Bull.* **88**, 280–299.

Marshall, A. G. (1985). Old World phytophagous bats (Pteropodidae) and their food plants: A survey. *Zool. J. Linn. Soc.* **83**, 351–369.

Martin, T. E., and Karr, J. R. (1986a). Temporal dynamics of neotropical birds with special reference to frugivores in second-growth woods. *Wilson Bull.* **98**, 38–60.

Martin, T. E., and Karr, J. R. (1986b). Patch utilization by migrating birds: Resource oriented? *Ornis Scand.* **17**, 165–174.

McFarland, D. C. (1986). The organization of a honeyeater community in an unpredictable environment. *Aust. J. Ecol.* **11**, 107–120.

Morrison, D. W. (1978). Foraging ecology and energetics of the frugivorous bat *Artibeus jamaicensis*. *Ecology* **59**, 716–723.

Morton, E. S. (1979). A comparative survey of avian social systems in northern Venezuelan habitats. *In* "Vertebrate Ecology in the Northern Neotropics" (J. F. Eisenberg, ed.), pp. 233–259. Smithsonian Institution Press, Washington, D. C.

Munn, C. A. (1985). Permanent canopy and understory flocks in Amazonia: Species composition and population density. *In* Buckley, P. A., *et al.* (eds.) pp. 683–712 1985.

Murray, K. G. (1988). Avian seed dispersal of three neotropical gap-dependent plants. *Ecol. Monogr.* **58**, 271–298.

Murray, K. G., Feinsinger, P., Busby, W. H., Linhart, Y. B., Beach, J. H., and Kinsman, S. (1987). Evaluation of character displacement among plants in two tropical pollination guilds. *Ecology* **68**, 1283–1293.

Nelson, J. E. (1965). Movements of Australian flying foxes (Pteropodidae: Megachiroptera). *Aust. J. Zool.* **13**, 53–73.

Newland, C. E., and Wooller, R. D. (1985). Seasonal changes in a honeyeater assemblage in *Banksia* woodland near Perth, Western Australia. *N. Z. J. Zool.* **12**, 631–636.

Oates, J. F. (1987). Food distribution and foraging behavior. *In* Smuts, B. B., *et al.* (eds.) pp. 197–209 1987.

Paton, D. C. (1985a). Honeyeaters and their plants in south-eastern Australia. *In* "The Dynamic Partnership: Birds and Plants in Southern Australia" (H. A. Ford and D. C. Paton, eds.), pp. 9–19. SA Government Printer, Adelaide, Australia.

Paton, D. C. (1985b). Food supply, population structure, and behaviour of New Holland honeyeaters *Phylidonyris novaehollandieae* in woodland near Horsham, Victoria. *In* Keast, A., *et al.* (eds.) pp. 219–230 1985.

Peters, R. H. (1983). "The Ecological Implications of Body Size." Cambridge University Press, Cambridge, Massachusetts.

Phillips, A. R. (1975). The migrations of Allen's and other hummingbirds. *Condor* **77**, 196–205.

Pleasants, J. M. (1990). Null-model tests for competitive displacement: The fallacy of not focusing on the whole community. *Ecology* **71**, 1078–1084.

Powell, G. V. N. (1985). Sociobiology and adaptive significance of interspecific foraging flocks in the neotropics. pp. 713-732 *in* Buckley, P. A., *et al.* (eds.) 1985.

Pratt, T. K. (1984). Examples of tropical frugivores defending fruit-bearing plants. *Condor* **86**, 123–129.

Pratt, T. K., and Stiles, E. W. (1985). The influence of fruit size and structure on composition of frugivore assemblages in New Guinea. *Biotropica* **17**, 314–321.

Primack, R. B. (1987). Relationships among flowers, fruits, and seeds. *Annu. Rev. Ecol. Syst.* **18**, 409–430.

Pyke, G. H. (1980). The foraging behaviour of Australian honeyeaters: A review and some comparisons with hummingbirds. *Aust. J. Ecol.* **5**, 343–369.

Pyke, G. H. 1983. Seasonal patterns of abundance of honeyeaters and their resources in heathland areas near Sydney. *Aust. J. Ecol.* **8**, 217–233.

Pyke, G. H. (1985). The relationships between abundances of honeyeaters and their food resources in open forests near Sydney. pp. 65–77 *in* Keast, A., *et al.* (eds.) 1985.

Pyke, G. H. (1988). Yearly variation in seasonal patterns of honeyeater abundance, flower density, and nectar production in heathlands near Sydney. *Aust. J. Ecol.* **13**, 1–10.

Pyke, G. H., and Recher, H. F. (1986). Relationship between nectar production and seasonal patterns of density and nesting of resident honeyeaters in heathland near Sydney. *Aust. J. Ecol.* **11**, 195–200.

Ramsey, M. W. (1989). The seasonal abundance and foraging behaviour of honeyeaters and their potential role in the pollination of *Banksia menziesii*. *Aust. J. Ecol.* **14**, 33–40.

Ratcliffe, F. N. (1932). Notes on the fruit bats (*Pteropus* spp.) of Australia. *J. Anim. Ecol.* **1**, 32–57.

Rathcke, B., and Lacey, E. P. (1985). Phenological patterns of terrestrial plants. *Annu. Rev. Ecol. Syst.* **16**, 179–214.

Remsen, J. V., Jr. (1985). Community organization and ecology of birds of high-elevation humid forest of the Bolivian Andes. pp. 733–756 *In* Buckley, P. A., *et al.* (eds.) 1985.

Richards, G. C. (1983a). Little red flying fox. *In* "The Complete Book of Australian Mammals" (R. Strahan, ed.), p. 277. Angus and Robertson, London.

Richards, G. C. (1983b). Queensland blossom bat. *In* "The Complete Book of Australian Mammals" (R. Strahan, ed.), p. 289. Angus and Robertson, London.

Robinson, J., and Janson, C. H. (1987). Capuchins, squirrel monkeys, and atelines: Socioecological convergence with Old World primates. Pp. 69–82 *in* Smuts, B. B., *et al.* (eds.) 1987.

van Roosmalen, M. G. M. (1980). "Habitat Preferences, Diet, Feeding Strategy, and Social Organization of the Black Spider Monkey (*Ateles p. paniscus* Linnaeus 1758) in Surinam." Rijksinstituut voor Natuurbeheer, Arnhem, The Netherlands.

Sandell, M. (1990). The evolution of seasonal delayed implantation. *Q. Rev. Biol.* **65**, 23–42.

Shreve, F., and Wiggins, I. R. (1964). "Vegetation and Flora of the Sonoran Desert." Vols. 1 and 2. Stanford University Press, Stanford, California.

Skeate, S. T. (1987). Interactions between birds and fruits in a northern Florida hammock community. *Ecology* **68**, 297–309.

Skutch, A. F. (1950). The nesting seasons of Central American birds in relation to climate and food supply. *Ibis* **92**, 185–222.

Smuts, B. B., Cheney, D. L., Seyfarth, R. M., Wrangham, R. W., and Struhsaker, T. T. (eds.) (1987). "Primate Societies." University of Chicago Press, Chicago.

Snow, B. K., and Snow, D. W. (1979). The ochre-bellied flycatcher and the evolution of lek behavior. *Condor* **81**, 286–292.

Snow, B. K., and Snow, D. W. (1984). Long-term defence of fruit by mistle thrushes *Turdus viscivorus. Ibis* **126**, 39–49.

Snow, D. W. (1962a). A field study of the black and white manakin, *Manacus manacus,* in Trinidad, W. I. *Zoologica* **47**, 65–104.

Snow, D. W. (1962b). A field study of the golden-headed manakin, *Pipra erythrocephala,* in Trinidad, W. I. *Zoologica* **47**, 183–198.

Snow, D. W. (1962c). The natural history of the oil bird, *Steatornis caripensis,* in Trinidad, W. I. Part 2. Population, breeding ecology, and food. *Zoologica* **47**, 199–221.

Snow, D. W. (1971). Evolutionary aspects of fruit-eating by birds. *Ibis* **113**, 194–202.

Snow, D. W. (1976). "The Web of Adaptations." Collins, London.

Spencer, H. J., and Fleming, T. H. (1989). Roosting and foraging behaviour of the Queensland tube-nosed bat, *Nyctimene robinsoni* (Pteropodidae): Preliminary radio-tracking observations. *Aust. Wildl. Res.* **16**, 413–420.

Start, A. N., and Marshall, A. G. (1976). Nectarivorous bats as pollinators of trees in West Malaysia. *In* "Tropical Trees: Variation, Breeding, and Conservation" (J. Burley and B. T. Styles, eds.), pp. 141–150. Academic Press, London.

Stiles, F. G. (1973). Food supply and the annual cycle of the Anna hummingbird. *Univ. Calif. Publ. Zool.* **97**, 1–109.

Stiles, F. G. (1977). Coadapted competitors: The flowering seasons of hummingbird-pollinated plants in a tropical forest. *Science* **198**, 1177–1178.

Stiles, F. G. (1978). Temporal organization of flowering among the hummingbird foodplants of a tropical wet forest. *Biotropica* **10**, 194–210.

Stiles, F. G. (1979). Reply to Poole and Rathcke. *Science* **203**, 471.

Stiles, F. G. (1980). The annual cycle in a tropical wet forest hummingbird community. *Ibis* **122**, 322–343.

Stiles, F. G. (1983). Birds. *In* "The Natural History of Costa Rica" (D. H. Janzen, ed.), pp. 502–543. University of Chicago Press, Chicago.

Stiles, F. G. (1985a). On the role of birds in the dynamics of neotropical forests. *In* "Conservation of Tropical Forest Birds" (A. W. Diamond and T. Lovejoy, eds.), pp. 49–59. I. C. B. P. Technical Publ. No. 4, Cambridge, England.

Stiles, F. G. (1985b). Seasonal patterns and coevolution in the hummingbird-flower community of a Costa Rican subtropical forest. pp. 757–787 *In* Buckley, P. A., *et al.* (eds.) 1985.

Stiles, F. G., and Wolf, L. L. (1979). Ecology and evolution of lek mating behavior in the long-tailed hermit hummingbird. *Ornithol. Monogr.* **27,** 1–78.

Symington, M. M. (1988). Food competition and foraging party size in the black spider monkey (*Ateles paniscus* Chamek). *Behaviour* **105,** 117–134.

Terborgh, J. (1983). "Five New World Primates." Princeton University Press, Princeton, New Jersey.

Terborgh, J. (1986). Community aspects of frugivory in tropical forests. *In* "Frugivores and Seed Dispersal" (A. Estrada and T. H. Fleming, eds.), pp. 371–384. Dr. W. Junk, Dordrecht, The Netherlands.

Terborgh, J., and Janson, C. H. (1986). The socioecology of primate groups. *Annu. Rev. Ecol. Syst.* **17,** 111–135.

Terborgh, J., Robinson, S. K., Parker, T. A., III, Munn, C. A., and Pierpont, N. (1990). Structure and organization of an amazonian forest bird community. *Ecol. Monogr.* **60,** 213–238.

Terborgh, J., and Stern, M. (1987). The surreptitious life of the saddle-backed tamarin. *Am. Sci.* **75,** 260–269.

Thomas, D. W. (1983). The annual migrations of three species of West African fruit bats (Chiroptera: Pteropodidae). *Can. J. Zool.* **61,** 2266–2272.

Thomas, D. W., and Marshall, A. G. (1984). Reproduction and growth in three species of West African fruit bats. *J. Zool. Lond.* **202,** 265–281.

Thompson, J. N., and Willson, M. F. (1979). Evolution of temperate fruit/bird interactions: Phenological strategies. *Evolution* **33,** 973–982.

Vaughan, T. A. (1976). Nocturnal behavior of the African false vampire bat (*Cardioderma cor*). *J. Mammal.* **57,** 227–248.

Vaughan, T. A., and Vaughan, R. P. (1986). Seasonality and the behavior of the African yellow-winged bat. *J. Mammal.* **67,** 91–102.

Waser, P. M. (1987). Interactions among species. pp. 210-226 *In* B. Smuts *et al.* (eds.) 1987.

Wheelwright, N. T. (1983). Fruits and the ecology of resplendent quetzals. *Auk* **100,** 286–301.

Wheelwright, N. T., and Orians, G. H. (1982). Seed dispersal by animals: Contrasts with pollen dispersal, problems of terminology, and constraints on coevolution. *Am. Nat.* **119,** 402–413.

White, F. J., and Wrangham, R. W. (1988). Feeding competition and patch size in the chimpanzee species *Pan paniscus* and *Pan troglodytes. Behaviour* **105,** 148–164.

Wickler, W., and Seibt, U. (1976). Field studies of the African fruit bat, *Epomophorus wahlbergi,* with special references to male calling. *Z. Tierpsychol.* **40,** 345–376.

Wiens, J. A. (1976). Population responses to patchy environments. *Annu. Rev. Ecol. Syst.* **7,** 81–120.

Wilson, D. E. (1979). Reproductive patterns. *In* "Biology of Bats of the New World Family Phyllostomatidae, Part 3" (R. J. Baker, J. K. Jones, Jr., and D. C. Carter, eds.), pp. 317–378. *Spec. Publ. Museum Texas Tech Univ.* No. 16, Lubbock, Texas.

Wolf, L. L. (1970). The impact of seasonal flowering on the biology of some tropical hummingbirds. *Condor* **72,** 1–14.

Wolf, L. L., Stiles, F. G., and Hainsworth, F. R. (1976). Ecological organization of a tropical highland hummingbird community. *J. Anim. Ecol.* **45,** 349–379.

Worthington, A. (1982). Population sizes and breeding rhythms of two species of manakins in relation to food supply. *In* "The Ecology of a Tropical Forest" (E. G. Leigh, Jr., A. S. Rand, and D. M. Windsor, eds.), pp. 213–225. Smithsonian Institution Press, Washington, D. C.

Wrangham, R. W. (1987). Evolution of social structure. pp. 282–296 *in* Smuts, B. B., *et al.* (eds.) 1987.

13

Inter- and Intraspecific Morphological Variation in Bumblebee Species, and Competition in Flower Utilization

Tamiji Inoue
Laboratory of Entomology
Faculty of Agriculture
Kyoto University, Kyoto 606, Japan[1]

Makoto Kato
Biological Laboratory
Yoshida College
Kyoto University, Kyoto 606, Japan

[1] Current address: Division of Tropical Ecology, Center for Ecological Research, Kyoto University.

I. Introduction

Morphological characters of animals that function in foraging have been intensively studied in relation to character divergence and displacement between community members interacting with each other (Hespenheide, 1974; Endler, 1986). Competition theory as described in the 1970s predicted that a limited number of species could coexist in a community, and that morphological characters would diverge so as to reduce competitive interactions between community members (reviewed by Schoener, 1987). Evidence for character displacement has been found in several animal species (Grant, 1972; Grant *et al.*, 1976; Fenchel, 1975; Dunham *et al.*, 1979). Recently, evidence has emerged that character convergence is possible, even in communities with strong interspecific competition (Vadas, 1990), and competition theory seems to fail to explain interspecific interactions in quite species-rich communities, such as tropical forests (Hubbell and Foster, 1986).

How morphological characters of foragers function in relation to the characteristics of the food that they utilize is important for an understanding of character divergence or convergence among foragers. However, very few studies (Davidson, 1978; Fenchel, 1975; Harder, 1983, 1985; Barrow and Pickard, 1984, 1985) have closely measured and matched the morphologies of both foragers and food. In most studies, there are three problems in approach and methodology. First, species averages of morphological characters are used for analysis, even in groups that exhibit large intraspecific variation in morphology, such as bumblebees. This is despite the fact that natural selection acts on individuals, not on species (see Chapter 2), and that the theory of competitive exclusion is formulated at the individual level. Second, in the majority of studies, the use of only one character for the analysis, assumes that it can represent multiple functions of morphology. In plant–pollinator interactions, only tongue length of insects and flower depth of plants are usually reported, although some studies (Harder, 1985; Feinsinger, 1987) have shown that other characters may influence plant–pollinator interactions. Third, many if not all studies are restricted to some portions or guilds of the whole community. This assumes that guilds are rather independent units that can be analyzed separately. Determining the membership of particular guilds is, in many cases, an arbitrary procedure, and concentrating on guilds may overlook interesting interactions between unexpected groups of animals (Waser, 1983; Roubik, 1989, Chapter 11).

In a volume devoted to investigating the consequences of variability among plants and animals for interactions between herbivores and their resources (Chapter 1), our message is a simple one. We argue that the foraging strategy of an animal species is actually the sum of heterogeneous strategies of individuals within that species. This is important because the responses of foragers to plant variability will be best understood by accounting for variability within the population of animals. Moreover, the influence that foragers have on their host plants (the other side of the animal–plant interaction) will also be affected by heterogeneity among individual foragers.

In this paper, we shall analyze multiple morphological characters of individual bumblebees in relation to the characters of the flowers that they visit, based on a 4-year study at three locations in Kyoto, central Japan (Kato *et al.*, 1990; Inoue *et al.*, 1990; Kakutani *et al.*, 1990). We shall show that characters other than tongue length are important in assessing the accessibility of flowers to bees, and that utilization patterns of bees can be better understood at the individual level than at the species level of bumblebees. We shall then discuss the influence of morphological characters on the fitness of both pollinators and plants and examine possible character displacement in bumblebees. Other variable characters, e.g., the flowering phenology of plants, are also discussed in relation to plant–pollinator interactions.

II. Materials and Methods

A. Study Sites

Our three study sites in Kyoto Prefecture, central Japan, include typical vegetation types in this region. Ashu has an intact beech (*Fagus crenata*) and *Cryptomeria japonica* forest in the northern Tango mountains (620–925 m above sea level). A total of 780 angiosperm species have been recorded from here (Kato *et al.*, 1990). Kibune, at the southern boundary of the Tango mountains, contains a well-preserved evergreen coniferous (*Abies firma* and *Tsuga sieboldii*) and deciduous oak (*Quercus crispula*) forest among *Cryptomeria* plantations (300–740 m above sea level). There is also a rich angiosperm flora (1267 species, Inoue *et al.*, 1990). The botanical garden in the campus of Kyoto University is located at the northern urban area of Kyoto city (60 m above sea level). The original vegetation of this area is thought to be evergreen oak forest in a warm temperate climate, but now many exotic plants have been introduced, and the vegetation is disturbed (Kakutani *et al.*, 1990).

B. Field Census

In the three study sites, we collected flower-visiting insects on flowers along fixed census routes, weekly or bimonthly throughout the flowering seasons of 1984 to 1987 (Kato *et al.*, 1990 for details). Specimens were preserved individually, and the plant species on which they were collected was recorded. A total of 9171 insect individuals or 1487 species were collected on 265 plant species (Table 1). Bumblebees were dominant flower visitors in

Table 1　Outline of Census in Three Locations of Kyoto Prefecture in 1984–1987

	Kyoto	Kibune	Ashu	Total area
Number of plant species	113	115	91	265
Number of plant species visited by bumblebees	19	54	50	98
Number of insect species	320	889	715	1487
Number of insect individuals	2109	4603	2459	9171

our study sites; in total, 735 individuals of five bumblebee species were collected on 98 plant species (Table 2). Out of these plants, 32 species were visited by ≥5 bumblebee individuals (Appendix). In this chapter, we mainly analyze visiting patterns of bumblebees and honeybees.

We did not measure the morphology of individual flowers in the above census, and we have used the species averages of flower depth of the 32 plant species (shown in Fig. 8) because generally intraspecific variation in flower morphology is far smaller than intraspecific variation among bumblebees in our study sites.

C. Bumblebees

There are 15 bumblebee species in Japan (Table 3). The bumblebee fauna is richer in northern regions: 11 species in Hokkaido, or in the highlands; nine species in Yatsugatake, Nagano Prefecture. Western regions of Japan, including Kyoto, have only five species, and lack species with longer (>14 mm) proboscises, e.g., *Bombus consobrius*. Long-tongued species are confined to subalpine or northern cool regions as in Europe (Ranta and Lundberg, 1980; Ranta, 1982).

Bombus diversus has the longest proboscis among the five species found in Kyoto. This species is thought to have come to Japan from the Korean Peninsula in the Riss Glacial stage [13 m.y.a.], and is now distributed in temperate regions. *Bombus honshuensis*, with the second longest proboscis, is thought to have come to Japan from the northern islands, Sakhalin Kuril'skiye Ostrova, between the Riss and the Würn stages (2 m.y.a.), and is now distributed mainly in cooler regions. Kyoto city lacks this species. *Bombus ardens*, with its moderate-length proboscis, is another old-comer from the Korean Peninsula and is distributed in warmer regions. Both *Bombus ignitus* and *B. hypocrita*, with shorter proboscises, are newcomers from the Würn glacial stage (1 m.y.a). *Bombus ignitus* is thought to have come from the Korean Peninsula and is now distributed only in warmer lowland regions, whereas *B. hypocrita* is distributed in cooler regions. The biogeography and natural history of Japanese bumblebees is described in Ito (1985, 1991), Ito and Sakagami (1980), Sakagami and Ishikawa (1969, 1972), Sakagami (1976), Kato (1987), and Katayama (1989).

Colonies of *B. ardens* were active from March to middle May (Inoue *et al.*,

Table 2 Distribution of Five Bumblebee Species in Kyoto Prefecture

Species	Kyoto			Kibune			Ashu			Total		
	Workers	Queens	Males	Workers	Queens	Males	Workers	Queens	Males	Workers	Queens	Males
B. diversus	2	0	1	257	14	16	117	3	3	376	17	20
B. honshuensis	0	0	0	14	1	5	70	6	11	84	7	16
B. ardens	16	13	1	5	13	11	21	3	3	42	29	15
B. hypocrita	1	14	0	38	4	4	33	10	6	72	28	10
B. ignitus	6	10	0	2	0	0	0	1	0	8	11	0
Total	25	37	2	316	32	36	241	23	23	582	92	61	Total 735

Table 3 Species Compositions of Bumblebees at 37 Localities in Japan[a]

	Bumblebee species														
	SS								S			L		LL	SS
Code, Locality	BI	BH	VF	PB	PA	PH	TH	TS	TD	TP	DD	DU	MY	MC	FN
1 Hokkaido		+	+	+	+	+	+	+	+	+	+		+		
2 Cape Nosappu		+	+	+				+	+	+	+		+		
3 Shibetsu Penninsula		+	+	+	+						+		+		
4 Rausu		+		+						+	+		+		
5 Kiritappu		+		+	+	+		+	+	+	+		+		
6 Kushiro Marshland		+								+	+		+		
7 Rishiri Island		+				+				+	+		+		
8 Yagishiri Island		+								+	+		+		
9 Kita-uryuu		+						+		+	+		+		
10 Sapporo, Hokkaido Univ.		+		+	+				+	+	+		+		
11 Mt. Obira		+		+	+		+			+	+		+		
12 Honshu	+	+		+	+		+	+	+	+	+	+		+	+
13 Towada Lake, Aomori		+		+	+		+				+	+		+	+
14 Atsumi, Yamagata		+			+		+				+				
15 Mt. Hayachine, Iwate		+		+	+		+			+	+				
16 Nasu, Tochigi		+		+	+		+				+				
17 Mt. Yatsugatake, Nagano		+		+	+		+		+		+	+		+	+
18 Karuizawa, Nagano		+		+	+		+		+		+	+		+	
19 Mt. Shiraiwa, Nagano		+		+	+		+				+			+	
20 Motosu Lake, Yamanashi		+		+	+		+				+	+		+	
21 Mt. Kushigata, Yamanashi		+		+	+		+				+			+	

398

22 Mt. Senjou, Shizuoka	+	+							+				+	
23 Abe Pass, Shizuoka	+	+							+				+	
24 Asahi-mura, Gifu	+	+	+						+				+	+
25 Tokuyama-mura, Gifu	+	+							+					
26 Mt. Ibuki, Shiga	+	+							+					
27 Ashu, Kyoto	+	+	+						+					
28 Kibune, Kyoto	+	+	+						+					
29 Kyoto Univ., Kyoto	+			+					+					
30 Mt. Sanjou, Wakayama			+						+					
31. Inagawa-cho, Hyogo	+	+							+					
32 Arita, Wakayama	+								+					
33 Shikoku	+	+							+					
34 Mt. Tebako, Ehime	+	+							+					
35 Kyushu	+	+							+					
36 Tsushima, Nagasaki	+	+							+					
37 Yakushima, Kagoshima	+	+							+					

[a] Species are sorted by proboscis length: SS, <9 mm; s, 9–12; L, 12–14; LL, >14. BI, *Bombus (Bombus) ignitus*; BH, *B. (B.) hypocrita*; BF, *B. (B.) florilegus*; PB, *B. (Pyrobombus) beaticola*; PA, *B. (P.) ardens*; PH, *B. (P.) hypnorum*; TH, *B. (Thoracobombus) honshuensis*; TS, *B. (T.) schrencki*; TD, *B. (T.) deuteronymus*; TP, *B. (T.) pseudobaicalensis*; DD, *B. (Diversobombus) diversus*; DU, *B. (D.) ussurensis*; MY, *B. (Megabombus) yezoensis*; MC, *B. (M.) consobrinus*; FN, *Psithyrus (Fernaldaepsithyrus) norvegicus*.

1990). Queens were collected on flowers until the end of May. The other four species were active from April to the end of October. In this paper we ignore males because of a small sample size.

D. Measurement of Morphology

We measured seven morphological characters of individual bumblebees that might relate to foraging activity. The total of prementum (PL) and glossa (GL) lengths is called proboscis (or tongue) length (PG) (Fig. 1). Proboscis length has been thought to be important in relation to accessibility to deep flowers (Fig. 1), and has been used in almost all studies of bumblebee–plant interactions (see Harder, 1982 for discussion of functional proboscis length). We measured prementum and glossa lengths separately. Mouthpart width (MW, the distance between the bases of the mandibles), head length (HL), and head width (HW) are three other head characters that may also be related to the accessibility of flowers to bumblebees (Fig. 1). Wing length (WL) is measured as the distance between the M–Cu bifurcation and the basal tip of the marginal cell. This may be related to flight ability. Corbicula length (CL) is measured to show the capacity of pollen loads.

E. Statistical Analysis

We used canonical discriminant analysis (CDA) to discriminate among the 32 plant species visited by ≥ 5 bumblebee individuals by the seven morphological characters of bumblebees that visited them. Note that species of bumblebees were not distinguished in the analysis. As CDA provides the

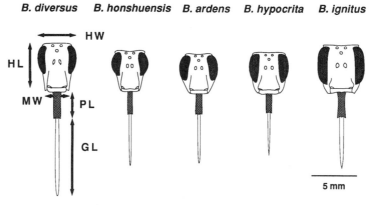

Figure 1   Average-sized worker's heads of five bumblebee species (from left to right, descending order of glossa length, GL). Head width (HW), head length (HL), mouthpart width (MW), prementum length (PL) and glossa length (GL) are measured for morphological analysis.

best direction to discriminate among groups (plant species in this case), it follows that plants that do not overlap on the plane of the canonical variables are visited by bees of different morphology (SAS User's Guide, 1985). In contrast, plant species that do overlap after CDA analysis are thought to be visited by bees of the same morphology. Furthermore, CDA tells us which morphological characters are important in determining visits to flowers by bees of different morphologies. In this paper, we pooled all the samples collected at different seasons from the three study sites to clarify overall relationships between the bee morphologies and flower-utilization patterns. Analyses separating seasonal, geographical, and morphological factors will be shown in a separate paper, but we have found that the results are similar to the pooled analysis even when we separately analyzed Ashu and Kibune (Kyoto contained only 9% of the total samples).

III. Results

A. Morphological Variation among Bumblebees

Figure 1 shows the heads of average-sized workers of five bumblebee species. *Bombus diversus* had the longest glossa, and its head resembled a horse face due to the elongation of the malar space. Mouthpart width was also slender compared with head width. *Bombus ignitus* had the shortest glossa but largest head and mouthpart width of the five species. *Bombus ignitus* is a facultative nectar robber and uses its mandibles to perforate flowers; mouthpart width may therefore relate to the power of the mandible muscles.

Frequency distributions of proboscis length greatly overlapped among four of the five species (except *B. diversus*, with the longest proboscis) (Fig. 2). There was some overlap even between workers of *B. diversus* and the second longest, *B. honshuensis*. However, this pattern was not static. Body sizes changed seasonally, as demonstrated by measurements of the head width and proboscis length of workers (Fig. 3). In *B. ignitus* and *B. honshuensis*, average proboscis lengths in September were 120% of those of early summer. If it were not that *B. ardens* ended colony activity by middle July, its proboscis lengths would largely overlap with *B. ignitus* and *B. honshuensis* late in the season. In every month, there was a vacant range between *B. diversus* and the other four species.

B. Discrimination among Plant Species by Bee Morphology

Figure 4(a) shows the relationship between the first two canonical variables of 32 plant species, based on the seven bumblebee characters. The first two canonical variables represented, respectively, 64% and 16% of the total variation, and thus 80% of the total variation was expressed by this figure.

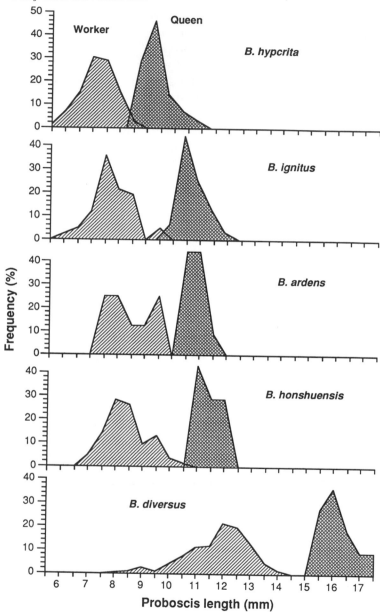

Figure 2   Frequency distributions of proboscis length (PG) of workers and queens of five bumblebee species (from top to bottom, ascending order of averages of workers).

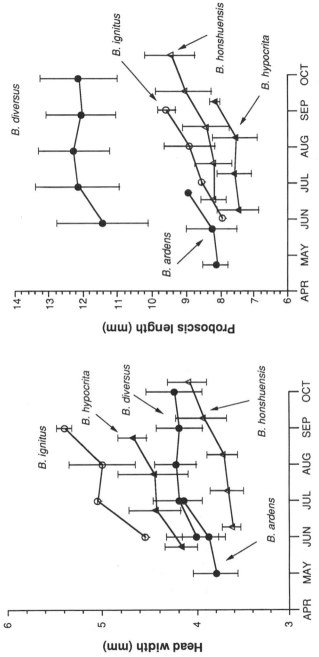

Figure 3 Seasonal changes in head width (HW) and proboscis length (PG) of workers of five bumblebee species.

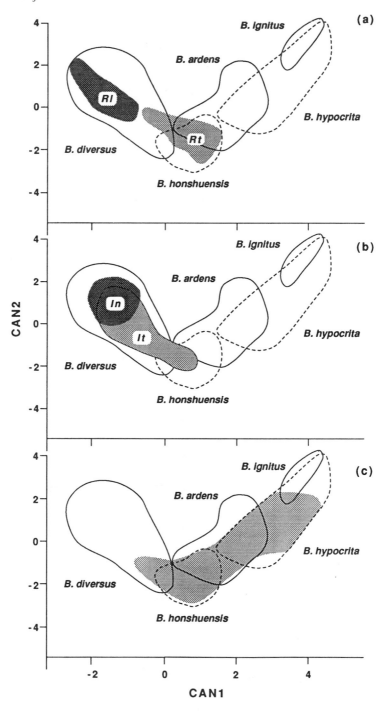

The ranges of plant individuals visited by *B. diversus* overlapped slightly with those visited by *B. honshuensis,* but not with the others. The range of plant individuals visited by *B. ardens* overlapped with visits by *B. honshuensis* and *B. hypocrita,* but not with the others. Thus, overlaps on the plane of the first two canonical variables occurred only between neighboring bumblebee species in Figure 4 (a). Canonical variables are linear functions of seven bumblebee characters and the contribution of each character is expressed by its standardized canonical coefficient (Fig. 5). The two most important characters determining the first canonical variable were GL and MW. This implies that bumblebees with longer glossa and narrower mouthparts are located at the left side of Figure 4(a). The two most important characters determining the second canonical variable were GL and PL, followed by MW. This implies that bumblebees with longer glossa, shorter prementa and wider mouthparts are located at the top of Figure 4(a). An interesting point is that, although PG is a simple total of GL and PL, standardized canonical coefficients of the two characters had different signs in each of the first two canonical variables (Fig. 5). This may be related to phylogenetic constraints. The other interesting point is that mouthpart width was also important in determining the canonical variables. *Bombus ignitus* had a longer proboscis (mean = 8.55 mm) than that of *B. honshuensis* (8.50), and is located at the upperright in Figure 4(a). Yet, as shown below, accessibility of long tubular flowers was higher for *B. honshuensis* owing to its narrower mouthparts. The implication of these results, that characters other than tongue length are important in separating plant–pollinator associations, is emphasized by the following case studies of morphological matching.

C. Case Studies of Flowers

1. *Rabdosia*

Rabdosia longituba and *R. trichocarpa* (Labiatae) bloom synchronously in September in the same habitat and thus, there is neither seasonal nor spatial niche segregation between the two plants. *Rabdosia longituba* was visited mainly by *B. diversus,* and less by *B. honshuensis,* a long-tongued hawkmoth, and a long-tongued fly (Table 4). In contrast, Asian honeybees (*Apis cerana*),

---

Figure 4 (a) Distribution ranges of the first two canonical variables (CAN1, CAN2) of individual bees that visited 32 plant species are shown by continuous or dotted lines, distinguishing five bumblebee species. The canonical variables of individual bees are calculated by the linear function of seven morphological characters, of which coefficients are shown in Figure 5. The distribution range of each bumblebee species is drawn so as to include all individuals that belonged to respective species within the distribution range. Distribution ranges of *Rabdosia longituba* (*Rl*) and *R. trichocarpa* (*Rt*) are also shown by shaded areas. (b) Distribution ranges of *Impatiens noli-tangere* (*In*) and *I. textori* (*It*). (c) A distribution range of *Aesculus turbinata*. See the text for details.

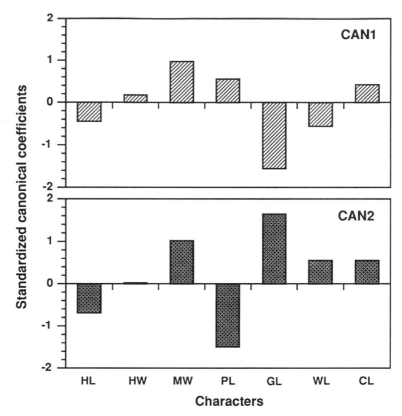

Figure 5   Standardized canonical coefficients (c_i) of 7 morphological characters ($i = 1$ to 7) for the first two canonical variables (CAN1, CAN2); *CAN 1* or *2* $= \sum_{i=1}^{7} c_i \frac{(x_i - m_i)}{s_i}$, where x_i is the value of i'th character of an individual bee, and m_i and s_i are respectively the mean and the s. d. of all the individuals of five species. HL, head length; HW, head width; MW, mouthpart width; PL, prementum length; GL, glossa length (shown in Fig. 1); WL, wing length, and CL, corbicula length.

native to Japan, were the main visitors to *R. trichocarpa* (Table 4) and were the main pollinators (confirmed by examination of pollen transfer by Kakutani, unpublished). European honeybees (*Apis mellifera*) rarely visited *R. tricocarpa*. *Bombus honshuensis* also visited *R. trichocarpa* flowers at a rather high frequency. Despite the fact that *B. honshuensis* visited both *Rabdosia* species, the distribution ranges of bumblebee morphologies that visited *R. longituba* and *R. tricocarpa* did not overlap with each other on the plane of the first two canonical variables (Fig. 4a). This suggests that individual bee morphology rather than species averages is the appropriate way to detect whether or not plants share pollinators.

When an average-sized worker of *B. diversus* visits a flower of *R. longituba*,

Table 4 Numbers of Visits to Five Plant Species by Bumblebees, Honeybees, and Other Insect Groups

	Plant species				
Insect groups	*R. longituba*	*R. trichocarpa*	*I. noli-tangere*	*I. textori*	*A. turbinata*
B. diversus	7	3	19	94	9
B. honshuensis	1	17	0	1	26
B. ardens	0	0	0	0	26
B. hypocrita	0	0	0	0	21
Apis cerana	0	34	0	0	2
Apis mellifera	0	2	0	0	38
Other bees	0	13	2	0	38
Other insects	2[a]	21	0	2	117

[a] *Macroglossum pynhosticta* (Sphingidae) and *Episyrphus balteatus* (Syrphidae).

the glossa tip just touches the nectar stored in the innermost part of the flower. Deeper penetration is prevented by a sharp tapering at the flower tip (Fig. 6) because the mouthparts of *B. diversus* are wider than the narrowed flower part. An elongated malar space also contributes to extend the reach of bees. The distribution of proboscis lengths of *B. diversus* individuals that visited *R. longituba* was slightly biased to the longer side, compared with the distribution of the total *B. diversus* population that was collected in September (z value of the normal distribution was 0.76). The proboscis of *B. honshuensis* was far shorter than the flower length of *R. longituba* (Fig. 6).

Small flowers of *R. trichocarpa* that are suitable for Asian honeybees are too small even for *B. honshuensis*, and handling time of *R. tricocarpa* flowers is longer in *B. honshuensis* than in Asian honeybees; bumblebees never hover for foraging and must land on plants, and thus a long proboscis reduces their efficiency (Kakutani, unpublished). *Bombus honshuensis* individuals that visited *R. tricocarpa* had slightly shorter proboscis lengths, compared to the September population of *B. honshuensis*. *Bombus diversus* rarely visited flowers of this plant and such bees were significantly smaller than the September population of *B. diversus* ($z = -1.83$, $p < 0.05$).

2. Impatiens

Two *Impatiens* species (Balsaminaceae) exhibited staggered flowering seasons in the same habitat. *Impatiens noli-tangere* flowers from July to middle August, and *I. textori* flowers from middle August to September (Kato *et al.*, 1990; Inoue *et al.*, 1990). Spurs of *I. noli-tangere* are slightly longer than those of *I. textori*. Again, the morphology of tapering flower spurs basically determined the accessibility of this plant genus to bees. *Impatiens noli-tangere* was exclusively visited by *B. diversus* (Table 4) (exceptions were two small bees that rob pollen without touching the stigma). *Impatiens textori* was also visited mainly by *B. diversus*. One *B. honshuensis* worker visited *I. textori*, but it is probably an insignificant pollinator (Heinrich, 1979a,b). The CDA distribution ranges of *B. diversus* that visited the two species of *Impatiens* were overlapped (Fig. 4b). We might speculate, therefore, that congenerics that stagger flowering periods can share bees of the same morphology (unlike *Rabdosia* above). That the range of *I. noli-tangere* is located toward the upper-left side of *I. textori* in Figure 4b may reflect the difference of spur length of the two species (Fig. 6).

B. diversus workers that visited *I. noli-tangere* and *I. textori* had significantly

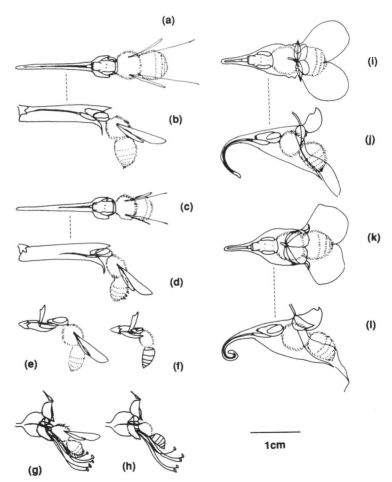

Figure 6 Morphological matching between bees (all workers) and flowers. (a) Top view, (b) side view, the average-sized *B. diversus* in *Rabdosia longituba*. (c), (d), *B. honshuensis* in *R. longituba*. (e) *B. honshuensis*, (f) *Apis cerana* in *R. tricocarpa*. (g) *B. ardens*, (h) *A. mellifera* in *Aesculus turbinata*. *B. diversus* in *Impatiens noli-tangere* (i, j) and in *I. textori* (k, l).

longer proboscises compared with other members of the worker population of *B. diversus* that were collected during the flowering seasons of the two species ($z = 3.48$, $p < 0.001$, $z = 2.21$, $p < 0.05$, respectively). Thus, the two *Impatiens* species share larger workers of *B. diversus* as pollinators (confirmed by Kakutani, unpublished).

3. *Aesculus*

The Japanese horse-chestnut, *Aesculus turbinata* (Hippocastanaceae) flowers for 2 weeks from the end of May. Although the rather open flowers (Fig. 6) of this species were visited by various bee groups (Table 4), only bumblebees were effective pollinators (Kakutani, unpublished). Others were essentially nectar or pollen robbers. *Aesculus turbinata* is one of the most important nectar sources for the apiculture of European honeybees in Japan, and flowers in the study sites were visited by them from an apiary nearby. European honeybees are considered nectar robbers of *A. turbinata* because they rarely touch stigmas and stamens (Fig. 6). Interestingly, Asian honeybees very rarely visited *A. turbinata*, although they should be able to take nectar from *A. turbinata*, as their morphology is similar to European honeybees. We found them on different flowers nearby during the flowering season of *A. turbinata*.

As *A. turbinata* was visited by four bumblebee species of different sizes (Fig. 4c), superficially there was no clear morphological matching between flowers and bees. However, careful examination of bee morphology reveals a clear pattern (Fig. 7). In Figure 7 we compared proboscis lengths of bumblebees that visited *A. turbinata* with those from the total population in the flowering season of *A. turbinata*. Only 7% of *B. diversus* workers were collected on *A. turbinata*, and the rest were collected at other flowers. They had significantly shorter proboscises than did the seasonal population ($z = -4.35$, $p < 0.0001$; Fig. 7). One *B. diversus* queen that was collected on *A. turbinata* was the smallest one found. *Bombus honshuensis* queens collected on *A. turbinata* also had shorter proboscises. There were no significant differences in proboscis length between *Bombus honshuensis* and *B. ardens* workers collected on *A. turbinata*, and those in the rest of their seasonal populations. Nearly half of the workers of these two bee species were collected on *A. turbinata*. Of the *B. hypocrita* workers, the frequency of visits to *A. turbinata* was 17%. This indicates that average-sized workers (7–8.5 mm proboscis length) of *B. honshuensis* and *B. ardens* matched flowers of *A. turbinata*. Morphological matching (Fig. 6) agreed with the above results.

We draw three major conclusions from these (and other) case studies. First, plant species can share bee species without actually sharing bee morphologies, and this is likely to reduce competition among plants for pollinators. Second, bee species that vary significantly in morphology can increase the number of plant species that the colony as a whole can utilize. Third, a plant species apparently visited by a wide variety of pollinator

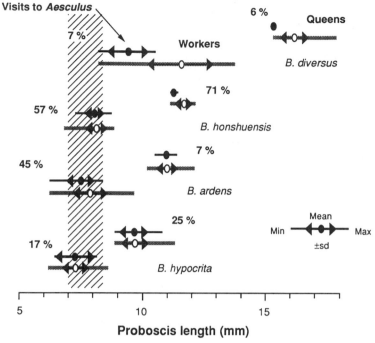

Figure 7 Distributions of proboscis length (PG) of queens and workers that visit *Aesculus turbinata* (top in each category) are compared with the distributions of the seasonal populations (bottom in each category). Percentage of individuals that visit *A. turbinata* are shown above respective categories.

species may actually be accessible to only a small proportion of the workers from some bee species, while accessible to the majority of workers of another bee species.

D. Community Structure

Thirty-two plant species visited by ≥5 bumblebee individuals were clustered by percentage visit frequency of five bumblebee and two honeybee species (Fig. 8). We added honeybees for the analysis because their floral resources largely overlapped with those of bumblebees.

The most abundant species in each cluster varied from *B. diversus* in cluster 1, *B. honshuensis* in cluster 5, *B. ardens* in cluster 2, *B. hypocrita* in cluster 3, and Asian honeybees in cluster 4 (Fig. 9). European honeybees were found in 4 clusters (not cluster 5) but percentage visits were always low. Clear separation is found only between cluster 1 and the others. Cluster 1 includes plants visited mainly by *B. diversus* (Fig. 9), such as *Rabdosia longituba* (Rab1), *Impatiens noli-tangere* (Imp1), *I. textori* (Imp2) and three *Cirisium*

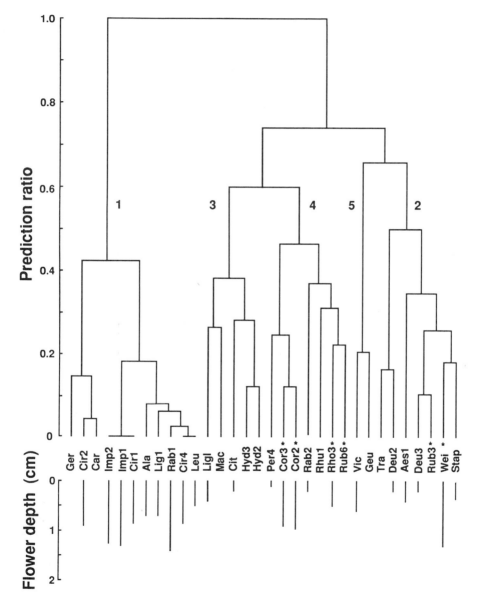

Figure 8 Clustering (by Ward method) of 32 plant species by percentage visit frequencies of five bumblebee and two honeybee species. Plant names and their codes are shown in appendix. Average floral depths are shown at bottom. * indicates spring flowers visited by bumblebee queens. Average visit frequencies are shown for each cluster in Figure 9.

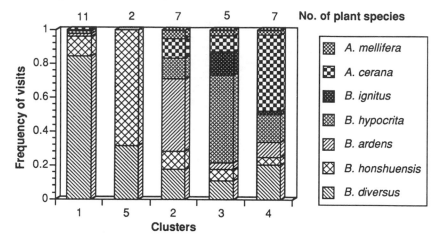

Figure 9 Average visit frequency of bumblebees and honeybees to plants in respective clusters shown in Fig. 8. Numbers of plant species that belong to individual clusters are shown at the top.

(Compositae) species (Cir1, Cir2, Cir4). However, other bumblebees also visited cluster 1 at a frequency up to 18%. As exemplified by *Rabdosia longituba* and *Impatiens, B. honshuensis* workers that visited plants in cluster 1 were significantly larger than the population average for that flowering season. Even plants in cluster 4, e.g., *A. tricocarpa*, were visited by *B. diversus*. *Bombus diversus* individuals that visited clusters 2, 3, and 4 had shorter-than-average proboscises. Changes in the spectrum of bee visits in clusters 2 to 5 are gradual, without distinct boundaries.

IV. Discussion

A. Causes of Morphological Variation in Bumblebees

Bumblebees rear multiple offspring in a single pollen pocket, and larvae located at the periphery of the pocket feed less and grow up as smaller adults (Michener, 1974, Morse, 1982). This communal brood chamber is the basis of the extensive size variation in bumblebees, compared with honeybees and stingless bees in the family Apidae, which employ the one-offspring-in-one-cell system. Size variation in the latter group is negligible (Michener, 1974). Large morphological variations frequently appear also in ants, which have no specific chambers for brood rearing (Wilson, 1971; Hölldobler and Wilson, 1990).

The average size of bumblebees in a colony is both nutritionally and genetically based. By changing the amount of pollen given to captive colonies, Plowright and Jay (1977) and Sutcliff and Plowright (1988) showed that the average size of bees increased with increasing amounts of food, but the width of the size distribution range remained rather constant. Heritability of size of queens and males is 0.2 to 0.5 (Owen, 1988, 1989), with a magnitude similar to environmental factors (mainly food). Thus, body size is a character that could be influenced by natural selection. Bumblebee size changes seasonally (Harder, 1985; this study) and geographically (Pekkarinen, 1979; Ito, personal communication). The seasonal increase in worker body size may reflect the increase in pollen income collected by the increased worker population in a colony.

B. Morphology and Competitive Divergence

Among morphological characters of insects, tongue length has been frequently employed in the study of insect–flower interactions, in a manner similar to bill size in avian studies. This is because tongue length has been thought exclusively to determine the accessibility of flowers of different depths to bees. After several attempts to measure the functional length of pollinator tongues (Taniguchi, 1954; Medler, 1962; Morse, 1977), Harder (1982) found that glossa length was the best estimate of how deep insects could insert their tongues into (artificial) flowers. The prementum was extended only into extremely deep flowers in Harder's (1982) experiment.

Positive correlations between the tongue lengths of insects and the depth of flowers that they utilize have been reported by many authors (Brian, 1957; Teräs, 1976; Pekkarinen, 1979; Ranta and Lundberg, 1980; Barrow and Pickard, 1984 for bumblebees, Ken Inoue, 1983 for moths, among others). These studies considered both average tongue lengths and average flower depths, ignoring the large individual variation within species. Macior (1968, 1973) did not find such a positive correlation and concluded that flowers of particular plant species (*Pedicularis*) were open to use by every bumblebee species in the local community. Our study is unique because we investigated individual variation in detail while including more than one morphological character.

Inouye (1977) found that tongue length ratios among bumblebee species in local communities in North America and Great Britain fitted well to the 1 : 3 ratio predicted by competitive exclusion theory (Hutchinson 1978). Inouye concluded that the number of sympatric species would be limited to four or five. Ranta (1982, 1984), Ranta and Lundberg (1980), and Ranta and Vepsäläinen (1981), however, found that up to 11 species coexisted locally in the mainland of Europe, and most of them had short tongues of similar lengths. Bumblebees with longer tongues were rare and/or distributed only locally, although long tongues were thought to be adaptive, because they allowed individuals to utilize both shallow and deep flowers (discussed below). Ranta considered that because of spatio–temporal heterogeneity among flower resources, competition was intermittent, and coexistence of bees with similar tongue lengths was possible (Chapter 11 considers competition among bees in more detail). These comparisons were done by using species averages, not individual measurements, even though natural selection acts on individuals, not on species (Endler, 1986), and competitive exclusion theory is based on competition among individuals (reviewed by Schoener, 1987). This is especially true for bumblebees, because size variation is quite large within a single species and caste (Pekkarinen, 1979; this volume).

In addition to individual variation in tongue length, other morphological characters may influence foraging and other ecological processes in bumblebees. Although Pekkarinen (1979) measured other morphological characters for biogeographical studies, and experiments with dead bees have shown that other characters influence their accessibility to flowers (Barrow and Pickard, 1985), there have been no multivariate analyses (but Harder, 1985).

In this study, we measured, individually, seven morphological characters that influence foraging activity. Canonical discriminant analysis demonstrated that such multivariate analysis could discriminate among individuals in multidimensional space, even in cases in which the distribution of a single character (e.g., tongue length, Figs. 2, 3) largely overlapped among bumblebee species. In the communities studied by us, glossa and prementa were the first two characters that had a strong influence on canonical variables. Although the total of the two characters is the proboscis (tongue) length, signs of standardized canonical coefficients of the two characters were always different (Fig. 5); individuals with longer glossa always had shorter prementa. Morphology of mouthparts was also important. Case studies on *Rabdosia* and *Impatiens* showed that not only tongue length but also head morphology determines how deep bees can insert their heads into flowers (see also Kato *et al.*, 1991). In addition to flower depth, the morphology of flower tips can interact with bee morphology to limit the phenotype of visitors (examples shown below).

Morphological characters of bees are related not only to their accessibility to flowers but also to other ecological traits. During foraging, morphological characters can be related to their dominance in flower territories (bees with wider head are winners, Kikuchi, 1965), and the cost of flight between flowers (body weight, Pyke, 1978a,b). Beyond effects on foraging, larger bumblebees are thought to be adapted to low temperatures and windy environments (Lundberg and Ranta, 1980), and large size may facilitate overwintering by queens (Owen, 1988). Natural selection for other morphological traits, and their correlation with tongue length, may result in superficial patterns of selection of tongue length (Endler, 1986). We suggest that future work should be based upon the careful analyses of multiple morphological characters and should exploit the tools of quantitative genetics to explore various ecological traits, although at present we completely lack information about covariances between morphological characters.

C. Plant Characters That Relate to Pollination

From the viewpoint of plants, plant morphology is generally thought to affect fertilization success, by increasing pollinator visitation and by decreasing improper pollen transfer (Rathcke, 1983). Visitation intensity of insects is basically determined by the attractiveness of plants (Chapter 5), or the net energetic intake from the plant per unit time (Pleasants, 1980). Rare plants produce much more nectar per flower than do abundant plants to increase their attractiveness (Heinrich, 1976a, b, c; Pleasants, 1980). Flower morphology adds a level of specificity to attractiveness by excluding noneffective pollinators, or robbers. Long flowers protect nectar from short-tongued bees and, as exemplified by *Rabdosia longituba*, characters other than flower depth may also determine the accessibility of flowers to bees. Such exclusion mechanisms have been thought not to be present in short and dish-shaped flowers, but this is not necessarily true (Kakutani *et al.*, 1989). Short flowers of the Japanese orchid, *Platanthera nipponica*, for example, that are pollinated by short-tongued noctuids, exclude long-tongued sphingids by having twisted spurs, because only straight spurs can be probed by the long tongues of hovering sphingids (Inoue, 1983; another example in *Impatiens* in Kato *et al.*, 1991). Neither type can escape from piercing by primary nectar robbers (Inouye, 1980; Roubik, 1989, 1990, see Chapter 11).

Intraspecific size variation is also found in flowers, and different-sized flowers within a single plant species attract different-sized bees. Larger flowers of cow vetch, *Vicia cracca*, were visited by larger workers of *Bombus vagans*, and *vice versa* (Morse, 1978). Assortive mating among similar size classes was found in this plant. Galen and Newport (1987) investigated optimal flower design in sky pilot, *Polemonium viscosum*. Larger flowers were advantageous because they increased the number of visits by pollinators (higher reward in large flowers), but suffered from a lower frequency of precise pollen transfer. Optimization of flower size results from stabilizing selection in this plant, which balances the opposing forces. By changing spur lengths of orchids *experimentally*, Inoue (1987) found that shortened spurs decreased both pollinia removal rate (paternal success) and fertilization rate (maternal success). There were no significant deleterious effects of elongating spurs and, in this case, longer flowers would be favored. Our study did not deal with intraspecific size variation of plants. We suggest that simultaneous studies of size variation of both plants and bees would prove fruitful.

The positions at which pollen or pollinia are attached to pollinator bodies are determined by flower morphology. Sharing of a single pollinator by multiple plant species becomes possible if pollen deposition sites vary among the plants (Waser, 1983; Macior, 1982; Feinsinger, 1987; Kato *et al.*, 1991). Sharing pollinators in this way is especially important for rare plants (Rathcke, 1983) to receive the service of trapliners (Janzen, 1971).

However, inter- and intraspecific morphological variation among plants does not always imply specialization of pollinator species specific to individual plant species or morphs (Chapter 11). There was considerable interspecific variation, for example, in the flower morphology of North American *Pedicularis,* but each species was visited by several common bumblebee species (Macior, 1982, but he did not measure bee morphology). Even specific morphological characters that seem specialized for a single pollinator are frequently pillaged by robbers, parasites, and others (Roubik, 1989; Chapter 11).

Color and odor are other floral characters that advertise flowers to pollinators (Keven, 1978; Macior, 1978). Intraspecific color morphs can promote assortive matings (Brown and Clegg, 1984), while intraspecific odor morphs, each of which is pollinated differentially by bumblebees and flies, are also known (Galen *et al.,* 1987). The classification of pollinator syndromes based on color and odor is found in classical works of pollination ecology.

Flowering phenology has been studied frequently in relation to competition for pollination (Rathcke, 1983). Both staggering of flowering periods among plant members in the same guild that share a single pollinator (Pleasants, 1980; Heinrich, 1976b; Yumoto, 1986, 1987), and random associations (Rathcke, 1984; Tepedino and Stanton, 1981, 1982) are reported from different communities. Such wide variation among the results, which depends on the communities studied and the researchers' prejudice, demands more careful examination of the patterns. For this purpose, methodologies for the statistical testing of data with null models have been developed by various authors (Pleasants, 1980; Rathcke, 1984). Pleasants (1990) examined by several methods the flowering times of plant species that were hypothetically staggered, and found that his original analysis of the plant–bee communities in the Rocky Mountains (Pleasants, 1980), by mean pairwise overlap, was statistically correct. But in Pleasants (1980), the problem is not in the statistical analysis, but in the rather arbitrary procedure used to assign plants to specific guilds. Boundaries between guilds are generally not distinct, but gradual, even in Pleasants (1980) data. We need a statistically rational standard to discriminate among guilds (Jaksić and Medel, 1990). Such a classification of plants according to *pollinator syndromes* tends, in many cases, to exaggerate specialization (Waser, 1983).

In the present study, we used cluster analysis for the classification of plants by visitation, but we could not detect distinct guilds of plants visited by four out of five bumblebee species. As shown below, in combinations of two or three plant species, overlap of flowering periods decreases seed set (see Chapter 5 for a further discussion of apparent competition among plants for pollinators generated by differences in nectar production rates). At the community level, however, staggering of flowering periods among plant species appears rare, and is incomplete if present. In addition, the phylogeny of plants (Kochmer and Handel, 1986) and habitat conditions (Macior, 1983) limit the free allocation of flowering periods by plants that is a basic assumption of null models. Simultaneous analyses of both morphology and phenology have been suggested (Murray *et al.*, 1987).

D. Foraging Efficiency and Patterns of Competition

From the point of view of the pollinator, foraging efficiency is the crucial parameter to consider in the matching of flower and pollinator characters. Foraging efficiency of bees is usually defined by net energy gain per unit time in nectar collection or by the amount of pollen collected per unit time (or per unit energy expenditure) (Heinrich, 1979b; Pyke, 1978a, and Krebs and Davies, 1987 for general discussion). Below we use the former definition. Grant (1972) proposed two models of foraging efficiency in relation to size. In Model 1, larger foragers (bill size in Grant, 1972) can utilize both small and large foods (e.g., seed grain width), while the smaller ones are confined to small foods (the same assumption as Wilson's 1975 model). In Model 2, there is an effective forager size for each food size, and thus larger foragers can use only large foods. In spite of its importance, only a few studies have measured foraging efficiency in this context (Fenchel, 1975 for shells, Davidson, 1978 for ants). In studies of bumblebee foraging, both models are supported in different cases; Model 1 by Holm (1966) and Inouye (1980) and Model 2 by Ranta and Lundberg (1980), Ranta (1983), and Harder (1983). Ranta and Lundberg (1980), for example, found that a wider range of flower sizes was used by larger bees. However, these studies analyzed only the relationships between tongue length and flower depth. Harder (1983) showed that, in addition to tongue length, body weight was also related to foraging efficiency, through changing traveling time between flowers. Harder also found that there was no species-specific differences in foraging efficiency, and it was the same if the morphology of the bees was the same, irrespective of species.

Foraging efficiency has been studied for a limited set of plant species in our study sites (Kakutani, unpublished data), but relationships between visit frequency (Fig. 9) and flower depth (Fig. 8) support Model 2 of Grant (1972). *Bombus diversus*, with the longest tongue on average, visited mainly deep flowers in cluster 1, but also visited plants in clusters 2 to 5 at lower frequencies. As already shown in the Results section above, *B. diversus* individuals that visited plants in clusters 2 to 5 were smaller than the species average. There were no significant differences between the flower depths of plants visited mainly by *B. honshuensis*, *B. ardens*, and *B. hypocrita*. Visit frequencies of bumblebee species changed gradually from cluster to cluster, and there was no distinct guild structure. Thus, generally, most plants (except for cluster 1) were visited by all bumblebee species and even by honeybees and other insects, as in Macior (1982). But as we demonstrated for *Aesculus turbinata*, there were fine-tuned patterns within each bee group (species and caste) that were basically determined by morphological matching between flowers and visitors (Fig. 7). Average-sized foragers of *B. honshuensis* and *B. ardens* were morphologically matched to *A. turbinata* flowers, and their visit frequency was high. On the other hand, average-sized foragers of *B. diversus* in that season were thought to be larger than the appropriate size for *A. turbinata* flowers, and only smaller workers visited them and at lower frequency. This pattern of foraging efficiency essentially determines the pattern of flower utilization, although it allows utilization by a wider range of insect species than expected from the competitive exclusion model of Pleasants (1980, 1990). At the same time, this does not mean that flowers are equally available for all visitors as suggested by Macior (1982). Learning and individual specialization would be the mechanism for the fine-tuned pattern observed in each group (Heinrich, 1979a,b).

Plant choice by bees of different sizes can be explained by morphological matching and foraging efficiency (Fig. 10). First, suppose that there is only one colony of Bee species 1 and there is size variation in the colony members. As discussed, foraging efficiency is thought to be a concave function of forager size. Assume that there are two plant species that have different foraging efficiency functions of forager size; Plant 1 fits larger foragers (Bee 1) and Plant 2 fits smaller foragers (Bee 2) on average (Fig. 10). At a low colony density of Bee 1, we can neglect consumption of floral resources and the resulting decrease in foraging efficiency. In this case, colony members smaller than the intersection (Point A) of the foraging efficiency functions of Plants 1 and 2 will visit Plant 2, because they get higher efficiencies by doing so. And colony members larger than Point A will visit Plant 1. When

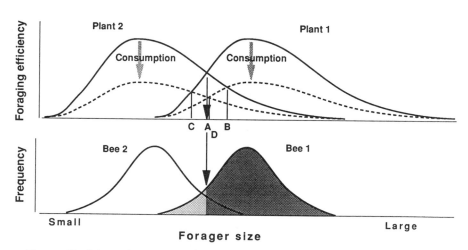

Figure 10 Schematic representation to show how plant choice by bees is determined by foraging efficiency and colony densities of bees. If consumption of floral resouces is negligible (shown by continuous lines of the top graph), foragers of Bee 1 smaller than Point A visit Plant 2 and foragers larger than Point A visit Plant 1. The intersections of the foraging efficiency functions of Plants 1 and 2 change, depending on the relative intensity of resource consumption (shown by dotted lines) on both plants. See the text for details.

colony density of Bee 1 increases (assume same frequency distributions of forager size among colonies), resource consumption occurs only in Plant 1 owing to a higher visit frequency (Fig. 10). Then, the intersection moves to the right side (Point B), and more colony members start foraging on Plant 2. This switching reduces competition for Plant 1 within the colony members and the total net gain of the colony will increase. When Bee 2 is added into this system, the intersection returns to the left side (Point D) owing to decreased foraging efficiency on Plant 2, because Bee 2 mainly consumes the floral resources of Plant 2. Thus, the intersection that decides plant choice is adjusted, depending on the foraging efficiency functions and colony densities of both bee species. This system is flexible and stable for both bees and plants. For example, Bee 1 can survive even when Plant 1 completely disappears from the habitat, by devoting all colony members' foraging to Plant 2. Plant 1 can produce some seeds by the service of Bee 2, and is capable of escaping extinction even when Bee 1 disappears from the habitat.

Evidence for competition among pollinators for plants is sporadic (Chapter 11), while competition among plants for pollinators may be common (Chapter 5). This may reflect the general rule that many pollinators, especially social bees, depend on a wide spectrum of plants during their relatively long lifespan, whereas plants bloom in a rather short period and are pollinated by a specific group of animals (the reverse is true for trees, especially in the tropics, Chapter 11). The Japanese orchids, *Platanthera*, for example, have on average about two moth pollinator species each. The moths, on the other hand, utilize quite a wide spectrum of plant species and are even distributed outside the distribution range of the orchids (Inoue, 1983). Conversely, plants may be pollinated by different insects in different parts of their range. *Rabdosia tricocarpa*, for example, is pollinated mainly by Asian honeybees in our study sites in Kyoto, but in Hokkaido, outside the distribution range of Asian honeybees, it is pollinated by *B. honshuensis*. Similarly, *B. consobrinus* was thought to be the obligate pollinator for *Aconitum* (Løken, 1962), but it is absent in Kyoto and *Aconitum japonicum* is pollinated by *B. diversus*. Competition patterns also change seasonally in the temperate regions; competition occurs among pollinators for flower resources in early spring when the flower density is low, and in autumn when the pollinator density increases. However, competition occurs among plants for pollinators in summer when flowering plants are abundant, relative to the pollinator density (Mosquin, 1971; Heinrich, 1976b).

There is evidence for competition among plants for pollination, both from comparisons between local communities, and from experiments. Coexistence between *Phlox cuspidata* and *P. drummondii*, for example, depends upon the red color morph of *P. drummondii*, which appears in sympatric populations. *P. cuspidata* and *P. drummondii* are both pink in allopatry, and seed set by *P. drummondii* is reduced, apparently due to hybridization, when the pink morph is planted artificially in sympatry with *P. cuspida* (Levin, 1985). Galen (1989) reported that sky pilot, *Polemonium viscosum*, distributed on alpine slopes, is pollinated by bumblebees at higher elevations, and by flies at lower elevations. Flower morphology changes locally, depending on the morphology of respective pollinators. Both length and diameter of flowers are differentiated. Flowers of the Japanese orchid, *Platanthera metabifolia*, that are usually pollinated by sphingids, are smaller in a locality where sphingids are absent, and smaller noctuids become the main pollinators (Inoue, 1986). *Campanula punctata* is pollinated exclusively by *B. diversus* on the mainland of Honshu, Japan. It is pollinated by *B. ardens* and by halictids on islands, which lack *B. diversus*, near the mainland and by halictids on remote islands. Flower size decreases in response to pollinator size on islands. At the same time, reproductive systems change from self-incompatibility on the mainland to self-compatibility and autogamy on remote islands, via intermediate patterns on close islands. The longevity of the male stage of flowers is extended, probably to compensate for a low visitation rate by halictids on remote islands (Inoue and Amano, 1986; Inoue, 1986).

Some natural removal experiments provide evidence for competition among pollinators for floral resources (Inouye, 1978; Bowers, 1985a). In allopatry, *B. flavifrons* and *B. rufocintus* show similar spectra of flower utilization, whereas, in sympatry, *B. rufocintus* is less abundant, and utilizes less-rewarding flowers, while the flower utilization spectrum of *B. flavifrons* remains constant (Bowers, 1985a). Morphological displacement in sympatric pollinators in relation to food size, to our knowledge has not been reported. In the polymorphic granivorous ant, *Veromessor pergandei*, however, the abundance of workers in different size classes in a colony changes locally in response to the morphology of other sympatric ant species, apparently to reduce overlap among resources (Davidson, 1978).

In our study, as with many other studies across a wide range of ecosystems (Chapter 10), competition, in this case for floral resources, was keen but intermittent. Kakutani (unpublished data), measured nectar consumption rates from more than a dozen plant species in Ashu. In fine weather, more than 90% of the nectar produced each day was consumed (mostly within a few hours). However, flower visitation was quite frequently interrupted by rain, especially during the monsoon in June and September, even though these 2 months represent peak flowering periods (Kato *et al.*, 1990). For example, seed set of *Aesculus turbinata* fluctuated greatly (1–10%) from year to year, depending on weather conditions during the short flowering period. This suggests that the amount of *A. turbinata* nectar collected by bees also fluctuated from year to year, and is consistent with competition for floral resources.

Competition among bumblebees during flower utilization, as shown above, is a short-term phenomenon, while the actual consequences of the competition for the colony is expressed in the longer term. Each colony integrates resource returns from many foragers over periods of a few or several months, and the low foraging efficiency of a specific forager on a specific plant species is not directly connected to the colony performance. Inclusive fitness of individual bumblebee workers should be measured by (the degree of relatedness) × (the number of reproductives in the next generation) and, in addition to natural selection, social selection (in the sense of West-Eberhard, 1979, 1983, 1989) operates in this process. This is quite different from competition theory in community ecology, in which the decrease of foraging efficiency of individuals directly links to a decrease in individual reproductive success, and natural selection favors individuals having higher efficiency. In spite of its importance, studies of colony performance of bumblebees in relation to foraging activity are quite rare (but Bowers, 1985a, b). One of us (T.I.) has studied the colony population dynamics of stingless bees (Apidae) for 6 years, and found that competition for nest sites with arboreal ants is crucial in determining colony density, whereas competition for floral resources is insignificant (Inoue, unpublished data). We need similar studies of bumblebee colonies to determine how large, intraspecific size variation, and the resulting large size overlap among species, are selected for.

E. Character Divergence

In our study sites in Kyoto, overlap between the frequency distributions of morphological characters of *B. diversus* and other bee species was minimal or absent (e.g., head width and proboscis length in Fig. 3). As already shown, *B. diversus* utilized the widest plant spectrum of the bumblebee species. In the highlands of Honshu, these *empty* size classes are filled by *B. consobrinus* (larger size classes) and by *B. deutoconymus* and *B. pseudobaicalensis* (smaller size classes) (Table 3). Character divergence or release might, therefore, be expected in our sites as on islands (Schluter, 1988; Schluter *et al.*, 1985), but the CV% of proboscis length was not significantly expanded in *B. diversus* (9.21% compared with 8.08 to 8.94% in the other species in Kyoto. Strictly speaking, however, the comparison should be made between *B. diversus* at localities with and without the empty size classes). Only *B. ardens* occurs on the southern island of Yaku, and character release is expected there (Yumoto, personal communication). But we need to investigate such patterns carefully to distinguish between character release and biogeographical clines (anti-Bergman law, Ito, personal communication), and we have started a field census to do this in the highlands of Honshu.

In Japan, plant guilds that are mainly pollinated by bumblebees are also visited by honeybees (although European honeybees are nectar robbers rather than pollinators of many plants). As shown in Figure 9, European honeybees even visit plants that are pollinated mainly by *B. diversus*, although native Asian honeybees mainly visit plants visited by short-tongued bumblebees. For morphological reasons, we cannot understand why Asian honeybees do not visit *Aesculus turbinata,* from which European honeybees rob nectar. Despite their high competitive ability, there are virtually no feral colonies of European honeybees in Japan. This contrasts strongly with the successful invasion of European and Africanized honeybees into North (Seeley, 1985) and South America, respectively (Roubik, 1989, 1990, Chapter 11). The reason is not direct competition with Asian honeybees, but attacks of vespid wasps. These vespids are specialist predators of social bees and wasps, and European honeybees do not have effective defense mechanisms. This results in the complete elimination of their colonies in autumn (Matsuura and Yamane, 1990). Asian honeybees have developed special defensive behavior against these vespids (Ono *et al.,* 1987). Thus, factors other than floral resources determine the possibility of invasion by European honeybees in Japan.

Acknowledgments

We express our sincere thanks to Mr. H. Muramatu and Mr. T. Kakutani, who helped with field collections and specimen measurement. We extend our thanks to Dr. S. F. Sakagami, Dr. M. Ito, and Dr. D. W. Roubik for their reading of the earlier manuscript and their valuable comments, and to Dr. M. D. Hunter and an anonymous reviewer for their kind reviewing of the manuscript and refining of expression. This study is partly supported by a Japan Ministry of Education, Science and Culture Grant-in-Aid for Scientific Research on Priority Areas (#319) (Project: "Symbiotic biosphere—an ecological interaction network promoting the coexistence of many species")

Appendix

A List of Plant Species on Which ≥5 Bumblebees Were Collected

FAMILY	Species	Species code
PAPAVERACEAE	*Macleaya cordata*	Mac
FUMARIACEAE	*Corydalis pallida*	Cor2
	C. incisa	Cor3
POLYGONACEAE	*Persicaria thunbergii*	Per4
ERICACEAE	*Rhododendron oomurasaki*	Rho3
ROSACEAE	*Geum japonicum*	Geu
	Rubus crataegifolius	Rub2
	R. palmatus	Rub3
	R. buergeri	Rub5
SAXIFRAGACEAE	*Cardiandra alternifolia*	Card
	Deutzia maximowicziana	Deu2
	D. gracilis	Deu3
	Hydrangea macrophylla	Hyd2
	H. paniculata	Hyd3
LEGUMINOSAE	*Vicia venosa*	Vic

ALANGIACEAE	*Alangium platanifolium*	Ala
STAPHYLEACEAE	*Staphylea bumalda*	Stap
HIPPOCASTANACEAE	*Aesculus turbinata*	Aes
ANACARDIACEAE	*Rhus javanica*	Rhul
RUTACEAE	*Citrus tachibana*	Cit
BALSAMINACEAE	*Impatiens noli-tangere*	Imp1
	I. textori	Imp2
OLEACEAE	*Ligustrum obtusifolium*	Lig1
LABIATAE	*Leucosceptrum stellipilum*	Leu
	Rabdosia longituba	Rab1
	R. trichocarpa	Rab2
CAPRIFOLIACEAE	*Weigela hortensis*	Wei
COMPOSITAE	*Cirsium japonicum*	Cir1
	C. kagamontanum	Cir2
	C. microspicatum	Cir3
	Ligularia fischerii	Lig1
COMMELINACEAE	*Tradescantia ohiensis*	Tra

References

Barrow, D. A., and Pickard, R. S. (1984). Size-related selection of food plants by bumblebees. *Ecol. Ent.* **9,** 369–373.

Barrow, D. A., and Pickard, R. S. (1985). Estimating corolla length in the study of bumble bees and their food plants. *J. Apic. Res.* **24,** 3–6.

Bowers, M. A. (1985a). Experimental analyses of competition between two species of bumble bees (Hymenoptera: Apidae). *Ecology.* **66,** 914–927.

Bowers, M. A. (1985b) Bumble bee colonization,extinction, and reproduction in subalpine meadows in northeastern Utah. *Ecology.* **66,** 914–927.

Brian, A. D. (1957). Differences in the flowers visited by four species of bumble-bees and their causes. *J. Anim. Ecol.* **26,** 71–98.

Brown, B. A., and Clegg, M. T. (1984). Influence of flower color polymorphism on genetic transmission in a natural population of the common morning glory, *Ipomoea purpurea.* *Evolution* **38,** 796–803.

Davidson, D. W. (1978). Size variability in the worker caste of a social insects (*Veromessor pergandei* Mayr) as a function of the competition environment. *Am. Nat.* **112,** 523–532.

Dunham, A. E., Smith, G. R., and Taylor, J. N. (1979). Evidence for ecological character displacement in western American catostomid fishes. *Evolution* **33,** 877–896.

Endler, J. A. (1986). "Natural Selection in the Wild." Princeton University Press, Princeton, New Jersey.

Feinsinger, P. (1987). Effects of plant species on each other's pollination: Is community structure influenced? *Trends Ecol. Evol.* **2,** 123–126.

Fenchel, T. (1975). Character displacement and coexistence in mud snails (Hydrobiidae). *Oecologia* **20,** 19–32.

Galen, C. (1989). Measuring pollinator-mediated selection on morphometric floral traits: Bumblebees and the alpine sky pilot, *Polemonium viscosum. Evolution* **43,** 882–890.

Galen, C., and Newport, M. E. A. (1987). Bumble bee behavior and selection on flower size in the sky pilot, *Polemonium viscosum. Oecologia* **74,** 20–23.

Galen, C., Zimmer, K. A., and Newport, M. E. (1987). Pollination in floral scent morphs of

Polemonium viscosum: A mechanism for disruptive selection on flower size. *Evolution* **41**, 599–606.

Grant, P. R. (1972). Convergent and divergent character displacement. *Biol. J. Linn. Soc.* **4**, 39–68.

Grant, P. R., Grant, B. R., Smith, J. N. M., Abbott, I. J., and Abbott, L. K. (1976). Darwin's finches: Population variation and natural selection. *Proc. Natl. Acad. Sci. U.S.A.* **73**, 257–261.

Harder, L. D. (1982). Measurement and estimation of functional proboscis length in bumblebees (Hymenoptera: Apidae). *Can. J. Zool.* **60**, 1073–1079.

Harder, L. D. (1983). Flower handling efficiency of bumble bees: Morphological aspects of probing time. *Oecologia* **57**, 274–280.

Harder, L. D. (1985). Morphology as a predictor of flower choice by bumble bees. *Ecology* **66**, 198–210.

Heinrich, B. (1976a). Resource partitioning among some eusocial insects: Bumblebees. *Ecology* **57**, 874–889.

Heinrich, B. (1976b). Flowering phenologies: Bog, woodland, and disturbed habitats. *Ecology* **57**, 890–899.

Heinrich, B. (1976c). The foraging specializations of individual bumblebees. *Ecol. Monogr.* **46**, 105–128.

Heinrich, B. (1979a). Majoring and minoring by foraging bumblebees, *Bombus vagans:* An experimental analysis. *Ecology* **60**, 245–255.

Heinrich, B. (1979b). "Bumblebee Economics." Harvard University Press, Cambridge, Massachusetts.

Hespenheide, H. A. (1974). Ecological inferences from morphological data. *Annu. Rev. Ecol. Syst.* **4**, 213–229.

Hölldbler, B., and Wilson, E. O. (1990). "The Ants." Harvard University Press, Cambridge, Massachusetts.

Holm, S. N. (1966). The utilization and management of bumble bees for red clover and alfalfa seed production. *Annu. Rev. Entomol.* **11**, 155–182.

Hubbell, S. P., and Foster, R. B. (1986). *In* "Community Ecology" (J. Diamond and T. Case, eds.), pp. 314–329. Harper & Row, New York.

Hutchinson, G. E. (1978). "An Introduction of Population Ecology." Yale University Press, New Haven, Connecticut.

Inoue, K. (1983). Systematics of the genus *Platanthera* (Orchidaceae) in Japan and adjacent regions with special reference to pollination. *J. Fac. Sci. Univ. Tokyo Sect. III* **13**, 285–374.

Inoue, K. (1986). Different effects of sphingid and noctuid moths on the fucundity of *Platanthera metabifolia* (Orchidaceae) in Hokkaido. *Ecol. Res.* **1**, 25–36.

Inoue, K. (1987). Coevolution between orchids and moths. *In* "Insect Communities in Japan." (S. Kimoto and H. Takeda, eds.), pp. 109–115. Tokai Univ. Press, Tokyo (in Japanese).

Inoue, K., and Amano, M. (1986). Evolution of *Campanula punctata* Lam. in the Izu Islands: Change of pollinators and evolution of breeding systems. *Plant Spec. Biol.* **1**, 89–97.

Inoue, T., Kato, M., Kakutani, T., Suka, T., and Itino, T. (1990). Insect–flower relationships in the temperate deciduous forest of Kibune, Kyoto: An overview of the flowering phenology and the seasonal pattern of insect visits. *Contr. Biol. Lab. Kyoto Univ.* **27**, 377–463.

Inouye, D. W. (1977). Species structure of bumblebee communities in North America and Europe. *In* "The Role of Arthropods in Forest Ecosystems." (W. J. Wattson, ed.), pp. 35–40. Springer-Verlag, Berlin, Heidelberg and New York.

Inouye, D. W. (1978). Resource partitioning in bumblebees: Experimental studies of foraging behavior. *Ecology* **59**, 672–678.

Inouye, D. W. (1980). The effect of proboscis and corolla tube lengths on patterns and rates of flower visitation by bumblebees. *Oecologia* **45**, 197–201.

Ito, M. (1985). Supraspecific classification of bumblebees based on the characters of male genitalia. *Low Temp. Sci. Ser. B* **20**, 1–143.

Ito, M. (1991). Taxonomy and biogeography of bumblebees in and around Japan. *In* "Evolution of Insect Societies." (T. Inoue and S. Yamane, eds.), in press, Shisakusha, Tokyo (In Japanese).

Ito, M., and Sakagami, S. F. (1980). The bumblebee fauna of the Kurile Islands (Hymenoptera: Apidae). *Low. Temp. Sci. Ser. B* **38,** 23–51.

Jaksić, F. M., and Medel, R. G. (1990). Objective recognition of guilds: Testing for statistically significant species clusters. *Oecologia* **82,** 87–92.

Janzen, D. H. (1971). Euglossine bees as long-distance pollinators of tropical plants. *Science* **171,** 203–205.

Kakutani, T., Inoue, T., and Kato, M. (1989). Nectar secretion pattern of the dish-shaped flower, *Cayratia japonica* (Vitaceae), and nectar-utilization patterns by insect visitors. *Res. Popul. Ecol.* **31,** 381–400.

Kakutani, T., Inoue, T., Kato, M., and Ichihashi, H. (1990). Insect–flower relationships in the campus of Kyoto University, Kyoto: An overview of the flowering phenology and the seasonal pattern of insect visits. *Contr. Biol. Lab. Kyoto Univ.* **27,** 465–521

Katayama, E. (1989). "Comparative Studies on the Egg-laying Habits of some Japanese Species of Bumblebees (Hymenoptera: Apidae)." Entomological Society of Japan, Tokyo, Japan.

Kato, M. (1987). Coevolution systems of angiosperm flora and bumblebees. *Shu-Seibutugaku Kenkyu* **11,** 1–13 (in Japanese).

Kato, M., Itino, T., Hotta, M., and Inoue, T. (1991). Pollination of four Sumatran *Impatiens* species by hawkmoths and bees. *Tropics* **1,** 59–73.

Kato, M., Kakutani, T., Inoue, T., and Itino, T. (1990). Insect-flower relationships in the primary beech forest of Ashu, Kyoto: An overview of the flowering phenology and the seasonal pattern of insect visits. *Contr. Biol. Lab. Kyoto Univ.* **27,** 309–375.

Keven, P. G. (1978). Floral coloration, its colorimetric analysis and significance in anthecology. *In* "The Pollination of Flowers by Insects." (A. J. Richards, ed.), pp. 51–78. Academic Press, London.

Kikuchi, T. (1965). Role of interspecific dominance–subordination relationship on the appearance of flower-visiting insects. *Sci. Rep. Tohoku Univ. Ser. 4* **31,** 275–296.

Kochmer, J. P., and Handel, S. N. (1986). Constraints and competition in the evolution of flowering phenology. *Ecol. Monogr.* **56,** 303–325.

Krebs, J. R., and Davies, N. B. (1987). "An Introduction to Behavioural Ecology, 2nd Ed." Blackwell, Oxford.

Levin, D. A. (1985). Reproductive character displacement in *Phlox. Evolution* **39,** 1275–1281.

Løken, A. (1962). *Bombus consobrinus* Dahlb., an oligolectic bumble bee (Hymenoptera, Apidae). *Verh. Int. Kongr. Entomol.* 11th. Vol. **1,** 598–603.

Lundberg, H., and Ranta, E. (1980). Habitat and food utilization in a subarctic bumblebee community. *Oikos* **35,** 303–310.

Macior, L. W. (1968) Pollination adaptation in *Pedicularis groenlandica. Am. J. Bot.* **55,** 927–932.

Macior, L. W. (1973). The pollination ecology of *Pedicularis* on Mount Rainier. *Am. J. Bot.* **60,** 863–871.

Macior, L. W. (1978). The pollination ecology of vernal angiosperms. *Oikos* **30,** 452–460.

Macior, L. W. (1982). Plant community and pollinator dynamics in the evolution of pollination mechanisms in *Pedicularis* (Scrophulariaceae). *In* "Pollination and Evolution" (J. A. Armstrong, J. M. Powell, and A. J. Richards, eds.), pp. 29–45. Royal Botanic Gardens, Sydney, Australia.

Macior, L. W. (1983). The pollination dynamics of sympatric species of *Pedicularis* (Scrophulariaceae). *Am. J. Bot.* **70,** 844–853.

Matsuura, M., and Yamane, S. K. (1990). "Biology of the Vespine Wasps." Springer-Verlag, Berlin.

Medler, J. T. (1962). Morphometric studies on bumble bees. *Ann. Entomol. Soc. Am.* **55,** 212–218.

Michener, C. D. (1974). "The Social Behavior of the Bees." Belknap Press of Harvard University Press, Cambridge, Massachusetts.

Morse, D. H. (1977). Estimating proboscis length from wing length in bumblebees (*Bombus* spp.). *Ann. Entomol. Soc. Am.* **70**, 311–315.

Morse, D. H. (1978). Size-related foraging differences of bumble bee workers. *Ecol. Entomol.* **3**, 189–192.

Morse, D. H. (1982). *In* "Social Insects" Vol. III. (H. R. Herman, ed.), pp. 245–322. Academic Press, New York.

Mosquin, T. (1971). Competition for pollinators as a stimulus for the evolution of flowering time. *Oikos* **22**, 398–402.

Murray, K. G., Feinsinger, P., and Busby, W. H. (1987). Evaluation of character displacement among plants in two tropical pollination guilds. *Ecology* **68**, 1283–1293.

Ono, M., Okada, I., and Sasaki, M. (1987). Heat production by balling in the Japanese honeybee, *Apis cerana japonica* as a defensive behavior against the hornet, *Vespa simillima xanthoptera* (Hymenoptera: Vespidae). *Experientia* **43**, 1031–1032.

Owen, R. E. (1988). Body size variation and optimal body size of bumble bee queens (Hymenoptera: Apidae). *Can. Entmol.* **120**, 19–27.

Owen, R. E. (1989). Differential size variation of male and female bumblebees. *J. Hered.* **80**, 39–43.

Pekkarinen, A. (1979). Morphometric, colour and enzyme variation in bumblebees (Hymenoptera, Apidae, *Bombus*) in Fennoscandia and Denmark. *Acta. Zool. Fennica.* **158**, 1–60.

Pleasants, J. M. (1980). Competition for bumblebee pollinators in rocky mountain plant communites. *Ecology* **61**, 1446–1459.

Pleasants, J. M. (1990). Null-model tests for competitive displacement: The fallacy of not focusing on the whole community. *Ecology* **71**, 1078–1084.

Plowright, R. C., and Jay, S. C. (1977). On the size determination of bumble bee castes (Hymenoptera: Apidae). *Can. J. Zool.* **55**, 1133–1138.

Pyke, G. H. (1978a). Optimal foraging in bumblebees and coevolution with their plants. *Oecologia* **36**, 281–293.

Pyke, G. H. (1978b). Optimal body size in bumblebees. *Oecologia* **36**, 255–266.

Ranta, E. (1982). Species structure of North European bumblebee communities. *Oikos* **38**, 202–209.

Ranta, E. (1983). Foraging differences in bumblebees. *Ann. Ent. Fenn.* **49**, 17–22.

Ranta, E. (1984). Proboscis length and the coexistence of bumblebee species. *Oikos* **43**, 189–196.

Ranta, E., and Lundberg, H. (1980). Resource partitioning in bumblebees: The significance of differences in proboscis length. *Oikos* **35**, 298–302.

Ranta, E., and Vepsäläinen, K. (1981). Why are there so many species? Spatio-temporal heterogeneity and northern bumblebee communities. *Oikos* **36**, 28–34.

Rathcke, B. J. (1983). Competition and facilitation among plants for pollination. *In* "Pollination Ecology" (L. Real, ed.), pp. 305–329. Academic Press, Orlando, Florida.

Rathcke, B. J. (1984). Patterns of flowering phenologies: Testability and causal inference using a random model. *In* "Ecological communities: Conceptual isssues and the evidence" (D. R. Strong, D. Simberloff, and A. B. Thistle, eds.), pp. 383–393. Princeton University Press, Princeton, New Jersey.

Roubik, D. W. (1989). "Ecology and Natural History of Tropical Bees." Cambridge University Press, Cambridge, Massachusetts.

Roubik, D. W. (1990). Niche preemption in tropical bee communities: A comparison of neotropical and Malesian faunas. *In* "Natural History of Social Wasps and Bees in Equatorial Sumatra." (S. F. Sakagami, R. Ohgushi, and R. Roubik, eds.), pp. 245–257. Hokkaido University Press, Sapporo, Japan.

Sakagami, S. F. (1976). Specific differences in the bionomic characters of bumblebees. A comparative review. *J. Fac. Sci. Hokkaido Univ. Zool.* **20**, 391–447.

Sakagami, S. F., and Ishikawa, R. (1969). Note préliminaire sur la répartiton géographique des bourdons Japonais, avec descriptions et remarques sur quelques formes nouvelles ou peu connues. *J. Rac. Sci. Hokkaido Univ. Zool.* **17,** 152–196.

Sakagami, S. F., and Ishikawa, R. (1972). Note supplémentaire sur la taxonomie et répartition géographique de quelques bourdons japonais, avec la description d'une nouvelle sous-espéce. *Bull. Nat. Sci. Museum* **15,** 607–616.

SAS (1985). "SAS User's Guide, Statistics, Ver. 5." SAS Institute, Cary, North Carolina.

Schluter, D. (1988). Character displacement and the adaptive divergence of finches on islands and continents. *Am. Nat.* **131,** 799–824.

Schluter, D., Price, T. D., and Grant, P. R. (1985). Ecological character displacement in Darwin's finches. *Science* **227,** 1056–1058.

Schoener, T. W. (1987). Resource partitioning. *In* "Community Ecology." (J. Kikkawa and D. J. Anderson, eds.), pp. 91–126. Blackwell, Melbourne, Australia.

Seeley, T. D. (1985). "Honeybee Ecology." Princeton University Press, Princeton, New Jersey.

Sutcliffe, G. H., and Plowright, R. C. (1988). The effects of food supply on adult size in the bumble bee *Bombus terricola* Kirby (Hymenoptera: Apidae). *Can. Entmol.* **120,** 1051–1058.

Taniguchi, S. (1954). Biological studies on the Japanese bees. I. Comparative study of glossa. *Sci. Rep. Hyogo Univ. Agric. Ser. Biol.* **1,** 81–89.

Tepedino, V. J., and Stanton, N. L. (1981). Diversity and competition in bee-plant communities on short-grass prairie. *Oikos* **36,** 35–44.

Tepedino, V. J., and Stanton, N. L. (1982). Estimating floral resources and flower visitors in studies of pollinator-plant communities. *Oikos* **38,** 384–386.

Teräs, I. (1976). Flower visits of bumblebees, *Bombus* Latr. (Hymenoptera, Apidae), during one summer. *Ann. Zool. Fenn.* **13,** 200–232.

Vadas, R. L. (1990). Competitive exclusion, character convergence, or optimal foraging: Which should we expect?. *Oikos* **58,** 123–128.

Waser, N. M. (1983). Competition for pollination and floral character differences among sympatric plant species: A review of evidence. *In* "Handbook of Experimental Pollination Biology" (C. E. Jones and R. J. Little, eds.), pp. 277–293. Van Nostrand Reinhold, New York.

West-Eberhard, M. J. (1979). Sexual selection, social competition, and evolution. *Proc. Am. Phil. Soc.* **123,** 222–234.

West-Eberhard, M. J. (1983). Sexual selection, social competition and speciation. *Q. Rev. Biol.* **58,** 155–183.

West-Eberhard, M. J. (1989) Phenotypic plasticity and the origins of diversity. *Annu. Rev. Syst.* **20,** 249–278.

Wilson, D. S. (1975). The adequacy of body size as a niche difference. *Am. Nat.* **109,** 769–784.

Wilson, E. O. (1971). "Insect Societies." The Belknap Press of Harvard University Press, Cambridge, Massachusetts.

Yumoto, T. (1986). The ecological pollination syndromes of insect-pollinated plants in an alpine meadow. *Ecol. Res.* **1,** 83–95.

Yumoto, T. (1987). Pollination systems in a warm temperate evergreen broad-leaved forest on Yaku Island. *Ecol. Res.* **2,** 133–145.

14

The Thermal Environment as a Resource Dictating Geographic Patterns of Feeding Specialization of Insect Herbivores

J. Mark Scriber and Robert C. Lederhouse

Department of Entomology
Michigan State University
East Lansing, Michigan

I. Introduction

A. The Voltinism–Suitability Hypothesis

A quarter century ago, Hairston *et al.* (1960) and Ehrlich and Raven (1964) presented fundamentally different views of ecological and evolutionary relationships between plants and herbivores. Despite vigorous research during the past 25 years, the relative importance of predation and chemical coevolution in affecting host range remains elusive and continues to be debated. In a recent feature article in *Ecology,* Bernays and Graham (1988) argue that chemical coevolution has been overemphasized and is of limited value in understanding herbivore specialization. They suggested that generalist natural enemies are the major selection pressure restricting host ranges of herbivorous insects. Ten respondents to this article agreed that in addition to phytochemistry, factors such as local host abundance, host microenvironments, restricted mating sites, sensory limitations, competition, and enemy-free space influence the evolution of host range in herbivores. However, none agreed that generalist predators are the most important. They further disagreed on the relative importance of specialist parasites versus generalist predators, and whether natural enemies should favor narrowing or broadening of the herbivore's host range (Lawton, 1986; Price *et al.,* 1986; Fox, 1988; Jermy, 1988; Thompson, 1988d). This indicates that the relative importance of various factors affecting the evolution of insect host range (Bernays and Graham, 1988; Futuyma and Moreno, 1988) can be resolved only through careful multifactor studies (Strong *et al.,* 1984; Barbosa, 1988; Courtney, 1988; Thompson, 1988a,d). Although field studies of herbivores are essential in determining the effects of variation in host abundance, phytochemistry, and host-specific natural enemies, complementary laboratory and genetic studies will be required to determine the interpopulation and intrapopulation variation in oviposition preference and larval performance, and their genetic relationships (Wiklund, 1981; Thompson, 1988b,c; Bossart and Scriber, 1992).

The geographic distribution of host plants can obviously limit the distribution of associated herbivorous insects. For any nonmigratory herbivorous insect, its maximal distribution is limited to the composite distribution of suitable host plants. Environmental conditions, especially temperature, are known to limit the geographic distribution of plant species, which, in turn, limits the distribution of associated insects.

The length of the *growing season* is limited generally by temperature. Accumulated thermal units above a base temperature are fundamental in predicting the number of generations of an insect herbivore possible at a given location (Apple, 1952). A commonly overlooked factor of similar importance affecting larval growth rate and voltinism patterns (number of generations) is the quality of the foodplant (Slansky, 1974; 1976; Scriber and Feeny, 1979; Scriber and Slansky, 1981). Faster larval growth rates on very nutritious host plants at locations where the available thermal units are marginal would permit an extra generation compared to poorer-quality hosts. These observations led one of us (RCL) to suggest that natural selection might favor local specialization by ovipositing females on the plant species, permitting the fastest larval development in an area where thermal units were generally marginal, for completion of an additional generation. In contrast, it was hypothesized that in other areas where an additional generation was not possible on any host plant, the herbivore species would be free of such a selection for specialization.

We propose that the degree of local specialization observed for a polyphagous species may be determined by the interaction of phenological limits and host plant suitability. Our evaluation of this voltinism–suitability hypothesis has involved extensive studies of the relative host plant quality and insect preferences across wide geographic areas. We have focused on two closely related polyphagous swallowtail species, *Papilio glaucus* and *P. canadensis* as our test species (Hagen *et al.*, 1991). The evaluation of our hypothesis has been aided by delineating the average position of latitudinal bands of critical thermal unit accumulations that determine insect generation limits for our test species (Hagen and Lederhouse, 1985; Ritland and Scriber, 1985; Scriber and Hainze, 1987). In this chapter, we describe the preliminary evaluation of local suitabilities of plants for larval growth and preliminary tests of actual patterns of host selection across this zone. The interaction of temperature and host quality in limiting the number of herbivore generations and in determining the geographical distribution of specialized and generalized patterns of host use represents a new hypothesis to explain local diet breadth.

B. Swallowtail Butterflies and Host Plants

Since swallowtails are large, showy butterflies of great popularity with collectors, considerable detail of their geographic ranges has been known since the 1800s (Edwards, 1884; Scudder, 1889). Members of the *Papilio glaucus* species group virtually cover the North American continent (Fig. 1), whereas the *P. troilus* group species have more limited ranges in eastern North America and Mexico (Fig. 2). Also, the taxonomy and phylogenetic relationships within the genus *Papilio* are relatively well established (Edwards, 1884; Rothschild and Jordan, 1906; Brower, 1959a; Munroe, 1961; Hagen and Scriber, 1991).

For both the *P. glaucus* and *P. troilus* species groups, considerable detail is known about which host plants are naturally used (Brower, 1959b; Scriber *et al.*, 1982; Scriber, 1984b), and which plant species could support larval development (Scriber, 1988; Dowell *et al.*, 1990; Scriber *et al.*, 1991b). Host records for the *P. troilus* group species are primarily Lauraceae species such as sassafras, *Sassafras albidum;* spicebush, *Lindera benzoin,* and redbay, *Persea borbonia* (Tietz, 1972). The *P. glaucus* species group is much more polyphagous. All *P. glaucus* group taxa use various cherries (*Prunus,* Rosaceae) and ashes (*Fraxinus,* Oleaceae) quite successfully. In addition, *P. glaucus* regularly uses tuliptree and sweetbay (Magnoliceae), hoptree (Rutaceae), sassafras (Lauraceae), and basswood (Tiliceae). Hoptree species (*Ptelea,* Rutaceae) are also important hosts of *P. multicaudatus.* Both *P. rutulus* and *P. canadensis* use various poplars and willows (Salicaceae), with *P. rutulus* also using sycamore (Platanaceae) and *P. canadensis* using alders and birches (Betulaceae). Unique among the *P. glaucus* group species, is the use of *Rhamnus* and *Ceanothus* species (Rhamnaceae) by *P. eurymedon.*

Clearly, the most obvious environmental factor limiting the geographic range of these swallowtail species is the distribution of actual and potential host plants. Our objectives have been to determine the nature of relationships between swallowtails and their potential hosts and how these have

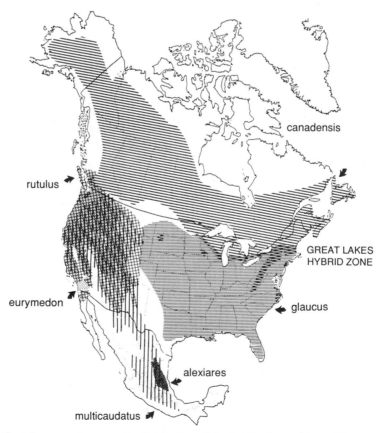

Figure 1   The current geographic distribution of the *Papilio glaucus* group of tiger swallowtail butterfly species in North America. After Brower (1959a); Beutelspacher and Howe (1984).

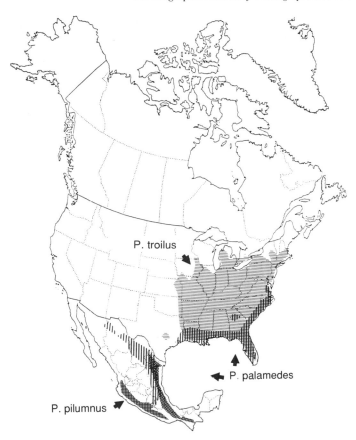

Figure 2 The current geographic distribution of the *Papilio troilus* species group of swallowtail butterfly species in North America. After Beutelspacher and Howe (1984); Opler and Krizek (1984).

been modified by abiotic and biotic factors to produce their current ranges. With detailed information, therefore, on swallowtail distribution, taxonomy, and host plant associations, North American *Papilio* species provide an excellent system with which to test our hypothesis.

II. Temperature and Host Plant Distributions

A. Temperature Extremes and Seasonal Accumulations

Since temperature is such a critical environmental factor and also relatively easy to measure, a great deal of information exists on the effects of temperature on organisms. Plant development and growth is limited by total seasonal thermal unit accumulations as well as by the temperature extremes. A

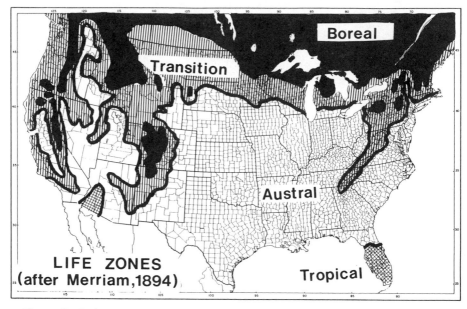

Figure 3   Redrawn copy of C. H. Merriam's (1894) life zones for the United States and Canada, illustrating the relationship between climate and vegetation.

developmental base temperature (T_0) of 8 to 10°C is common for many temperate zone plants. Agricultural crop cultivars are in fact often described in terms of the number of days for maturity (e.g., *90-day* versus *110-day* corn). Plant hardiness zones (U.S.D.A.) are determined largely by the length and severity of the *cold* period (Fowells, 1965, p. 752). The *taiga*, a predominantly forested ecosystem that covers North America to the arctic tree line from roughly north of 50°N latitude, is, for example, delineated by the winter and summer position of the arctic front (and the arctic permafrost; Oechel and Laurence, 1985). This plant community includes the boreal forests (with Salicaceae and Betulaceae), which largely delineate the distribution of *Papilio canadensis*. South of the taiga, Merriam's *transition zone*, based on temperature and rainfall (Fig. 3), also basically corresponds to the area where the average number of days between the last killing spring frost and the first killing fall frost equals 120–160 days. This also closely corresponds to plant hardiness zones (Fowells, 1965, p. 753) in which the average annual minimal temperature is −25 to −15°F.

With photosynthetic bark as well as leaves to facilitate its existence at high latitudes throughout the Canadian boreal forest, quaking aspen (*Populus*

Table 1 Selected Lepidoptera with Major Disjunctions in the Great Lakes Transition Zone from Boreal Forests to Temperate Deciduous Forests and Agriculturally Disturbed Areas

Genus	Species with disjunction at the transition zone & host plant affiliations	Potential isolating factor[a]	References
Hyalophora (giant silkmoths)	*H. cecropia* (generalist)	a) No evidence of specific limitations b) No evidence of specific limitations *Cecropia* is a southern generalist	Waldbauer & Sternburg (1973). Scriber (1983).
	H. columbia (larch specialist)	c) Only on larch (?) d) Only in bogs (?) e) No evidence of different predation	Ferge (1983). Ferge (1983).
Callosamia (promethea silk moth group)	*C. promethea* (generalist)	a) No evidence of specific limitations b) *L. angulifera* is a southern specialist; *C. promethea* feeds on black cherry and Ash further north	Sternburg & Waldbauer (1984). Scriber (1983).
	C. angulifera (tuliptree specialist)	c) Only lack of oviposition on paper birch limits *promethea* across Canada, since larvae survive & grow excellently; *angulifera* only on tuliptree (?) d) Mating times differ/habitat preferences (?) e) Male *promethea* may mimic *Battus philenor* which goes only to 41–42° N. latitude	Scriber & Hainze (1987). Brown (1972); Scriber (1983). Waldbauer & Sternburg (pers. comm.)
Limenitis (purple butterflies)	*L. astyanax* (black cherry and others)	a) No evidence of specific limitations b) Differential larval specificities and use abilities possible c) Oviposition preferences likely	Opler & Krizek (1984). Scriber, unpublished. Opler & Krizek (1984).
	L. arthemis (*Salix* and *Populus*)	d) Minimal barriers to hybridization e) Differential predation on adults likely (*astyanax* mimics *Battus philenor*) Northern limits to *philenor* are approximately 41°–42° N. latitude	Platt (1983). Platt & Brower (1968); Waldbauer *et al.* (1988).

(continued)

Table 1 (Continued)

Genus	Species with disjunction at the transition zone & host plant affiliations	Potential isolating factor[a]	References
Diaryctria (pine shoot moths)	*D. zimmermani* (pine generalist)	a) No evidence of specific limitations b) Host shift to red pine (new species) c) Oviposition in red pine shoots	Scriber & Hainze (1987). Mutuura (1982). Hainze & Benjamin (1985).
	D. resinosella (red pine specialist on shoots)	d) Mating (?) e) No mortality disjunctions	Hainze & Benjamin (1985); Scriber & Hainze (1987).
Hydraecia (noctuid borers)	*H. micacea* (generalist: introduced & spreading high latitudes)	a) Different thermal thresholds, obligate univoltine likely (both lack capacity for multivoltinism) b) Different survival (recent host shift) c) Different oviposition preferences	Giebink, et al. (1985); Scriber & Hainze (1987). Giebink & Scriber (in prep.).
	H. immanis (grass/hop spec. w/recent grass/corn host shift)	d) Same habitats e) No significant predation rates	Giebink et al. (1984). Deedat et al. (1983).
Ostrinia (European corn borer)	*nubilalis* (multivoltine strain)	a) Bivoltine north to the same latitude as *P. glaucus* Univoltine in northern range and New York, possibly central Michigan	Showers (1981). Ruppel (pers. comm.). Eckenrode et al. (1983). Straub et al. (1986). McLeod (1981). Roelofs et al. (1987).
	nubilalis (univoltine strain)	b) Different host races likely c) Oviposition likely to be selective d) Pheromone differences (mating; X-linked control) e) No consistent differences in predation known between races	

[a] a) diapause/voltinism limits b) larval host specificity c) oviposition specificity d) sexual selection/habitat differences e) differential predation on adults (mimicry).

tremuloides) is also able to extend its geographic range southward into the western montane forests with dry soils and open areas throughout the Rocky Mountains because of its vegetative reproduction by sprouting, as opposed to seed germination with sexual reproduction. Although its climatic needs and geographic range are very similar to those of paper birch across Canada, aspen has higher densities in the Rocky Mountains, yet paper birch reaches its highest densities north of the Great Lakes (Halliday and Brown, 1943). The aspen–birch forests are an essential resource for *P. canadensis*. Only since the Pleistocene glaciations have the eastern North American plant populations reestablished contact with Alaskan plant populations, creating an extensive range for *P. canadensis* (Scriber, 1988). In the east, aspens merge with the northern hardwoods of the temperate deciduous forests at the Great Lakes region (Hicks and Chabot, 1985). This area of the Midwest is of particular interest, not only because it represents a rather distinct plant ecotone (Curtis, 1959), but because it represents a suture zone of hybrid interaction for animals as well (Remington, 1968; Scriber and Hainze, 1987; Table 1).

B. The Great Lakes Ecotone

The major plant ecotone between the boreal and eastern deciduous forest biomes in the Great Lakes region has been described in detail previously by Curtis (1959) and Braun (1974). This ecotone area across central Wisconsin, south-central Michigan and into New England also represents the overlap in plant ranges for the hostplants of *Papilio glaucus* and *Papilio canadensis* (Scriber, 1988). In this area, the southern distribution limits are reached for a large number of northern plant species including potential *Papilio* hosts (e.g., paper birch, *Betula papyrifera;* balsam poplar, *Populus balsamifera;* and quaking aspen, *P. tremuloides*), and the northern distribution limits are reached for a number of temperate deciduous forest host species (e.g., tuliptree, *Liriodendron tulipifera;* hoptree, *Ptelea trifoliata;* sassafras, *Sassafras albidum;* white ash, *Fraxinus americana;* black cherry, *Prunus serotina;* and basswood, *Tilia americana*). Range limits of 32 potential *Papilio* hosts in this Great Lakes ecotone were plotted on a county basis in order to illustrate the geographic location of the transition zone (Fig. 4). This area also corresponds almost precisely with the geographic zone of average seasonal thermal unit accumulations of 1400 to 1500 degree days (above a developmental base temperature of 10°C based on 20-year averages; Fig. 5). This correspondence of thermal unit accumulations of 1400 to 1500°C degree days with the host plant transition zone from boreal to temperate deciduous represents covariables that are critical to our goal of determining causal factors for disjunct *Papilio* distributions in the Great Lakes region.

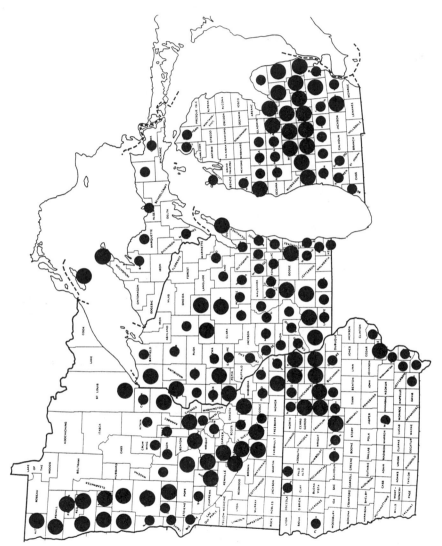

Figure 4 The number of *Papilio* host plants with range limits reached in each county of Minnesota, Iowa, Wisconsin, and Michigan. The total number of hosts included in this study was 32. ●:5–10 plant species reached their range limit in that county; ●:3–4 plant species reached their range limit in that county; ●:3–4 plant species reached their range limit in that county.

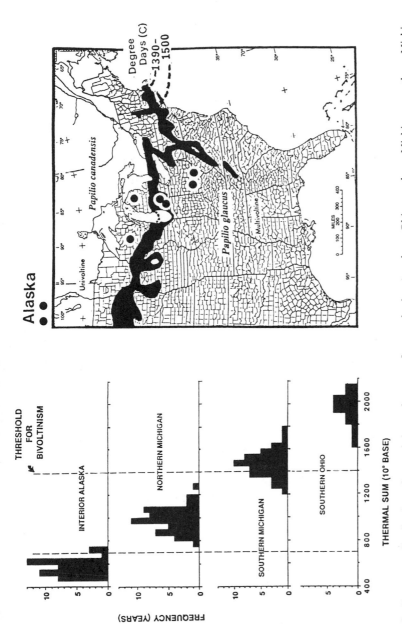

Figure 5 Frequency distribution of annual thermal sums at four study sites (southern Ohio, southern Michigan, northern Michigan, and interior Alaska) based on 10 to 50 years of recent weather records. The bivoltine *P. glaucus* in southern Ohio and the univoltine *P. canadensis* in northern Michigan have a relative surplus of thermal units in which to complete development. In contrast, *P. glaucus* in southern Michigan and *P. canadensis* in interior Alaska are under intense selection in many years to complete development before leaf senescence. In the Alaska population, larger eggs, reduced adult size, and elevated development potential at low temperatures, allow completion of development in 500 to 600 degree-days compared to the estimated 700 degree-days for northern Michigan populations (Ayres *et al.*, in preparation). Otherwise, we doubt *P. canadensis* would be able to sustain a population in this region of Alaska.

III. Patterns of Swallowtail Distributions

A. *P. troilus* Group Species

The *Papilio troilus* species group provides good examples of close relationships between host plant ranges and the geographic distribution of herbivores. Most host records for the three species, *P. troilus*, *P. palamedes*, and *P. pilumnus*, are trees or shrubs in the family Lauraceae. These include sassafras, *Sassafras albidum*, redbay, *Persea borbonia*, spicebush, *Lindera benzoin*, and the introduced ornamental camphortree, *Cinnamomum camphora*. No-choice feeding studies conducted in our laboratory confirm that various plant species in families other than the Lauraceae are unacceptable to *P. troilus* and *P. palamedes* (Scriber, 1986; Scriber *et al.*, 1991b). Neonate larvae apparently starve rather than initiate feeding on non-lauraceous plants (Scriber *et al.*, 1991b). Literature records of *P. troilus* or *P. palamedes* larvae on species of Magnoliaceae, Rutaceae, Rosaceae, Oleaceae, and Leguminosae are probably in error. Although "Laurel" or *Litsea* is the only reported host of *P. pilumnus* in Mexico (Brower, 1959b; Scott 1986); *P. pilumnus* can complete development on sassafras in our laboratory (Scriber and Lederhouse, 1988).

The northern geographic distribution of *P. troilus* is congruent with a composite of the distribution of spicebush and sassafras in eastern North America (Opler and Krizek, 1984; Scott, 1986; Petrides, 1972; Little, 1980; Fig. 6). Spicebush and sassafras are most commonly used by *P. troilus* wherever they occur. The rare and endangered silky spicebush, *Lindera melissifolia*, has also been recorded as a host for *P. troilus* (Morris, 1989). From central Florida southward where spicebush and sassafras are absent (Bell and Taylor, 1982), redbay serves as the primary host for *P. troilus* populations. Larvae from southern Florida populations perform much better on redbay than those from populations outside of its range (Nitao *et al.*, 1991).

In contrast with *P. troilus*, the geographic distribution of *P. palamedes* closely resembles that of its primary host, redbay, *Persea borbonia* in the United States, possibly adding other *Persea* species such as avocado, *Persea americana*, in Mexico. Larvae show varying abilities to use other lauraceous plants under laboratory conditions (Brooks, 1962; Scriber *et al.*, 1991b; Lederhouse *et al.*, 1991). Indeed, Harris (1972) comments on the rapid decline in *P. palamedes* numbers as one leaves the coastal plain in Georgia. Although redbay drops out on the piedmont, it seems unlikely that other abiotic or biotic factors would prevent *P. palamedes* from effectively exploiting sassafras there. Oviposition behavior of *P. palamedes* females may be the key. In east Texas, where the two *Papilio* species co-occur with both sassafras and redbay, *P. troilus* uses primarily sassafras and *P. palamedes* primarily redbay (Lederhouse *et al.*, 1991). In the laboratory, *P. palamedes* females

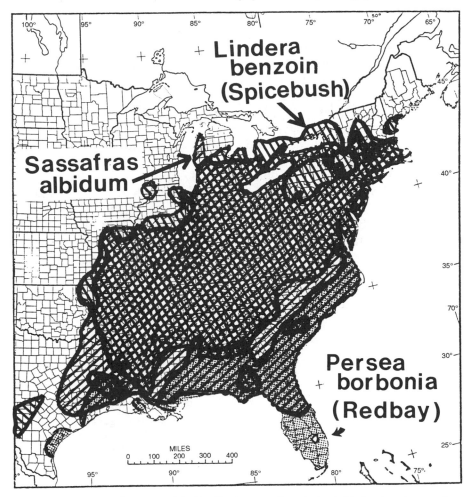

Figure 6 Composite geographic distribution map for spicebush *Lindera benzoin*, sassafras, *Sassafras albidum,* and redbay, *Persea borbonia.* Based on Hightshoe (1988).

strongly prefer to oviposit on redbay (Brooks, 1962; Lederhouse *et al.*, 1991). Our observations suggest a very tight relationship of insect and host plant distributions in *P. troilus* group species.

B. *P. glaucus* Group Species

Unlike the Lauraceae-specialized *P. troilus* species group, the *P. glaucus* species group is much more polyphagous. Numerous plants from ten different plant families are used as hosts (Scriber, 1975, 1984b, 1988). Also, unlike the rather restricted southeast distribution of Lauraceae hosts, the ten primary host families of the *P. glaucus* group (Rosaceae, Rutaceae, Lauraceae, Magnoliaceae, Oleaceae, Betulaceae, Salicaceae, Tiliaceae, Platanaceae, and Rhamnaceae) are geographically spread across the entire North American continent. Deciphering the relative importance of these plant distributions in the determination of the geographic distribution limits of the *glaucus* species group is more complex than with the Lauraceae specialists of the *troilus* group, but is more interesting. A long series of laboratory bioassays with the *glaucus* group taxa has allowed us to define some of the foodplants that are fundamentally unacceptable to certain tiger swallowtail species.

Restricted in natural distribution primarily to the southeastern United States from the southern Great Lakes zone, the Magnoliaceae (e.g., tuliptree and sweetbay) and the Lauraceae (e.g., spicebush and redbay) are generally toxic to all of the western species (*P. rutulus, P. eurymedon, P. multicaudatus,* and the northern *P. canadensis* (Scriber *et al.,* 1991b; Dowell *et al.,* 1990). Although used by the northern *P. canadensis* and western *P. rutulus,* plants of the Salicaceae (especially *Populus*) are toxic to all of the eastern *P. glaucus,* and most of the Mexican *P. alexiares* (Scriber *et al.,* 1988) as well as most of the western *P. multicaudatus* and *P. eurymedon* populations.

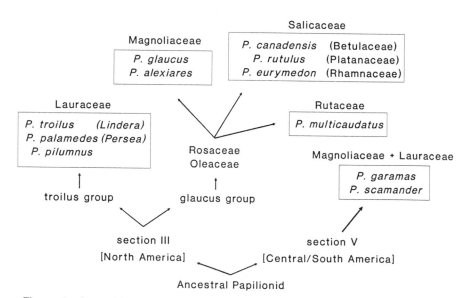

Figure 7 General host use patterns of Section III and Section V *Papilio* species. Hosts written in parentheses indicate additional feeding specialization. We thank James Nitao for crystallizing this figure.

The Rhamnaceae are used only by the western pale swallowtail, *P. eurymedon* (all other *glaucus* group taxa do very poorly on *Rhamnus* and *Ceanothus* species in this family). Similarly, the ability of *P. rutulus* to use sycamore (Platanaceae) is not shared by any other *glaucus* group taxon. These are presumably evolutionarily derived preferences and abilities (Fig. 7).

Although these rather distinct host plant affinities among different *glaucus* group species for the Magnoliaceae and Lauraceae, the Salicaceae, the Rhamnaceae, and the Platanaceae might imply a closely corresponding insect–host plant distribution, several additional host plants from different families are also used by each *Papilio* species. For example, several of the Rosaceae (black cherry, *Prunus serotina* or choke cherry, *P. virginiana*) are naturally used by all *glaucus* group taxa (*rutulus, eurymedon, multicaudatus, glaucus, alexiares,* and *canadensis*), in addition to their specific host plant families mentioned above. Therefore, it is not surprising that *glaucus* species distributions do not necessarily follow that of any single plant or plant family as was observed with the *troilus* species group.

IV. Environmental Determinants of Insect Distribution Limits

A. Tolerance of Extremes and Shorter Seasons

The combination of cold and desiccation during higher-latitude winters results in environmental extremes that limit a large number of insect species to lower latitudes. Gradients in species diversity along latitudes have many potential causes (Strong *et al.*, 1984), but inability to adapt to temperature extremes is certainly among the leading reasons for fewer insect species at higher latitudes. In addition, phytophagous insects find fewer host species at high latitudes since the plant species cannot tolerate the environmental extremes.

Insects have evolved diapause and migration as means of surviving such higher-latitude harsh environments during unfavorable winter periods (Danks 1981; Tauber *et al.*, 1986). Diapausing insects often exhibit special behavioral and/or physiological adaptations associated with freeze-tolerance (Ring 1981; Bale 1987; Kukal *et al.*, 1988a, b; Cannon and Black, 1988; Storey and Storey, 1988). Diapause regulation involving environmental cues is an important adaptation for certain *Papilio*. Facultative and obligate diapause regulation may be one of the key variables differentiating *Papilio* biogeography (Rockey *et al.*, 1987b).

In addition to thermal extremes at higher latitudes, insect species face shorter, cooler growing seasons that may provide inadequate thermal unit accumulation to complete a given number of generations. Among the adaptations of Lepidoptera to shorter seasons are very long life cycles with extended multiyear larval periods (Kukal, 1988). Additional adaptations include faster larval growth rates at a given temperature (i.e., lower developmental thresholds), the ability to molt at lower temperatures (Ayres and MacLean, 1987; M. P. Ayres, unpublished), thermal maximization by larvae (Rawlins and Lederhouse, 1981; Porter, 1982), and shortened larval periods producing smaller pupae and adults. Also females of some species select favorable microhabitats for oviposition (Williams, 1981; Grossmueller and Lederhouse, 1985), and larvae select the most nutritious host plant leaves (Scriber and Slansky, 1981) to maximize growth rates.

B. Plant Phenology and Changing Nutritional Quality

Changing thermoperiods and photoperiods serve as cues in initiation of physiological processes in many plant species that results in declining leaf water and nitrogen content. These declines in leaf water and leaf nitrogen content, along with other correlated changes in nutritional quality of host plant leaves, result in significant reductions in growth rates for many insect species of the leaf-chewing guild (Scriber, 1984a; Mattson and Scriber, 1987). Since different plant species have vastly different nutritional and allelochemical composition and their own seasonal pattern of phenological responses (with considerable individual variation), we might expect that a significant and wide-ranging potential exists for selection by insects of the most suitable hosts for growth and/or survival (Scriber, 1984a).

C. The Voltinism–Suitability Hypothesis of Feeding Specialization

We suggest that the length and quality of the growing season can determine the generality or specificity of host use in polyphagous insect herbivores (see Voltinism–Suitability Hypothesis of the introduction). Since different hosts permit different rates of larval development, a polyphagous herbivore may be able to complete an additional generation on its more favorable hosts in a particular area but one less generation in the same area on hosts that permit slower growth. To our knowledge, the interaction of host suitability with thermal units available during the growing season has not been previously considered to affect host choice by phytophagous insects (Ehrlich and Raven, 1964; Wiklund, 1974; Chew, 1975; Rausher, 1980, 1981; Fox and Morrow, 1981; Tauber and Tauber, 1981; Futuyma, 1983; Singer, 1983; Williams, 1983; Diehl and Bush, 1984; Strong *et al.*, 1984; Futuyma and Peterson, 1985; Tauber *et al.*, 1986; Bernays and Graham, 1988; Singer *et al.*, 1988).

An important aspect of the life history of an insect is the number of generations it can complete per year (voltinism). Since many insect species can diapause successfully in only one stage, individuals that initiate a generation that cannot be completed may suffer complete reproductive failure (Taylor, 1980a,b). However, individuals that do not start a generation that can be completed have lower relative fitness than those that do (Rolf, 1980; Taylor, 1980a). Therefore, selection should favor fitting the maximal number of generations into the available time. Because the host species can determine developmental rates, it can also limit the number of possible generations for herbivorous insects (Hare, 1983). Therefore host suitability can select for specialist–generalist host preferences, and selective oviposition should be most advantageous in phenologically limiting areas.

However, advantages of selective oviposition may decrease in areas where the favorable season easily supports a given number of generations, but is almost always too short for an additional one (Scriber and Lederhouse, 1983; Hagen and Lederhouse, 1985). In such areas, the same number of generations can be produced regardless of host, and the host range may expand to include lower-quality host plant species that support only slower development. Use of hosts with lower physiological suitability might be selected for by lower enemy loads on these hosts (Brower, 1958; Root and Kareiva, 1984; Feeny *et al.*, 1985; Bernays and Graham, 1988). Differential enemy loads could result from disproportionate numbers of specialized predators or parasites on a specific host, or from greater populations of generalized predators or parasites attracted to higher prey levels on specific host species.

V. Testing the Voltinism–Suitability Hypothesis

A. The Hypothesis and a Graphical Model

We hypothesize that local host use patterns for *Papilio glaucus* and *P. canadensis* result from predictable interactions between phytochemical constraints, phenological restrictions, and predation pressures (Fig. 8). Work already completed has largely documented the role of phytochemistry in determining the subset of available plant species that are potential hosts (Lindroth *et al.*, 1988; Scriber, 1983, 1986a, 1986b, 1988; Scriber *et al.* 1991b). This range is further restricted by phenological constraints, especially in regions where the season is barely long enough to complete development. The remaining hosts are of adequate nutritional suitability, but their overall suitability (probability that an egg will yield a reproductive adult in the next generation) may be further modified by the pressures of natural enemies. To test our voltinism–suitability hypothesis we need to

1. Identify areas where the seasonal thermal accumulations are barely adequate to permit an *extra* generation, and areas where no extra generation is possible on any host in any year (i.e., areas where selection is *intense* versus *relaxed* for fast larval growth);
2. Determine the suitability of various host plants for larval growth rates at each test location within and across these thermal accumulation zones;
3. Determine the actual patterns of host plant use in the field, within, and across the zones;
4. Evaluate the role of natural enemies (predators, parasites, or pathogens)

that might widen the host use patterns in areas of relaxed selections for fast growth; and

5. Evaluate the role of other factors affecting the selection and larval use of host plants (e.g., plant ecotones and insect species differences and genetic introgression in the hybrid zone).

B. The Evidence

1. Delineating Thermal Accumulation Zones

Growth and development in *Papilio glaucus* and *P.canadensis* is temperature dependent, as in most insects (Scriber and Lederhouse, 1983). Accumu-

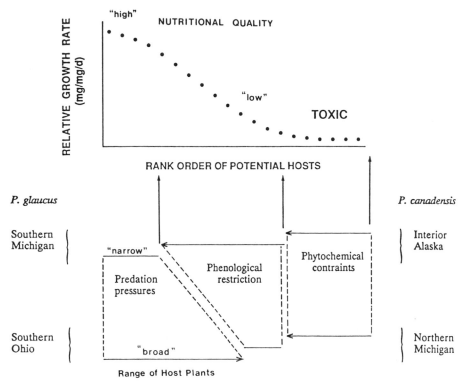

Figure 8 Graphical model illustrating the hypothesized role of phenological restrictions and predation pressures in influencing overall host suitability. All four plant communities contain plant species that are unsuitable owing to phytochemical constraints (characterized by very low growth rates and poor survival). Plant species that allow intermediate growth rates are unsuitable for populations in southern Michigan and interior Alaska because larval development cannot be completed before leaf senescence (phenological restriction). The suitability of remaining hosts will be influenced by host-specific rates of predation and parasitism. In southern Ohio and northern Michigan, where phenological restrictions are relaxed, even nutritionally marginal hosts may allow completion of larval development; in these regions, there are more potential host species subject to the influences of predation pressures. We thank Matt Ayres for assistance in developing this figure.

lation of a given number of degree-days above a developmental threshold temperature accurately predicts the completion of a particular life stage. Using an approximate developmental threshold of 10°C for *Papilio glaucus* and *canadensis* (Scriber and Lederhouse, 1983; Ritland and Scriber, 1985), we calculate that about 700°C degree-days are needed to complete one generation on the best food plants. We have summarized seasonal thermal accumulations for the eastern United States and Great Lakes region (Fig. 5), using 20–50 year averages from various state and regional reports (see Scriber and Hainze, 1987). This has sufficed to delineate an important latitudinal (and altitudinal) band of mean thermal accumulation along which numerous general geographic areas of interest can be selected to test our hypothesis. Based on differential developmental rates of *Papilio glaucus* and *P. canadensis* larvae on different food plants (Scriber and Feeny, 1979; Scriber, 1984a; Scriber *et al.*, 1982), it is possible to predict the northernmost limits for completion of a given number of generations on any particular host plant (Hagen and Lederhouse, 1985; Scriber and Hainze, 1987).

However, annual variation in thermal-unit accumulation is probably more critical to populations in thermally marginal zones than are long-term averages. As an example, we have illustrated seasonal thermal sums for key locations for recent decades (interior Alaska and northern Michigan) for *Papilio canadensis* and (southern Michigan and southern Ohio) for *Papilio glaucus* (Fig. 5; Ayres and Scriber, unpublished). Alaskan *P. canadensis* are under extreme selection to grow fast (or they could not complete even a single generation) compared to populations of *P. canadensis* in northern Michigan (or northern Wisconsin), where one generation can easily be achieved. A parallel situation exists for the multivoltine *Papilio glaucus*. In Ohio, *P. glaucus* should easily complete two generations on any nutritionally suitable host. But in south-central Michigan, *P. glaucus* can complete a second generation on only the most nutritionally suitable hosts (Figs. 5, 8). Another zone of interest occurs in the southern Smokey Mountains, where the threshold for three generations is reached; however, the delineation of this zone from NOAA weather records has not yet been completed, and our hypothesis testing would be further complicated by the complexity of overlap of two very slow generations with three very fast generations.

2. Determining Differential Suitability of Potential Hosts

As a preliminary test of our hypothesis, we have used data of M. P. Ayres and Scriber *et al.* (1982) to calculate overall growth rates (hatching to pupation) for *P. canadensis* on potential hosts in interior Alaska and in northern Wisconsin, and for *P. glaucus* on potential hosts in southern Michigan. We have plotted these values in decreasing rank order to compare the suitability profiles of available hosts in the three areas (Fig. 9). Total duration (hatching to pupation) includes everything that affects the general vulnerability of larvae through time. However, the duration of specific

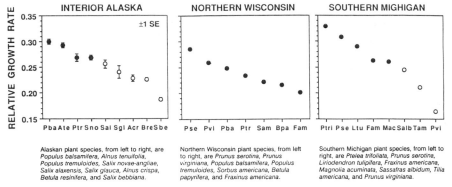

Alaskan plant species, from left to right, are
*Populus balsamifera, Alnus tenuifolia,
Populus tremuloides, Salix novae-angliae,
Salix alaxensis, Salix glauca, Alnus crispa,
Betula resinifera,* and *Salix bebbiana.*

Northern Wisconsin plant species, from left
to right, are *Prunus serotina, Prunus
virginiana, Populus balsamifera, Populus
tremuloides, Sorbus americana, Betula
papyrifera,* and *Fraxinus americana.*

Southern Michigan plant species, from left to
right, are *Ptelea trifoliata, Prunus serotina,
Magnolia acuminata, Sassafras albidum, Tilia
americana,* and *Prunus virginiana.*

Figure 9   Relative growth rate, mg · mg^{-1} · d^{-1} (where d = day), of *P. canadensis* on potential hosts from interior Alaska; *P. canadensis* on potential hosts from northern Wisconsin; and *P. glaucus* on potential hosts from southern Michigan. Darkened symbols represent plants with published records of host use by these populations. Relative growth rates encompass the entire larval period and were calculated as $[\ln(W_f) - \ln(W_i)]/t$, where W_f = dry mass of pupae, W_i = dry mass at hatch, and t = development time. N = 10 to 20 larvae per host. We thank Matt Ayres for providing the Alaska data.

stages may be disproportionately important because of stage-specific mortality factors that are particularly intense. For example, the vulnerability of the larval molting stages are each potentially as significant as total larval duration, and are known to be quite temperature sensitive (Ayres and MacLean, 1987). The first instar growth rates are perhaps the most critical, since difficulty in fast growth at this stage is likely to result in desiccation much more easily than for larger and more mobile larvae (Mattson and Scriber, 1987). Penultimate and final instar larvae of many *Papilio* species lose their brown and white color patterns (that apparently mimic bird-droppings) and become mostly cryptic green, with false thoracic eyespots (presumably mimicking tree snakes or frogs; Feeny 1976; Lederhouse, 1990). Differential predation pressures at different locations may select for a premium on fast growth in these late stages. Host plant nutritional quality changes seasonally and at different rates, depending on each plant's phenology (Scriber, 1984a). Thus, host suitability must, in the broadest sense, represent more than just the nutritional adequacy for rapid growth at any point in time.

3. Determining Actual Host-Use Patterns

As a first approximation of natural host-use patterns, we must use literature records and our own field observations of oviposition (Tietz, 1972; Scriber *et al.*, 1982). Since both common and rare occurrences are reported, we cannot quantify the relative frequency of host use using these records. Field records indicate that only four of the nine interior Alaska plants tested are ever used by natural populations of *P. canadensis* (Scriber and Ayres, 1990), and these are the four best host plants for overall growth (*Populus balsamifera, Alnus tenuifolia, P. tremuloides,* and *Salix novae-angliae;* Fig. 9). In contrast, in northern Michigan and Wisconsin, *P. canadensis* has been found on plant species of low as well as high quality for larval growth (Fig. 9). For example, white ash, paper birch, pin cherry, and mountain ash are some of the poorest plants tested for larval growth rates, yet all are used as natural hosts by *P. canadensis* populations in this area. Thus, we observe that essentially every potential host plant for larvae is naturally utilized in this zone of relaxed thermal selection (where one generation is possible on any host but where two generations are impossible on any host; Scriber *et al.,* 1982; Hagen and Lederhouse, 1985). In New York State, sufficient degree-days occur for completion of one generation on almost any host but never for two generations (Hagen and Lederhouse, 1985). Natural egg distribution (60% on white ash and 40% on black cherry; $n = 45$; Scriber 1975, and unpublished) was essentially identical in proportion to the distribution of leaf area at a Tompkins County, N. Y. site (58% ash, 42% cherry).

Similarly, we predict *Papilio glaucus* populations in south-central Michigan to be more specialized on those hosts that would permit the second generation to be completed before severe weather or leaf abscission. Our own preliminary observations and the literature records generally support our predictions (Fig. 9). In southern Ohio, we would expect the natural host range to be broad, but literature records of actual host use from this area are too few for even a preliminary assessment. Aspects of the life histories of a *P. glaucus* population from northern New Jersey are consistent with the predictions for southern Michigan. In that area, as in southern Michigan, the tiger swallowtail is bivoltine, but normally there are barely enough degree-days per summer to permit completion of two generations even on the most favorable hosts (Grossmueller, 1984; Fig. 5). The *P. glaucus* population in northern New Jersey exhibited traits consistent with maximizing the probability of completing two generations. Of the three most common host plants (tuliptree, black cherry, and white ash), females placed 88% of their eggs on tuliptree (*Liriodendron tulipifera*), the host on which larvae developed fastest (Grossmueller, 1984), although this host species averaged only a third (34%) of all the available host leaf area (black cherry averaged 17%, and white ash, 49%).

Females also oviposited in thermally favorable microsites on individual trees in New Jersey (Grossmueller and Lederhouse, 1985). In these sunward sites, eggs developed more quickly, and thermal maximization (i.e., *basking*) by the larvae was favored (Rawlins and Lederhouse, 1981; Porter, 1982). In fact, larvae of the *P. glaucus* group spin silken mats on the upper leaf surfaces such that the larvae rest off the surface of the leaf. One of us (JMS) suggested that this could be the focal point for reflected light and a thermally warmer place to rest (see figure in Mattson and Scriber, 1987).

Grossmueller and Lederhouse (1985) subsequently showed that larvae basking on curled leaves averaged 3.5°C warmer than larvae basking on uncurled leaves. We expect to find that females in locations of marginal thermal accumulations will select warmer microenvironments for their eggs. Faster growth in these sites would enhance the probability of successfully completing the second generation and would also reduce the time larvae would be exposed on leaves and vulnerable to predators, parasites, and disease (see reviews by Dempster, 1983, 1984).

4. The Role of Natural Enemies in Determining Breadth of Herbivore Host Use

Preliminary evidence supporting the role of predators and parasites in shaping host-use patterns of *P. glaucus* and *canadensis* is more limited. However, hosts that support slower development may be favored in areas where the premium on rapid development is decreased by a broad window of favorable conditions, especially if risks of predation and parasitism are lower on nutritionally marginal hosts, as some of our preliminary evidence suggests. For example, survivorship of *P. canadensis* caterpillars in New York varies greatly by host species, area, and year (Scriber, 1975; Hagen, 1986), nevertheless, field survival on white ash, which supports relatively slow growth, was as high as or higher than that on hosts supporting more rapid development (Scriber, 1975; Fig. 10).

5. Plant Ecotones and Insect Species Hybrid Zones

Plant community composition variability is doubtedlessly an important influence on host plant selection, as are genetically based differences in oviposition preference. A full test of our hypothesis requires a careful analysis of these variables.

The Great Lakes boreal forest/temperate deciduous forest ecotone (Section II, B) basically delineates the range limits for *P. canadensis* and *P. glaucus*. The region marks a climatic transition between northern areas dominated by relatively cool, dry arctic air masses and southern areas dominated by warmer, wetter tropical air masses (Bryson and Hare, 1974). Moreover, it corresponds to the southern edge of Pleistocene continental glaciation, and thus marks a transition in topography, soil types, and history of occupancy by animals and plants. Range limits and hybrid zones involving a variety of animal taxa coincide with this ecotone (e.g., Platt and Brower, 1968; Remington, 1968; Howard, 1986; Scriber and Hainze, 1987).

Evidence of natural hybridization between *P. glaucus* and *P. canadensis* has come from studies of morphology (Rothschild and Jordan, 1906; Luebke *et al.*, 1988), host plant use (Scriber, 1986b, 1988; Hagen, 1990), inheritance of mimetic female color (Scriber *et al.*, 1987; J. M. Scriber, R. H. Hagen, and

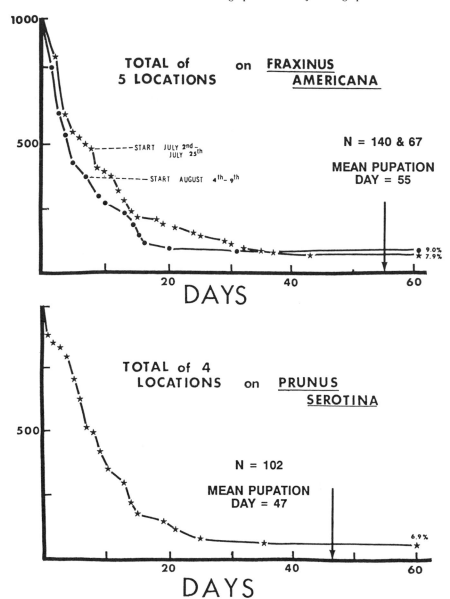

Figure 10 Survival of tiger swallowtail larvae on a slow-growth plant (white ash; *Fraxinus americana*) and on a fast-growth plant (black cherry; *Prunus serotina*) in Ithaca, New York. Nonetheless, these survivorship functions are very similar. Each function represents an average of study sites at five locations of different elevations for ash, four locations for cherry from neonate to final instar. The survival of planted larvae (i.e., placed on hosts by us) compares nearly identically with larvae from natural oviposition (Scriber, 1975).

R. C. Lederhouse, in preparation), and allozyme electrophoresis (Hagen, 1990). However, hybridization is confined to a relatively narrow zone (Fig. 5) within the broader transition zone (Fig. 3) extending from New England through the Great Lakes region, which falls generally between 41 and 44°N latitude, except in the Appalachian Mountain region.

In the upper Midwest, the position of the hybrid zone tracks closely the seasonal degree-day isotherm corresponding to the predicted northern limit for a multivoltine life cycle in *P. glaucus* (Fig. 11; Scriber 1983, 1988). Allele proportions for *Hk*, *Ldh*, and *Pgd* from population samples of males we collected in Michigan from 1986 to 1989 (Fig. 11; Hagen *et al.*, 1991) supports the morphometric evidence (Luebke *et al.*, 1988) that the hybrid zone appears to be quite narrow; only 50–150 km separates sites with *glaucus* alleles in high frequency from those with *canadensis* alleles. In none

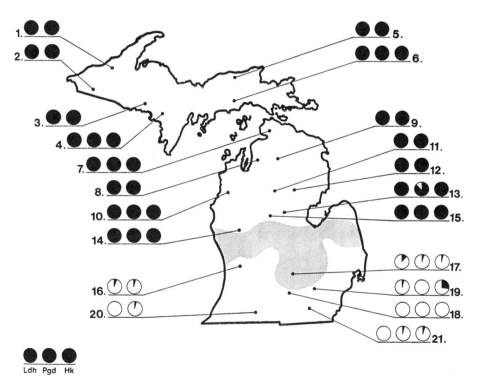

Ldh Pgd Hk

Figure 11 Proportion of all *canadensis* alleles for *Ldh*, *Pgd*, and *Hk* loci in samples of butterflies from 21 Michigan counties (Hagen *et al.*, 1991). *Hk* was not scored for samples for all counties; only two circles are shown in those cases. Filled circle, 100% *canadensis;* open circle, 0% *canadensis;* (100% *glaucus*); intermediate frequencies of *canadensis* alleles are indicated by the proportion of each circle filled. The shaded region indicates the position of 1400 to 1500°C seasonal isotherm in Michigan, the predicted northern limit for a multivoltine life cycle.

of the samples were both *glaucus* and *canadensis* alleles present in equal frequency.

In the northern Appalachian region, the isotherm corresponds somewhat less closely to the hybrid zone defined by *Ldh* and *Pgd* allele frequencies (Hagen, 1990). Approximately 250–300 km separates samples clearly *P. glaucus* from those that are clearly *P. canadensis.* Also unlike Michigan samples, the three central samples in this Appalachian transect had both types of *Pgd* alleles present at intermediate frequency, while *canadensis Ldh* alleles occurred at high frequency. Larvae from these three hybrid zone populations showed intermediate survival rates on both *canadensis* and *glaucus* host plants (Hagen, 1990), but, as expected from their locations north of the isotherm, enter an obligate pupal diapause (Hagen and Lederhouse, 1985).

The apparent difference in hybrid zone populations from the upper Midwest and the Appalachians is probably because of the more-complex distribution of habitats through the northern Appalachian region, as suggested for a ground cricket hybrid zone (Howard, 1986). Additional swallowtail samples from the northern Appalachian region are needed to determine whether this habitat complexity has produced a mosaic hybrid zone (Harrison and Rand, 1989), or a broad, but still clinal zone.

Since there is no obvious geographic barrier to prevent dispersal of *P. glaucus* and *P. canadensis* across the hybrid zone, maintenance of the sharp boundary between them must be attributed to other factors that prevent extensive interbreeding or the establishment of sympatric populations. One possibility may be inability of *P. glaucus* to adapt to the univoltine life cycle that is essential for survival north of the hybrid zone (Hagen and Lederhouse, 1985; Rockey *et al.*, 1987a,b). This could prevent northward extension of the range of *P. glaucus*, though it should not prevent *P. canadensis* from extending its range further south. Other sources for selection gradients can be readily envisioned, including differences in tolerance for extremes of high—or low—temperature encountered on either side of the hybrid zone, differences in host plant availability, or frequency-dependent mating preference. The latter two may be correlated with the central lowlands of Michigan (Scriber, 1992), which has been extensively deforested for vegetable production (drybeans, sugarbeets, carrots, potatoes, etc.).

Alternatively, the boundary between the species may be maintained as a tension zone, produced by disruption of development in the offspring of interspecies matings. A possible contributor to this may be the incompatibility that is responsible for reducing viability among the hybrid daughters of a *P. glaucus* female × a *P. canadensis* male (Haldane effect). A process that could result in a combination of selection gradients and a tension zone is suggested by the two-part *sex-chromosome–speciation* hypothesis outlined by Charlesworth *et al.* (1987) and Coyne and Orr (1989). For the first part, adaptive divergence between *P. glaucus* and *P. canadensis* with respect to diapause control and Batesian mimicry may have been facilitated by sex-linkage of the recessive (or female-limited) alleles responsible for the new phenotypes. For the second part, the same sex-linked alleles could have resulted in deleterious effects on females when hybridization places their sex chromosome in a genetic background different from their normal one. This genetic differentiation and evolutionary divergence appears to be greater for *Papilio glaucus* and the western species (*rutulus*, *eurymedon*, and *multicaudatus*) than for *P. canadensis*. Regardless of the mechanisms maintaining genetic integrity of species at hybrid zones, the introgression that does occur (e.g., Scriber, 1990), can affect the host-use preferences and our interpretation of the voltinism–suitability hypothesis. Consequently the oviposition preferences of individual females across the hybrid zone were analyzed with respect to their genetic make-up (i.e., their key diagnostic alleles; Scriber *et al.*, 1991a).

Successful growth of *Papilio* on the Salicaceae (aspen, poplars, willows) and Betulaceae (birch, alders) is prevalent north of the hybrid zone, but these hosts are toxic to most individuals south of the narrow zone. Conversely, use of the Magnoliaceae (e.g., tuliptree) is prevalent in the southern half of the United States, but the plant is toxic to essentially all individuals north of the hybrid zone. These detoxification differences are genetically based (Scriber 1986b; Scriber *et al.*, 1989), and geographically quite sharp (Scriber *et al.*, 1991a). However, while the quaking aspen and tuliptree neonate survival clines are very sharply defined across this zone (at approximately the same latitude and locations as the plant transition zone), there are several natural foodplants used by both *P. canadensis* and *P. glaucus*, which transcend this zone (e.g., black cherry, white ash, chokecherry).

Despite an occasional mistake, butterflies are usually quite precise in placing their eggs. This most often results in selectivity for host species in general (Chew, 1977; Smiley, 1978; Rausher, 1981; Singer, 1982, 1983; Mackay, 1985), but may also produce selection of individual plants (Rausher *et al.*, 1981; Rausher and Papaj, 1983; Thompson, 1988c,d), or parts of individual plants (Ives, 1978; Williams, 1981; Grossmueller and Lederhouse, 1985). Ovipositional patterns may also reflect choice of habitat (Shapiro and Carde, 1970; Singer, 1971; Wiklund, 1977; Rausher, 1979a), choice of host-dispersion pattern (Cromartie, 1975; Shapiro, 1975; Rausher *et al.*, 1981; Root and Kareiva, 1984), long-distance cues (Courtney, 1982; Papaj, 1986), or influences of factors such as location of nectar sources (Murphy, 1983; Murphy *et al.*, 1984, Grossmueller and Lederhouse 1987) or conspecific eggs (Gilbert, 1975; Rausher, 1979b; Shapiro, 1980). These variables are obviously also important in further evaluations of our voltinism–suitability hypothesis for feeding specialization.

In order to assess the genetically based variation in host preferences across the Great Lakes ecotone, a large series of individuals from various geographic populations were tested in a three-choice oviposition study (Scriber *et al.*, 1991a). Although our laboratory arena tests for close-range visual discrimination and contact chemoreception, it does not assess female ability to recognize long-range visual or volatile cues. Nonetheless, our laboratory studies readily revealed strong and ecologically consistent trends in preference along latitudinal clines from Alaska and south across the Great Lakes region to Florida. This indicates that the laboratory protocol is sufficiently robust, and that we are measuring meaningful traits. Results of three-choice oviposition studies using quaking aspen, tuliptree, and black cherry show a reciprocal latitudinal cline in oviposition preferences for aspen and tuliptree (especially in the ecotone region of Michigan), with insignificant low-level variation in selection of black cherry (Fig. 12). This emphasizes the need to evaluate the phytochemical, as well as the nutritional suitability, of each potential host at each geographic site chosen as a test location for our voltinism–suitability hypothesis.

6. Implications of the Voltinism–Suitability Hypothesis of Feeding Specialization: An Example with the European Corn Borer

Field corn (*Zea mays*), grown yearly in monocultures across the Midwest, has experienced serious biological stress from a variety of insects. Over the last

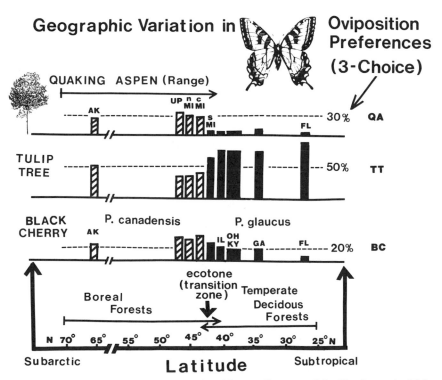

Figure 12 Latitudinal variation in oviposition preferences of *Papilio glaucus* (solid bars) and *P. canadensis* (hatched bars) in three-choice lab studies (see Scriber *et al.*, 1991a).

three decades, the most devastating pest has probably been the European corn borer, *Ostrinia nubilalis* (ECB). Recently, at latitudes corresponding to the juncture of the univoltine and bivoltine populations (i.e., south central Wisconsin, central Michigan, and central New York), significant ECB damage has occurred in peppers (McLeod, 1981), potatoes, green beans (Webb *et al.*, 1987), apples (Straub *et al.*, 1986), and other vegetable crops. It is also intriguing that the NC-180 Regional Research Committee (USDA, Cooperative State Research Service) has observed a recent shift of univoltine populations to bivoltine populations (in particular South Dakota, Michigan, and Minnesota). A complex mixture of ecotypes and voltinism patterns has been known for central New York for a longer period of time, with a mixture of univoltine and bivoltine types, both Z and E pheromone types, and certain strains—ecotypes preferentially selecting apples and other vegetables (Roelofs *et al.*, 1985; Showers *et al.*, 1989). The genetic basis of strain differences are known for pheromone detection and perception (Roelofs *et al.*, 1987) and diapause regulation (Showers, 1981), but not for host plant preferences.

Predictability of potential insect damage to specific crop plants is a highly desired feature in essentially all integrated pest management (IPM) or crop management programs (Kogan, 1986; Horn, 1988; Pedigo, 1989). Economic injury thresholds or decision-making levels regarding pest management are based primarily on pest population growth rates and the phenological stage of the plant or plant part attacked at a given locality. However, many pest management models are almost exclusively thermally driven (based on degree-day accumulations). Many pest species are polyphagous; the crop of interest may be only one of several potential hosts used locally. Management models failing to consider varying nutritional quality of alternative host species or even different plant tissues may result in large and sometimes perplexing errors in predicting insect population development (Scriber and Slansky, 1981; Slansky and Rodriguez, 1987).

For example, in Wisconsin, second flights of adult ECBs, have occasionally been hundreds of degree-days earlier than predicted (Showers *et al.* 1989; 10°C base temperature). This had serious consequences for sweet corn producers that were not expecting adults and egg deposition to begin for 1 to 2 weeks (based on development rate models for corn, *Zea mays*). Although it has been suggested that these borers migrated or were blown in from more southern states (e.g., Iowa or Illinois), it is equally likely that polyphagous ECB moths were emerging from nutritionally more favorable adjacent hosts (potatoes, peppers, beans, or other wild or weed species) earlier than those from corn. Indeed, our voltinism—suitability hypothesis predicts the selection of nutritionally superior host plants (rather than corn) in areas of marginal growth degree accumulation. The possibility of introgression from bivoltine populations into historically univoltine areas of South Dakota, Minnesota, and Michigan may be an additional explanation for some of these unusual observations, especially in conjunction with recent warming trends. A 20-state regional research program will address these possibilities across the Great Lakes region (NC-205 Project) and also farther south where 3–4 generations per year occur. Variable roles of generalized and specific biological control agents have been noted at different locations and are part of management plans. Whatever the case with the European corn borer, any phenological models of insect population growth should consider the very significant effects that differential plant quality has on insect development rates. In some cases, this may represent a simple fine tuning of the model, but in others it can transcend the entire range of temperatures in its magnitude of effect (Slansky and Scriber, 1985; Tabashnik and Slansky, 1987; Mattson and Scriber, 1987).

VI. Summary and Conclusions

We have introduced a new voltinism–suitability hypothesis to explain patterns of herbivore specialization. This hypothesis is predicated on the assumption that thermal unit accumulation is a critical selective force (resource) in determining insect voltinism patterns. Furthermore, it presumes that differential host plant quality (i.e., different plant species) can provide the basis for strong selection for specialization on the most suitable plants (where the larval growth rate is most rapid) in regions where seasonal thermal unit accumulations may be marginal for an additional generation. If so, we predict alternating latitudinal bands of specialization and generalization. Such subtle variation in plant species phenologies (e.g., plant water content) has been shown to occur in treehoppers, possibly enhancing the genetic divergence and speciation of host-specific insects (Wood *et al.*, 1990). Of course, other selective pressures such as host-specific natural enemies may alter these patterns (Denno *et al.*, 1990). The complex interactions of abiotic and biotic factors in determining distribution is exemplified by the narrow *P. canadensis* and *P. glaucus* hybrid zone in the Great Lakes region, which has certain very distinct clines in host preference and detoxification abilities on certain hosts, but not on others.

In summary, the sharply defined *Papilio* hybrid zone of south-central Wisconsin and Michigan is the result of a combination of both biotic (plant community type, plant availability), and abiotic (thermal) selective factors. Our hypothesis suggests that the interaction of temperature-imposed and host plant–imposed limitations for larval growth rates may determine the distribution of specialists and generalists across this ecotone, but more significantly, across the continent. The similarity of base thresholds (near 10°C) for the development of many insects (Taylor, 1981) to those we have determined for *Papilio*, suggests that other insect herbivores may also have thermally imposed limits on the potential for multivoltinism (multiple generations), maybe at the same geographic location (Table 1). We predict that the differential host plant quality for these insects would provide selection pressure for specialization in host use, in areas where their thermal units are marginal for completion of an extra generation, and the evolution of generalists north and south of these latitudinal zones. Only careful analysis at the interpopulation and intrapopulation level will permit us to know for sure.

Acknowledgements

This research was supported by Michigan State University College of Natural Sciences and the Agricultural Experiment Station (Projects 8051and 8072), and in part by NC-180 and 205 regional research, the National Science Foundation (BSR 8718448, BSR 8801184, BSR 9001391), and USDA Grants #87-CRCR-1-2851 and #90-37153-5263. We would particularly like to thank the following people for valuable discussion and/or their assistance in field collections for this study: Matthew Ayres, Janice Bossart, Robert Dowell, Mark Evans, Bruce Giebink, David Grossmueller, Robert Hagen, Kelly Johnson, James Maudsley, James Nitao, and Doozie Snider.

References

Apple, J. W. (1952.) Corn borer development and control on canning corn in relation to temperature accumulation. *J. Econ. Entomol.* **45,** 877–879.

Ayres, M. P., and MacLean, S. F. (1987). Molt as a component of insect development: *Galerucella sagittariae* (Chrysomelidae) and *Epirrita autumnata* (Geometridae). *Oikos* **48,** 273–279.

Bale, J. S. (1987). Insect cold-hardiness—an ecophysiological perspective. *J. Insect Physiol.* **33,** 899–908.

Barbosa, P. (1988). Some thoughts on the evolution of host range. *Ecology* **69,** 912–015.

Bell, C. R., and Taylor, B. J. (1982). Florida wild flowers and roadside plants. Laurel Hill Press, Chapel Hill, North Carolina.

Bernays, E., and Graham, M. (1988). On the evolution of host specificity in phytophagous arthropods. *Ecology* **69,** 886–892.

Beutelspacher, C. R., and Howe, W. H. (1984). Mariposas de Mexico. Fasc. 1. La Prensa Medica Mexicana. S.A. Mexico, D.F.

Bossart, J. L., and J. M. Scriber. (1992). Genetic variation in oviposition preference in the tiger swallowtail butterfly: Interspecific, interpopulation, and interindividual comparisons. *In:* (J. M. Scriber, Y. Tsubaki and R. C. Lederhouse, eds.) The Swallowtail Butterflies: Their Ecology and Evolutionary Biology. Cornell University Press. Ithaca, New York. In press.

Braun, E. L. (1955). The phytogeography of unglaciated eastern United States and its interpretations. *Botan. Rev.* **21,** 297–375.

Braun, E. L. (1974). "Deciduous forests of eastern North America" (Reprint of 1950 Ed.). Free Press, New York.

Brooks, J. C. (1962). Foodplants of *Papilio palamedes* in Georgia. *J. Lepid. Soc.* **16,** 198.

Brower, L. P. (1958). Bird predation and foodplant specificity in closely related procryptic insects. *Am. Nat.* **92,** 183–187.

Brower, L. P. (1959a). Speciation in butterflies of the *Papilio glaucus* group. I. Morphological relationships and hybridization. *Evolution* **13,** 40–63.

Brower, L. P. (1959b). Speciation in butterflies of the *Papilio glaucus* group. II. Ecological relationships and interspecific sexual behavior. *Evolution* **13,** 212–228.

Brown, L. N. (1972). Mating behavior and life habits of the sweet-bay silk moth (*Callosamia carolina*) *Science* **176,** 73–75.

Bryson, R. A., and Hare, F. K. (1974). The climates of North America. *In* "Climates of North America (World survey of climatology, Vol. II)" (R. A. Bryson and F. K. Hare, eds.), pp. 1–47. Elsevier, Amsterdam.

Cannon, R. J. C., and Black, W. (1988). Cold tolerance of microarthropods. *Biol. Rev.* **63,** 23–77.

Chabot, B. F., and Mooney, H. A. (1985). "Physiological Ecology of North American Plant Communities." Chapman and Hall, New York.

Charlesworth, B., Coyne, J. A., and Barton, N. H. (1987). The relative rates of evolution of sex chromosomes and autosomes. *Am. Nat.* **130,** 113–146.

Chew, F. S. (1975). Coevolution of pierid butterflies and their cruciferous foodplants. I. The relative quality of available resources. *Oecologia (Berl.)* **20,** 117–127.

Chew, F. S. (1977). Coevolution of pierid butterflies and their cruciferous foodplants. II. The distribution of eggs on potential foodplants. *Evolution* **31,** 568–579.

Chumbley, C. A., Baker, R. G., and Bettis, E. A. (1990). Midwestern Holocene paleoenvironments revealed by floodplain deposits in northeastern Iowa. *Science* **249,** 272–274.

Courtney, S. P. (1982). Coevolution of pierid butterflies and their cruciferous foodplants. IV. Crucifer apparency and *Anthocharis cardamines* (L.) oviposition. *Oecologia (Berl.)* **52,** 258–265.

Courtney, S. P. (1988). If it's not coevolution, it must be predation? *Ecology* **69,** 910–911.

Coyne, J. A., and Orr, H. A. (1989). Two rules of speciation. *In* "Speciation and Its Consequences" (D. Otte and J. A. Endler, eds.), pp. 180–207. Sinauer, Sunderland, Massachusetts.

Cromartie, W. J. (1975). The effect of stand size and vegetational background on the colonization of cruciferous plants by herbivorous insects. *J. Appl. Ecol.* **12,** 517–533.

Curtis, J. T. (1959). "The Vegetation of Wisconsin." University of Wisconsin Press, Madison, Wisconsin.

Danks, H. V. (1981). "Arctic Arthropods." Entomol. Soc. Canada, Ottawa.

Deedat, Y. D., Ellis, C. R., and West, R. J. (1983). Life history of the potato stem borer (Lepidoptera: Noctuidae) in Ontario. *J. Econ. Entomol.* **76,** 1033–1037.

Delcourt, P. A., and Delcourt, H. R. (1979). Late Pleistocene and Holocene distributional history of the deciduous forest of southeastern United States. *Veroffentilichungen des Geogotanischen Institutes der Eidgenossische Technische Hockschule, Stiftung Rubel* **68,** 79–107.

Dempster, J. P. (1983). The natural control of populations of butterflies and moths. *Biol. Rev.* **58,** 461–481.

Dempster, J. P. (1984). The natural enemies of butterflies. *In* "The Biology of Butterflies" (R. I. Vane-Wright and P. R. Ackery eds.), pp. 97–104. Academic Press, New York.

Denno, R. F., Larsson, S., and Olmstead, K. L. (1990). Role of enemy free space and plant quality in host plant selection by willow beetles. *Ecology* **71,** 124–137.

Diehl, S. R., and Bush, G. L. (1984). An evolutionary and applied perspective of insect biotypes. *Annu. Rev. Entomol.* **29,** 471–504.

Dowell, R. V., Scriber, M. J., and Lederhouse, R. C. (1990). Survival of *Papilio rutulus* Lucus (Lepidoptera: Papilionidae) on 42 potential host plants. *Pan-Pacific Entomol.* **66,** 140–146.

Eckenrode, C. J., Robbins, P. S, and Andaloro, J. T. (1983). Variations in flight patterns of European corn borer (Lepidoptera: Pyralidae) in New York. *Environ. Entomol.* **12,** 393–396.

Edwards, W. H. (1884). "The Butterflies of North America" (Vol. II). Houghton Mifflin, Boston, Massachusetts.

Ehrlich, P. R., and Raven, P. H. (1964). Butterflies and plants: A study in coevolution. *Evolution* **18,** 586–608.

Feeny, P. P. (1976). Plant apparency and chemical defense. *Rec. Adv. Phytochem.* **10,** 1–40.

Feeny, P., Blau, W. S., and Kareiva, P. M. (1985). Larval growth and survivorship of the black swallowtail butterfly in central New York. *Ecol. Monogr.* **55,** 167–187.

Ferge, L. (1983). Distribution and hybridization of *Hyalophora columbia* (Lepidoptera: Saturniidae) in Wisconsin. *Great Lakes Entomol.* **16,** 67–71.

Fowells, H. A. (1965). "Silvics of Forest Trees of the United States". Agric. Handbook No. 271. U.S.D.A. Forest Service, Washington, D.C.

Fox, L. R. (1988). Diffuse coevolution within complex communities. *Ecology* **69**, 906–907.

Fox, L. R., and Morrow, P. A. (1981). Specialization: Species property or local phenomenon? *Science* **211**, 887–893.

Futuyma, D. J. (1983). Selective factors in the evolution of host choice by phytophagous insects. *In* "Herbivorous Insects Host Seeking Behavior and Mechanisms" (S. Ahmad, ed.), pp. 227–244. Academic Press, New York.

Futuyma, D. J., and Peterson, S. C. (1985). Genetic variation in the use of resources by insects. *Annu. Rev. Entomol.* **30**, 217–238.

Futuyma, D. J., and Moreno, G. (1988). The evolution of ecological specialization. *Annu. Rev. Ecol. Syst.* **19**, 207–233.

Giebink, B., Scriber, J. M., and Wedberg, J. L. (1984). Biology and phenology of the hop vine borer, *Hydraecia immanis* Guenee and detection of the potato stem borer, *H. micacea* (Esper) (Lepidoptera:Noctuidae) in Wisconsin. *Environ. Entomol.* **13**, 1216–1224.

Giebink, B. L., Scriber, J. M., and Hodd, D. (1985). Developmental rates of the hop vine borer and potato stem borer (*Hydraecia immanis* and *H. micacea*) (Lepidoptera: Noctuidae): Implications for insecticidal control. *J. Econ. Entomol.* **78**, 311–315.

Gilbert, L. E. (1975). Ecological consequences of coevolved mutualism between butterflies and plants. *In* "Coevolution of Animals and Plants (L. E. Gilbert and P. H. Raven, eds.), pp. 210–240. University of Texas Press, Austin, Texas.

Graham, A. (1964). Origin and evolution of the biota of southeastern North America: Evidence from the fossil record. *Evolution* **18**, 571–585.

Grossmueller, D. W. (1984). Factors affecting voltinism in the tiger swallowtail *Papilio glaucus*. Dissertation. Rutgers University, Newark, New Jersey.

Grossmueller, D. W., and Lederhouse, R. C. (1985). Oviposition site selection: An aid to rapid growth and development in the tiger swallowtail butterfly, *Papilio glaucus*. *Oecologia (Berl.)* **66**, 68–73.

Grossmueller, D. W., and Lederhouse, R. C. (1987). The role of nectar source distribution in habitat use and oviposition by the tiger swallowtail butterfly. *J. Lepid. Soc.* **41**, 159–165.

Hagen, R. H. (1986). The evolution of host plant use by the tiger swallowtail butterfly, *Papilio glaucus*. Dissertation. Cornell University, Ithaca, New York.

Hagen, R. H. (1990). Population structure and host use in hybridizing subspecies of *Papilio glaucus* (Lepidoptera: Papilionidae). *Evolution* **44**, 1914–1930.

Hagen, R. H., and Lederhouse, R. C. (1985). Polymodal emergence of the tiger swallowtail, *Papilio glaucus* (Lepidoptera: Papilionidae): Source of a false second generation in central New York State. *Ecol. Entomol.* **10**, 19–28.

Hagen, R. H., and Scriber, J. M. (1989). Sex-linked diapause, color, and allozyme loci in *Papilio glaucus:* Linkage analysis and significance in a hybrid zone. *Heredity* **80**, 179–185.

Hagen, R. H., and Scriber, J. M. (1991). Systematics of the *Papilio glaucus* and *P. troilus* groups: Inferences from allozymes. *Ann. Ent. Soc. Am.*, **84**, 380–395.

Hagen, R. H., Lederhouse, R. C., Bossart, J., and Scriber, J. M. (1991). *Papilio canadensis* is a species. *J. Lepid. Soc.*, in press.

Hainze, J. H., and Benjamin, D. M. (1985). Partial life tables for the red pine shoot moth, *Diaryctria resinosella* (Lepidoptera: Pyralidae) in Wisconsin red pine plantations. *Environ. Entomol.* **14**, 545–551.

Hairston, N. G., Smith, F. E., and Slobodkin, L. B. (1960). Community structure, population control, and competition. *Am. Nat.* **94**, 421–425.

Halliday, W. E. D., and Brown, A. W. A. (1943). The distribution of some important trees in Canada. *Ecology* **24**, 353–373.

Hare, J. D. (1983). Seasonal variation in plant-insect associations: Utilization of *Solanum dulcamara* by *Leptinotarsa decemlineata*. *Ecology* **64**, 345–361.

Harris, L. (1972). "Butterflies of Georgia." University of Oklahoma Press, Norman, Oklahoma.

Harrison, R. G., and Rand, D. M. (1989). Mosaic hybrid zones. *In* "Speciation and Its Consequences" (D. Otte and J. A. Endler, eds.), pp. 111–113. Sinauer, Sunderland, Massachusetts.

Hicks, D. J., and Chabot, B. F. (1985). Deciduous forests. *In* "Physiological Ecology of North American Plant Communities" (B. F. Chabot and H. A. Mooney, eds.), pp. 257–277. Chapman and Hall, New York.

Hightshoe, G. L. (1988). "Native Trees, Shrubs and Vines for Urban and Rural America". Van Nostrand-Reinhold, New York.

Hopkins, D. M., P. A. Smith, and J. V. Mathews (1981). Dated wood from Alaska and the Yukon: Implications for forest refugia in Beringia. *Q. Res.* **15**, 217–249.

Horn, D. J. (1988). Ecological approach to pest management. Guilford Press, New York.

Howard, D. J. (1986). A zone of overlap and hybridization between two ground cricket species. *Evolution* **40**, 23–43.

Ives, P. M. (1978). How discriminating are cabbage butterflies? *Aust. J. Ecol.* **3**, 261–276.

Jermy, T. (1988). Can predation lead to narrow food specialization in phytophagous insects? *Ecology* **69**, 902–904.

King, J. E., and Allen, W. H. (1977). A Holocene vegetation record from the Mississippi river valley, southeastern Missouri. *Q. Res.* **8**, 307–323.

Kogan, M. (1986). "Ecological theory and integrated pest management practice." Wiley, New York.

Kukal, O. (1988). Caterpillars on ice. *Natural History* **59(1)**, 36–40.

Kukal, O., Heinrich, B., and Duman, J. G. (1988a). Behavioral thermoregulation in the freeze-tolerant arctic caterpillar, *Gynaephora groenlandica. J. Exp. Biol.*, in press.

Kukal, O., Serianni, A. S., and Duman, J. G. (1988b). Glycerol metabolism in a freeze-tolerant arctic insect: An *in vivo* 13-C NMR study. *J. Comp. Physiol.* **158**, 175–183.

Lawton, J. (1986). The effect of parasites on phytophagous insect communities. *In* "Insect Parasitoids" (J. Waage and D. Greathead, eds.), pp. 265–287. Academic Press, London, England.

Lederhouse, R. C. (1990). Avoiding the hunt: Primary defenses of Lepidoptera caterpillars. *In* "Insect Defenses" (D. L. Evans and J. O. Schmidt, eds.), pp. 175–189 SUNY Press, Albany, New York.

Lederhouse, R. C., Ayres, M. P., Nitao, J. K., and Scriber, J. M. Differential use of lauraceous hosts by *Papilio troilus* and *P. palamedes* Manuscript submitted.

Lindroth, R. L., Scriber, J. M., and Hsia, M. T. S. (1988). Chemical ecology of the tiger swallowtail: Mediation of host use by phenolic glycosides. *Ecology* **69**, 814–822.

Little, E. L. (1971). "Atlas of United States Trees. Vol. 1. Conifers and important hardwoods." Misc. Publ. No. 1146. U.S. Forest Service, Washington, D.C.

Little, E. L. (1976). "Atlas of United States Trees. Vol. 3. Minor Western Hardwoods." Misc. Publ. No. 1314. U.S. Forest Service, Washington, D.C.

Little, E. L. (1980). "The Audubon Society Field Guide to North American Trees." Knopf, New York.

Luebke, H. J., Scriber, J. M., and Yandell, B. S. (1988). Use of multivariate discriminate analysis of male wing morphometrics to delineate the Wisconsin hybrid zone for *Papilio glaucus glaucus* and *P. g. canadensis. Am. Midl. Nat.* **119**, 366–379.

Mackay, D. A. (1985). Prealighting search behavior and host plant selection by ovipositing *Euphydryas editha* butterflies. *Ecology* **66**, 142–151.

Martin, P. S., and Harrell, B. E. (1957). The Pleistocene history of temperate biotas in Mexico and eastern United States. *Ecology* **38**, 468–480.

Mattson, W. J., and Scriber, J. M. (1987). Nutritional ecology of insect folivores of woody plants: Water, nitrogen, fiber, and mineral considerations. *In* "Nutritional Ecology of Insects, Mites and Spiders" (F. Slansky, Jr. and J. G. Rodriguez, eds.), pp. 105–146. John Wiley, New York.

McLeod, D. G. R. (1981). Damage to sweet pepper in Ontario by three strains of European corn borer, *Ostrinia nubilalis*. *Proc. Entomol. Soc. Ont.* **112,** 29–32.

McVaught, R. (1952). Suggested phylogeny of *Prunus serotina* and other wide-ranging phylads in North America. *Brittonia* **7,** 317–346.

Merriam, C. H. (1894). Laws of temperature control of the geographic distribution of terrestrial animals and plants. *Nat. Geogr. Mag.* **6,** 229–238.

Morris, M. W. (1989). *Papilio troilus* L. on a new and rare larval food plant. *J. Lepid. Soc.* **43,** 147.

Munroe, E. (1961). The generic classification of the Papilionidae. *Can. Entomol. Suppl.* **17,** 1–51.

Murphy, D. D. (1983). Nectar sources as constraints on the distribution of egg masses by the checkerspot butterfly, *Euphydryas chalcedona* (Lepidoptera: Nymphalidae). *Environ. Entomol.* **12,** 463–466.

Murphy, D. D., Menninger, M. S., and Ehrlich, P. R. (1984). Nectar source distribution as a determinant of oviposition host species in *Euphydryas chalcedona*. *Oecologia (Berl.)* **62,** 269–271.

Mutuura, A. (1982). American species of *Diaryctria* (Lepidoptera: Pyralidae) VI. A new species of *Diaryctria* from eastern Canada and northeastern United States. *Can. Entomol.* **114,** 1069–1076.

Nitao, J. K., Ayres, M. P., Lederhouse, R. C., and Scriber, J. M. (1991). Larval adaptation to lauraceous hosts: Geographic divergence in the spicebush swallowtail butterfly. *Ecology.* **72,** 1428–1435.

Oechel, W. C., and Laurence, W. T. (1985). Tiaga. *In* "Physiological Ecology of North American Plant Communities" (B. F. Chabot and H. A. Mooney, eds.), pp. 66–94. Chapman & Hall, New York.

Opler, P. A., and Krizek, G. O. (1984). Butterflies east of the Great Plains: An illustrated natural history. Johns Hopkins University Press, Baltimore, Maryland.

Papaj, D. R. (1986). Interpopulation differences in host preference and the evolution of learning in the butterfly, *Battus philenor*. *Evolution* **40,** 518–530.

Pearcy, R. W., and Robichaux, R. H. (1985). Tropical and subtropical forests. *In* "Physiological Ecology of North American Plant Communities" (B. F. Chabot and H. A. Mooney, eds.), pp. 278–295. Chapman & Hall, New York.

Pedigo, L. P. (1989). "Entomology and Pest Management." Macmillan, New York.

Petrides, G. A. (1972). "A Field Guide to Trees and Shrubs." Houghton Mifflin, Boston, Massachusetts.

Platt, A. P. (1981). Evolution of North American admiral butterflies. *Bull. Ent. Soc. Am.* **29,** 10–22.

Platt, A. P., and Brower, L. P. (1968). Mimetic versus disruptive coloration in intergrading populations of *Limenitis arthemis* and *astyanax* butterflies. *Evolution* **22,** 699–718.

Porter, K. (1982). Basking behavior in larvae of the butterfly *Euphydryas aurinia*. *Oikos* **38,** 308–312.

Price, P. W., Westoby, M., Rice, B., Atsatt, P. R., Fritz, R. S., Thompson, J. W., and Mobly, K. (1986). Parasite mediation in ecological interactions. *Annu. Rev. Ecol. Syst.* **17,** 487–505.

Rand, D. M., and Harrison, R. G. (1989). Ecological genetics of a mosaic hybrid zone: Mitochondrial, nuclear, and reproductive differentiation of crickets by soil type. *Evolution* **43,** 432–439.

Rausher, M. D. (1979a). Larval habitat suitability and oviposition preference in three related butterflies. *Ecology* **60,** 503–511.

Rausher, M. D. (1979b). Egg recognition: Its advantage to a butterfly. *Anim. Behav.* **27,** 1034–1040.

Rausher, M. D. (1980). Host abundance, juvenile survival, and oviposition preferences in *Battus philenor*. *Evolution* **34,** 342–355.

Rausher, M. D. (1981). Host plant selection by *Battus philenor* butterflies: The roles of predation, nutrition, and plant chemistry. *Ecol. Monogr.* **51,** 1–20.

Rauscher, M. D., and Papaj, D. R. (1983). Demographic consequences of discrimination among conspecific host plants by *Battus philenor* butterflies, *Ecology* **64**, 1402–1410.

Rauscher, M. D., Mackay, D. A., and Singer, M. C. (1981). Pre- and post-alighting host discrimination by *Euphydryas editha* butterflies: The behavioral mechanisms causing clumped distributions of egg clusters. *Anim. Behav.* **29**, 1220–1228.

Rawlins, J. E., and Lederhouse, R. C. (1981). Developmental influences of thermal behavior on monarch caterpillars (*Danaus plexippus*): An adaptation for migration (Lepidoptera: Nymphalidae: Danainae). *J. Kans. Entomol. Soc.* **54**, 185–193.

Remington, C. L. (1968). Suture-zones of hybrid interaction between recently joined biotas. *Evol. Biol.* **2**, 321–428.

Ring, R. A. (1981). The physiology and biochemistry of cold tolerance in arctic insects. *J. Therm. Biol.* **6**, 219–229.

Ritland, D. B., and Scriber, J. M. (1985). Larval developmental rates of three putative subspecies of tiger swallowtail butterflies, *Papilio glaucus,* and their hybrids in relation to temperature. *Oecologia (Berl.)* **65**, 185–193.

Rockey, S. J., Hainze, J. H., and Scriber, J. M. (1987a). Evidence of a sex-linked diapause response in *Papilio glaucus* subspecies and their hybrids. *Physiol. Entomol.* **12**, 181–184.

Rockey, S. J., Hainze, J. H., and Scriber, J. M. (1987b). A latitudinal and obligatory diapause response in three subspecies of the eastern tiger swallowtail, *Papilio glaucus* (Lepidoptera: Papilionidae). *Am. Midl. Nat.* **118**, 162–168.

Roelofs, W. L., Du, J. W., Jang, X. H., Robins, P. S., and Eckenrode, C. J. (1985). Three European corn borer populations in New York based on sex pheromones and voltinism. *J. Chem. Ecol.* **11**, 829–836.

Roelofs, W., Glover, T., Tang, X–H., Sreng, I., Robbins, P., Eckenrode, C., Lofstedt, C., and Bengtsson, B. (1987). Sex pheromone production and perception in European corn borer moths is determined by both autosomal and sex-linked genes. *Proc. Natl. Acad. Sci. U.S.A.* **84**, 7585–7589.

Rolf, D. (1980). Optimizing development time in a seasonal environment: The 'ups and downs' of clinal variation. *Oecologia (Berl.)* **45**, 202–208.

Root, R. B., and Kareiva, P. M. (1984). The search for resources by cabbage butterflies (*Pieris rapae*): Ecological consequences and adaptive significance of Markovian movements in a patchy environment. *Ecology* **65**, 147–165.

Rothschild, R., and Jordan, K. (1906). A revision of the American papilios. *Novitates Zoologicae* **13**, 411–744.

Scott, J. A. (1986). "The Butterflies of North America." Stanford University Press, Stanford, California.

Scott, J. A., and Shepard, J. H. (1976). Simple and computerized discriminant functions for difficult identifications: A rapid nonparametric method. *Pan-Pacific Entomol* **52**, 23–28.

Scriber, J. M. (1975). "Comparative Nutritional Ecology of Herbivorous Insects: Generalized and Specialized Feeding Strategies in Papilionidae and Saturniidae (Lepidoptera)." Dissertation. Cornell University, Ithaca, New York.

Scriber, J. M. (1983). Evolution of feeding specialization, physiological efficiency, and host races in selected Papilionidae and Saturniidae. *In* "Variable Plants and Herbivores in Natural and Managed Systems" (R. F. Denno and M. S. McClure, eds.), pp. 373–412. Academic Press, New York.

Scriber, J. M. (1984a). Host-plant suitability. *In* "Chemical Ecology of Insects" (W. J. Bell and R. T. Carde, eds.), pp. 159–202. Chapman & Hall, London.

Scriber, J. M. (1984b). Larval foodplant utilization by the world Papilionidae (Lep.): Latitudinal gradients reappraised. *Tokurana* (Acta Rhaloceralogica) **2**, 1–50.

Scriber, J. M. (1986a). Origins of the regional feeding abilities in the tiger swallowtail butterfly: Ecological monophagy and the *Papilio glaucus australis* subspecies in Florida. *Oecologia (Berl.)* **71**, 94–103.

Scriber, J. M. (1986b). Allelochemicals and alimentary ecology: Heterosis in a hybrid zone? *In* "Molecular Mechanisms in Insect–plant Associations." (L. Brattsten and S. Ahmad, eds.), pp. 43–71. Plenum Press, New York.

Scriber, J. M. (1988). Tale of the tiger: Beringial biogeography, bionomial classification, and breakfast choices in the *Papilio glaucus* complex of butterflies. *In* "Chemical Mediation of Coevolution" (K. C. Spencer, ed.), pp. 240–301. Academic Press, New York.

Scriber, J. M. 1990. Interaction of introgression from *Papilio glaucus canadensis* and diapause in producing "spring form" eastern tiger swallowtail butterflies, *Papilio glaucus* (Lepidoptera: Papilionidae). *Great Lakes Entomol.* **23,** 127–138.

Scriber, J. M. (1992). Pollution and global climate change: Plant ecotones, butterfly hybrid zones, and biodiversity. *In* "The Swallowtail Butterflies: Their Ecology and Evolutionary Biology." (J. M. Scriber, Y. Tsubaki, and R. C. Lederhouse, eds.), Cornell University Press, Ithaca, New York. In press.

Scriber, J. M., and Feeny, P. (1979). Growth of herbivorous caterpillars in relation to feeding specialization and to the growth form of their food plants. *Ecology* **60,** 829–850.

Scriber, J. M., and Slansky, Jr., F. (1981). The nutritional ecology of immature insects. *Annu. Rev. Entomol.* **26,** 183–211.

Scriber, J. M., and Lederhouse, R. C. (1983). Temperature as a factor in the development and feeding ecology of tiger swallowtail caterpillars, *Papilio glaucus* (Lepidoptera). *Oikos* **40,** 95–102.

Scriber, J. M., and Hainze, J. H. (1987). Geographic invasion and abundance as facilitated by different host plant utilization abilities. *In* "Insect outbreaks: Ecological and Evolutionary Processes" (P. Barbosa and J. C. Schultz, eds.), pp. 433–468. Academic Press, New York.

Scriber, J. M., and Lederhouse, R. C. (1988). Hand-pairing of *Papilio glaucus glaucus* and *Papilio pilumnus* (Papilionidae) and hybrid survival on various food plants. *J. Res. Lepidopt.* **27,** 96–103.

Scriber, J. M., and Ayres, M. P. (1990). New foodplant and oviposition records for the tiger swallowtail butterfly, *Papilio glaucus canadensis* in Alaska (Lepidoptera: Papilionidae). *Great Lakes Entomol.* **23,** 145–147.

Scriber, J. M., Lintereur, G. L., and Evans, M. H. (1982). Foodplant suitabilities and a new oviposition record for *Papilio glaucus canadensis* (Lepidoptera: Papilionidae) in northern Wisconsin and Michigan. *Great Lakes Entomol.* **15,** 39–46.

Scriber, J. M., Evans, M. H., and Ritland, D. (1987). Hybridization as a causal mechanism of mixed color broods and unusual color morphs of female offspring in the eastern tiger swallowtail butterflies, *Papilio glaucus.* In "Evolutionary Genetics of Invertebrate Behavior" (M. Huettel, ed.) pp. 119–134. Plenum Press, New York.

Scriber, J. M., Evans, M. H., and Lederhouse, R. C. (1988). Hybridization of the Mexican tiger swallowtail, *Papilio alexiares* garcia (Lepidoptera: Papilionidae) with other *P. glaucus* group species and survival of pure and hybrid larvae on potential host plants. *J. Res. Lepidopt.* **27,** 222–232.

Scriber, J. M., Lindroth, R. L., and Nitao, J. (1989). Differential toxicity of a phenolic glycoside from quaking aspen to *Papilio glaucus* subspecies, hybrids, and backcrosses. *Oecologia (Berl.)* **81,** 186–191.

Scriber, J. M., Giebink, B. L., and Snider, D. (1991a). Reciprocal latitudinal clines in oviposition behavior of *P. glaucus* and *P. canadensis* across the Great Lakes hybrid zone: possible sex-linkage of oviposition preferences. *Oecologia (Berl.)* **87,** 360–368.

Scriber, J. M., Lederhouse, R. C., and Hagen, R. H. (1991b). Foodplants and evolution within *Papilio glaucus* and *Papilio troilus* species groups (Lepidoptera: Papilionidae). *In* "Plant-animal Interactions: Evolutionary Ecology in Tropical and Temperate Regions." (P. W. Price, T. M. Lewinsohn, G. W. Fernandes, and W. W. Benson, eds.), pp. 341–373. John Wiley, New York.

Scudder, S. H. (1889). "The Butterflies of the Eastern United States and Canada." Vol. II. Published by the author, Cambridge, Massachusetts.

Shapiro, A. M. (1975). Ecological and behavioral aspects of coexistence in six crucifer-feeding pierid butterflies in the central Sierra Nevada. *Am. Midl. Nat.* **98,** 424–433.

Shapiro, A. M. (1980). Egg load assessment and carryover diapause in *Anthocharis. J. Lepid. Soc.* **34,** 307–315.

Shapiro, A. M., and Carde, R. T. (1970). Habitat selection and competition among sibling species of satyrid butterflies. *Evolution* **24,** 48–54.

Showers, W. B. (1981). Geographic variation of the diapause response in the European corn borer. *In* "Insect Life History Patterns: Habitat and Geographic Variation" (D. F. Denno and H. Dingle, eds.), pp. 97–111. Springer-Verlag, New York.

Showers, W. B., Witkowski, J. F., Mason, C. E. Mason, Calvin, D. D., Higgins, R. A., and Dively, G. P. (1989). European corn borer: Development and management. North Central Regional Extension Publication No. 327. Iowa State University, Ames, Iowa.

Singer, M. C. (1971). Evolution of food plant preferences in the butterfly *Euphydryas editha.* Evolution **25:** 383–389.

Singer, M. C. (1982). Quantification of host preference by manipulation of oviposition behavior in the butterfly *Euphydryas editha. Oecologia (Berl.)* **52,** 224–229.

Singer, M. C. (1983). Determinants of multiple host use by a phytophagous insect population. *Evolution* **37,** 389–403.

Singer, M. C., Ng, D., and Thomas, C. D. (1988). Heritability of oviposition preference and its relationship to offspring performance within a single insect population. *Evolution* **42,** 977–985.

Slansky, F. 1974. Relationship of larvae food-plants and voltinism patterns in temperate butterflies. *Psyche* **81,** 243–253.

Slansky, F. 1976. Phagism relationships among butterflies. *J. N.Y. Entomol. Soc.* **84,** 91–105.

Slansky, F., and Scriber, J. M. (1985). Food consumption and utilization. *In* "Comprehensive Insect Physiology, Biochemistry, and Pharmacology," Vol. 4. (G. A. Kerkut and L. I. Gilbert, eds.), pp. 87–163. Pergamon Press, Oxford, England.

Slansky, F., Jr., and Rodriguez, J. G. (1987). "Nutritional Ecology of Insects, Mites, Spiders and Related Invertebrates." Wiley, New York.

Smiley, J. (1978). Plant chemistry and the evolution of host specificity: New Evidence from *Heliconius* and *Passaflora. Science* **201,** 745–747.

Sternburg, J. G., and Waldbauer, G. P. (1984). Diapause emergence patterns in univoltine and bivoltine populations of promethea (Lepidoptera: Saturniidae). *Great Lakes Entomol.* **17,** 155–162.

Straub, R. W., Weiures, R. W., and Eckenrode, C. J. (1986). Damage to apple cultivars by races of European corn borer (Lepidoptera: Pyralidae). *J. Econ. Entomol.* **79,** 359–363.

Storey, K. R., and Storey, J. M. (1988). Freeze tolerance in animals. *Physiol. Rev.* **68,** 27–84.

Strong, D. R., Lawton, J. H., and Southwood, R. (1984). "Insects on Plants. Community Patterns and Mechanisms." Harvard University Press, Cambridge, Massachusetts.

Tabashnik, B. E., and Slansky, F. (1987). Nutritional ecology of forb foliage-chewing insects. *In* "Nutritional Ecology of Insects, Mites, Spiders and Related Invertebrates (F. Slansky and J. G. Rodriguez, eds.), pp. 71–103. Wiley, New York.

Tauber, C. A., and Tauber, M. J. (1981). Insect seasonal cycles: Genetics and evolution. *Annu. Rev. Ecol. Syst.* **12,** 281–308.

Tauber, M. J., Tauber, C. A., and Masaki, S. (1986). "Seasonal Adaptations of Insects." Oxford University Press, New York.

Taylor, F. (1980a). Timing in the life histories of insects. *Theor. Popul. Biol.* **18,** 112–124.

Taylor, F. (1980b). Optimal switching in diapause in relation to the onset of winter. *Theor. Popul. Biol.* **18,** 125–133.

Taylor, F. (1981). Ecology and evolution of physiological time in insects. *Am. Nat.* **117,** 1–23.

Thompson, J. N. (1988a). Coevolution and alternative hypotheses on insect/plant interactions. *Ecology* **69,** 893–895.

Thompson, J. N. (1988b). Evolutionary ecology of the relationship between oviposition preference and performance of offspring in phytophagous insects. *Ent. Expt. Appl.* **47,** 3–14.

Thompson, J. N. (1988c). Variation in preference and specificity in monophagous and oligophagous swallowtail butterflies. *Evolution* **42**, 118–128.

Thompson, J. N. (1988d). The evolutionary genetics of oviposition preference in swallowtail butterflies. *Evolution* **42**, 1223–1234.

Thompson, J. N., Wehling, W., and Palowsky, R. (1990). Evolutionary genetics of host use in swallowtail butterflies. *Nature* **344**, 148–150.

Tietz, H. M. (1972). "An Index to the Described Life Histories, Early Stages, and Hosts of the Macrolepidoptera of the Continental United States and Canada." Allyn, Sarasota, Florida.

Tyler, H. (1975). "The Swallowtail Butterflies of North America". Naturegraph, Healdsburg, California.

Waldbauer, G. P., and Sternburg, J. B. (1973). Polymorphic termination of diapause by cecropia: Genetic and geographical aspects. *Biol. Bull.* **145**, 627–641.

Waldbauer, G. P., Sternburg, J. B., and Ghent, A. W. (1988). Lakes Michigan and Huron limit gene flow between the subspecies of the butterfly *Limenitis arthemis*. *Can. J. Zool.* **66**, 1790–1795.

Watts, W. A. (1980). The late quaternary vegetation history of the southeastern United States. *Annu. Rev. Ecol. Syst.* **11**, 387–409.

Webb, D. R., Eckenrode, C. J., and Dickson, M. H. (1987). Variation among green and wax beans in survival of larvae of a bivoltine-E race of the European corn borer (Lepidoptera: Pyralidae). *J. Econ. Entomol.* **80**, 521–524.

Whitney, S. (1985). "Western Forests." National Audubon Society and A. A. Knopf, New York, New York.

Whittaker, R. H. (1975). "Communities and Ecosystems." 2nd Ed. Macmillan, New York.

Wiklund, C. (1974). Oviposition preferences in *Papilio machaon* in relation to the host plants of the larvae. *Ent. Exp. Appl.* **17**, 189–198.

Wiklund, C. (1977). Oviposition, feeding, and spatial separation of breeding and foraging habitats in a population of *Leptidea sinapis* (Lepidoptera). *Oikos* **28**, 56–68.

Wiklund, C. (1981). Generalist vs. specialist oviposition behavior in *Papilio machaon* and functional aspects on the hierarchy of oviposition preferences. *Oikos* **36**, 163–170.

Williams, E. H. (1981). Thermal influences on oviposition in the montane butterfly *Euphydryas gillettii*. *Oecologia (Berl.)* **50**, 342–346.

Williams, K. S. (1983). The coevolution of *Euphydryas chalcedona* butterflies and their larval host plants: III. Oviposition behavior and host plant quality. *Oecologia (Berl.)* **56**, 336–340.

Wood, T. K., Olmstead, K. L., and Guttman, S. I., (1990). Insect phenology mediated by host-plant water relations. *Evolution* **44**, 629–636.

Author Index

Subject Index

Taxnomic Index